AF695847

PROGRESS in High-Temperature Superconductors

Published by
The American Ceramic Society
735 Ceramic Place
Westerville, Ohio 43081 USA
www.ceramics.org

For information on ordering titles published by The American Ceramic Society, or to request a publications catalog, please contact our Customer Service Department at 614-794-5890 (phone), 614-794-5892 (fax), <info@ceramics.org> (e-mail), or write to Customer Service Department, P. O. Box 6136, Westerville, OH 43086-6136, USA.

Printed in the United States of America.

4 3 2 1–07 06 05 04

ISBN 1-57498-194-3

www.ceramics.org

www.ceramicbulletin.org

www.ceramicjournal.org

CONTENTS

CERAMIC TRANSACTIONS

INTRODUCTION

The discovery of high-temperature superconductivity in the material LaBaCuO by Bednorz and Mueller almost two decades ago has been hailed as one of the most exciting advancements in modern physics. There has been much progress in the area of high-temperature superconductors since. For example, only a few months after the discovery of superconductivity in the La-Ba-Cu-O system, another copper oxide based ceramic was found, $Yba_2Cu_3O_7$, with a three-fold increase in the superconducting transition temperature to over 90 K. Today there are more than 50 types of copper oxide superconductors presently known.

There is still much to be learned however; there continues to be no universally accepted theory for why these materials are superconductors, and the full impact of high-temperature superconductivity will only be realized with the success of large-scale applications. For most large-scale bulk applications of high-temperature superconductors, low cost, long-length, flexible wires that can carry a large amount of supercurrent are required. While tremendous progress has been made, no current high-temperature superconductors yet have the ideal combination of these needed properties.

The American Ceramic Society (ACerS) continues to contribute to the progress of high-temperature superconductors by organizing and sponsoring forums for information exchange and disseminating information through its various periodicals and books. In 1987, the year following the discovery of the high-temperature superconductivity in the La-Ba-Cu-O system, The American Ceramic Society (ACerS) held special symposiums on ceramic superconductors at their Annual and Electronic Division meetings, and symposiums on this topic have been held every year since.

This *Progress in Ceramic Technology* series book contains a compilation of recent articles and papers published by ACerS on high-temperature superconductors. The lead article, Oxide Superconductors by Robert J. Cava, was published as an ACerS centennial feature article in the *Journal of the American Ceramic Society* and provides an excellent overview. This article, along with the additional 34 articles and papers on current research and development, provides a valuable resource for those involved in this field.

A special thanks is given to Amit Goyal, Oak Ridge National Lab, for helping with the selection of articles included in this book.

References to additional articles not included in this collection may be located by searching the following ACerS websites:

- The *Journal of the American Ceramic Society*: www.ceramicjournal.org/
- *Ceramic Transactions*: www.ceramics.org/ctindex.asp

journal

of the American Ceramic Society

journal

J. Am. Ceram. Soc., 83 [1] 5–28 (2000)

Oxide Superconductors

Robert J. Cava*

Department of Chemistry and Materials Institute, Princeton University, Princeton, New Jersey 08544

This article briefly reviews ceramic superconductors from historical and materials perspectives. It describes the factors that distinguish high-temperature cuprate superconductors from most electronic ceramics and places them in the context of other families of superconducting materials. Finally, it describes some of the scientific issues presently being actively pursued in the search for the mechanism for high-temperature superconductivity and the directions of research into new superconducting ceramics in recent years.

I. Introduction

During the long history of electronic oxide ceramics before 1986, even the most astute observer would have found few clues predicting the revolution in materials science and physics that those materials would eventually unleash. Oxide ceramics are an important part of electronics technologies because of their electrically insulating, optical, and dielectric properties, but very rarely for their ability to conduct electrical current. Insulators such as steatite, dielectrics such as SiO_2 and Ta_2O_5, piezoelectrics such as lead zirconate titanate (PZT), optical materials such as $LiNbO_3$, and the transparent semiconductor indium tin oxide (ITO) can be considered as representative of the large class of metal oxides with real and potential applications in electronics technologies. Although such materials are sophisticated and complex when considered from any point of view, their technological usefulness in electronics generally results (with the exception of ITO and a handful of other materials) from their ability to block or highly resist the passage of electrical current and, as an important consequence in some cases, their transparency at optical wavelengths.

These important characteristics of most oxides are related to a fundamental chemical difference between the metallic elements and oxygen: the difference in their electronegativities. These differences in electronegativity result in semiconducting or insulating materials, with forbidden energy gaps between the highest occupied orbitals in the solid derived from oxygen electronic states and the lowest unoccupied orbitals derived from the metal electronic states. The technical applications of electronic ceramics depend very highly on the materials scientist's manipulations of these energies, the introduction or elimination of defects that alter the distribution of energy states and/or their occupancy by charge carriers, and other factors such as the polarizabilities of the component atoms.

Not all metal oxides are transparent. Transition-metal oxides have been used since ancient times as pigments because of strong optical-frequency localized electronic absorptions between different excited states of the electrons in the transition-metal *d* orbitals, which lead to color. As electronic conductors, however, i.e., as competitors for copper wire, with few exceptions, they have never been of significant technological interest. Their principal consideration as electrical conductors has come primarily from their stability under high temperature, chemically extreme conditions where metals are not expected to be thermodynamically stable, as, for example, is found in electrodes for magneto-hydrodynamic generators.

This whole picture was stood up side down in late 1986 with the discovery of superconductivity at surprisingly high temperatures in an electronically conducting oxide based on lanthanum, barium, copper, and oxygen. Superconductivity is the passage of electrical current with *zero* resistance below a certain critical temperature (T_c). The best superconductors known at the time, to be described below, were metallic compounds, not ceramics, as one might expect as a consequence of the (no longer obvious) general fact that metals are good electrical conductors and oxides are not. Good superconductors do not have to be good metallic conductors above their superconducting transition temperature, however, and the fundamental characteristics that result in superconductivity at high temperatures in the copper oxides have little to do with any type of conventional understanding of the electronic properties of materials. In fact, more than a decade after their discovery, there continues to be no universally accepted theory for why these materials are superconductors. No one working in the field in the early days could possibly have foreseen that this would be the case after so much work and thought had been devoted to the problem.

All reviews on topics that are the subject of active on-going research are necessarily snapshots in time. The same has been true for reviews of the status of research in high-temperature (high-T_c) superconductivity. Many of the reviews of this topic published in

H. U. Anderson—contributing editor

Manuscript No. 189179. Received August 5, 1999; approved October 15, 1999.
The authors research program in superconducting materials at Princeton University is supported by the U.S. Department of Energy under Grant No. DE-FG02-98-ER45706.

*Member, American Ceramic Society.

Centennial Feature

the past decade are of considerable value, and they are not be duplicated here. The references in those reviews,[1–10] especially Ref. 1, serve to provide the primary sources for much of the information summarized in the present article. References to more-recent work or work complementary to what is found in those reviews are included here. The occasion of this centennial review series in ceramic materials is used to present a more tutorial and historical perspective of the topic, with strong emphasis on the structural and chemical aspects of the story. Although widely viewed as a revolution in condensed matter physics, it is equally valid to view the breakthrough of high-T_c superconductivity in copper oxides as a revolution in materials science and ceramic materials. The pace of discovery of new materials has slowed from its peak in the late 1980s, making a catalog of the known oxide superconductors and their "history" as understood at the turn of the millenium somewhat longer lived than it would have been a decade ago, and that comprises a significant proportion of this review. In this context, a set of uniform representations of oxide superconductors emphasizing their relation to the important perovskite structure type, in terms of the packing of metal–oxygen polyhedra, is presented.

The scientific issues under study at the present time primarily involve the precise determination of the details of the electronic and magnetic characteristics of the oxide superconductors in an attempt to get at the vexing microscopic heart of the problem of why superconductivity occurs in the cuprates. The appearance of superconductivity at high temperatures in such oxides has revealed a glaring hole in our understanding of the way magnetism and electrical conductivity are intertwined and related in transition-metal-based compounds in general—both an embarrassing and wonderful event in materials science and condensed matter physics. Magnetism and superconductivity continue to be the "Jekyll and Hyde" of oxide superconductors—they are clearly intimately related and may be different sides of the same personality—but, until we catch them directly in the act of changing from one to the other, we may not know exactly what their relationship is. The issues for the nonexpert may appear obscure, but they can be presented in a way that shows why they are interesting, and they are addressed in this review. To understand what a surprise the high-T_c copper oxide superconductors were, a brief summary of the superconducting materials under active research at the time of discovery of the cuprates also is included, as is a description of the oxide superconductors known before 1986—the precursors to the revolution. The progress in materials processing of ceramics and the progress toward the ultimate application of oxide superconductors, topics central to continuing investment in the field, are not reviewed here, because that is better done by experts in that area.

II. Distinguishing Characteristics of Copper Oxides

There are several chemical, structural, and electronic aspects of the copper oxide superconductors that distinguish them from other electronic oxides, even those that are electrically conducting. The great majority of conducting oxides studied in the past have as their basis the motion of electrons in energy bands that are formed from the interactions of the *d* orbitals of the transition metals present. The individual transition-metal atomic orbitals overlap and interact, forming a band of allowed energy states that is partially filled by the electrons available. The energy states of the oxygens play only a minor or insignificant role, a reflection of the same type of electronegativity difference that results in the bandgap in the majority of oxides. Examples of this type of conducting oxide are compounds such as V_6O_{13}, whose electronic conductivity is essential in its use as a possible cathode in experimental secondary batteries. In the copper oxides, however, the difference in energy between the oxygen and metal orbitals is very small, resulting in electronic energy bands in which the oxygen orbitals play a major role in the vicinity of the highest occupied electronic states. Thus *oxygen* is as highly involved in the conductivity as is copper, a rare situation in conducting oxides.

The second, and most interesting factor, especially from the physicists' point of view, derives from the electronic configuration of the Cu^{2+} ions that form the basis for the superconducting compounds. The electronic configuration for Cu^{2+} is $3d^9$: nine of the ten available *d*-orbital energy states are filled. In the Cu–O coordination polyhedra (octahedra, pyramids, and squares) that form in the cuprate superconductors, these energy levels are nondegenerate. The t_{2g} orbitals, which are directed between the oxygen atoms, are at low energy and, therefore, are completely filled with electrons (six). The shapes of the Cu–O coordination polyhedra are such that there are four near in-plane oxygen neighbors and one (pyramidal) or two (octahedral) apical oxygens more distant when present. This makes the energies of the orbitals with "z" components (i.e., toward the apices) lower because of lower repulsion from the oxygen orbitals. The result is that the nine-electron configuration has a single unpaired electron in the $dx^2 - y^2$ orbital (Fig. 1b), which points toward the in-plane oxygens. The single unpaired electron carries a spin of 1/2. This low spin value allows for the possibility of nonclassical (e.g., quantum mechanical) interactions between spins in the solid, which is different from the more familiar case of the spins in ferrites, for example, where the interactions between the larger spins (e.g., spin 5/2 for Fe^{3+} in Fe_3O_4) can be described by more conventional physics.

In isolated atoms, these orbitals are discrete energy states, but, in solids, where the atoms are close together, the orbitals interact, and the sharp atomic energy states become bands of energies. The similar energy of the O $2p$ states and the Cu $3d$ states leads to the situation shown in Fig. 1(c). The electrons fill half of the Cu $dx^2 - y^2$-derived band, and the high-energy part of the oxygen-derived band has almost the same energy as the highest occupied copper states. This is more complicated than in the usual electronic oxides, but the complexity continues. This type of picture would predict that Cu^{2+} oxides with the superconducting structure types

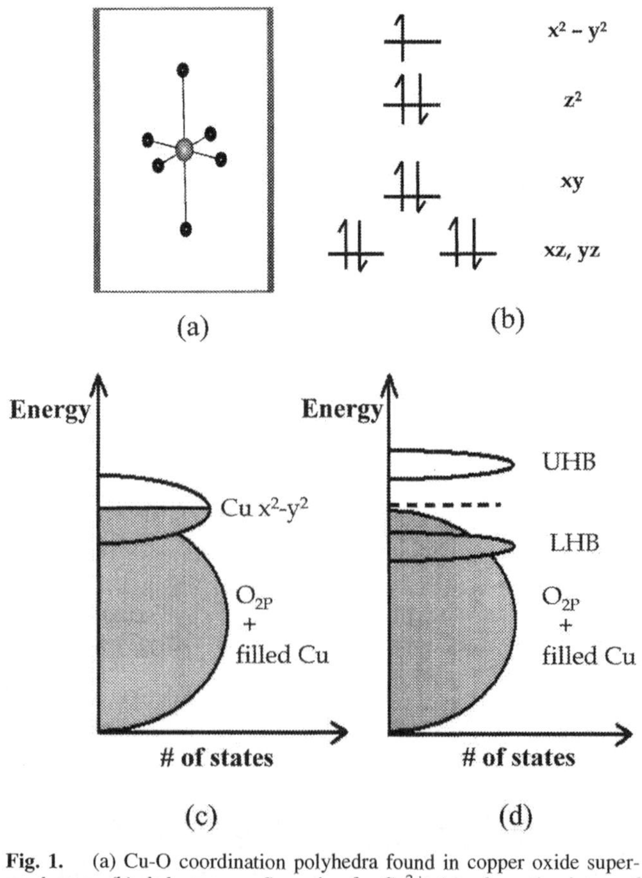

Fig. 1. (a) Cu-O coordination polyhedra found in copper oxide superconductors. (b) *d* electron configuration for Cu^{2+}, (c) schematic picture of the energy states in copper oxide superconductors, and (d) split $dx^2 - y^2$ band for interacting electrons in cuprates (UHB is upper Hubbard band and LHB is lower Hubbard band).

should be *metallic* conductors (e.g., the electrons are in a partially filled band with available energy states close by). They are, instead, electrical *insulators*. The standard picture for the electronic properties of solids assumes that the electrons are not interacting with each other, only with the underlying atomic lattice. In the copper oxides, this is not the case—they *do* interact. This realization came very early in the history of the field, causing great joy, continuing headaches, and great opportunities for theoretical physicists. It is this unexpected interaction that is one of the reasons why it is taking so long to understand the mechanism of high-T_c superconductivity. This interaction had been observed in other materials before the high-T_c superconductors; it was the focus of the work of Sir Nevill Mott.[11] Probably because there was no real compelling reason to determine the details, the ideas were incompletely formed and incompletely explored from the experimental viewpoint before the advent of high-T_c superconductors.

The interaction between electrons is greatest in half-filled bands, which is exactly the case for the cuprates. The addition of the second electron to the $dx^2 - y^2$ orbital takes a significant amount of extra energy, because it is repelled by the electron already present (called by the physicists the "on-site repulsion energy"). The result is that the energy states for the second electron are higher than those for the first electron, and there is a gap of nonallowed energies between the two. The bandgap between the highest occupied oxygen states and the empty part of the now split-in-two Cu $dx^2 - y^2$ band (Fig. 1(d)) is what leads to the insulating behavior.

What exactly happens when electrons are added to or taken away from the electronic ground state shown in Fig. 1(d), for the particular geometry of the copper–oxygen lattice in cuprate superconductors, is the big problem that, to date, has not been solved. A generalized view of the structure of cuprate superconductors is shown in Fig. 2. At their electronic heart are infinite CuO_2 planes (Fig. 2(a)) made from a checkerboard-like pattern of the in-plane basal squares of the CuO_x coordination polyhedra (such as the octahedra shown in Fig. 1(a)) sharing corners with each other. Each of the four oxygens in the CuO_4 square is shared with another copper, resulting in 180° (or ~180°) Cu–O–Cu bonds and an overall stoichiometry of CuO_2. Between these CuO_2 layers are other layers, known as "charge reservoir layers" (Fig. 2(b)). These layers serve to control, through chemistry, the number of electrons in the available electronic states in the CuO_2 planes and to electronically connect or isolate the CuO_2 planes in the third dimension. The key to determination of the superconducting transition temperatures within the family of cuprates depends in large part on the chemistry of these charge reservoir layers.

In the copper oxide superconductor parent materials, the Cu^{2+} spin 1/2 ions (one unpaired electron per copper in the $dx^2 - y^2$ orbital) in this CuO_2 plane are ordered antiferromagntically at a high temperature, and the material is insulating, as previously described. The high antiferromagnetic ordering temperatures indicate that the copper spins are coupled very strongly. Superconductivity is induced when the electron count on the CuO_2 plane is changed from one electron per copper site: i.e., the compounds are doped to make the formal copper valence different from Cu^{2+}, typically higher. This is accomplished through manipulation of the charge reservoir layer, either by adding oxygen, by partial substitution of one atom of higher or lower valence for another, or by the naturally occurring electron count present because of the valences of the atoms in the compound. Examples of these three cases are the inducement of superconductivity on oxygen intercalation from $YBa_2Cu_3O_6$ to $YBa_2Cu_3O_7$ in the "123" compound, the partial substitution of strontium for lanthanum in a solid solution in $La_{2-x}Sr_xCuO_4$, and the naturally occurring copper valence in the stoichiometric compound $Tl_2Ba_2CaCu_3O_8$. In semiconductor language, the parent compound is doped by either electrons or holes, and the antiferromagnetic ordering changes to superconductivity. This happens at an excess electron doping of ~0.2 electrons per copper (the formal reduction of the Cu to $Cu^{1.8+}$) or on the introduction of electron deficiency (hole doping; the formal oxidation of the Cu to $Cu^{2.2+}$). At higher doping concentrations, the materials become "normal" metallic conductors and are non-superconducting. A generic electronic phase diagram based on many experiments has been developed over the past decade to represent what happens as a function of electron concentration, and it is shown in Fig. 3.

Understanding superconductivity becomes complicated when we take the next step, and this is where the intellectual action has been concentrated in this field for some time. Hole doping is, by far, the most common case, and we must consider now what exactly happens to the holes in the electronic bands in Fig. 1(d). It appears they should be going into oxygen *p* states. If this is the case, then we must consider what apparently happens to a particular copper ion on doping. The Cu^{2+} keeps its 1/2 spin (the copper orbitals are not at the Fermi level and would not change electron count on hole doping), and the hole introduced is centered on oxygen *p* states: The spin (the unpaired electron on the copper) and the charge (the doped hole on the oxygens) are *separated*. The most prevalent belief is that the doped hole is in the electronic states in the square of four oxygens surrounding the copper. The system becomes less magnetic on doping, because the hole in the oxygen states also has a spin, in the opposite direction as the copper spin, and effectively cancels it out.[12,13]

We now consider what happens when this entity with zero net spin and one deficient electron (called a Zhang–Rice singlet) moves to carry current. It has to move through a lattice of strongly

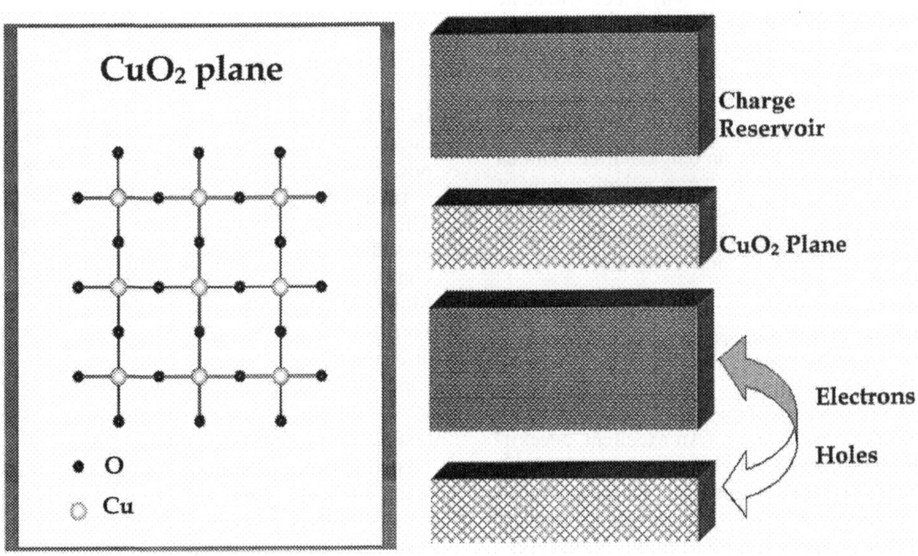

Fig. 2. (a) CuO_2 planes made from CuO_4 squares sharing corner oxygens and (b) schematic view of the electronic layers in copper oxide superconductors.

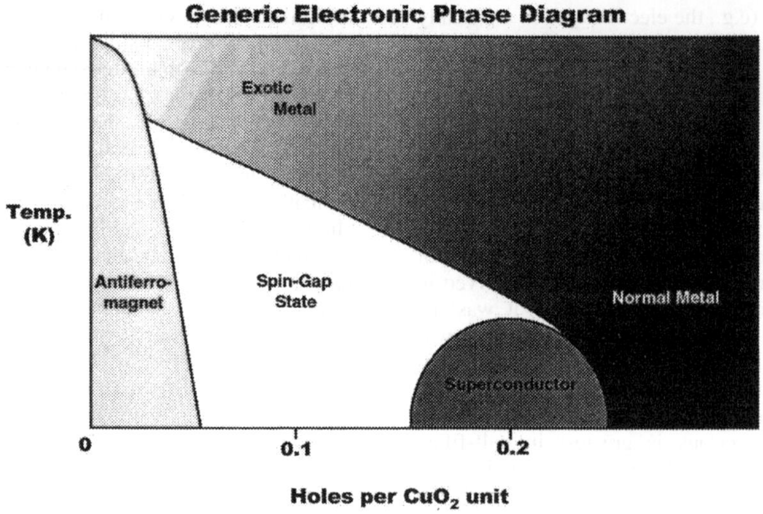

Fig. 3. Generic electronic phase diagram for copper oxide superconductors.[29]

antiferromagnetically coupled copper spins, a problem that condensed matter physicists previously had not considered. Many believe that this exotic conduction process is the key to the whole problem. Somehow, at just the right concentration of these entities wandering around in the CuO_2 plane, their complex interactions result in superconductivity. Sophisticated neutron-scattering measurements have confirmed that the interactions between spins remain present in the superconducting materials, even though they no longer show conventional magnetic ordering. Complicating the whole issue are spectroscopic results that suggest that the overall electronic picture moves toward the case shown in Fig. 1(c) on doping, i.e., a conventional electronic band structure. Some believe strongly that it is one peculiar characteristic of this recovered, otherwise normal looking, band structure (known as a van Hove singularity) that results in the superconductivity.

This whole story is completely different from and much more complex than what people were thinking a decade ago. It is certainly not believed (sometimes vigorously) by all the theoretical physicists working on the problem, and it is difficult to judge whether it will ultimately prove to be correct. It does seem to set out many of the issues that need to be addressed before the problem can be completely understood. There is evidence for other factors to consider as well, such as the possible influence of electronic excitations and the coupling of the charge carriers to the lattice. Sometimes one wonders whether it is similar to the story of the person who loses a ten dollar bill in the street at night and looks for it under the lamppost because that is the only place where he can see: Some of the ideas about why superconductivity happens in the cuprates are much easier to explore experimentally than are other ideas and necessarily get all the attention. It is wonderful, really, that something so remarkable as superconductivity at very high temperatures in relatively simple transition-metal oxides, such as the cuprates, could continue to be an enigma after so much thought has been devoted to the problem. An excellent review of the detailed state of our current understanding of cuprates in specific and of transition-metal compounds in general can be found in the review by Imada, Fujimori, and Tokura.[14]

III. Intermetallic and Oxide Superconductors Known before 1986

A very brief explanation of superconductivity and some of the phenomena associated with it are in order. (An excellent, accessible description of the topic can be found in the book by Vialdi.[15]) Superconductivity is the ability of some materials, at temperatures below a certain critical temperature (T_c), to conduct electrical current with zero electrical resistance. There are some very important limitations on this zero resistance state—to be described shortly—particularly when it comes to the practical applications of superconducting materials.

Superconductivity was discovered in the laboratory of Kamerlingh Ohnes in Leiden, Holland, in 1911. Ohnes was testing a theory about the resistivities of metals at very low temperatures, only then possible because of the liquefaction of helium. It was believed that the electron gas in a solid might freeze out at absolute zero and that metals would become insulating. Instead, their resistivities were found to decrease as the temperature was decreased, and experiments indicated that more-pure metals would attain lower resistivities. Mercury was the metal that could be most easily purified at that time, and, therefore, it was of particular interest. It was found to suddenly lose all resistance to current flow just above 4 K. The practical implications of the discovery were immediately obvious, but it was quickly realized that elemental superconductors would never be of any practical value. Ordinary resistivity reappeared when certain current limits (the critical current, J_c) and certain magnetic fields (the critical field, H_c) were exceeded. Both were too low for superconductors to be useful. Lead and niobium have the highest superconducting transition temperatures (9.3 and 7 K), tungsten has the lowest (0.01 K), and many are not superconducting.

Two important discoveries were made during the 1950s. Compound superconductors had been known for some time, with particular attention paid to materials based on niobium. One set of such materials, the A15 compounds (described below), were found to have relatively high T_c values, 10–20 K. More importantly, they were found to be able to conduct very large amounts of current and withstand high magnetic fields without becoming normal metals. In these materials, the magnetic field first penetrates the superconductor at low magnetic fields (the lower critical field, H_{c1}) but does not destroy the superconductivity. The field penetration causes normal metal regions within the superconductor, called vortexes. There is a wide region of field strengths where superconductivity and the vortexes coexist in the material, and zero resistance continues to be displayed, as long as the vortexes do not move. Materials chemistry and processing are critical in "pinning" the vortexes in their places, a very important area of research. Finally, at a much higher applied field (the upper critical field, H_{c2}), the superconductivity is fully destroyed. These compounds and alloys—called "type II superconductors" because of this important distinction in behavior from the superconducting elements (called "type I superconductors"), where the field penetrates all at once—were realized to hold the key to practical applications of superconductivity. They form the basis for current uses of superconductors, most importantly in the high-field magnets used in magnetic resonance imaging (MRI). (Copper oxide superconductors are extreme examples of type II superconductors, with H_{c1} values on the order of hundreds of Oersteds, and H_{c2} values on the order of millions of Oersteds.)

The other important event of this period was the development of a theoretical model that finally successfully explained how superconductivity could occur. The theory was proposed by John Bardeen (the inventor of the transistor), Leon Cooper, and J. Robert Schrieffer (the BCS theory) in the late 1950s. BCS theory explains that, below the critical temperature in superconducting materials, pairs of electrons are formed, for which the possible energy states are quantized. The energies of the possible states are sufficiently high that the electron pairs cannot absorb the low energy from the lattice vibrations and other low-energy events in the solid. They cannot absorb the energy of the normal scattering mechanisms experienced by ordinary conducting electrons and, therefore, move completely freely through the solid. (In a sense, this is analogous to window glass passing visible light without absorbing it—light is insufficiently energetic to excite the electrons from the ground state to the lowest excited state.) Experiments soon showed that the electrons in the pairs, expected to repel each other because of their like electrical charge, were held together by an attractive interaction mediated through coupling of the electrons to vibrations of the underlying crystal lattice ("electron–phonon coupling"). This coupling is at the basis of what is now called "conventional" superconductivity. It is operating in most known superconducting materials but is clearly not responsible for superconductivity in the copper oxides.

Although the field of superconductivity research in the 1980s before the advent of high-T_c superconductivity was very small by later standards, it was an intensely intellectually active field and not "sleepy" as is often described in the popularizations of the high-T_c story. Although there had been no new families of very-high-T_c materials discovered since the 1960s, there was significant activity in more esoteric, but forefront, areas. These areas included, for example, the interplay between magnetism and superconductivity, later to become an even more urgent issue in the cuprates.

The pre-1986 high-T_c standards were intermetallic superconductors of the A15 structure type. These were alloys of niobium or vanadium with the metals aluminum, silicon, germanium, tin, gallium, etc., with the formula Nb_3X. The crystal structure, shown in Fig. 4, consists of a primitive cubic array of X atoms with niobium or vanadium atoms in pairs in each of the faces of the cube. The niobium (vanadium) atom network forms nonintersecting one-dimensional chains. The highest T_c for these materials is 23.4 K, for the metastable alloy Nb_3Ga, which can be synthesized only under nonthermodynamic conditions, i.e., in thin-film form.

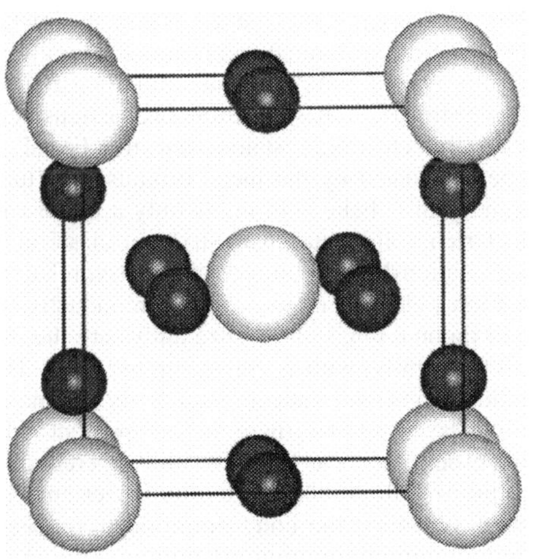

Fig. 4. The crystal structure of Nb_3Ge[30] ((●) niobium and (○) germanium).

These materials completely defined, by example, what the community believed as the important characteristics of high-T_c superconductors.

(1) Good superconductors should be metal alloys.

(2) High symmetry is necessary (preferably cubic, with simple structures).

(3) The electronic density of states must be high.

(4) The highest-T_c superconductors are those that are structurally unstable or metastable (because the forces that result in superconductivity tend to tear crystal structures apart or make them distort into lower symmetry and unacceptable structures if they get too strong).

(5) Structural distortions that lower the electronic density of states at the Fermi level and superconductivity compete with each other.

(6) The physics that leads to superconductivity at high temperatures is very well understood, and the highest possible T_c can be calculated (30 K); therefore, T_c might go higher with more research, but not too much higher (perhaps the most dangerous intellectual limitation).

Materials that were not believed to function by a conventional pairing mechanism were, in fact, being studied in the early 1980s, but their T_c values were very low. These were based on alloys of uranium and cerium with other metals.[16] One of the more famous of these materials, $CeCu_2Si_2$, with $T_c < 1$ K, is shown in Fig. 5. The superconductivity in these materials is derived from the hybridization at the Fermi energy of the cerium or uranium *f* orbitals, which are very narrow in energy, with the more-extended *d* orbitals of other metallic atoms in the crystal structure. The conduction electrons then travel in an electronic band that is very narrow in energy, resulting in that they act as if they have very large effective masses. The effective mass of the conduction electrons in $CeCu_2Si_2$, for example, is 1500 times the mass of a free electron. These materials are consequently called "heavy Fermion superconductors." Electrons traveling in narrow bands typically lead to magnetism, not superconductivity, and the study of these materials was intensely pursued for that reason. The low T_c values of heavy Fermion superconductors seemed to indicate that unconventional mechanisms for superconductivity would lead

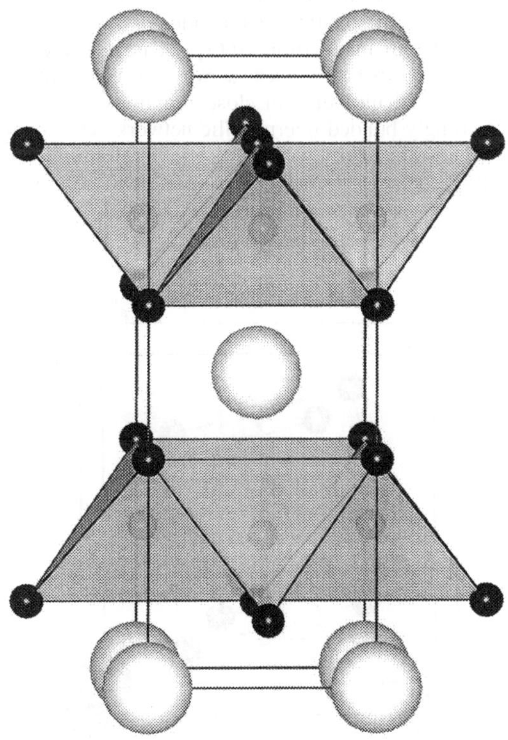

Fig. 5. Crystal structure of $ThCr_2Si_2$ ((○) thorium, (◍) chromium, and (●) silicon; $CrSi_4$ tetrahedra are shown).[31]

only to lower T_c values than what was known for the more-conventional alloys.

Intermetallic compounds were not the only materials studied. Another family of superconductors studied in the early 1980s was that whose structural and chemical complexity rival those of the cuprates. These were the "Chevrel phase" chalcogenide superconductors.[17] The highest T_c was known for the compound $PbMo_6S_8$ (14 K). Again, the T_c value was not as high as that of the A15 structure materials, and, therefore, the basic understanding of superconductivity was not challenged. Similar to the cuprates, they are part of a very large chemical and structural family. Several members of the family are superconducting. The crystal structure of $PbMo_6S_8$ is shown in Fig. 6. At the electronic heart of the structure is an Mo_6 cluster in the shape of an octahedron. This Mo_6 cluster can be considered to consist of atoms on the faces of a cube whose corners are occupied by sulfur atoms. This octahedron within a cube is a strongly bonded electronic entity, the Mo_6S_8 cluster, and it is the basis for many inorganic compounds. The lead atoms lie between the cubes. The Mo_6S_8 cubes are linked electronically to each other because some of the molybdenum on the Mo_6 octahedra are bonded to sulfur at the corners of neighboring Mo_6S_8 cubes. There are many interesting chemical variants of this structure, especially derived from putting different atoms in the positions of the lead atoms, ranging in size and chemical character from lithium through lead. These materials distinguish themselves by having very large H_{c2} values for the disappearance of superconductivity (as do the cuprates). The first detailed studies linking the superconducting properties in this family to the existence of chemical defects in the structures were announced at the time high-T_c superconductivity was announced, but the subject was quickly dropped. Despite considerable research into their structures and properties, these materials may be a potential source of interesting new materials science, because the relationship between structure, defect chemistry, and superconductivity was never fully explored.

Few superconducting oxides were known before 1986. For the most part, they had transition temperatures that were low compared with the known intermetallic superconductors, and, although they were of considerable scientific interest, they did not strongly challenge the status quo. One of the earliest known superconducting oxides was NbO, shown in Fig. 7. Its structure presages in some ways that of the cuprate superconductors, in that it can be considered as being made from NbO_4 squares. These squares, however, share corners in such a way to bring the highly chemically reduced niobiums in close proximity to each other, forming a strongly bonded intermetallic network. The conduction electrons are in a band of essentially metallic niobium character, and, therefore, the oxygen plays only a minor, if any, role.

Two types of oxide superconductors studied before the cuprates had classical oxide structure types, one based on the spinel structure and one based on the tungsten–bronze structure. Oxide spinels are well-known for their magnetic properties and have been extensively studied. The lithium titanate spinel $LiTi_2O_4$ was found in 1973 to be superconducting at 13 K. It is a normal spinel, with lithium in the tetrahedral sites and titanium in the octahedral sites. The effective titanium valence is $Ti^{3.5+}$, and all titanium sites are equivalent. The titanium–oxygen octahedra share edges in three sets of intersecting chains running along the cubic [111]-type directions. Although the crystal structure (Fig. 8) is that of a classical oxide, the electronic analysis of the compound shows that the conduction electrons are in a band of strong titanium-dominated character. Although many magnetic spinels are known, the number of superconducting spinel oxides is very limited. Some normal chalcogenide spinels, such as $CuRh_2S_4$ and $CuRh_2Se_4$, however, are known to be superconducting. The copper in these latter materials is formally monovalent and not involved in the superconductivity.

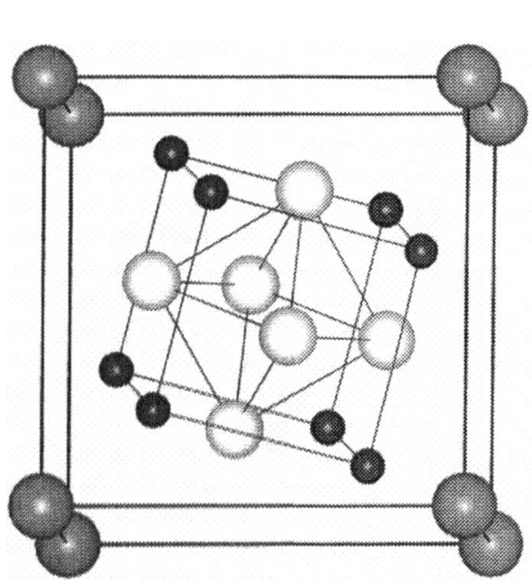

Fig. 6. Crystal structure of $PbMo_6S_8$ ((◎) lead, (○) molybdenum, and (●) sulfur).[32]

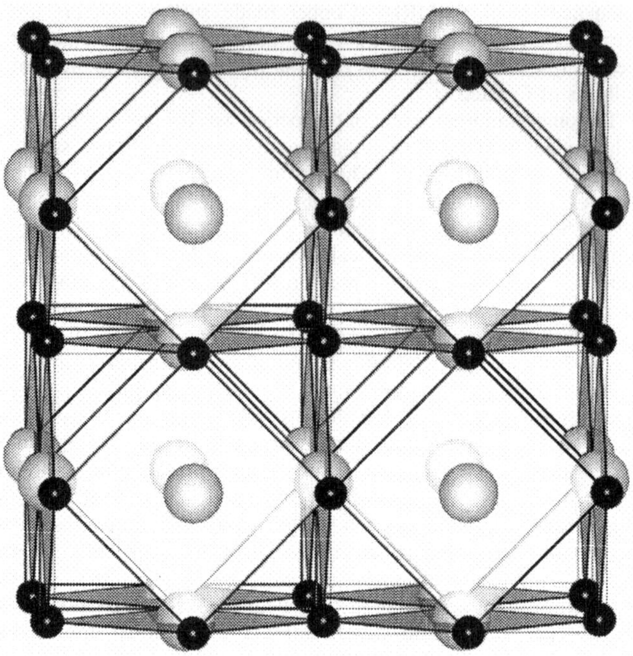

Fig. 7. Crystal structure of NbO ((○) niobium and (●) oxygen; NbO_4 squares are shown).[33]

The tungsten–bronze family of oxides is structurally related to perovskites. These materials are best known in ceramic technology as excellent ferroelectrics. $Ba_2NaNb_{10}O_{30}$ and $Ba_{0.5}Sr_{0.5}Nb_2O_6$ (BSN) are two good examples. Compounds with alkali metals in tunnel sites and tungsten–oxygen octahedra are superconducting and can adopt cubic, tetragonal, and hexagonal structures. The crystal structures of the tetragonal and hexagonal tungsten–bronze alloys are shown in Figs. 9 and 10. Although the octahedra share corners in all the structure types, resulting in an extensive W–O–W bonding array, the electronic states near the Fermi level are believed to be dominated by the metal orbitals—the tungsten 5*d* states. Oxygen again is believed to play only a minor role. There is a wealth of rich solid-state chemistry and classical materials science in the relationships among stoichiometry, relative sizes of the interstitial ions, electron count, and superconductivity in these phases. The hexagonal tungsten–bronze alloys are the best superconductors in the family, with T_c values up to ~7 K. There is an apparent relationship between superconductivity and magnetism in the copper oxides; similarly, there is an apparent relationship between charge density wave behavior and superconductivity in the tungsten–bronze alloys. Charge density waves are a collective electronic state, in which the conduction electrons become partially localized in particular chemical bonds and, therefore, are no longer conducting. The wavelength and direction of the charge density waves are directly determined by the wavelength and

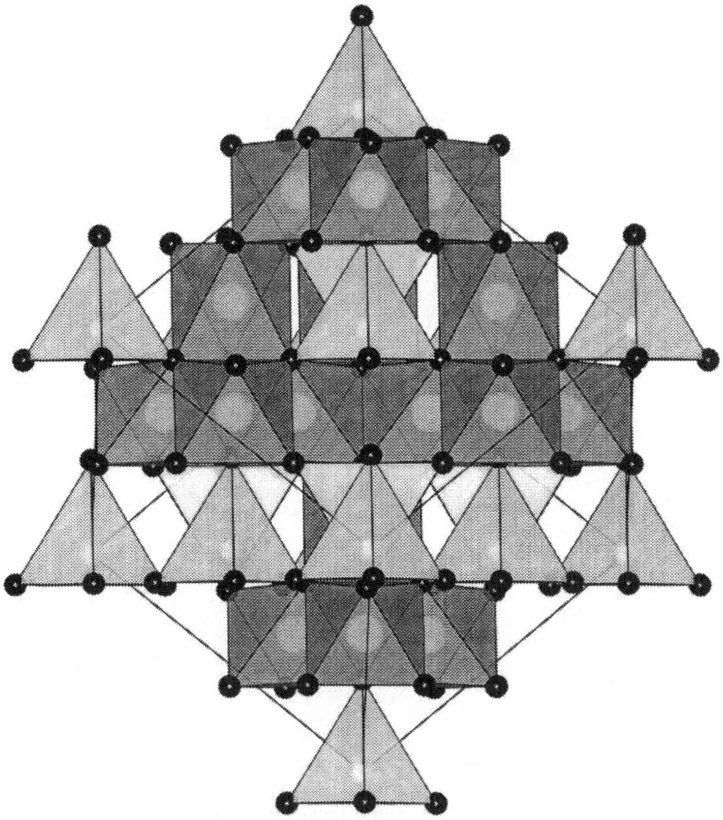

Fig. 8. Crystal structure of $LiTi_2O_4$ ((●) oxygen; TiO_6 octahedra and LiO_4 tetrahedra are shown).[34]

direction of the electrons at the Fermi level. The localized charge is coupled to the lattice, causing periodic lattice distortions through movements of the bonded atoms from their mean positions. This type of collective electron state directly competes with superconductivity. Although these materials were extensively studied in the 1960s and 1970s, they may deserve further study, because new ideas have been formed in the past decade concerning the behavior of complex electronic solids.

More influential as precursors to the cuprates are perovskite-related superconductors. In the 1970s, it was found that $SrTiO_3$, a classical ferroelectric perovskite, can be made superconducting through small amounts of electron doping—generally obtained by doping a small percentage of niobium on the titanium sites. The superconducting transition temperature is <1 K. This might not be significant, except that the creation of superconductivity by doping a semiconducting ferroelectric, even with a T_c at an extremely low temperature, piqued the interest of J. G. Bednorz and K. A. Müller, suggesting to them that oxides might someday be the source of new types of superconductors from a scientific standpoint.

The real precursor of the cuprate revolution was barium lead bismuth oxide, found to be superconducting in the 1970s.[18] This superconductor was made from the perovskite solid solution of the insulating compound $BaBiO_3$ in the metallic compound $BaPbO_3$, e.g., with bismuth and lead randomly mixed in the perovskite $Ba(Pb_{1-x}Bi_x)O_3$. There is one important feature of the metallic compound $BaPbO_3$ that was apparent to the community at large in the early 1980s. This was that the similarity in energy of the lead and oxygen valence orbitals meant that the conduction electrons were traveling in a highly hybridized electronic band, one which had significant oxygen character as well as significant lead character: a different case from what is usually encountered in conducting oxides. Moreover, $BaBiO_3$ is an exotic insulator. It is a classical perovskite, but instead of having Bi^{4+}, as would be suggested by simple electron counting, there are actually two different types of bismuth in the crystal structure, with localized valences near Bi^{3+} and Bi^{5+}. The Bi^{4+} is said to be "disproportionated" into Bi^{3+} and Bi^{5+}. This is simply understood from a chemical point of view, in that Bi^{4+} would have a single electron in a 6*s* orbital, a highly energetically unfavorable configuration ($6s^2$ is a lone-pair configuration). These bismuths are ordered in alternating B sites, forming a double perovskite. In other words, the formula can be written as $Ba_2Bi_2O_6$, in analogy with compounds similar to Ba_2LaNbO_6, a more familiar type of ordered 3+/5+ perovskite. The crystal structure of $BaBiO_3$ is shown in Fig. 11.

The idea behind attempting to find superconductivity in the mixed $Ba(Pb_{1-x}Bi_x)O_3$ perovskite solid solution was to smoothly vary the electron count across the metal-to-insulator transition as a function of bismuth content. Superconductivity appears at the boundary between the metallic and insulating phases, at ~12 K, at the relatively low bismuth concentration of $x = 0.25$. This superconducting transition temperature could be completely explained by the classical electron–phonon coupling mechanism. Nonetheless, the relatively high T_c in a perovskite structure and the admixture of metal and oxygen states at the Fermi energy captured the interest of a few physicists around the world, all becoming an important part of the early work in copper oxides, but especially stimulating Bednorz and Müller to pursue oxide perovskites further.

Although the vast majority of workers in the field believe $BaPb_{0.75}Bi_{0.25}O_3$ to be a conventional superconductor, I, for one, am not convinced. An alternative explanation is that the charge disproportionation of bismuth is at the heart of it and that an unconventional electronic mechanism is at work. Against my view is that the oxygen isotope effect (the shift of superconducting transition temperature with mass of the components, described in more detail below) is large, indicating strong coupling of the electrons to the lattice, consistent with a conventional mechanism for superconductivity. In favor of my point of view is the dramatic increase in T_c that occurs when the electronically active ion is completely bismuth, and the electron count adjustment is performed entirely by solid solution on the large-atom site: $Ba_{0.6}K_{0.4}BiO_3$ was found in 1988 to have a T_c of 30 K.[19] Another member of this family, made with antimony in solid solution with

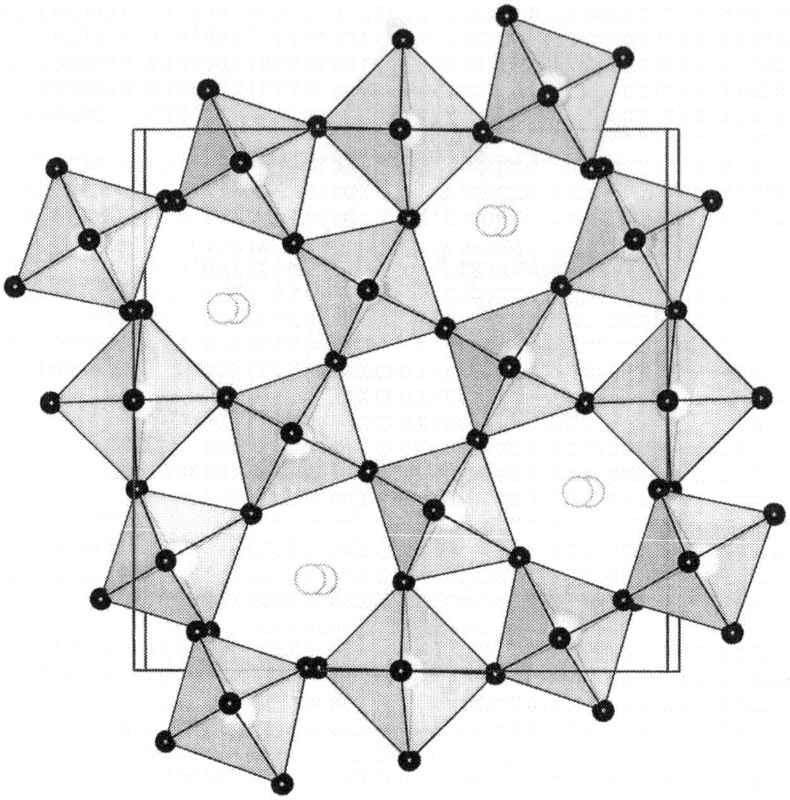

Fig. 9. Crystal structure of $Na_{0.28}WO_3$.[35] In this figure and the structural figures to follow, metal–oxygen coordination environments are shown as polyhedra. Oxygen are represented as small black circles and transition metals as small gray circles. Larger atoms, such as lanthanides, alkaline earths, and alkalis are shown, for clarity, as large circles without coordination polyhedra. For the important less-electropositive elements in the charge reservoir layers in the cuprates, such as lead, bismuth, tellurium, and mercury, the coordination polyhedra are shown.

lead on the B site, $BaPb_{0.75}Sb_{0.25}O_3$ has a T_c of 3.5 K. The order of magnitude difference in T_c values for these compounds never has been explained. These materials are not the subject of active research at the time of this writing, and they are perhaps most widely viewed as being of interest primarily because of their influence on the researchers to later discover high-T_c superconductivity.

IV. Discovery and Development of Copper Oxide Superconductors

The excitement that electrified the physical science community after the discovery of superconductivity above the temperature of liquid nitrogen in early 1987 spread into the popular press and from there to the public at large. There were many newspaper stories, TV news reports, and several best-selling books written about the events surrounding the initial discoveries. These popularizations made some into heroes, some into villains, some minor players into geniuses, and some major players invisible. None of them ever matched my view of what actually happened. The accounts in this country seemed always to be a blend of the actual facts with American cultural values and stereotypes—such as our national pride, our desire to root for the underdog, and our need to associate remarkable historical events with the great deeds of individuals rather than the collective efforts of many different people.

Hopefully, by considering the above summary of the world of superconductivity as it was in 1986, the reader has enough background to appreciate what a remarkable discovery the two scientists Georg Bednorz and Alex Müller made and published in late 1986.[20] They broke all the rules about what electronic ceramics could do and what good superconductors should look like. Pursuing their own idea about how superconductivity might occur in oxides, they were looking for superconductivity in perovskite structure oxides based on nickel and copper. They attempted for some time to find superconductivity in nickel oxides, without success. They then turned to copper oxides, carefully considering the reports of metallic conductivity in the ternary cuprate $La_4BaCu_5O_{13}$ by the group led by the French chemist Bernard Raveau. Their initial report, cautious and objective, published in *Zeitschrift für Physik*, reported the possible existence of superconductivity in the La-Ba-Cu-O chemical system, an oxide ceramic, at the surprisingly high temperature of 28 K—not that much higher than was known for the best intermetallic superconductor (23 K), but in a completely different type of material. The actual chemical compound responsible for the observed superconductivity was not known. The results of their work became known in the very small circle of scientists interested in superconductivity in November 1986.

While Bednorz and Müller continued to work on their discovery in Switzerland, two groups on opposite sides of the world began to follow up on their work immediately: one at the University of Tokyo and one at the University of Houston. Both groups reported some of their preliminary results at the annual meeting of the Materials Research Society (MRS) in early December 1986. The group at the University of Tokyo was under the leadership of the visionary materials scientist and physicist Shoji Tanaka. Tanaka, like Bednorz and Müller, believed that there was something very unusual about the superconductivity in barium lead bismuth oxide and was very quick to realize the possible importance of the discovery. The group at the University of Houston was under the leadership of C. W. (Paul) Chu, well-known in the community of scientists working on superconductivity in the United States and an expert in the measurement of the properties of materials under very high applied pressures. Similar to the others in the story, he believed that superconductivity in exotic materials at unexplainable temperatures was possible, and his group moved quickly to look into the Bednorz and Müller announcement.

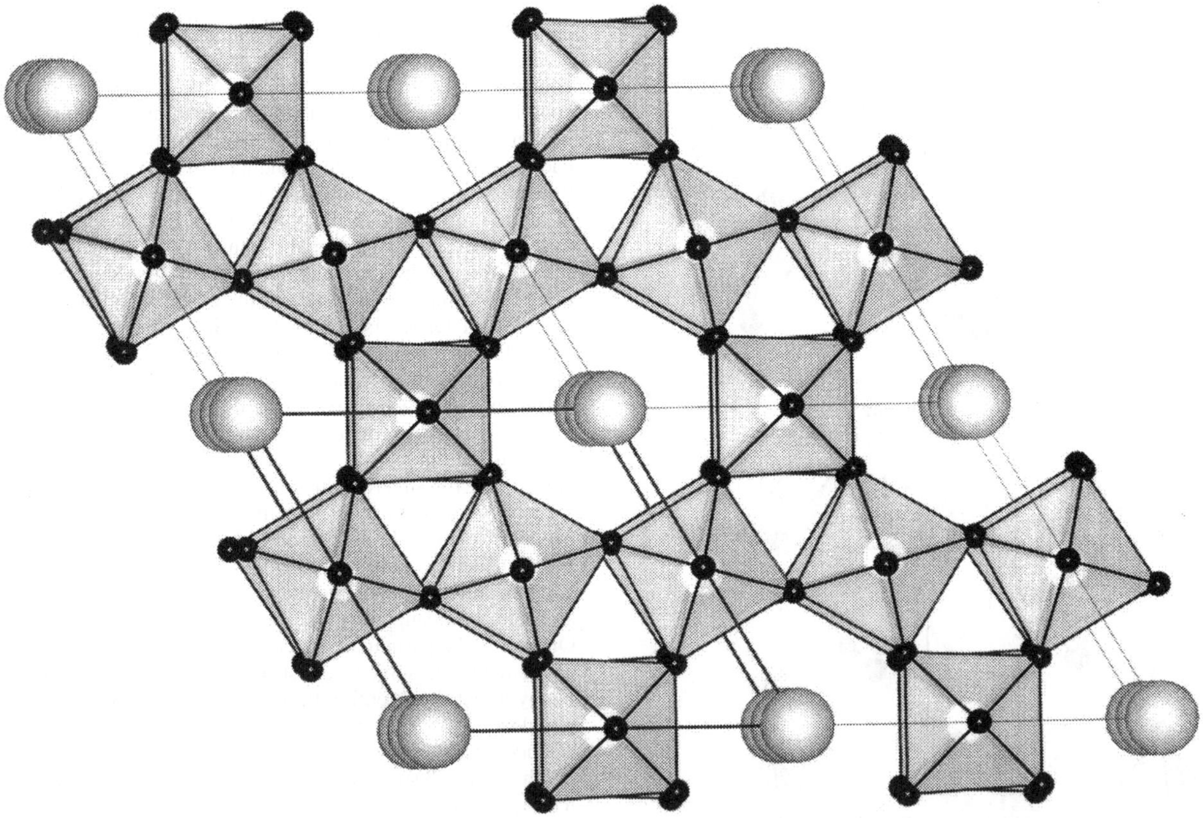

Fig. 10. Crystal structure of $Rb_{0.27}WO_3$ (see Fig. 9 for legend).[36]

Two important things were revealed in the early work of these two groups, aside from their confirmation that the surprising announcement of the European team was indeed correct. Chu's group found that the superconducting transition temperature in the original formulation increased significantly under the application of pressure; i.e., T_c increased as the crystal lattice got smaller. The Japanese group was attempting to determine which compound in the La-Ba-Cu-O chemical system was actually superconducting. I was sitting in the seminar room at Bell Laboratories a few days after the MRS meeting at an impromptu talk given by Koichi Kitazawa from the University of Tokyo when a telephone was brought into the room and he called from the United States to the laboratory at the University of Tokyo (where it was 3 a.m.) to get the report that they had just determined that the superconducting compound was the layered K_2NiF_4-type oxide solid-solution $La_{2-x}Ba_xCuO_4$. The crystal structure is shown in Fig. 12. A few more groups, including ours at Bell Laboratories, began its research after the MRS meeting, and, before Christmas 1986, the strontium analog of the K_2NiF_4-type compound, $La_{1.8}Sr_{0.2}CuO_4$, was found to be superconducting at the then astonishingly high temperature of 38 K, consistent with the high-pressure results that smaller lattices (strontium is smaller than barium) might lead to higher transition temperatures.

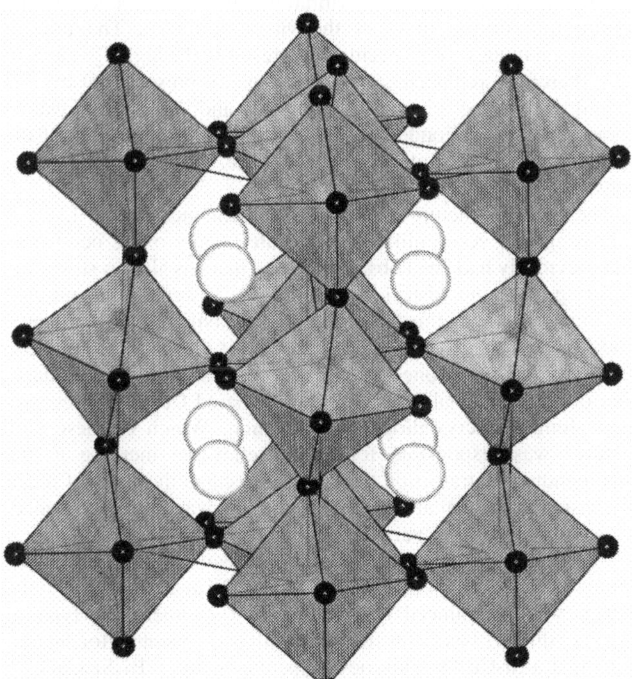

Fig. 11. Crystal structure of $BaBiO_3$ (see Fig. 9 for legend).[37]

Almost everyone in the scientific community remained oblivious to what was happening. Although many of those who noticed argued that the new materials could easily be explained by a conventional picture for superconductivity, theoretical physicist and Nobel laureate Philip Anderson published a paper very early arguing that an exotic and completely unconventional mechanism must be behind the superconductivity. The few groups actively pursuing the problem worked through the holidays and through New Years 1987. It was clear that something remarkable might happen any time. A report then appeared from the group of Z. X. Zhao at the University of Beijing that very tiny amounts of superconductivity occurred in the original Bednorz and Müller formulation at a temperature of 58 K. I can remember that copies of the article announcing the result in the Beijing Peoples Daily newspaper (in Chinese of course) were very hot items for a week or so and that the only recognizable character in the article was the chemical symbol "Pb," prompting considerable effort on our part to find Chinese colleagues who could tell us what lead had to do with it (they had used superconducting lead metal as a standard for

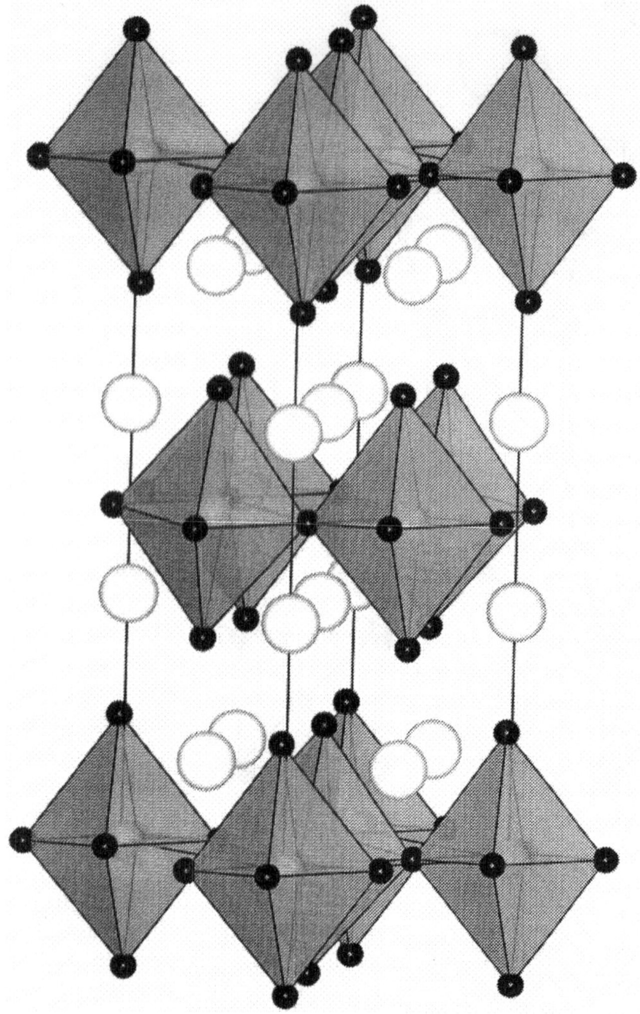

Fig. 12. Crystal structure of La_2CuO_4 (see Fig. 9 for legend).[38]

calibration). This result could not be reproduced in other laboratories, was soon eclipsed, and, therefore, is often omitted in the telling of the story. In retrospect, it was correct, because later research would show that there was indeed a higher-temperature superconductor in the La-Ba-Cu-O system (of the 123 type). Things were already getting almost out of control—there were stories circulating about spies and secret late-night telephone calls; we probably will never know exactly what was going on.

Then came the event that attracted everyone's attention. The story began to circulate that M. K. Wu at the University of Alabama and Paul Chu at the University of Houston (Wu was one of Chu's former students) had found a superconducting copper oxide with a transition temperature of >90 K. The superconductor was also rumored to be green, by people who had actually *seen* it, which was a bizarre claim, inconsistent with what most believed to be possible for a conducting oxide, but given that, if true, everything we knew was wrong anyway. They had substituted the smaller atom yttrium for the lanthanide lanthanum in the original Bednorz and Müller formula in an attempt to shrink the lattice, as was suggested by their high-pressure data, and the T_c had gone up by an incredible 50 K!

This is the part of the story that has been told many times in the popular accounts. The original paper was submitted with the incorrect chemical components preventing the leaking of the results by the referees (always claimed to be an inadvertent error by the authors, but something that would have been a very good idea anyway). The initial report did not identify the superconducting compound and suggested that this was such a departure from the ordinary that the superconductivity might not be due to a compound at all, but rather due to complex electronic interactions at chemical interfaces. The identification of the superconducting compound (the 123 superconductor $YBa_2Cu_3O_7$) soon revealed a completely different crystal structure than that of the Bednorz and Müller superconductor, and the era of high-T_c superconductivity had begun. The 123 superconductor, the first found to have a transition temperature above the temperature of the boiling point of liquid nitrogen (77 K), remains one of the few that is considered to have the potential for real commercial application. (As of the time of this writing, more than a dozen years after the determination of its composition and the filing of the patent applications, the U.S. Patent Office is yet to determine who rightfully holds the patent to this material. Aside from the ancient Zen question: "What is the sound of one hand clapping?" no other single question, it seems, has ever been more difficult to answer.)

Now that superconductivity above the temperature of liquid nitrogen was a reality, all types of new applications might be envisioned for superconducting materials, based on the fact that expensive, difficult to handle liquid helium would no longer be needed to cool the superconductors to temperatures close to absolute zero. Many were quite enthusiastic about what could happen, and there was much speculation about how our lives might someday be changed by the discoveries of the past few months. One had to be particularly stoic or bitter not to be caught up in the excitement, and some scientists outside of the maelstrom were not happy about the attention these developments were getting in the popular press. This was not, however, business as usual for science, and there was no reason to treat it as such.

The new 123 superconductor, similar to many electronic ceramics before it, was found to have a widely variable oxygen stoichiometry, The oxygen content determined whether it was superconducting. A few scientists quickly realized that this ability of the 123 compound to breathe oxygen in and out would allow them to test whether the superconductivity in copper oxides could be due to the same mechanism operating in the intermetallic superconductors. They remembered that the "conclusive evidence" for the conventional understanding of superconductivity by coupling of the electrons to the vibrations of the crystal lattice was proved by the observation of a change in superconducting transition temperature with the mass of the constituent atoms—an isotope effect. The transition temperature in the electron–phonon picture is proportional to the frequency of the lattice vibrations, and the frequency of lattice vibrations, in turn, depends on the square root of the mass of the vibrating atoms. The oxygen vibrational modes were speculated to result in the high T_c values in 123; therefore, replacement of the normally present ^{16}O by ^{18}O should lower the phonon frequencies and T_c if this was a conventional superconductor. ^{18}O did lower the frequency of the vibrational modes, but T_c changed little. These could not be conventional superconductors!

The crystal structure of the 123 superconductor (Fig. 13) was surprising and very unlike the structure of any oxide perovskite that previously had been observed: a fact that has always struck me as fitting for such a historical material. There were two significant differences between this material and the K_2NiF_4-type forerunners. The first was that there were two layers of CuO_2 planes in close proximity instead of one. The second was that there were remarkable one-dimensional chains of corner-shared CuO_4 squares running along one crystallographic direction. Which of these new features was responsible for the unbelievable increase in the superconducting transition temperature became the topic of much discussion and speculation. This speculation was soon to be ended, however, by the discovery of the next new class of materials.

The French group under Bernard Raveau, which knew more about the chemistry of this type of material than anyone else on the world, soon announced the existence of another new type of superconductor, based on the substitution of bismuth for lanthanum in La-Sr-Cu-O. This new superconductor ("$Bi_2Sr_2CuO_6$"), which was found to have a crystal structure different from 123 and the K_2NiF_4 materials, ultimately turned out to be the first of a gigantic class of new materials to follow. Its T_c was <10 K,

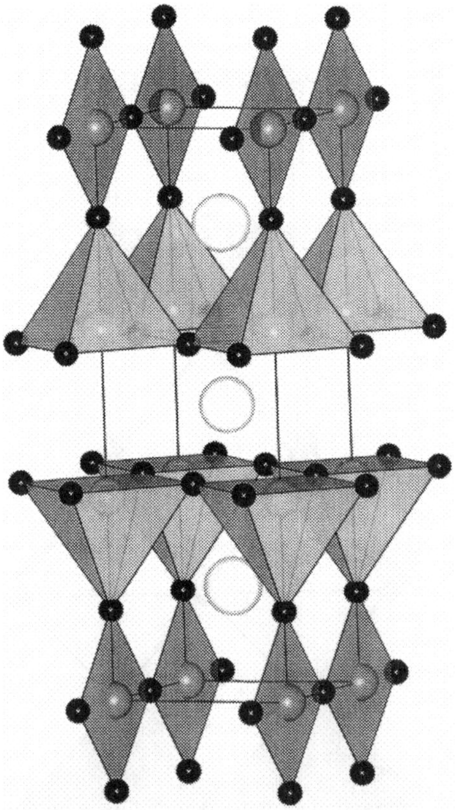

Fig. 13. Crystal structure of $YBa_2Cu_3O_7$ (see Fig. 9 for legend).[39]

however, and it attracted relatively little attention at first, because most were struggling to understand what could be going on in the 123-based materials. Excited by the developments in the new superconducting materials, T. Maeda, from NIRM in Tskuba, Japan, began experimenting with the French superconductor by adding other elements to the mixture. What he found changed everything. By adding calcium to the mixture, the superconducting temperature of Bi-Sr-Cu-O increased to >80 K, and there was good evidence that parts of the sample were superconducting at temperatures near 100 K. In the now often repeating pattern, the composition and structures of the superconductors were not known, and another race to identify the superconducting compound or compounds was initiated.

The crystal structure of the 80 K superconductor in the Bi-Sr-Ca-Cu-O system is shown in Fig. 14. The approximate formula is $Bi_2Sr_2CaCu_2O_8$ (the real formula has deviations from these simple integer ratios of constituent atoms). There are no copper oxide chains and, therefore, that ended all discussion about whether those chains could be the reason for the high T_c in the 123 compound. What this does have in common with the 123 compound is a double layer of CuO_2 planes. Eventually the superconductor in this system with an even higher T_c (~110 K) was isolated. Its ideal formula is $Bi_2Sr_2Ca_2Cu_3O_{10}$: It had *three* sets of nearby CuO_2 planes. This compound, stabilized by partial substitution of lead for bismuth and the 123 superconductor $YBa_2Cu_3O_7$ are to this day the high-T_c superconductors most likely to see practical application. This is the point in history (1988) when it was realized that higher T_c values were more likely in compounds with a larger number of CuO_2 planes in proximity. This remains the general rule, although several materials with single CuO_2 layers are now known to have superconducting transition temperatures near 90 K. This was also the time when the common function of the chains and the bismuth oxide layers in these new materials was realized to be that of a charge reservoir, serving to control the amount of electron or hole doping in the CuO_2 planes.

1988 was a very good year for the discovery of new superconducting materials. It became clear that the way to find a new

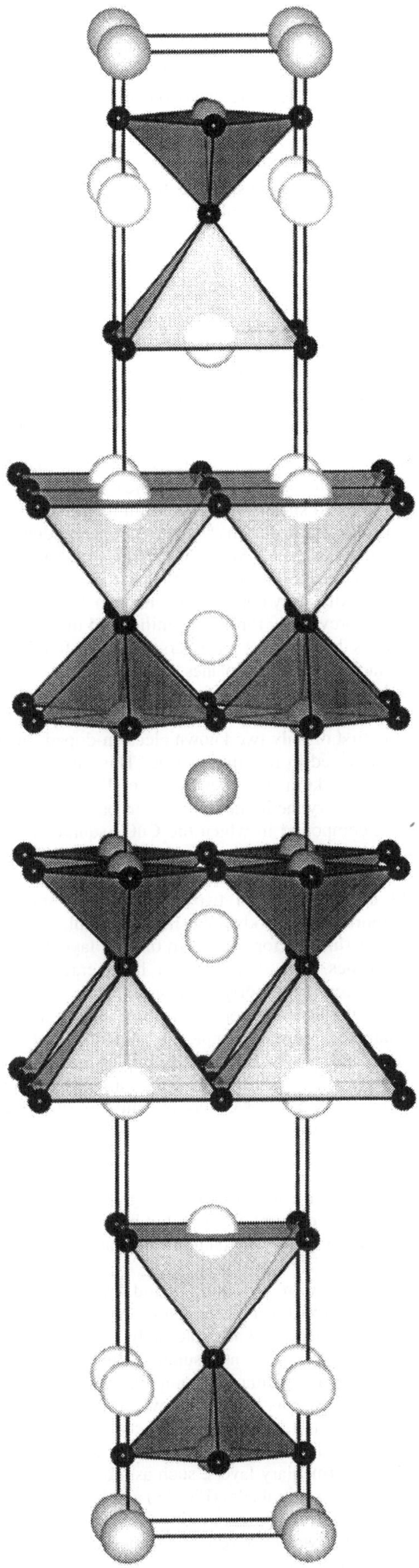

Fig. 14. Crystal structure of $Bi_2Sr_2CaCu_2O_8$ (see Fig. 9 for legend).[40]

superconductor was to find a new atomic constituent to put into the charge reservoir layer and then to change the number of CuO_2 planes. This was easier said than done, but soon to follow were materials with thallium and lead in place of bismuth in the intermediary layer, stacked with different numbers of CuO_2 planes. Thallium and lead are analogous in size, electronegativity, and chemistry to bismuth, but, because of serious complications in the synthesis of the compounds, they are nontrivial extensions of the bismuth-based materials. The key to their accommodation in the CuO_2-based perovskites is that their size is such that edge-shared octahedra and pyramids in bismuth, lead, and thallium oxygen layers have the same in-plane size as corner-shared CuO_4 squares; this is the same size criterion that determines whether the perovskite structure is stable in any chemical system. The thallium-based materials had as their distinguishing characteristic very high superconducting transition temperatures, holding the record for a very long time; they also were "self-doped." Because of the relative energies of thallium and copper orbitals, there is natural doping of the CuO_2 layers at the stoichiometric compositions. One of these materials, $Tl_2Ba_2Ca_2Cu_3O_{10}$, is shown in Fig. 15. The lead-based compounds had as their distinguishing characteristic a more complex charge reservoir layer than had been previously seen, involving Pb^{2+} pyramids and Cu^{+} sticks, both chemically reduced species in equilibrium in the same compound with chemically oxidized $Cu^{2.2+}$ in the superconducting layers. One of these materials, $Pb_2Sr_2Y_{1-x}Ca_xCu_3O_8$, is shown in Fig. 16. These lead-based materials also had the characteristic that, because of their complexity, they could not be made in as many variants as had been seen previously for other families and that their T_c values never were as high as the members of other families with an equivalent number of CuO_2 planes.

Given the whole history so far, it should not be obvious that electron doping of a copper oxide should lead to superconductivity as well. The first of only two known electron-doped superconductors was discovered in this time period. This discovery, from the group of Yoshi Tokura at the University of Tokyo, came as a very big surprise to everyone in the field. It was based on the doping of Nd_2CuO_4, a compound in which the CuO_2 squares in the planes are not capped with further oxygen in their apical positions (Fig. 17). The doping to reduce the formal valence of copper was eventually accomplished in two ways: partial substitution of cerium or thorium for neodymium in a solid solution and partial substitution of fluorine for oxygen in oxygen layers isolated from the CuO_2 planes. This material has been enigmatic since its discovery and remains poorly understood. It is not sufficient to prepare the material at the correct electron-doping level to obtain a superconductor. Certain complex treatments must be used to induce superconductivity, just slightly milder than conditions that clearly lead to decomposition of the phase. Moreover, for all superconducting preparations, there is a mixture of phases present with the same overall structure but with different short-range-ordered structures. No one knows which one is the superconductor or what the difference between them is. Years later, a second *n*-type superconductor was discovered, again apparently based on CuO_2 planes with no apical oxygens, but it can be synthesized only by a high-pressure synthesis technique and has not been the subject of detailed study.

The early 1990s were a grueling period if your business was the discovery of new oxide superconductors. All the simple combinations of elements that could be explored by conventional synthesis had been attempted, and the new superconductors that were found were structurally and chemically complex, did not have T_c values that rivaled the materials already known, and did not, in general, appear to be pointing to any new aspects of the physics of the materials. Nonetheless, the truly hardy did find new materials with new types of intermediary layers, such as those based on intermediary layers of GaO_4 tetrahedra (Fig. 18) and NbO_6 octahedra (Fig. 19). There was, however, a continuing undercurrent of research in new materials that was to have serious consequences in the years to follow: one of the most interesting stories associated with the discovery of oxide superconductors.

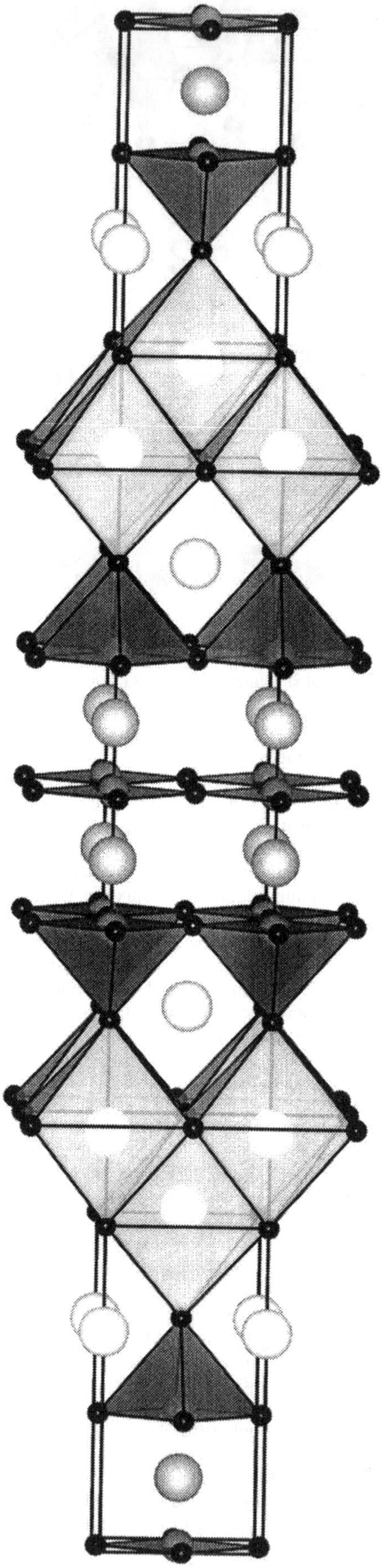

Fig. 15. Crystal structure of $Tl_2Ba_2Ca_2Cu_3O_{10}$ (see Fig. 9 for legend).[41]

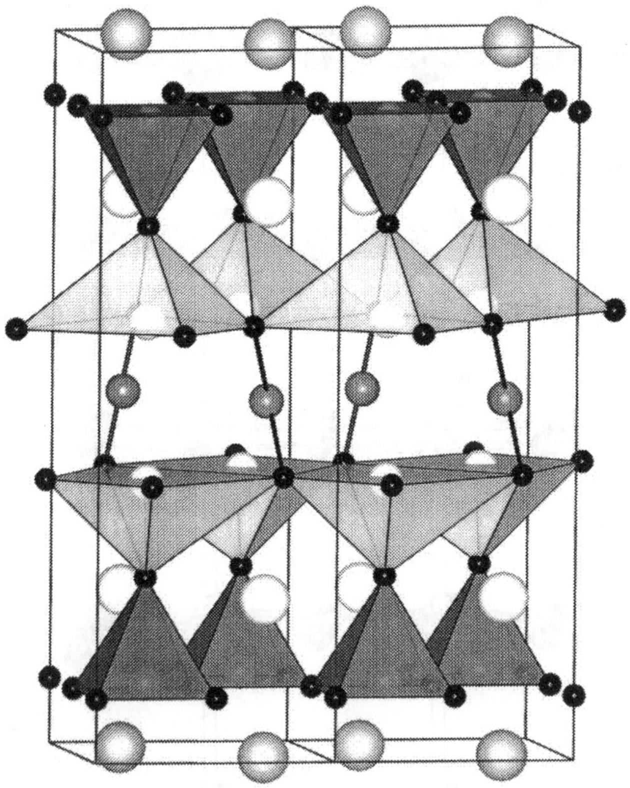

Fig. 16. Crystal structure of $Pb_2Sr_2YCu_3O_8$ (see Fig. 9 for legend).[42]

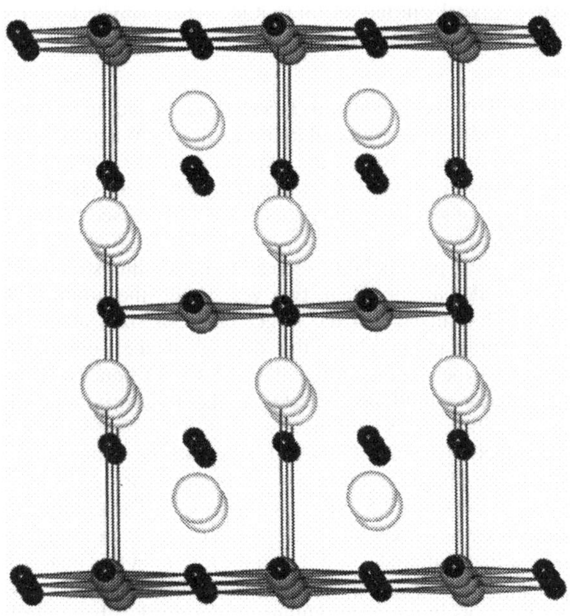

Fig. 17. Crystal structure of Nd_2CuO_4 (see Fig. 9 for legend).[43]

The story begins in 1988, when Bob Roth and co-workers, studying the phase-equilibrium diagram for the Sr-Ca-Cu-O chemical system as part of a much larger project to characterize the important (and complicated!) Bi-Sr-Ca-Cu-O chemical system, found a remarkable chemical compound of composition $Ca_{0.84}Sr_{0.16}CuO_2$. The composition of the compound was very precise, with no measurable variability, at this peculiar Ca:Sr ratio. The determination of the crystal structure by Theo Siegrist showed that this was a compound made entirely of CuO_2 layers separated by alkaline-earth layers made of a solid solution of calcium and strontium mixed *randomly* but at one particular ratio.[21] This

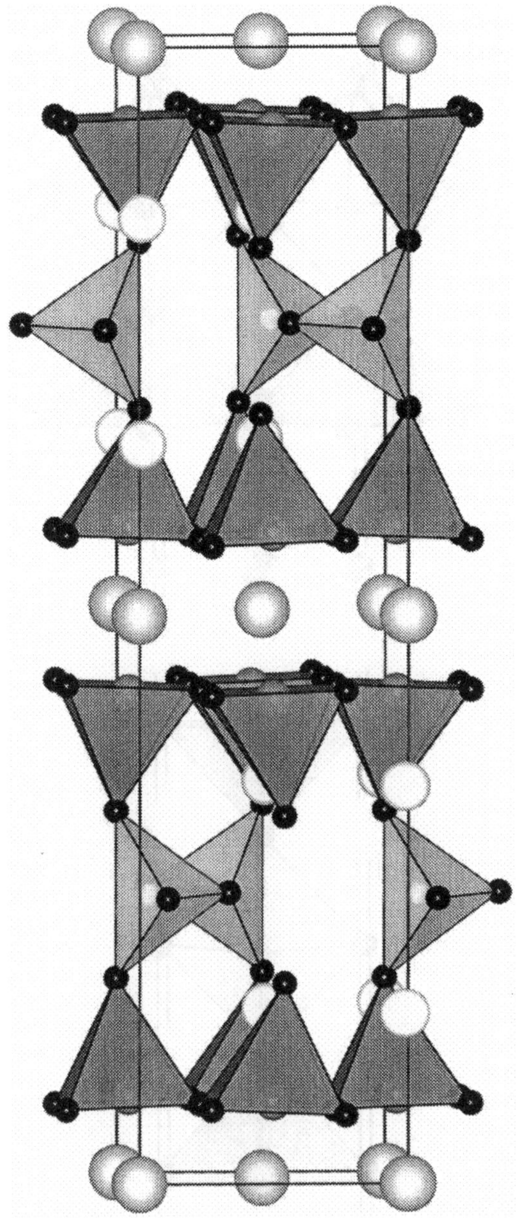

Fig. 18. Crystal structure of $GaSr_2(Y,Ca)Cu_2O_7$ (see Fig. 9 for legend).[44]

important material is variously known as the "parent structure of the cuprate superconductors," or "the infinite-layer phase." It consists only of the electronically active CuO_2 planes responsible for superconductivity (Fig. 20). If it could be chemically doped, then its T_c might be extremely high!

However, it apparently could not be done easily. The precise Ca:Sr (whose sizes differ by only 10%) ratio needed to form the compound indicates that the compound is sensitive to about the 0.1% level to the precise matching of size of the CuO_2 plane to the Sr + Ca spacing layer. All attempts at doping by conventional means failed, apparently because of this extreme size sensitivity. Then came an important insight from the group of M. Takano in Japan: Because of the great differences in compressibility of the alkaline-earth–oxygen and copper–oxygen bonds, synthesis under high pressures might allow the range of relative sizes of the atoms in the compound to be expanded. Synthesis under high-pressure/high-temperature conditions often leads to the formation of phases that are metastable only under ordinary conditions of pressure (e.g., diamond!). Thus, in 1989, a year after the announcement of the existence of the phase, the Japanese group reported that, during synthesis at 40 000 or more atmospheres (4 GPa) of pressure at 1000°C, the stability of the all-layer phase could be expanded to a

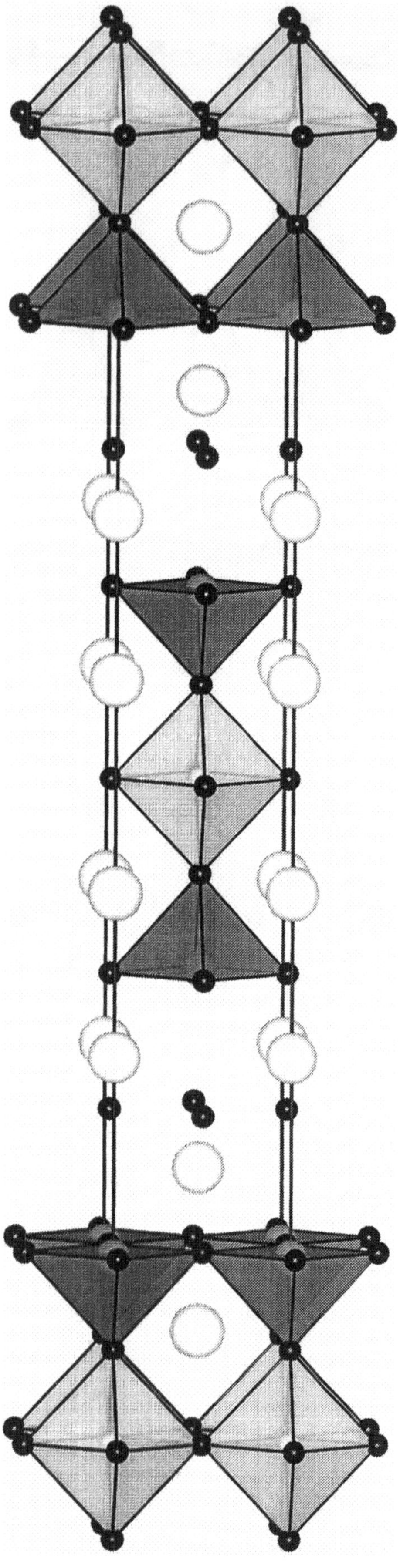

Fig. 19. Crystal structure of $NbSr_2(Nd,Ce)_2Cu_2O_{10}$ (see Fig. 9 for legend).[45]

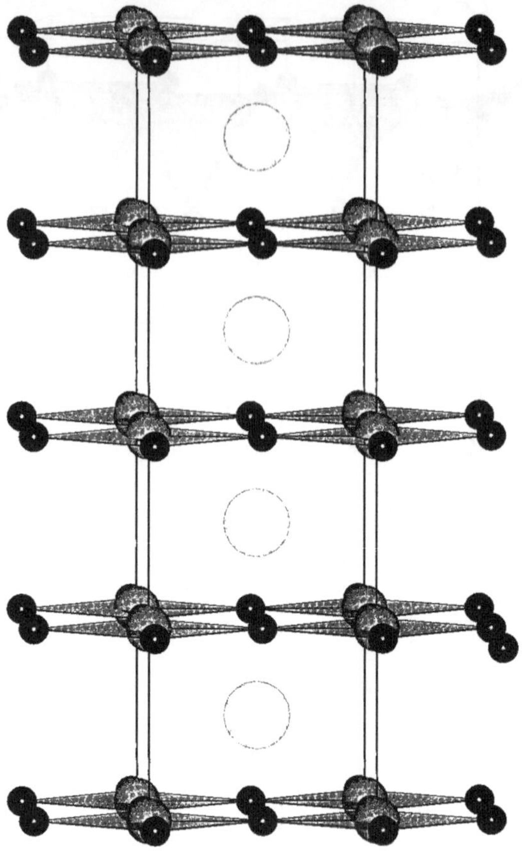

Fig. 20. Crystal structure of $(Sr,Ca)CuO_2$—the "infinite layer phase" (see Fig. 9 for legend).[46]

wider range of sizes, notably to *larger* sizes, including mixtures of strontium and barium on the alkaline-earth site. When these larger atoms went in, they forced the expansion of the CuO_2 plane to allow them to fit. At the time, there were few groups in the world with the type of synthesis apparatus that could be used for this type of experiment.

For the most part, nobody seemed to notice this structural and synthetic undercurrent. Then, two years later, the group of John Goodenough realized that the larger range of sizes attainable at high pressures might allow the doping of the compound, in particular *n*-type doping, for which the Cu–O bond with lower formal charge would be expected to be relatively long. They announced, three years after the discovery of the phase, that *n*-type superconductivity with T_c = 40 K could be induced in the infinite-layer phase in materials of formula $Sr_{1-x}Nd_xCuO_2$ by synthesis at 25 kbar (2.5 MPa) and 1000°C.[22] The community at the time was, however, notably one-dimensional in thought (the only thing that mattered was T_c), and the low T_c of this material did not at first generate a perturbation in the direction of the field in proportion to its importance. Things changed radically in late 1991 and in 1992 when a Japanese group announced the existence of superconductivity at temperatures up to 110 K in the *p*-type-doped infinite-layer phase.[23] The doping was accomplished through the introduction of strontium deficiency in randomly arranged defect layers in the all-layer phase of composition $Sr_{1-x}CuO_2$ synthesized at 60 kbar (6 MPa) and 1000°C. At a T_c of 110 K, people took notice. The most important consequence of this notice was that it became evident to many that high-pressure synthesis was the road to the discovery of new superconducting materials and could break the logjam plaguing the field for the past four or more years.

Again, this is easier said than done, and it took several years for groups to gear up to this new expensive and elaborate synthetic method. The result was a great new flurry of discovery of superconducting compounds, starting in 1994, based almost exclusively on the synthesis of materials under high pressures. The most notable of these are the Ba-Ca-Cu-O and Ba-Ca-Cu-C-O multiple-layer cuprates. Although not initially discovered by high-pressure methods, the synthesis of the large family of superconductors with mercury intermediary layers has been greatly facilitated by high-pressure synthesis, needed to prevent mercury from volatilizing. This is now virtually the only way these important materials are synthesized in bulk form. This family of superconducting compounds is based on intermediary layers of a single plane of O–Hg–O sticks, as, for example, is shown in Fig. 21. (Double mercury layer materials also have been synthesized.) The triple layer variant of these, $HgBa_2Ca_2Cu_3O_9$, is currently the high-T_c record holder at ambient pressure (134 K) and when measured at high pressure (164 K).

In the same time period, intermediary layers that involved the mixing of CO_3 groups and CuO_x polyhedra in the intermediary layers, with the charge on the superconducting CuO_2 planes controlled by the Cu:C ratio, were discovered to result in a family of superconductors. High-pressure synthesis has been used extensively to make these materials. An example of one of them is shown in Fig. 22. The highest T_c values (>100 K) in this family are obtained in the Ba-Ca-Cu-C-O system when synthesized at high pressures. A very interesting structural phenomena in the carbonate family is the discovery of a series of complex multiple-layer cuprates by the Raveau group, in which TlO- and CO_3-based intermediary layers alternate.[24] Aside from the more straightforward $Y_2Ba_4Cu_7O_{15}$ superconductors, in which single-chain CuO and double-chain CuO intermediary layers alternate, these are the only materials known with more than one type of intermediary layer in the same compound; a new type of structural principle not yet fully explored.

The influence of high-pressure synthesis and measurement continues to be profoundly felt in the field, and it is discussed in context of current directions. However, one of the most interesting parts of the story of the all-layer phase came years after the discovery that it could apparently be made superconducting. Further study of the p-type doping of the all-layer phase in 1997 by an independent group of researchers revealed that the superconductivity above 100 K in that material was not from the all-layer phase but, rather, due to inadvertent second phases based in the Sr-K-Cu-O-Cl system![25] The potassium and chlorine in the superconducting impurity phases are because $KClO_3$ is used in the synthesis at high pressures as a source of extra oxygen (it decomposes at high temperatures to KCl + oxygen). Ostensibly separated from the main reactants, it nonetheless managed to get into the sample and form superconducting compounds. The superconducting compounds, later studied in more detail, are $Sr_{3-x}K_xCu_2O_4Cl_2$ and $Sr_{2-x}K_xCuO_2Cl_2$. Sodium-doped variants are also known. The all-layer phase cannot be made into a p-type high-T_c superconductor, only into an n-type superconductor.

Only a few groups around the world continue to look for new superconductors among the copper oxides. For the most part, research in "new materials" has consisted of adding a variety of elements to already known superconductors, sometimes with the intention of exploring one aspect of the superconductivity or another and sometimes for no obvious reason. From a chemical or structural consideration, they do not, in general, represent new directions. Nonetheless, there have been some interesting discoveries in recent years. These are described in further detail in a later section.

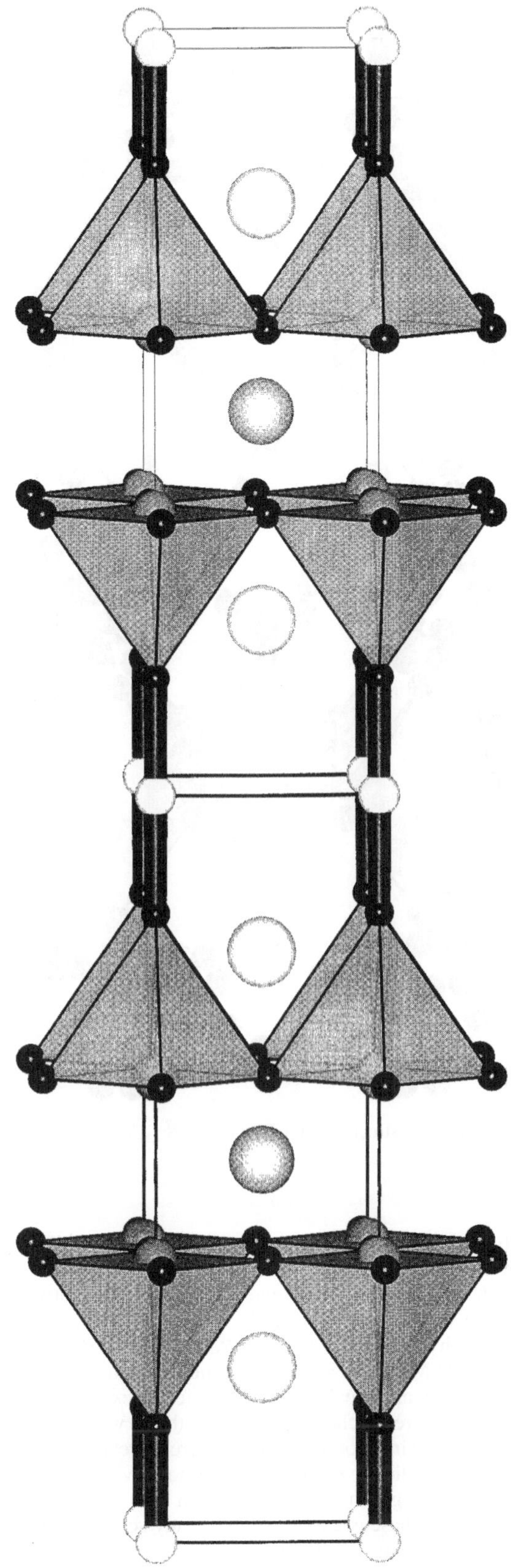

Fig. 21. Crystal structure of $HgBa_2CaCu_2O_6$ (see Fig. 9 for legend).[47]

V. Historical Sideshows and Final Count of Cuprate Superconductors

The sidetracks, sideshows, false alarms, and downright fraudulent claims in the history of the field of high-T_c superconductivity have been an ever-present and often important part of the story since the field's inception. Some of the claims of the observations of very small amounts of very-high-T_c materials (e.g., at the recurring apparent T_c value of 240 K) remain real mysteries to this day, because no obvious flaw in logic or mistake in the experimental procedure is evident. These observations have never been reproducible in other laboratories following the reported procedure—this is logical, in a sense, because the extreme superconductivity is always present at the 1 part in 10 000 level or lower. It is impossible to estimate, however, how many hours have been

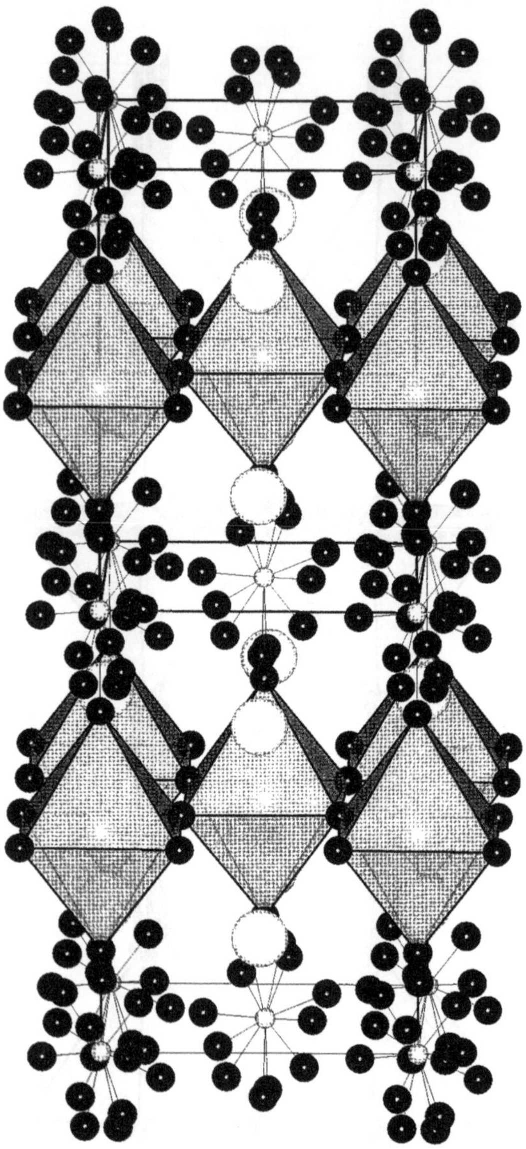

Fig. 22. Crystal structure of $(C_{1-x}Cu_x)(Ba,Sr)_2CuO_y$ (see Fig. 9 for legend).[48]

spent by the materials community attempting to duplicate what appeared to be spectacular results and which were mistakes in judgment, inadvertent error, overzealousness, or outright lies. During the early years of high-T_c research, I came to call the prestigious physics journal *Physical Review Letters* the "Journal of Irreproducible Materials Science." In the first years of research in the field, there was virtually no negative feedback by the scientific community for researchers who made these wild claims. The attention such claims got in the popular press created considerable havoc and never did the retraction of the story eventually shown to be incorrect ever get the type of attention the original announcement received. These events frustrated many scientists, especially those working outside of the field, and left some with a negative attitude for high-T_c research. The great majority of workers in the field, however, were not working in that mode. The relationship between the media and scientists that evolved during the heady days in high-T_c research may have had an influence in the disastrous "cold fusion" folly that followed, because the media were primed with the expectation of spectacular announcements in science and are never nearly as critical of scientists' statements as they are, for example, about the assertions of politicians. In a very small way, primarily through exaggeration or error, incorrect claims about new superconductors or their properties continue to this day. I am sure it is difficult for the nonexpert to tell what is real and what is not.

Koichi Kitazawa coined the perfect term for these announcements of outrageously high transition temperatures in 1987: USOs. This is actually a triple entendre: USO stands for "unidentified superconducting object." The obvious analogy to UFO conjures up the appropriate implications about their verisimilitude, and the character of the people who chase them, and, finally, USO means "lie" in Japanese. USO sightings continue to this day, for example, the recent report of "superconductivity" at 91 K in the surface layer of a sodium tungsten bronze.

Some of the origins of USOs and confusing claims are presented in the following, as examples of the variety of reports that can obfuscate the issues when a group reports the existence of a "new superconductor," making keeping track of things difficult.

(1) Experimental error or lack of scientific judgment is always the most spectacular claim; there are many examples of them in the literature. It would not be appropriate to point out particular examples. They continue to this day. Many are determined to be obvious nonsense after looking at the data for one minute, but, for some, it is more difficult to discover the mistake. An all too typical error is lack of good thermal contact between the sample and the thermometer measuring its temperature, resulting in the sample being colder than the claimants believe.

(2) There are sincerely reported and apparently experimentally sound but confusing results. The most famous example is the 240 K USO sighting caused by the "reverse Josephson effect" (even most physicists cannot easily recall exactly what that is) in Y-Ba-Cu-O. The authors claim there can be no other explanation for their observation other than parts per million of superconductivity present in their sample, turning on at 240 K. This 240 K value has appeared in many USO sightings by other groups since. I have never heard a convincing argument as to why these experiments might be incorrect, although one may have escaped me.

(3) There continue to be reports of a "new" superconductor in a chemical system that already has a known superconductor in it, made from some subset of the elements present. There are almost countless examples of these claims in the literature. One example is the report of a new layered-structure superconductor in the Ba-K-Pb-Bi-O system—the Ba-K-Bi-O and the Ba-Pb-Bi-O systems have superconductors in them, and the combination of all the elements is also superconducting, but from the known superconducting phases. It is possible for more than one superconducting compound to exist in the same chemical system, e.g., $Tl_2Ba_2CaCu_2O_8$ and $Tl_2Ba_2Ca_2Cu_3O_{10}$, but the case has to be carefully proved. Errors of this type are difficult for nonexperts to identify. The culture of contemporary science makes it inappropriate to publish papers disagreeing with these types of reports; therefore, they reside in the literature unopposed.

(4) The superconducting phase actually comes from a combination of elements or an element not believed to be present by the experimenter. A good example is *p*-type superconductivity reported for the "all-layer" Sr-Cu-O phase, which actually came from Sr-K-Cu-O-Cl, as described previously. More insidious are, for example, reports of superconductivity in Sr_xNbO_y and related oxides. The superconductivity actually comes from the presence of filaments of niobium metal in the sample, or worse, from NbN, which forms very easily when there is even a small amount of residual air or nitrogen present during synthesis.

(5) A "new" superconductor is sometimes reported when the authors have merely substituted one minor constituent for another in a known superconductor. That the 123-type phase does not form in the Sr-Y-Cu-O system without being stabilized by a small amount of another metal ion mixed in on the copper chain site is responsible for an almost countless number of these announcements. A "new" superconductor is not announced every time $YBa_2Cu_3O_7$ is doped with a minor amount of some additional metal ion, and the same should be the case in the Sr-Y-Cu-O system. Partial substitution of 10–15 elements in the periodic table allows for the stabilization of the 123-type phase in this system.

Table I. Structurally or Chemically Distinct Copper Oxide Superconductors and Year of Discovery

Superconductor	Year
$(La,M)_2CuO_4$	1986
La_2CuO_{4+x}	1988
$La_2CuO_4F_x$	1988
$(Nd,Sr,Ce)_2CuO_4$	1989
$(Nd,Ce)_2CuO_4$	1989
$Nd_2CuO_{4-x}F_x$	1989
Sr_2CuO_{3+x}	1993
$Sr_2CuO_2F_{2+x}$	1994
$(Ca,Na)_2CuO_2Cl_2$	1994
$(Ca,Na)_3Cu_2O_4Cl_2$	1995
(Sr,K) versions of previous two also are superconducting	
$(La,Sr)_2CaCu_2O_6$	1990
$(Sr,Ca)_2(Sr,Ca)_{n-1}Cu_nO_x$ (n = 2, 3, 4)	1993
$PbBaSr(Y,Ca)Cu_3O_x$	1990
$Sr_{1-x}Nd_xCuO_2$	1991
$YBa_2Cu_3O_7$	1987
$YBa_2Cu_4O_8$	1988
$Y_2Ba_4Cu_7O_{15}$	1988
$(Cu,M)Sr_2(Y,Ca)Cu_2O_7$ M stabilized Sr 123 M = Pb, Ga, Fe, B, SO_4, CO_3, Al, (Bi + Cd)	1988
$Pb_2Sr_2(Y,Ca)Cu_3O_8$	1988
$Pb_2(Sr,La)_2Cu_2O_6$	1988
"$Bi_2Sr_2CuO_6$"	1987
$Bi_2Sr_2CaCu_2O_8$	1988
$Bi_2Sr_2Ca_2Cu_3O_{10}$	1988
$Bi_2Sr_2(Ln,Ce)_2Cu_2O_{10}$	1990
$Tl_2Ba_2Ca_{n-1}Cu_nO_{2n+4}$ n = 1, 2, 3, 4	1988
$TlBa_2Ca_{n-1}Cu_nO_{2n+3}$ n = 1, 2, 3	1988
$TlBa_2(Eu,Ce)_2Cu_2O_9$	1992
(Tl,Pb)SrCaCu variants of Tl, Ba, Ca compounds	1988
$GaSr_2(Y,Ca)Cu_2O_7$	1991
$GaSr_2Ca_{n-1}Cu_nO_x$ n = 3, 4	1994
$AuSr_2CaCu_2O_7$	1997
$NbSr_2(Nd,Ce)_2Cu_2O_{10}$	1992
$RuSr_2(Nd,Ce)_2Cu_2O_{10}$ and others in "1222" type with a mix of Cu and M on first Cu site (e.g., Ti, V, Cr) also known	1996
$Cu(Eu,Ce)_2(Eu,Sr)_2Cu_2O_9$ and others in this type with a mix of Cu and M on first Cu site (e.g., Pb, Ga) also known	1989
$HgBa_2Ca_{n-1}Cu_nO_{2n+2}$ n = 1, 2, 3, 4, 5, 6	1993
$Hg_2Ba_2(Y,Ca)Cu_2O_8$	1994
$(Hg_{0.5}Cr_{0.5})Sr_2CuO_5$	1995
$(Hg_{0.5}Cr_{0.5})Sr_4Cu_2O_7CO_3$	1995
$(Ba,Sr)_2Cu_{1+x}(CO_2)_{1-x}O_y$	1992
$(Cu_{1-x}(CO_2)_x)_m(Ba,Sr)_2Ca_{n-1}Cu_nO_y$ $m = 1$; n = 2, 3, 4, 5; $x \neq 0$ or $x = 0$; $m = 2$; n = 3, 4, 5; $x \neq 0$	1994
$Bi_2Sr_4Cu_2O_8CO_3$	1993
$Bi_2Sr_5Cu_3O_{10}(CO_3)_2$	1994
$(Tl,Pb)Sr_4Cu_2O_7CO_3$	1993
$Ca_{13.5}Sr_{.5}Cu_{24}O_{41}$ (60 kbar (6 MPa) applied pressure only)	1996

Considering all the above, the materials I believe to be unique superconducting copper oxides, with the date of their first report, are listed in Table I. Uniqueness is taken to mean a new structure type, a new intermediary layer, or a significant difference in chemistry from previously known examples with the same structure type. (Perhaps others using a different philosophy for what constitutes a truly unique superconductor would have a different listing.) There have been many reports of "new" superconductors in the past few years that are actually only substitutions of one type of stabilizing atom for another in a well-known structure type. This means, as far as I can tell, that only one unique cuprate superconductor was discovered in 1997 and none in 1998. The list presented here represents my best assessment of the current situation and is slightly different from earlier versions I have published. The compounds are organized by type of intermediary layer and number of CuO_2 planes in Table II, generally accepted to be the

Table II. Classification of Copper Oxide Superconductors by Intermediary Layer and Number of CuO_2 Planes

	$(CuO_2)_n$						
Intermediary layer	$n = 1$	$n = 2$	$n = 3$	$n = 4$	$n = 5$	$n = 6$	$M_2O_2 + 2CuO_2$
AO (A = Ln, Ca, Sr, Ba)	X	X	X	X			
CuO or (Cu,M)O		X					X
2CuO		X					
PbO–Cu–PbO	X	X					
2BiO	X	X	X				X
TlO	X	X	X				X
2TlO	X	X	X	X			
GaO		X	X				
NbO_2							X
RuO_2							X
AuO		X					
HgO	X	X	X	X	X	X	
2HgO		X					
(Cu,C)O	X	X	X	X	X		
2(Cu,C)O			X	X	X		
BiO–CO_3		X	X				
(Tl, Pb)O–CO_3		X					
MCl	X	X					

most useful way of looking at them and seeing the patterns. Many reviews have been written about such classification schemes, and I will not treat them here. Finally, the rate of discovery of superconducting copper oxides as a function of year is presented in Fig. 23. Figure 23 shows the incubation period of 1986–1987, the banner year of 1988, the doldrums of the early 1990s, and the dramatic impact of high-pressure synthesis beginning in 1994 and lasting until 1997.

VI. Some Current Directions in Oxide Superconductor Research

After more than 12 years of active research in high-T_c superconductors, much has been learned about them, and new insight into the behavior of other materials has been attained as well, as transition-metal compounds are being reconsidered in a new light. Current research directions primarily involve the posing and answering of complex, specific questions designed to attempt to clarify the superconducting mechanism, the fabrication of experimental electronic devices, and the processing of bulk materials for large-scale applications. This review is primarily of a crystal–chemical nature, and, therefore, those three topics are best left to expert reviews elsewhere. Two areas of research into new superconducting oxides are active at this time: the investigation of cuprates with copper–oxygen geometries that are more amenable to first-principles calculation of physical properties, and the investigation of a different type of oxide superconductor, based on ruthenium oxide, in which an implied relationship between magnetism and superconductivity is also present. These are described briefly in this section.

The first area of active research in oxide superconductors has been inspired by the predictions of theorists.[12,13] The CuO_2 planes responsible for superconductivity in the normal cuprate materials, even though geometrically simple, are very difficult to model theoretically. Theorists can model the electronic and magnetic behavior of complex one-dimensional systems exactly, but, in two dimensions, approximations must be made that to date apparently have been insufficient to allow the prediction of a superconducting state from first principles. The important insight that came to one group of theorists was that they might be able to understand CuO_2 planes, which are made from corner-shared CuO_4 squares, if they first modeled the behavior of one-dimensional chains of corner-shared CuO_4 squares. These one-dimensional chains could then be lined up side-by-side, parallel to each other, sharing the oxygen between them, forming "ladders." The theorists wondered whether they could model that system exactly, because the ladder is electronically and magnetically intermediate between one and two dimensions. Two chains side-by-side make a ladder with two legs. Three chains side by side make a ladder with three legs, and so on, until eventually a full plane is built. It turned out they could model them very well. When they began to look at the predicted electronic and magnetic properties of such ladder systems, they found some interesting surprises.

They found theoretically that a two-legged ladder made of Cu^{2+} and oxygen, and all ladders with an even numbers of legs, should display a "spin gap"—a range of forbidden energies in the set of allowed spin states beginning just above the ground state—exactly analogous to the electronic bandgap in a semiconductor. They also predicted that, if a system such as this could be electronically doped, then it should become superconducting. The predicted phenomenology of the behavior of the two-legged ladder was exactly analogous to that observed in the cuprate superconductors. For the three-legged ladders and all ladders with an odd numbers of legs, no magnetically or electronically unusual behavior could be observed.

Now that such behavior was predicted, could it be observed in a real material? Again high-pressure synthesis was the key. The compounds $SrCu_2O_3$ and $Sr_2Cu_3O_5$, with two-legged and three-legged ladder structures, respectively, were synthesized by solid-state chemists who had heard the predictions.[26] Their crystal structures are shown in Figs. 24 and 25. Magnetic and electronic interactions for copper–oxygen–copper arrays are very strong for Cu–O–Cu bond angles of ~180° and very weak for O–Cu–O bond angles of ~90°; the magnetic two-legged and three-legged nature of the compounds are shown in the figures. Consistent with the predictions, the two-legged ladder compound showed the presence of a spin gap and the three legged ladder compound did not. Half of the prediction held true. However, it turned out to be impossible to electronically dope these materials, and the theoretical picture was not fully verified.

Continuing to pursue these ideas, there were several false starts in finding ladder systems that could become superconducting when doped. Then, in 1997, a group announced the discovery of superconductivity in a ladder system that was made by high-pressure synthesis and had to be subjected to high pressure while cooling to low temperatures to become superconducting.[27] The

The Rate of Discovery of Copper Oxide Superconductors

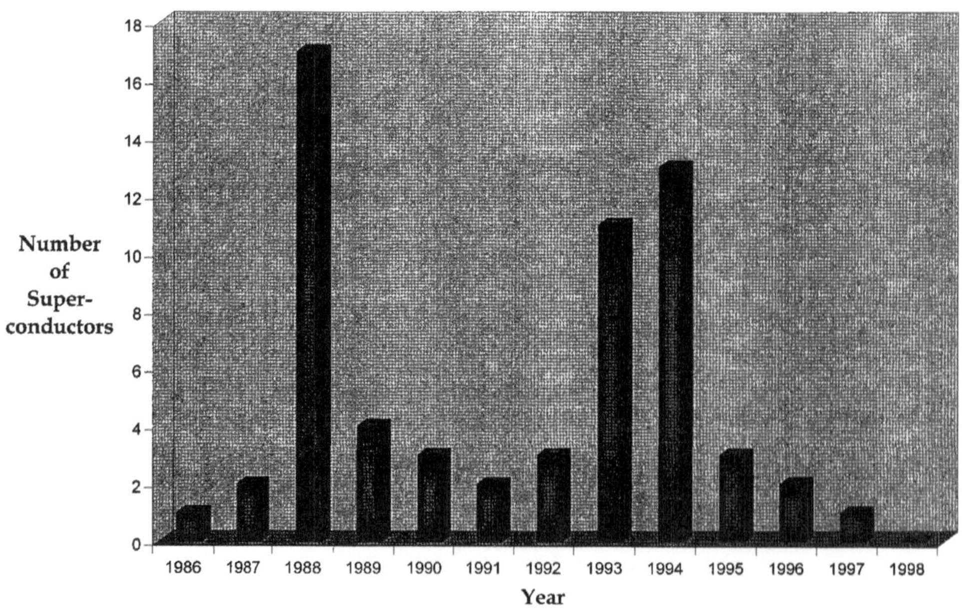

Fig. 23. Number of copper oxide superconductors discovered per year, 1986–present.

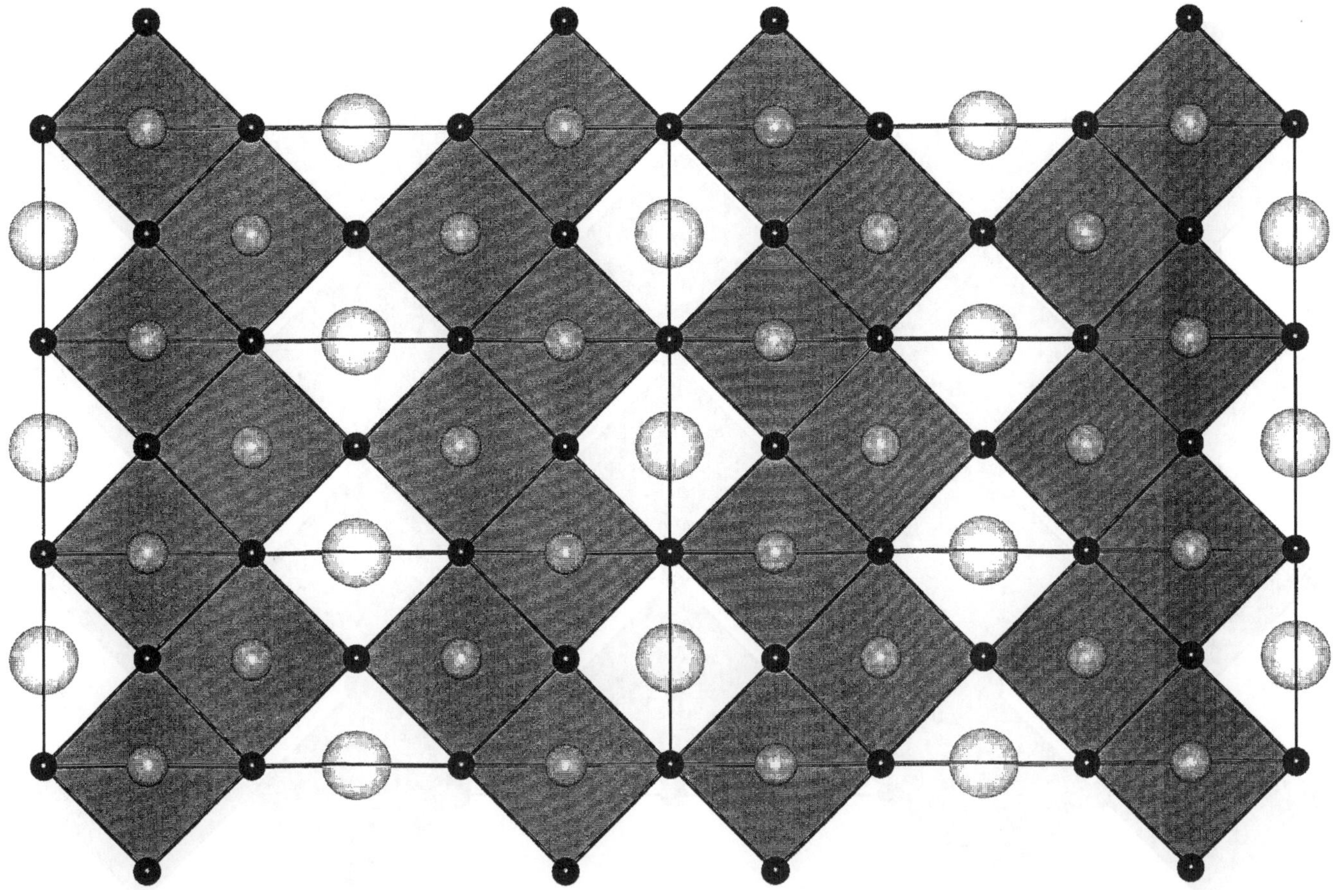

Fig. 24. Hypothetical crystal structure of $SrCu_2O_3$ (see Fig. 9 for legend).

superconductor announced was based on the compound $Sr_{14}Cu_{24}O_{41}$, which had been known since the early days of high-T_c research. The crystal structure of this complex material is shown in Fig. 26. It is made from severely skewed CuO_5 pyramids attached at their vertices by badly twisted CuO chains. The bases of the pyramids form the same type of plane as is found in the much simpler compound $SrCu_2O_3$, i.e., a two-legged ladder. Unlike the other materials considered as possible ladder superconductors, however, this one has a charge reservoir layer—the CuO chains—and, therefore, can be doped.

The doping process is complicated. At a stoichiometry of $Sr_{14}Cu_{24}O_{41}$, the formal copper valence is 2.25. The material is therefore hole doped as made. However, the holes are concentrated in the chains. When the solid solution $Sr_{14-x}Ca_xCu_{24}O_{41}$ is made, the unit-cell dimensions change significantly, and charge is transferred from the chains to the ladder plane. Under ambient pressure conditions, the calcium solid-solution limit is $x = 10$, and there is never sufficient charge transfer to reach a superconducting state. However, synthesis under high pressure extends the upper limit of calcium solubility to $x = 13.5$. This material, however, is not superconducting. Because the calcium substitution shrinks the lattice significantly, the next step would be to subject the $x = 13.5$ material to high pressure with cooling to attempt to obtain superconductivity. That was successful, and a superconducting transition was obtained in a narrow pressure range near an applied pressure of 30 kbar (3 MPa).

Therefore, has the theory been verified and has high-T_c superconductivity been understood theoretically? Some believe so, but there is no consensus in the community at large. Time will tell. It is fair to say that much-more-sophisticated experiments need to conducted to explore the detailed predictions of the theoretical picture. Competing theories might also be able to explain the existence of superconductivity in this compound. Also, the difficulty in making and testing the compound has severely limited its study. The lesson on the false report of superconductivity of the all-layer compound $Sr_{1-x}CuO_2$ synthesized by the same method due to inadvertent potassium and chlorine contamination should be remembered. Study of this material in more detail would be an interesting course of research in the future.

The second area of research centers around the very-low-temperature oxide superconductor Sr_2RuO_4. This is the first (n = 1) member of the Ruddlesden–Popper series, $Sr_{n+1}Ru_nO_{3n+1}$. In this series, common in ternary transition-metal oxides, n is the number of RuO_2 layers. The members $n = 1$, $n = 2$, and $n = \infty$ (the perovskite) are known in the Sr-Ru-O chemical system. Their crystal structures are shown in Fig. 27. The interest in this family comes from the manner in which ferromagnetism transforms into superconductivity as a function of the number of RuO_2 layers. The perovskite $SrRuO_3$ is a ferromagnetic metallic conductor with a large ordered magnetic moment and a ferromagnetic ordering temperature of 160 K. The double-layer compound $Sr_3Ru_2O_7$ has a local moment at high temperatures, consistent with the spin expected for Ru^{4+}, and is a metallic conductor. The Curie–Weiss theta, indicative of the strength of the magnetic interactions between ruthenium atoms, is low, ~10 K. No magnetic ordering transitions are observed in this compound down to low temperatures. Thus, on going from the fully three-dimensional compound to the double-layer compound, the magnetism has changed rather dramatically—from that of a high-T_c ferromagnet with strong interactions between spins, to a (very) low-T_n antiferromagnet, with weak spin interactions. Finally, the single RuO_2 layer compound Sr_2RuO_4 shows no local moment magnetic behavior whatsoever at high temperatures—the magnetic susceptibility indicates that the electrons are completely delocalized. No magnetic transitions are observed. Instead, this material becomes superconducting at temperatures near 1 K[28]!

Therefore, with a superconducting transition temperature so low, what is the interest in this material? The interest is purely

Fig. 25. Crystal structure of $Sr_2Cu_3O_5$ (see Fig. 9 for legend).[49]

scientific—this appears to be another example of an oxide for which the mechanism of the superconductivity is not conventional electron–phonon coupling. Many sophisticated experiments have been performed that show that to be the case. As in the cuprates, there appears to be a relationship between superconductivity and magnetism, in this case between *ferromagnetism* and superconductivity. This is a highly unexpected relationship, because the two effects should be mutually exclusive by conventional understanding. The transition from ferromagnetism to superconductivity as a function of increasing two-dimensionality of crystal structure is intriguing, especially in the context of the importance of two-dimensionality in the cuprate superconductors: The same is true for the analogous correlation of the disappearance of the localized magnetic moment and the appearance of superconductivity.

Given that transition-metal oxides frequently walk a fine line between localized moment magnetism and delocalized nonmagnetic electron states, there are likely to be other examples of exotic superconductivity of a similar nature in other oxides. Two aspects of the results on the ruthenates, however, suggest that these interesting materials will be very difficult to find. Unlike in other oxide superconductors, in the ruthenates the presence of superconductivity is very sensitive to the presence of impurities. The superconductivity in Sr_2RuO_4 is destroyed by the presence of impurity atoms at the level of tens of parts per million, as opposed, for instance, to the case in the cuprates, where a few percent impurity concentration is needed to suppress the superconductivity. Only single crystals of Sr_2RuO_4 grown in the traveling-solvent floating-zone furnace, which grows the crystal and refines it of impurities, have been found to be superconducting. Finally, testing materials for superconductivity at temperatures near 1 K is hardly something that can be done routinely. Liquid helium can be used easily to cool samples to 4.2 K and, with some effort, to 1.8 K. A more sophisticated and expensive cooling technology, based on ^{3}He, is needed to test materials at lower temperatures. Not many groups in the world have the means and the will to perform an exploratory materials program of that type.

VII. The Future

The pursuit of the microscopic mechanism for high-T_c superconductivity in the cuprate ceramics and their potential practical application has had a significant impact on the field of materials science in general. Examples of these types of influences are the development of laser ablation for the fabrication of thin films, the dramatic improvements in the energy and angular resolution in photoelectron spectroscopy, the widespread use of traveling-solvent floating-zone crystal-growth methods, and the fabrication of complex experimental magnetic and electronic devices based on oxides rather than conventional semiconductors. The impact of these developments is clearly being felt in areas outside of superconductivity research even now, and it will continue well into the future. From my point of view, the last of these—the development of the ability to fabricate new multilayer devices based on the magnetic and electronic properties of complex transition-metal oxides, including the development of the fabrication methods and the sophisticated analytical tools used in their characterization—may be the most far-reaching legacy of high-T_c superconductivity research. It has brought ceramics from the old world of technological application as bulk components into the modern world of integration into microelectronic systems. These innovations would have happened eventually in any case, but the very strong driving force of the understanding and application of oxide superconductors has greatly accelerated the pace of their development.

Constant progress is being made toward the development of fabrication techniques for making bulk conductors of high-T_c superconductors. Innovative processing methods that allow for the

Fig. 26. Crystal structure of $Sr_{14}Cu_{24}O_{41}$ (see Fig. 9 for legend).[50]

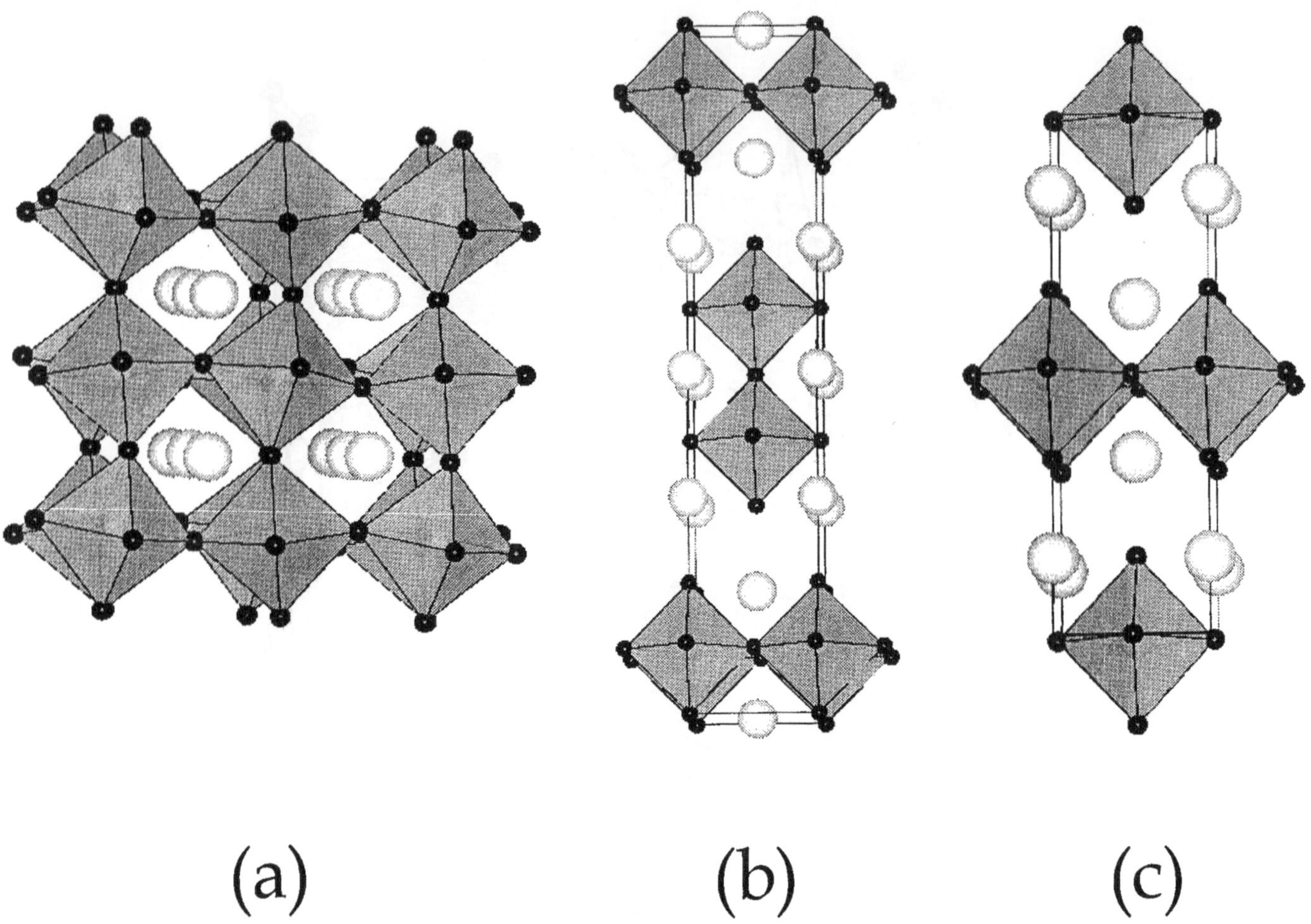

Fig. 27. Crystal structures of (a) $SrRuO_3$,[51] (b) $Sr_3Ru_2O_7$,[52] and (c) Sr_2RuO_4 (see Fig. 9 for legend).[53]

alignment of superconducting crystallites in bulk ceramics, such that the superconducting current travels in preferred directions across grain boundaries, are presently in a stage of active research. The processing of superconducting cuprates appears to me to be one of the most important areas for research in the future in this field, because the potential of high-T_c superconductors to change the way society functions cannot be realized until problems with making them into practical materials can be solved. It is important for funding agencies and corporate research managers to take a long-term view of the possibilities for success, although the progress may in many ways often appear slow. Perhaps several concentrated worldwide consortia working within different cultural contexts, with sufficient human investment to achieve teams of critical mass, is the best way to go about this. Although the materials-processing issues are difficult, nothing has appeared to date that would preclude the eventual large-scale use of high-T_c superconductors in technologies that would impact daily life.

As far as new superconductors with even higher transition temperatures are concerned, it now does not appear likely that they will come in the cuprate family in materials amenable to conventional synthesis. The issue of practical application is not a single-variable problem, and other factors—in particular the modification of the behavior of the superconductors in high fields and high currents through innovative processing and chemical manipulation—are more important than an increase in T_c at this time. Closed-cycle refrigeration systems can be designed and built to cool the materials now available to temperatures appropriate for their use if the current-carrying and magnetic-field response issues can be resolved. On the other hand, great potential exists for the discovery of oxides with different lattice geometries or other characteristics that contribute to the understanding of the superconducting mechanism.

It is a big periodic table. It is conceivable that there will be another new family of superconductors that will display exotic or exciting superconducting properties. Perhaps the cuprates are unique in having a mechanism for superconductivity that depends on the particular set of complex factors described above, and other superconductors may not have as wildly high superconducting transition temperatures, but there is no doubt in my mind that equally unexpected superconductors will be found elsewhere as well. Materials with much more modest T_c values, in the 30–40 K range, and different processing or critical field/critical current capabilities may ultimately prove to be the ones that change the way we live. Maybe the superconductors will be oxides and maybe not, but the superconducting oxides will always be considered the first to open our eyes to the fact that anything is possible in materials science if we just keep looking.

Acknowledgment

I thank Joanna Slusky for preparation of the figures for this review.

References

[1] C. Park and R. L. Snyder, "Structures of High-Temperature Superconductors," *J. Am. Ceram. Soc.*, **78,** 3171 (1995).

[2] J. Hauck and K. Mitka, "Structure Families of Superconducting Oxides and Interstitial Alloys," *Supercond. Sci. Technol.*, **11,** 614 (1998).

[3] J. Etourneau, "Superconducting Materials"; pp. 60–111 in *Solid State Chemistry: Compounds*. Edited by A. K. Cheetham and P. Day. Clarendon Press, Oxford, U.K., 1992.

[4] B. L. Chamberland, "Historical Introduction and Crystal Chemistry of Oxide Superconductors"; pp. 2–104 in *Chemistry of Superconductor Materials*. Edited by T. A. Vanderah. Noyes, Park Ridge, NJ, 1992.

[5]H.-K. Muller-Buschbaum, "The Crystal Chemistry of High-Temperature Oxide Superconductors and Materials with Related Structures," *Angew. Chem. Int. Ed. Engl.*, **28,** 1472 (1989).

[6]A. W. Sleight, "Chemistry of High-Temperature Superconductors," *Science (Washington, DC)*, **242,** 1519 (1988).

[7]R. J. Cava, "Structural, Chemistry, and the Local Charge Picture of Copper Oxide Superconductors," *Science (Washington, DC)*, **247,** 656 (1990).

[8]Y. Tokura and T. Arima, "New Classification Method for Layered Copper Oxide Compounds and Its Application for Design of New High-T_c Superconductors," *Jpn. J. Appl. Phys.*, **29,** 2388 (1990).

[9]E. Takayama-Muromachi, "High-Pressure Synthesis of Homologous Series of High-T_c Superconductors," *Chem. Mater.*, **10,** 2686 (1998).

[10]B. Raveau, C. Michel, M. Hervieu, and D. Groult, "The Complex Chemistry of Superconductive Layered Cuprates"; see Ref. 4, pp. 106–45.

[11]N. F. Mott, *Metal–Insulator Transitions*. Taylor and Francis, London, U.K., 1990.

[12]E. Dagotto and T. M. Rice, "Surprises on the Way from One to Two Dimensional Magnets: The Ladder Materials," *Science (Washington, DC)*, **271,** 618 (1996).

[13]T. M. Rice, "The Rich Variety of Cuprates," *Physica B (Amsterdam)*, **5,** 241–43 (1998).

[14]M. Imada, A. Fujimori, and Y. Tokura, "Metal Insulator Transitions," *Rev. Mod. Phys.*, **70,** 1039 (1998).

[15]G. Vialdi, *Superconductivity: The Next Revolution?* Cambridge University Press, Cambridge, U.K., 1993.

[16]N. B. Brandt and V. V. Moshchalkov, "Concentrated Kondo Systems," *Adv. Phys.*, **33,** 373 (1984).

[17]See, e.g., R. N. Shelton, "Ternary Phases: Superconductivity, Magnetism, and Crystallographic Stability"; pp. 123–32 in *Superconductivity in d- and f- Band Metals*. Edited by W. Buckel and W. Weber. Kern for Schungsentrum, Karlsruhe, Germany, 1982.

[18]A. W. Sleight, J. L. Gillson, and P. E. Bierstedt, "High-Temperature Superconductivity in the $BaBi_{1-x}Pb_xO_3$ System," *Solid State Commun.*, **17,** 27 (1975).

[19]R. J. Cava, B. Batlogg, J. J. Krajewski, R. L. Farrow, L. W. Rupp Jr., A. E. White, K. Short, W. F. Peck, and T. Kometani, "Superconductivity Near 30 K without Copper, The $Ba_{0.6}K_{0.4}BiO_3$ Perovskite," *Nature (London)*, **332,** 814 (1988).

[20]J. G. Bednorz and K. A. Müller, "Possible High-T_c Superconductivity in the La-Ba-Cu-O System," *Z Phys. B: Condens. Matter*, **64,** 188 (1986).

[21]T. Siegrist, S. M. Zahurak, D. W. Murphy, and R. S. Roth, "The Parent Structure of the Layered High-Temperature Superconductors," *Nature (London)*, **334,** 231 (1988).

[22]M. G. Smith, A. Manthiram, J. Zhou, J. B. Goodenough, and J. T. Mackert, "Electron-Doped Superconductivity at 40 K in the Infinite Layer Compound $Sr_{1-x}Nd_xCuO_2$," *Nature (London)*, **351,** 549 (1991).

[23]M. Azuma, Z. Hiroi, M. Takano, Y. Bando, and Y. Takeda, "Superconductivity at 110 K, in the Infinite Layer Compound $(Sx_{1-x}Ca_x)_{1-y}Cu_2$," *Nature (London)*, **356,** 775 (1992).

[24]S. Malo, C. Michel, D. Pelloquin, M. Hervieu, O. Toulemonde, and B. Raveau, "Compared Structural Behavior of $Hg_{0.5}Cr_{0.5}Sr_2CuO_{4.82}$ and $Hg_{0.5}Cr_{0.5}Sr_4Cu_2O_7CO_3$ Superconductors," *Physica C (Amsterdam)*, **304,** 213 (1998).

[25]Y. H. Wang, B. A. Scott, B. H. Chen, and D. Walker, "The Role of Contamination in Sr-Cu-O Reactions at High Pressure," *Physica C (Amsterdam)*, **275,** 52 (1997).

[26]M. Azuma, Z. Hiro, M. Takano, K. Ishida, and Y. Kitaoka, "Observation of a Spin Gap in $SrCu_2O_3$ Comprising Spin-1/2 Quasi-1*d* 2-Leg Ladders," *Phys Rev. Lett.*, **73,** 3463 (1994).

[27]H. Takahashi, N. Mori, T. Nakanishi, T. Nagata, M. Uehara, J. Akimitsu, and K. Kinoshita, "Superconductivity in Quantum Spin-Ladder Compound $Ca_{13.6}Sr_{0.4}Cu_{24}O_{41-d}$ at High Pressures," *Physica B (Amsterdam)*, **112,** 237–38 (1997).

[28]Y. Maeno, "Electronic States of the Superconductor Sr_2RuO_4," *Physica C (Amsterdam)*, **206,** 282–87 (1997).

[29]T. M. Rice, "Reviews, Prospects, and Concluding Remarks, High-T_c Superconductivity—Where Next?" *Physica C*, **282–289**, xix–xxiii (1997).

[30]J. H. Carpenter and A. W. Searcy, "Preparation, Identification, and Chemical Properties of the Niobium Germanides," *J. Am. Chem. Soc.*, **78,** 2079 (1956).

[31]Z. Ban and M. Sikiritsa, "The Crystal Structure of Ternary Silicides ThM_2Si_2 (M = Cr, Mn, Fe, Co, Ni, Cu)," *Acta Crystallogr.*, **18,** 594 (1965).

[32]J. Guillevic, H. Lestrat, and D. Grandjean, "Etude Structurale de Combinaisons Sulfurees et Seleniees du Molybdene, VI. Structures Cristallines de $Pb_xMo_3S_4$ and $Pb_xMo_3Se_4$," *Acta Crystallogr., Sect. B: Struct. Crysatllogr. Cryst. Chem.*, **32**, 1342 (1976).

[33]G. Brauer, "Die Oxyde des Niobs," *Z. Anorg. Allg. Chem.*, **248,** 1 (1941).

[34]A. Deschanvres, B. Raveau, and Z. Sekkal, "Mise en Evidence et Etude Cristallographique d'Une Nouvelle Solution Solide de Type Spinelle $Li_{1+x}Ti_{2-x}O_4$ $0 < x < 0.33$," *Mater. Res. Bull.*, **6,** 699 (1971).

[35]A. Garcia-Ruiz and B. Okhimi, "Superconductivity and Crystalline Structure in the Na-W-O system," *Physica C (Amsterdam)*, **204,** 79 (1992).

[36]A. Magneli, "Studies on the Hexagonal Tungsten Bronzes of Potassium, Rubidium, and Caesium," *Acta Chem Scand.*, **7,** 315 (1953).

[37]C. Chaillout, J. P. Remeika, A. Santoro, and M. Marezio, "The Determination of the Bi-Valence State in $BaBiO_3$ by Neutron Powder Diffraction Data," *Solid State Commun.*, **56,** 829 (1985).

[38]M. H. Tuilier, B. Chevalier, A. Tressaud, C. Brisson, J. L. Soubeyroux, and J. Etourneau, "EXAFS Study at the LaL_{III} X-ray Absorption Edge of Superconducting Materials Obtained by Fluorination of $La_{2-x}Sr_xCuO_4$ Oxides $(0 < x < 0.15)$," *Physica C (Amsterdam)*, **200,** 113 (1992).

[39]F. Beech, S. Miraglia, A. Santoro, and R. S. Roth, "Neutron Study of the Crystal Structure and Vacancy Distribution in the Superconductor $Ba_2YCu_3O_{9-d}$," *Phys. Rev. B: Condens. Matter*, **35,** 8778 (1987).

[40]J. K. Liang, S. S. Xie, G. C. Che, J. Q. Huang, Y. L. Zhang, and Z. X. Zhao, "Crystal Structure and Superconductivity of $Bi_2Sr_2CaCu_2O_8$ Compound," *Mod. Phys. Lett. B*, **2,** 483 (1988).

[41]J. K. Liang, Y. L. Zhang, J. Q. Huang, S. S. Xie, G. C. Che, and X. R. Cheng, "A Series of Crystal Structures of Superconducting Phases in Tl-Ba-Ca-Cu-O System," *Sci. Sin., Ser. A*, **32,** 826 (1989).

[42]H. Fujishita, S.-I. Yamagata, and M. Sato, "Crystal Structures of $Pb_2Sr_2YCu_3O_{8+d}$ (d = 0 and 1.67)," *J. Phys. Soc. Jpn.*, **60,** 913 (1991).

[43]J. Gopalkrishnan, M. A. Subramanian, C. C. Torardi, J. P. Attfield, and A. W. Sleight, "Properties and Structures of $R_{2-x}A_xCuO_4$ Phases: R = La, Pr, and Nd; A = Sr, Pb, and Cd," *Mater. Res. Bull.*, **24,** 321 (1989).

[44]T. G. N. Babu and C. Greaves, "Synthesis, Composition, and Structure of Superconducting $(Y,Ca)(Sr,Ba)_2Cu_2GaO_{7-d}$," *J. Supercond.*, **8,** 21 (1995).

[45]T. J. Goodwin, H. B. Radousky, and R. N. Shelton, "Superconducting, Magnetic, Electronic Transport, and Structural Properties of $(R_{1.5}Ce_{0.5})Sr_2Cu_2NbO_{10}$, R = Pr, Nd, Sm, and Eu," *Physica C (Amsterdam)*, **204,** 212 (1992).

[46]T.-Z. Cheng, J. Lu, L.-Z. Yao, L. Shi, Y.-T. Lei, G. Zhou, Y. Zhang, Z.-Y. Chen, Y. Qian, and Z.-R. Li, "Crystal Structures of $(Ca_{0.9}Sr_{0.1})CuO_2$ and $(Ca_{0.89}Sr_{0.11})_2CuO_3$," *Diwen Wuli Xuebao*, **12,** 272 (1990).

[47]B. A. Hunter, J. D. Jorgensen, J. L. Wagner, P. G. Radaelli, D. G. Hinks, H. Shaked, R. L. Hitterman, and R. B. Von Dreele, "Pressure-Induced, Structural Changes in Superconducting $HgBa_2Ca_{(n-1)}Cu_{(n)}O_{(2n+2+d)}$ (n = 1, 2, 3)," *Physica C (Amsterdam)*, **221,** 1 (1994).

[48]H. Shibata, K. Kinoshita, and T. Yamada, "Crystal Structure and Growth of $(Ba_{(1-x)}Sr_{(x)})_2Cu_{1.1}O_{(2.2+d)}(CO_3)_{0.9}$ Single Crystals," *Physica C (Amsterdam)*, **235,** 519 (1994).

[49]S. M. Kazakov, S. Pachot, E. M. Kopnin, S. N. Putilin, E. V. Antipov, C. Chaillout, J. J. Capponi, P. G. Radaelli, and M. Marezio, "Synthesis, Neutron Diffraction Study, and Cation Substitutions in $Sr_{(n-1)}Cu_{(n+1)}O_{(2n)}$ (n = 3, 5)," *Physica C (Amsterdam)*, **276,** 139 (1997).

[50]T. Siegrist, L. F. Schneemeyer, S. A . Sunshine, and J. V. Waszczak, "A New Layered Cuprate Structure Type, $(A_{1-x}A'_x)_{14}Cu_{24}O_{41}$," *Mater. Res. Bull.*, **23** 1429 (1988).

[51]C. W. Jones, P. D. Battle, P. Lightfoot, and W. T. A. Harrison, "The Structure of $SrRuO_3$ by Time-of-Flight Neutron Powder Diffraction," *Acta Crystallogr., Sect. C: Cryst. Struct. Commun.*, **45,** 365 (1989).

[52]I. B. Sharma and D. Singh, "Preparation and Study of Structure, Electric Transport, and Magnetic Properties of $Sr_3Ru_2O_7$ and $Sr_3Ru_{1.5}Fe_{0.5}O_7$," *Indian As-Chem. Sci.*, **108,** 415 (1996).

[53]L. Walz and F. Lichtenberg, "Refinement of the Structure of Sr_2RuO_4 with 100 and 295 K X-ray Data," *Acta Crystallogr., Sect. C: Cryst. Struct. Commun.*, **49,** 1268 (1993). □

J. Am. Ceram. Soc., **83** [7] 1675–80 (2000)

Influence of Ca_2PbO_4 on Phase Formation and Electrical Properties of $(Bi,Pb)_2Sr_2Ca_2Cu_3O_{10}$/Ag Superconducting Composites

Yuan Chang Guo, Hua Kun Liu, and Shi Xue Dou

Institute for Superconcduting and Electronic Materials, The University of Wollongong, Wollongong, New South Wales 2522, Australia

$(Bi,Pb)_2Sr_2Ca_2Cu_3O_{10}$ (Bi2223) precursor powders with large and small amounts of Ca_2PbO_4 phase were prepared and used to make superconductor/silver composite tapes. The melting behavior of the powders and tapes was examined by differential thermal analysis (DTA). The influence of Ca_2PbO_4 on the formation and microstructure of Bi2223, and electrical properties of the tapes, was investigated by X-ray diffraction (XRD), scanning electron microscopy (SEM), and critical current measurements. It was found that the melting onset temperature ($T_{m,onset}$) of precursor powders and composite tapes was strongly dependent on the amount of Ca_2PbO_4. Tapes with a small amount of Ca_2PbO_4 had a higher $T_{m,onset}$ and a higher optimum sintering temperature compared with tapes with a large amount of Ca_2PbO_4. Also because of the higher sintering temperature, the total sintering time required for the former tapes was drastically shortened compared with the latter ones (250 vs 110 h). Furthermore, the microstructure and the current-carrying capacity of the tapes were significantly improved by reducing the Ca_2PbO_4 content of the precursor powders. These results are of practical significance for the commercialization of Bi-based high-temperature superconductors.

I. Introduction

Bismuth-based high-temperature superconductors are new materials with great potential for application in electrical and energy engineering. Since their discovery,[1] intensive research efforts have been directed toward the fabrication of materials suitable for practical applications. Silver-sheathed Bi2223 composite tapes with critical current density (J_c) at a level of 10^4 A/cm^2 (77 K) and lengths up to 1000 m have been achieved.[2–6] These advances have brought Bi2223 to a stage where commercialization is under consideration.

To realize commercialization, materials are required to be not just technically feasible, but also economically profitable. The present fabrication process is, however, very long and complicated. An accumulative sintering time of 200 to 300 h, and two or more intermediate mechanical deformation steps, are required for the tapes to form a pure Bi2223 core with a dense and highly textured microstructure.[7,8] This is the major disadvantage of the present process from a fabrication-cost point of view. More work on fundamental material aspects is needed to improve Bi2223/Ag tape fabrication.

Phase composition in the precursor powders has been proved to play an important role in determining the microstructure and properties of Bi2223/Ag tapes. For instance, tapes made from a $Bi_2Sr_2Ca_1Cu_2O_8$ (Bi2212)-predominated precursor powder (partially reacted) usually have a higher J_c than those made from Bi2223-predominated powder (fully reacted) after heat treatment.[9,10] This is because in the partially reacted powders, microstructural features, such as grain alignment and grain connectivity, are enhanced during the tape heat treatment due to a phase transformation mechanism. However, this is not achieved in the case of fully reacted powders, in which the phase transformation is finished before the tape heat treatment.

Some minor phases are also believed to affect the microstructure and properties of Bi2223/Ag tapes. Shortly after the discovery of the Bi-based systems, it was found that partial substitution of Bi by Pb lowers the sintering temperature and enhances the Bi2223 phase formation.[11–14] It is believed that Pb substitution improves reaction kinetics via formation of a Ca_2PbO_4-related liquid phase.[15–17] However, excessive liquid results in the formation of large nonsuperconducting phases and causes deterioration of current-carrying capacity of the tapes.[18,19] To avoid the formation of excessive liquid, a relatively low sintering temperature has to be used in the Pb-substituted Bi2223 system; because of the low sintering temperature, Bi2223 phase formation is sluggish and the required sintering time is long.

Since the Ca_2PbO_4-related liquid phase is the key factor in formation of the Bi2223 phase, it may be possible to increase the sintering temperature and consequently shorten tape processing time through optimization of Ca_2PbO_4 content in precursor powders. Some experimental results on the effect of Pb on Bi2223 phase formation have been reported in the literature.[20–22] In this work, a systematic investigation on the influence of Ca_2PbO_4 phase on the melting behavior, optimum sintering temperature, Bi2223 phase formation, microstructure, and critical current density of Bi-2223/Ag was conducted by using Bi2223 precursor powders with large and small amounts of Ca_2PbO_4.

II. Experimental Procedure

Precursor powders were prepared by a spray drying process. Nitrates and oxides, weighed in a cation ratio of Bi/Pb/Sr/Ca/Cu = 1.84/0.34/1.91/2.03/3.06, were dissolved in dilute nitric acid, sprayed into a drying chamber, and dried at a temperature of ~200°C. The dried mixture was placed in a muffle furnace and calcined at 780°C for 15 h in air. The calcined powder was ground and sintered. To obtain precursor powders with different amounts of Ca_2PbO_4, the temperature and time of powder sintering were varied. The sintering temperature was varied from 820° to 840°C, and sintering time from 15 to 60 h. XRD was conducted after each sintering to monitor the content of Ca_2PbO_4 and other phases. The melting behavior of each precursor powder was studied by DTA in air with a heating rate of 10°C/min. Particle size and size distribution were analyzed by a low-angle laser light scattering (LALLS) method.

Two precursor powders, one with a large amount of Ca_2PbO_4 and the other with a small amount of Ca_2PbO_4, were selected and fabricated into silver-sheathed composite tapes by a powder-in-tube (PIT) technique. In this technique, precursor powders were packed into pure silver tubes of 10 mm outer diameter and 8 mm

H. U. Anderson—contributing editor

Manuscript No. 190343. Received March 9, 1998; approved January 28, 2000.
*Member, American Ceramic Society.

inner diameter. The oxide/silver composites were then drawn to a diameter of 1.8 mm and rolled into flat tapes with a thickness of ~0.25 mm. Short pieces (4 cm) were cut and subjected to a thermomechanical process consisting of alternate thermal sintering and mechanical pressing. Preliminary experiments were conducted to determine the optimum sintering temperature (at which the highest J_c was obtained) for each powder. Sintering was performed at each tape's optimum temperature for 50–80 h. Bi2223 phase fraction and J_c were measured after each sintering. Bi2223 fraction was estimated from the sample XRD patterns by comparing the peak heights. J_c was measured by the standard dc four-probe technique using a criterion of 1 μV/cm. The microstructure of each tape after final sintering was observed by SEM.

III. Results and Discussion

(1) Ca_2PbO_4 Content

Figure 1(a) shows the XRD pattern for the powder as calcined at 780°C for 15 h. The powder consisted of a major phase of $Bi_2Sr_2Cu_1O_6$ (Bi2201), together with some minor phases such as Bi2212, CuO, and Sr-Ca-Cu-O. In particular, a very large amount of Ca_2PbO_4, indicated by the height of its characteristic peak at $2\theta = 17.5°$, was present in this calcined powder. Figures 1(b–d) show the XRD patterns for the precursor powders, which were obtained by further sintering the calcined powder. The powders shown in Figs. 1(b) and (c) were sintered at the same temperature (820°C), but for different times (20 and 60 h, respectively). The powder in Fig. 1(d), however, was sintered at a higher temperature (830–840°C) for a shorter time (15 h). It is seen that all three powders had Bi2212 as the major phase. The main difference between these powders was the Ca_2PbO_4 content. The powder in Fig. 1(b) contained a large amount of Ca_2PbO_4, while those in Figs. 1(c) and (d) contained relatively little Ca_2PbO_4. Therefore, the amount of Ca_2PbO_4 in each sample was strongly dependent on the temperature and time used for powder sintering. The higher the sintering temperature and the longer the sintering time, the lower the Ca_2PbO_4 content in the resulting precursor powders. In terms of chemical reaction, Pb quickly reacted with Ca during initial calcination forming the Ca_2PbO_4 phase, and then eventually entered the Bi2212 phase during subsequent sintering.

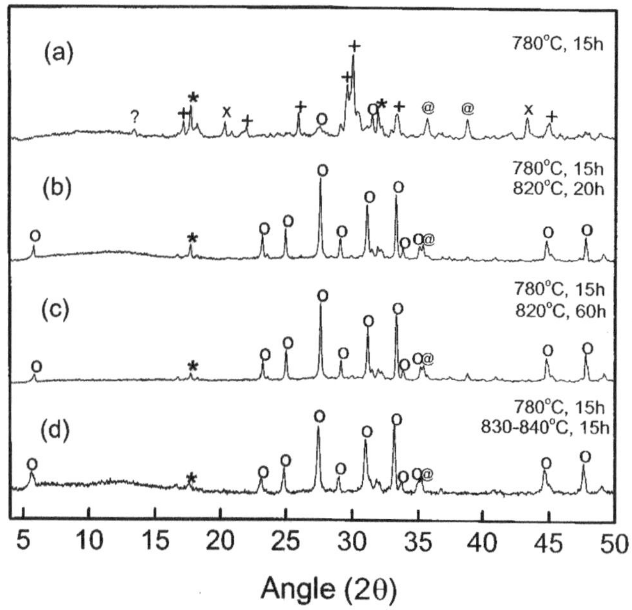

Fig. 1. X-ray diffraction patterns of (a) powder as calcined at 780°C for 15 h and precursor powders prepared by sintering the calcined powder (b) at 820°C for 20 h, (c) at 820°C for 60 h, and (d) at 830–840°C for 15 h: (*) Ca_2PbO_4, (+) $Bi_2Sr_2CuO_6$ (Bi2201), (O) $Bi_2Sr_2Ca_1Cu_2O_8$ (Bi2212), (@) CuO, (X) Sr-Cu-O, and (?) unknown phase.

(2) Particle Size and Size Distribution

Precursor powders shown in Figs. 1(b) and (d), hereafter identified as powders I and II, were used to make silver-sheathed tapes to investigate the influence of Ca_2PbO_4 on the properties of the superconductors. To isolate the Ca_2PbO_4 effect, other parameters for the two powders, such as particle size and size distribution, were made as similar as possible. In this work, both powders were ground under the same conditions, and their particle size and size distribution were analyzed by the LALLS method. The obtained results are summarized in Table I.

The results of particle size analysis by LALLS were given in three volume proportions: $D(v,0.1)$, $D(v,0.5)$, and $D(v,0.9)$, which indicate that the volume of particles with diameters below $D(v,0.1)$, $D(v,0.5)$, and $D(v,0.9)$ comprises 10%, 50%, and 90%, respectively, of the total volume of all particles. $D(v,0.5)$ is the volume median diameter. The results are also given using the derived diameters $D(m,n)$. According to an internationally agreed method of defining the mean and other moments of particle size,

$$D(m,n) = [\Sigma V_i d_i^{m-3} / \Sigma V_i d_i^{n-3}]^{1/(m-n)} \tag{1}$$

where V_i is the relative volume in class i with mean diameter of d_i. m and n are integer values describing the type of derived diameter, e.g., $D(4,3)$ is the volume-weighted mean diameter and $D(3,2)$ the surface-weighted mean diameter.

The other two important terms describing the distribution type of powders are distribution span and uniformity. The distribution span is defined as

$$\text{span} = [D(v,0.9) - D(v,0.1)]/D(v,0.5) \tag{2}$$

It gives a description of the width of the distribution, which is independent of the median size. The distribution uniformity is a measure of the absolute deviations from the median size and is defined as

$$\text{uniformity} = \Sigma V_i |D(v,0.5) - d_i| / [D(v,0.5) \Sigma V_i] \tag{3}$$

As seen in Table I, all of the measured parameters for powders I and II were very similar, with the latter having only slightly larger particle size and narrower size distribution over the former. Therefore, the difference in phase formation and other properties due to differences in particle size and size distribution between the two samples was negligible.

It should also be pointed out that the characteristic sizes shown in Table I (i.e., $D(v,0.1)$, $D(v,0.5)$, and $D(v,0.9)$) are volume-weighted particle sizes rather than linear sizes, which are normally measured by optical and electron microscopy.

(3) Melting Behavior

Heat flow as a function of temperature for powders I and II is shown in Fig. 2. There were two partially overlapping endothermic peaks for powder I in the temperature range of 800° to 900°C. The onset temperatures (T_{onset}) for the small and large peaks were 838° and 859°C, respectively. T_{onset} was defined as the point at which the downward tangent at the point of greatest slope on the leading edge of the peak intersects the extrapolated base line. Comparatively, there was only one large distinct endothermic peak in the same temperature range for powder II. Its T_{onset} was 852°C, i.e., 14°C higher than that of the first peak of powder I (838°C).

The large endothermic peak in both powders was attributed to the melting of the Bi2212 phase.[25] As indicated by the XRD results (Fig. 1), the main difference between powders I and II was the Ca_2PbO_4 content, i.e., a large amount of Ca_2PbO_4 present in the former and little in the latter. The small endothermic peak present in powder I but absent in powder II suggested that this peak was associated with the Ca_2PbO_4 phase. According to the CaO–PbO phase diagram, Ca_2PbO_4 melts incongruently at ~822°C, decomposing into CaO and a liquid phase.[26] It is therefore concluded that the melting temperature of the precursor powders was strongly dependent on the Ca_2PbO_4 content. The higher the Ca_2PbO_4 content, the lower the melting onset temperature of the precursor powder.

Table I. Results of Analysis of Particle Size and Size Distribution for Powders I and II

Sample	$D(v,0.1)$ (μm)	$D(v,0.5)$ (μm)	$D(v,0.9)$ (μm)	$D(4,3)$ (μm)	$D(3,2)$ (μm)	Span	Uniformity
Powder I	1.34	4.73	25.59	9.20	3.04	5.12	1.56
Powder II	1.85	5.76	28.85	10.82	4.05	4.68	1.32

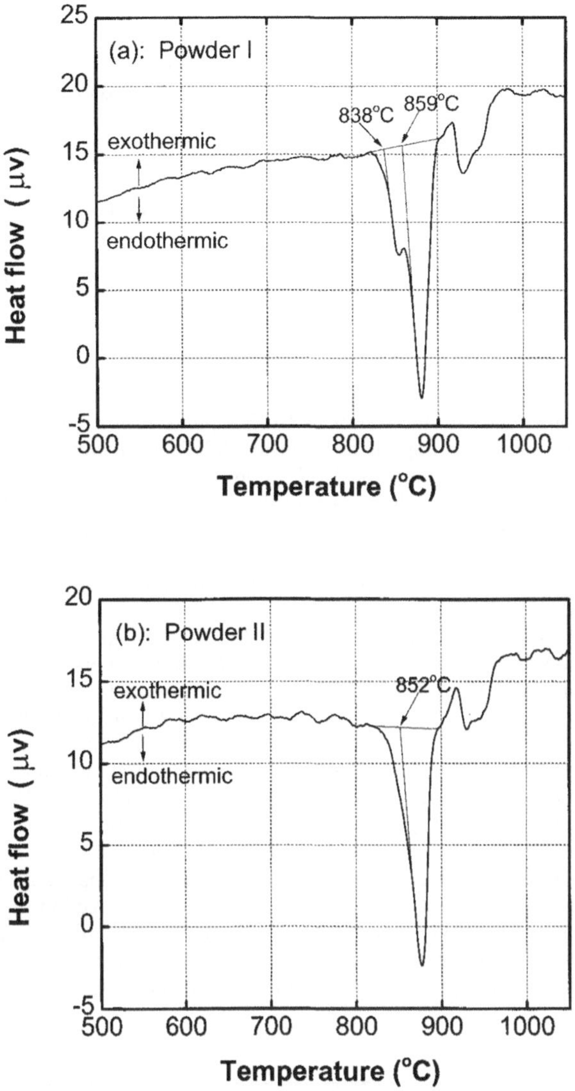

Fig. 2. Differential thermal analysis curves of (a) powder I and (b) powder II. In the temperature range of 800–900°C, powder I had two endothermic peaks with T_{onset} = 838° and 859°C, and powder II had only one distinct peak with T_{onset} = 852°C.

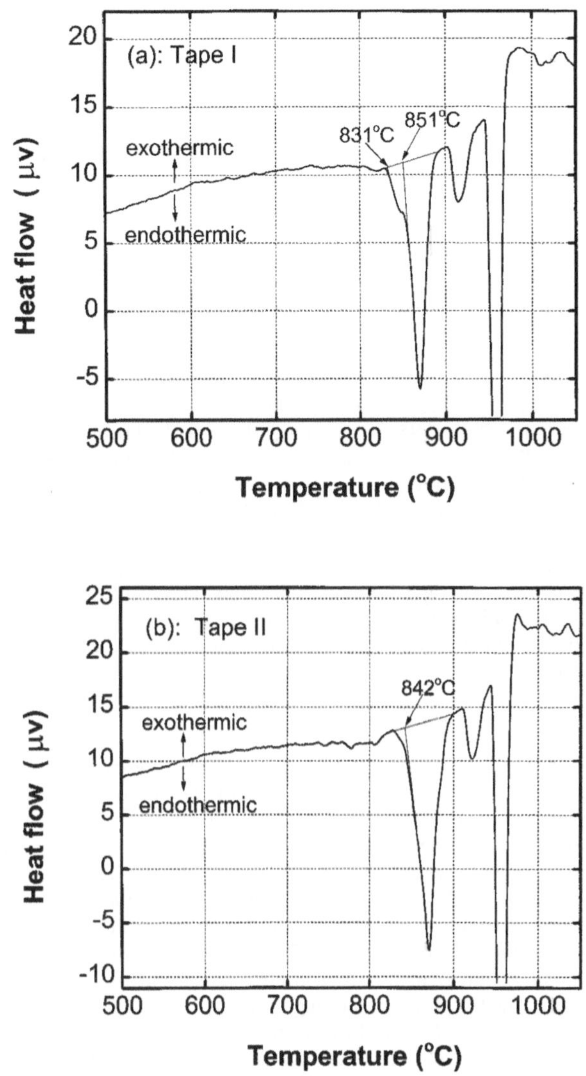

Fig. 3. Differential thermal analysis curves of (a) tape I and (b) tape II, showing that between 800° and 900°C, powder I had two endothermic peaks with T_{onset} = 831° and 851°C while powder II had only one peak with T_{onset} = 842°C. The large endothermic peak at ~850°C for both tapes was due to melting of the silver sheath.

DTA curves for Bi2223/Ag composite tapes are shown in Fig. 3. In the figure, tape I and tape II refer to tapes made from powders I and II, respectively. Melting behaviors similar to their corresponding precursor powders were observed for each tape. Between 800° to 900°C, tape I showed two partially overlapping endothermic peaks, with T_{onset} = 831° and 851°C, respectively. Tape II showed, however, only one large distinct endothermic peak in the same temperature range, with a T_{onset} of 842°C. The huge endothermic peak at ~950°C for both tapes was due to melting of the silver sheath.

By comparing Fig. 2 with Fig. 3, one can see that after sheathing of powders in silver, the melting temperature of the samples was lowered by 7–10°C. However, the difference in melting temperature between the two types of samples was not affected. That is, T_{onset} for tape II was 9°C higher than that for tape I. This was because the latter tape contained a large amount of Ca_2PbO_4, while the former tape contained relatively little Ca_2PbO_4. The difference in T_{onset} would result in a difference in optimum sintering temperature of the tapes. As is well known, the optimum temperature for tape sintering is near its melting onset temperature.[9,27] In this work, the optimum sintering temperatures for tapes I and II were determined to be 832° and 841°C, respectively, by preliminary experiments.

(4) *Bi2223 Phase Formation*

Tapes I and II were sintered at their optimum temperatures and Bi2223 fraction was measured by X-ray diffraction analysis after each step of sintering. Figure 4 shows the Bi2223 fraction as a function of sintering time for both tapes. The fraction was estimated by comparing the height of the Bi2223 (0010) peak to that of the Bi2212 (008) peak.[28,29] It is obvious that tape II had a

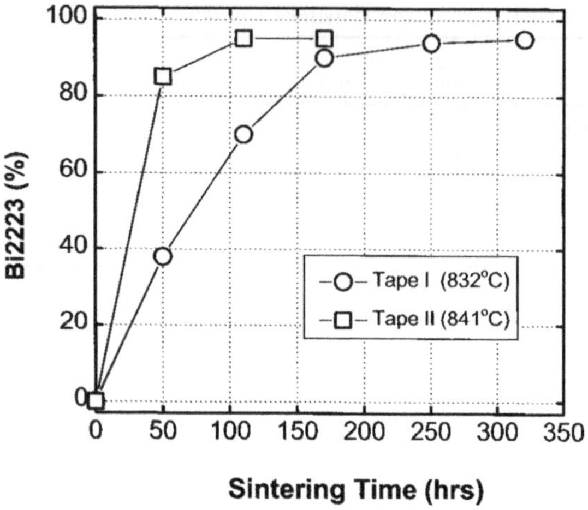

Fig. 4. Fraction of Bi2223 phase as a function of sintering time for tapes I and II, showing a much faster phase formation rate for the latter than the former, but an identical maximum fraction for both.

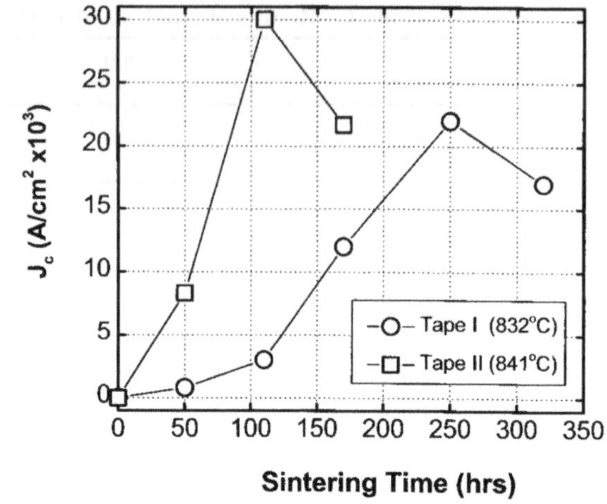

Fig. 5. Critical current density ($J_{c,77\ K}$) vs sintering time for tapes I and II. Tape II reached a J_c of 30 000 A/cm^2 after a total sintering time of 110 h and one step of pressing while tape I reached only 22 000 A/cm^2 after a total time of 250 h and three steps of pressing.

much faster Bi2223 formation rate than tape I. After a first sintering for 50 h, ~85% of the core of tape II had been converted into Bi2223 phase. With another 60 h of sintering (a total of 110 h), a Bi2223 fraction as high as 95% had been achieved. The Bi2223 fraction saturated at ~95% with further sintering. For tape I, however, the Bi2223 fraction did not reach its maximum until an accumulative time of 250 h. Furthermore, three pressing steps were needed for tape I to reach its maximum Bi2223 fraction, compared with only one pressing for tape II.

The chemical reaction kinetics are described by the Arrhenius equation,[30]

$$k(T) = A\ \exp(-E/RT) \tag{4}$$

where k is the reaction rate constant, A is the pre-exponential or frequency factor, and E, R, and T are the activation energy, the Boltzmann constant (8.3145 J·mol^{-1}·K^{-1}), and the absolute temperature, respectively. The activation energy, E, is typically in the range of a few to several hundred thousand J·mol^{-1}. Therefore, for most reactions the rate increases by 2 to 4 times when the temperature is raised by 10°C. In this work, the sintering temperatures for tapes I and II were 832° and 841°C, respectively. The values of E and A in this temperature range can be regarded as constant. The 9°C higher sintering temperature of tape II over tape I had resulted in a Bi2223 formation rate ~2.4 times greater in the former than the latter. In other words, the total sintering time for tapes to reach their maximum Bi2223 fraction was shortened from 250 h (for tape I) to 110 h (for tape II). These results are in agreement with the Arrhenius theory. The remarkable shortening of the sintering time and reduction of intermediate deformation steps simplify the process of Bi2223/Ag superconductor fabrication and will have an impact on the commercialization of the material.

It is also noted in Fig. 4 that the final Bi2223 fraction for both tapes was almost identical, meaning that Ca_2PbO_4 in the precursor powders seemingly only affected the phase formation kinetics, but did not change the nature of the formation reaction.

(5) *Critical Current Density (J_c)*

Critical current densities (J_c) for tapes I and II were measured after each step of sintering and the results are shown in Fig. 5. Compared with tape I, tape II showed a more rapid J_c increase on sintering and had a much higher maximum J_c value. For example, with two steps of sintering (a total of 110 h) and one intermediate pressing, J_c for tape II reached a maximum value of 30 000 A/cm^2 at 77 K and self-generated magnetic field. For tape I, however, J_c reached a maximum of only 22 000 A/cm^2 after four steps of sintering (a total of 250 h) and three intermediate pressings. This is consistent with the results of Bi2223 phase formation (Fig. 4), which shows that tape II had a faster Bi2223 formation rate than tape I. As reported previously, J_c improves as a result of enhancement of Bi2223 fraction, core density, and grain alignment.[7,31] J_c of both tapes dropped with sintering beyond the maximum due to unhealed cracks introduced by mechanical deformation, as well as formation of an amorphous phase with prolonged sintering time.[32–34]

Another important property of Bi2223/Ag tapes is the J_c behavior in an external magnetic field (H). In this work, the J_c–H behavior of tapes I and II was measured with a magnetic field applied parallel ($H \| ab$) and perpendicular ($H \perp ab$) to the broad surface of the tape. The results obtained are shown in Fig. 6. To simplify the comparison, J_c at a given magnetic field was normalized to its zero field value. It is clear that in both $H \| ab$ and $H \perp ab$ directions, J_c for tape II showed a weaker magnetic field dependence than tape I. As is well known, J_c–H behavior in the $H \| ab$ direction is dependent on both flux pinning strength and the degree of grain alignment of the tapes. In the $H \perp ab$ direction, J_c–H behavior is mainly governed by flux pinning.[35,36] The weaker J_c–H dependence for tape II indicates that it had stronger flux pinning and improved grain alignment compared with tape I.

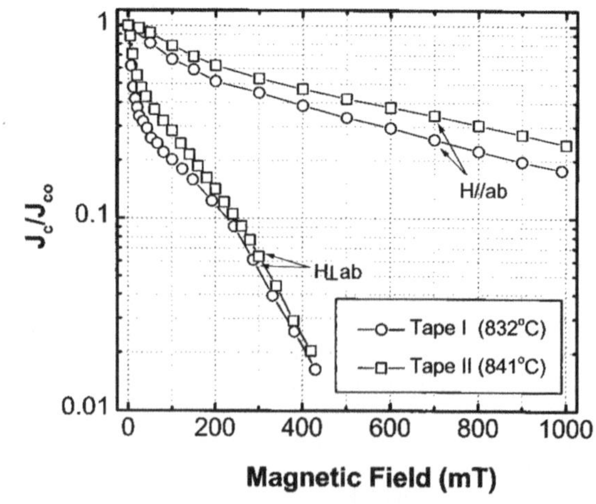

Fig. 6. Normalized critical current density ($J_{c,77\ K}$) against applied magnetic field for tapes I and II. In both $H \| ab$ and $H \perp ab$ directions, J_c of tape II showed less dependence on magnetic field compared with tape I, indicating the former had stronger flux pinning and better grain alignment than the latter.

(6) Microstructure

As with all ceramic materials, the properties of Bi2223/Ag composite superconductors are dependent on their microstructures. In this work, microstructures for tapes I and II were examined by SEM and are compared in Figs. 7(a) and (b), respectively. The Bi2223 grains have a platelike morphology, with their dimension in the *a–b* plane much larger than in the *c*-axis direction. They look like needles along the *a–b* planes. It is because of this platelike morphology that Bi2223 grains inside tapes are easily aligned with their *a–b* plane parallel to the interface of the superconductor core and silver sheath during mechanical deformation and thermal treatment.

Figure 7 shows that the needlelike Bi2223 grains, observed from the cross sections of the tapes, were intimately stacked along the core/silver interface with one on top of another. Electrical current passes through the tapes by flowing from one grain to another through either the wide *c*-axis boundaries (brick-wall model[37,38]), or the low-angle *a–b* boundaries (railway-switch model[39,40]). But, in both cases, a dense and highly aligned microstructure is crucial for the samples to be able to carry a large current. Compared with tape I, tape II had a denser and better aligned superconductor core. Furthermore, the Bi2223 grains in tape II were larger and better grown than those in tape I. All these microstructural features were responsible for the higher J_c and improved J_c–H behavior of tape II over tape I, as shown in Figs. 4 and 5.

IV. Conclusions

(1) Ca_2PbO_4 content was strongly dependent on the temperature and time used for powder sintering. The higher the temperature and the longer the time, the smaller the Ca_2PbO_4 content in the resulting precursor powders.

(2) The melting temperature of Bi2223/Ag composite tapes was lowered by the presence of a large amount of Ca_2PbO_4 in the precursor powder. Consequently, the optimum sintering temperature for the tape with a large amount of Ca_2PbO_4 (tape I) was $\sim$10°C lower than for that with little Ca_2PbO_4 (tape II).

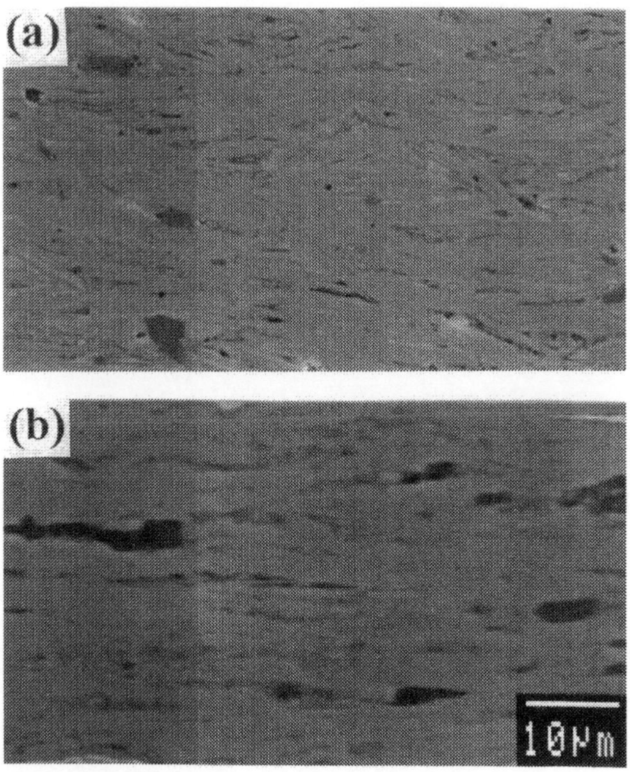

Fig. 7. SEM photographs of (a) tape I and (b) tape II, showing the latter had a denser and better aligned superconductor core than the former.

(3) The Bi2223 formation rate was significantly increased and the sintering time drastically shortened for tapes with little Ca_2PbO_4 compared with those with a large amount of Ca_2PbO_4 (110 to 250 h), as a result of the higher optimum sintering temperature of the former as compared with the latter.

(4) The critical current density (J_c) and J_c behavior in a magnetic field were significantly improved for the tapes containing little Ca_2PbO_4 compared with those with a large amount of Ca_2PbO_4, because of the higher density, larger grains, and better grain alignment in the former compared with the latter.

These results are of practical significance for the commercialization of Bi-based high-T_c superconductors.

References

[1]H. Meada, Y. Tanaka, M. Fukutomi, and Asano, "A New High-T_c Oxide Superconductor without a Rare Earth Element," *Jpn. J. Appl. Phys.*, **27**, L209–L210 (1988).

[2]M. Ueyama, T. Hikata, T. Kato, and K. Sato, "Microstructures and J_c–*B* Characteristics of Ag-Sheathed Bi-Based Superconducting Wires," *Jpn. J. Appl. Phys.*, **30**, L1384–L1386 (1991).

[3]S. X. Dou and H. K. Liu, "Silver-Sheathed Bi(Pb)SrCaCuO Superconducting Tapes," *Supercond. Sci. Technol.*, **6**, 297–314 (1993).

[4]Q. Li, K. Broderson, H. A. Hjuler, and T. Freltoft, "Critical Current Density Enhancement in Silver-Sheathed Bi-2223 Superconducting Tapes," *Physica C*, **217**, 360–66 (1993).

[5]G. Grasso, A. Jeremie, and R. Flukiger, "Optimization of the Preparation Parameters of Monofilamentary Bi(2223) Tapes and the Effect of the Rolling Pressure on J_c," *Supercond. Sci. Technol.*, **8**, 827–32 (1995).

[6]K. Sato, K. Hayashi, K. Ohkura, K. Ohmatsu, T. Hikata, T. Kaneko, T. Kato, T. Sashida, M. Ueyama, J. Fujikami, M. Ito, S. Kobayashi, N. Saga, S. Hahakura, T. Shibata, T. Masuda, and S. Isojima, "Bismuth-Based HTS Wires and Their Application Progress"; pp. 1347–52 in Proceedings of the 16th ICEC/ICMC Conference (Kitakyushu, Japan, May 1996). Edited by T. Haruyama, T. Mitsui, and K. Yamafuji. Elsevier Science, Oxford, England, 1997.

[7]Y.C. Guo, H. K. Liu, and S. X. Dou, "Development of Microstructure and Superconductivity of Silver-Clad Bi2223 Composite Tapes in the Process of Heat Treatment," *Mater. Sci. Eng. B*, **23**, 58–63 (1994).

[8]G. Grasso, F. Marti, Y. Huang, and R. Flukiger, "Long Lengths of Bi2223 Superconducting Tapes," *Cryogenics*, **37** [10] 597–99 (1997).

[9]Y. Yamada, B. Obst, and R. Flukiger, "Microstructural Study of Bi(2223)/Ag Tapes with J_c(77 K, 0 T) Values of up to 3.3 $\times$ 10^4 A/cm^2," *Supercond. Sci. Technol.*, **4**, 165–71 (1991).

[10]Y. C. Guo, "Investigation of Silver Clad $(Bi,Pb)_2Sr_2Ca_2Cu_3O_{10}$ High-T_c Superconducting Composites"; Ph.D. Thesis, pp. 110–12. University of New South Wales, Sydney, Australia, 1994.

[11]S. A. Sunshine, T. Siegrist, L. F. Schneemeyer, D. W. Murphy, R. J. Cava, B. Batlogg, R. B. van Dover, R. M. Fleming, S. H. Glarum, S. Nakahara, R. Farrow, J. J. Krajewski, S. M. Zahurak, J. V. Waszczak, J. H. Marshall, P. Marsh, L. W. Rupp Jr., and W. F. Peck, "Structure and Physical Properties of Single Crystals of the 84 K Superconductor $Bi_{2.2}Sr_2Ca_{0.8}Cu_2O_{8+d}$," *Phys. Rev. B*, **38** [1] 893–96 (1988).

[12]R. Ramesh, G. Thomas, S. Green, C. Jiang, Y. Mei, M. L. Rudee, and H. L. Luo, "Polytypoid Structure of Pb-Modified Bi-Ca-Sr-Cu-O Superconductor," *Phys. Rev. B*, **38** [10] 7070–73 (1988).

[13]B. W. Statt, Z. Wang, M. J. G. Lee, J. V. Yakomie, P. C. de Camargo, J. F. Maya, and J. W. Rutter, "Stabilizing the High-T_c Superconductor $Bi_2Sr_2Ca_2Cu_3O_{10+x}$ by Pb Substitution," *Physica C*, **157**, 251–55 (1988).

[14]J. L. Tallon, R. G. Buckley, P. W. Ilberd, M. R. Presland, I. W. M. Brown, M. E. Bowden, L. A. Christian, and R. Goguel, "High-T_c Superconducting Phases in the Series $Bi_{2.1}(Ca,Sr)_{n+1}Cu_nO_{2n+4+d}$," *Nature (London)*, **333**, 153–55 (1988).

[15]Y. L. Chen and R. Steven, "2223 Phase Formation in Bi(Pb)-Sr-Ca-Cu-O: The Role of Chemical Composition," *J. Am. Ceram. Soc.*, **75** [5] 1142–49 (1992).

[16]K. Aota, H. Hattori, T. Hatano, K. Nakamura, and K. Ogawa, "Growth of the 2223 Phase in Leaded Bi-Sr-Ca-Cu Oxide under Reduced Oxygen Partial Pressure," *Jpn. J. Appl. Phys.*, **28**[12] L2196–L2199 (1989).

[17]T. Uzumaki, K. Yamanaka, N. Kamehara, and K. Niwa, "The Effect of Ca_2PbO_4 Addition on Superconductivity in a Bi-Sr-Cu-O System," *Jpn. J. Appl. Phys.*, **28** [1] L75–L77 (1989).

[18]Y. C. Guo, H. K. Liu, and S. X. Dou, "Stability of High-T_c Phase in Ag-Clad Bi-Based Superconducting Wires," *Physica C*, **200**, 147–52 (1992).

[19]C. K. Rhee, C. J. Kim, H. G. Lee, I. H. Kuk, J. M. Lee, I. S. Chang, C. S. Rim, P. S. Han, S. I. Pyun, and D. Y. Won, "Effects of Pb Content on the Formation of the High-T_c Phase in the (Bi,Pb)-Sr-Ca-Cu-O System," *Jpn. J. Appl. Phys.*, **28** [7] L1137–L1139 (1989).

[20]J. C. Grivel, A. Jeremie, B. Hensel, and R. Flukiger, "Effects of Pb in the First Stages of the Bi(2223) Phase Formation," *Supercond. Sci. Technol.*, **6**, 725–29 (1993).

[21]A. Jeremie, K. Alami-Yadri, J. C. Grivel, and R. Flukiger, "Bi,Pb(2212) and Bi(2223) Formation in the Bi-Pb-Sr-Ca-Cu-O System," *Supercond. Sci. Technol.*, **6**, 730–35 (1993).

[22]J. S. Luo, S. E. Dorris, A. K. Fischer, J. S. LeBoy, V. A. Maroni, Y. Feng, and D. C. Larbalestier, "Mode of Lead Addition and Its Effects on Phase Formation and Microstructure Development in $Ag/(Bi,Pb)_2Sr_2Ca_2Cu_3O_x$ Composite Conductors," *Supercond. Sci. Technol.*, **9**, 412–21 (1993).

[23]N. Fukushima, H. Niu, S. Nakamura, S. Takeno, M. Hayashi, and K. Ando, "Structural Modulation and Superconducting Properties in $Bi_{2-x}Pb_xSr_2CaCu_2O_{8+d}$ and $Bi_{2-y}Pb_ySr_2YCu_2O_{8+d}$," *Physica C*, **159**, 777–83 (1989).

[24]Y. Iwai, Y. Hoshi, H. Sato, and M. Takata, "Influence of the Oxygen Partial Pressure on the Solubility of PbO in the $(Bi,Pb)_2Sr_2CaCu_2O_{8+y}$ Superconducting Oxides," *Physica C*, **170**, 319–24 (1990).

[25]K. Aota, H. Hattori, T. Hatano, K. Nakamura, and K. Ogawa, "Growth of the 2223 Phase in Leaded Bi-Sr-Ca-Cu Oxide under Reduced Oxygen Partial Pressure," *Jpn. J. Appl. Phys.*, **28** [12] L2196–L2199 (1989).

[26]R. S. Roth, T. Negas, and L. P. Cook; pp. 94–95 in *Phase Diagrams for Ceramists*, Vol. 4. Edited by G. S. Smith. American Ceramic Society, Columbus, OH, 1981.

[27]T. Hatano, K. Aota, S. Ikeda, K. Nakamura, and K. Ogawa, "Growth of the 2223 Phase in Leaded Bi-Sr-Ca-Cu-O System," *Jpn. J. Appl. Phys.*, **27** [11] L2055–L2058 (1988).

[28]C. H. Chan and S. H. Risbund, "Effect of Supercooling on the Growth Kinetics of the 2223 Phase in Bi-Pb-Sr-Ca-Cu-O Superconducting Glass-Ceramics," *Supercond. Sci. Technol.*, **6**, 736–42 (1993).

[29]D. M. Spiller, M. K. Al-Mosawi, Y. Yang, C. Beduz, and R. Riddle, "The Effect of the Heating Rate on the Core Morphology and Transport J_c of (Pb,Bi)2223 Superconducting Tapes," *Cryogenics*, **37** [10] 623–25 (1997).

[30]J. I. Steinfeld, J. S. Francisco, and W. L. Hase, *Chemical Kinetics and Dynamics*; pp. 14–17. Prentice-Hall, Englewood Cliffs, NJ, 1989.

[31]H. S. Edelman, J. A. Parrell, and D. C. Larbalestier, "Evolution of the Transport Critical Current Density and Irreversibility Field as a Function of Heat Treatment and Pressing Pressure During Processing of Ag-Sheathed $(Bi,Pb)_2Sr_2Ca_2Cu_3O_x$ Tapes," *J. Appl. Phys.*, **81** [5] 2296–301 (1997).

[32]J. Horvat, Y. C. Guo, and S. X. Dou, "Formation of Weak and Strong Links in Ag/Bi-2223 Superconducting Tapes," *Physica C*, **271**, 59–60 (1996).

[33]X. J. Wu, S. Horiuchi, L. Ben-dor, and H. Diab, "Structure Evolutionin High-T_c Bi-Based Superconductors with Pb Doping Revealed by Electron Microscopy," *J. Solid State Chem.*, **97**, 82–92 (1992).

[34]M. Catoni, Y. Tanaka, M. Ishizuka, and S. Horiuchi, "Microstructure Evolution in High-T_c Bi(Pb)2223/Ag(Hf) Tapes," *Physica C*, **276**, 259–69 (1997).

[35]J. W. Ekin, T. M. Larson, A. M. Hermann, Z. Z. Sheng, K. Togano, and H. Kumakura, "Double-Step Behavior of Critical Current vs. Magnetic Field in Y-, Bi- and Tl-Based Bulk High-T_c Superconductors," *Physica C*, **160**, 489–96 (1989).

[36]J. Horvat, W. G. Wang, R. Bhasale, Y. C. Guo, H. K. Liu, and S. X. Dou, "Critical Currents in Ag/Bi2223 Superconducting Tapes: Grain Connectivity and Flux Pinning," *Physica C*, **275**, 327–32 (1997).

[37]L. N. Bulaevskii, J. R. Clem, L. I. Glazman, and A. P. Malozemoff, "Model for the Low-Temperature Transport of Bi-Based High-Temperature Superconducting Tapes," *Phys. Rev. B*, **45**, 2545–48 (1992).

[38]L. N. Bulaevskii, L. L. Daemen, M. P. Maley, and J. Y. Coulter, "Limits to the Critical Current in High-T_c Superconducting Tapes," *Phys. Rev. B*, **48**, 13798–816 (1993).

[39]B. Hensel, J. C. Grivel, A. Jeremie, A. Perin, A. Pollini, and R. Flukiger, "A Model for the Critical Current in $(Bi,Pb)_2Sr_2Ca_2Cu_3O_x$ Silver-Sheathed Tapes: The Role of Small-Angle *C*-axis Grain Boundaries and of the Texture," *Physica C*, **205**, 329–37 (1993).

[40]B. Hensel, G. Grasso, and R. Flukiger, "Limits to the Critical Transport Current in Superconducting $(Bi,Pb)_2Sr_2Ca_2Cu_3O_{10}$ Silver-Sheathed Tapes: The Railway-Switch Model," *Phys. Rev. B*, **51** [21] 15456–73 (1995). □

journal

J. Am. Ceram. Soc., **83** [10] 2622–24 (2000)

Influence of Undercooling and $Nd_4Ba_2Cu_2O_{10}$ Content on the Growth Rate of Large-Grain Bulk Nd-Ba-Cu-O Superconductor

Nadendla Hari Babu, Wai Lo, and David A. Cardwell

IRC in Superconductivity, University of Cambridge, Cambridge CB3 0HE, United Kingdom

The dependence of growth rate on isothermal undercooling and composition has been investigated for Nd-Ba-Cu-O single grains containing various amounts of nonsuperconducting $Nd_4Ba_2Cu_2O_{10}$ (Nd-422) phase inclusions and fabricated under a 1% O_2 in N_2 atmosphere using a top seeded melt growth technique. The growth rate along the crystallographic *c*-direction is observed to exceed that along the *a/b* direction at all undercooling temperatures and exhibits a maximum for a Nd-422 content of ≈10 mol%. The samples have been examined by optical microscopy and the results interpreted within an established planar solidification model.

I. Introduction

HIGH-TEMPERATURE superconducting Y-Ba-Cu-O (YBCO) may be fabricated in the form of large, single grains which do not contain current-limiting grain boundaries via the so-called top seeded melt growth (TSMG) technique.[1–4] This process involves using a seed crystal of chemical and structural compatibility to YBCO as a heterogeneous nucleus to achieve controlled grain nucleation and growth. The irreversibility field of YBCO, however, is limited typically to around 5 T at 77 K for field applied parallel to the *c*-axis, which represents the maximum field trapping potential of this material at this temperature. Recently, the Nd-Ba-Cu-O (NdBCO) system has been successfully melt processed and observed to exhibit high critical temperature and current density (T_c and J_c), even in the presence of relatively high magnetic fields.[5] One important feature of melt-processed NdBCO is that it exhibits high irreversibility fields of ≈9 T and >40 T (extrapolated) at 77 K for fields applied parallel and perpendicular to the *c*-axis, respectively.[6,7]

Knowledge of the growth rate of any (RE)BCO material (RE = rare-earth element or Y) in the TSMG process is particularly important if it is to be fabricated in large-grain form by peritectic solidification with controlled microstructure and crystallographic orientation. Earlier studies have established that the grain growth rate of YBCO, for example, is influenced significantly by interface undercooling[8] and, to understand the growth kinetics, careful growth rate measurements have subsequently been performed on this material.[9] The growth rate of NdBCO, which is sensitive to process variables such as the solubility of the rare-earth atom in the in Ba-Cu-O melt in the peritectic,[10] on the other hand, is generally under-researched despite the potential advantages of this material for engineering applications compared with YBCO. In addition, there are only a limited number of reports of the fabrication of large (>2 cm diameter) single NdBCO grains by the TSMG technique.[11]

S. M. Mukhopadhyay—contributing editor

Manuscript No. 189031. Received October 8, 1999; approved May 30, 2000.
*Member, American Ceramic Society.

In this paper we report the dependence of the growth rate of large-grain, top seeded melt grown NdBCO processed under a 1% O_2 in N_2 atmosphere on undercooling and $Nd_4Ba_2Cu_2O_{10}$ (Nd-422) phase content.

II. Experimental Procedure

$NdBa_2Cu_3O_{7-\delta}$ (Nd-123) and Nd-422 precursor powders were synthesized by spray drying and subsequent calcination.[12] Powders of stoichiometric Nd-123 enriched by 0, 5, 10, 15, 20, 25, 40, 50, 60, 80, and 100 mol% Nd-422 were pressed uniaxially (1.5 tonnes) into 32 mm diameter and 12 mm thick pellets. These were sintered in air at 1060°C for 6 h to form solid precursor bodies. A MgO seed crystal with its (001) surface oriented parallel to the sample surface was placed on the sintered pellet before melt processing using the temperature profile shown in Fig. 1. No significant variation in the peritectic temperature, T_p, of the samples with increasing Nd-422 content was observed from differential thermal analysis (DTA) measurements under 1% O_2 in N_2. This was approximately constant at 1032°C. The growth temperature, T_g, for each sample was consequently chosen below T_p. All of the samples were grown isothermally (i.e., at a constant temperature) over a period of a few hours and then cooled rapidly to room temperature. The holding time for growth temperatures close to T_p was varied between 5 and 40 h, within which range the variation of growth length with holding time was observed to be linear. A hold time of 5 h was used for lower growth temperatures, on the other hand, to limit the formation of homogeneous nuclei. Finally, the melt-processed samples were mounted and polished for microstructural examination under polarized light using an optical microscope.

III. Results and Discussion

Figure 2(a) shows the top surface of a large NdBCO grain containing 20 mol% Nd-422 processed isothermally for 15 h under reduced oxygen pressure at a growth temperature of 1020°C. It can

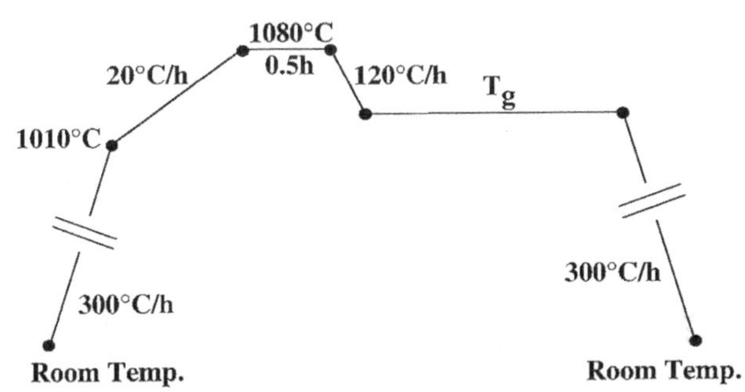

Fig. 1. Schematic diagram of the thermal profile used for the isothermal solidification of NdBCO. The growth temperature, T_g, and hold time were varied for different samples.

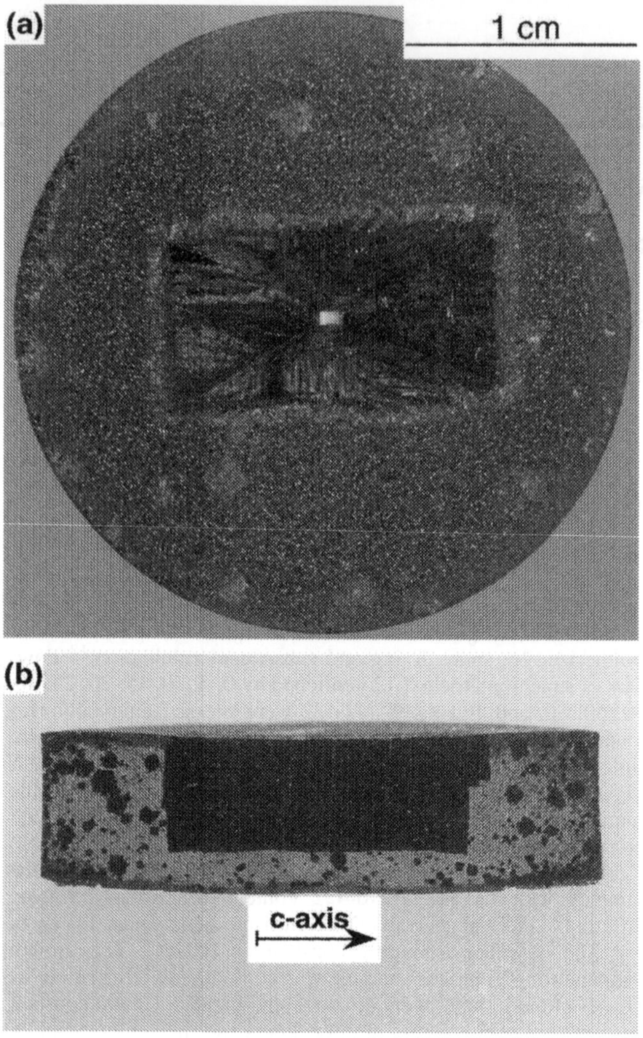

Fig. 2. Photograph of (a) the top surface and (b) the cross section normal to the sample surface of a NdBCO single grain containing 20 mol% Nd-422 processed isothermally at 1020°C (ΔT = 12°C) under a controlled 1% O_2 in N_2 atmosphere by the TSMG technique.

be seen clearly that the melt process yields a single grain which nucleates heterogeneously at the contact surface between the MgO seed and Ba-Cu-O melt. No homogeneous nucleation is apparent in the sample microstructure for this particular degree of undercooling ($\Delta T = T_p - T_g \approx$ 12°C) although satellite homogeneous nuclei were observed to form on the sample surface for all compositions as the undercooling increased beyond ≈1018°C. This clearly hinders the growth of single grains and inevitably yields multidomain material. Single grains of the various compositions were subsequently grown at temperatures at which only heterogeneous nucleation was observed.

(1) *Effect of Undercooling on Growth Rate*

The as-grown bulk samples were cut perpendicular to their top surface, as illustrated in Fig. 2(b), to measure the growth lengths of the grain along the *a*/*b* and *c* lattice directions. Grain growth rates were measured by this method for various growth times and temperatures for all of the compositions investigated. Figure 3 illustrates the growth rate of the sample containing 15 mol% Nd-422 as a function of undercooling, from which it can be seen that the grain size increases with increasing ΔT for both directions (i.e., as the driving force for nucleation increases). It is apparent from this figure that the *c*-axis growth rate, R_c, is higher than that for the *a*-direction, R_a, as is the case for YBCO melt processed in air at relatively low undercooling.[9] From Fig. 3, the variation of

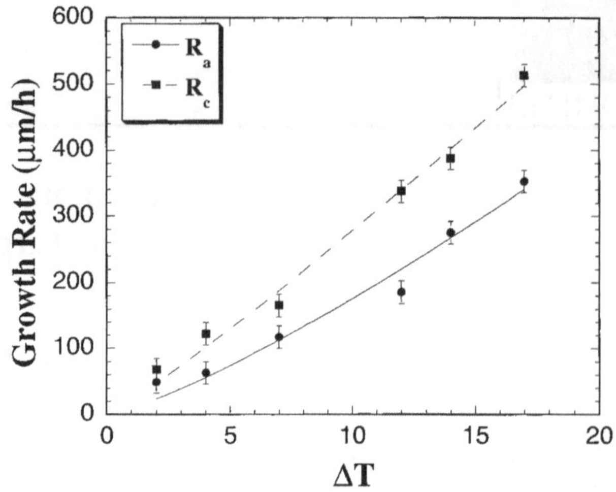

Fig. 3. Growth rate of single NdBCO grains containing 15 mol% Nd-422 as a function of undercooling for both *a* and *c* lattice directions.

the growth rate with undercooling may be expressed as

$$R_a = K_a \Delta T^{1.24}$$

$$R_c = K_c \Delta T^{1.09}$$

where K_a = 9.97 μm/h and K_c = 22.27 μm/h (i.e., R_c increases almost linearly with ΔT). It is very difficult to measure the growth rate beyond 18°C undercooling because of the formation of random homogeneous nuclei in the melt. Endo *et al.*[12] measured the growth rate of a $YBa_2Cu_3O_{7-\delta}$ grain processed in air by TSMG and reported the following growth rate relationships:

$$R_a = 4.5 \times 10^{-7} \Delta T^{1.9} \text{ (mm/s)}$$

$$R_c = 2.8 \times 10^{-6} \Delta T^{1.3} \text{ (mm/s)}$$

It can be concluded, therefore, that the growth rate of NdBCO processed under a 1% O_2 in N_2 atmosphere (which exhibits high T_c, J_c and irreversibility field[6]) is greater than that observed for YBCO by typically a factor of 3 at all undercooling temperatures and along both growth directions. The higher growth rate in the NdBCO system occurs primarily as a consequence of the lower liquidus slope, m_l, in the phase diagram and the higher solubility of Nd, compared with Y, in the Ba-Cu-O liquid.[10] The higher values of measured growth rate of NdBCO single grains is an important factor in the development of a practical bulk materials process. In addition, good transport properties of NdBCO samples directionally solidified at high rates have been reported which underline the potential of this material for engineering applications.[14]

(2) *Effect of Nd-422 Content on Growth Rate*

Single grains containing various amounts of Nd-422 were grown at constant undercooling of 12°C to investigate the dependence of growth rate on the Nd-422 phase content. The results of this study are shown in Fig. 4. It can be seen that the growth rate exhibits a maximum for a Nd-422 content of ~10 mol%, which is significant within the measurement error, and tends to zero as the Nd-422 content tends to 100 mol% for both growth directions. Cima *et al.*[8] reported the following equation describing the maximum growth rate for a planar solidification front, which may be used to explain the above observation:

$$R_{max} = \frac{D_L(\Delta T + G\ell)}{[\ell(C_s^{123} - C_\ell^p)m_\ell]} \quad (1)$$

where D_L is the coefficient of the diffusion of Nd, G is the thermal gradient, C_s^{123} is the Nd concentration of the Nd-123 phase, C_ℓ^p is the Nd concentration in the liquid at the Nd-123 interface, and ℓ is the distance between Nd-422 particles (proportional to the Nd-422 particle size). The inter-Nd-422 particle distance decreases as the

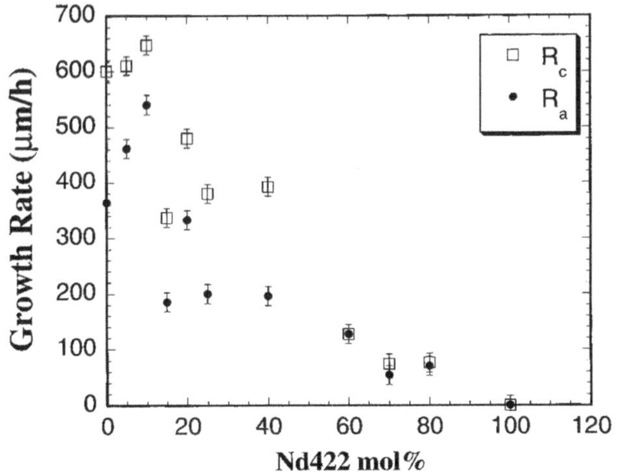

Fig. 4. Growth rate of NdBCO single grains grown at 1020°C (ΔT = 12°C) as a function of Nd-422 content for both *a* and *c* lattice directions.

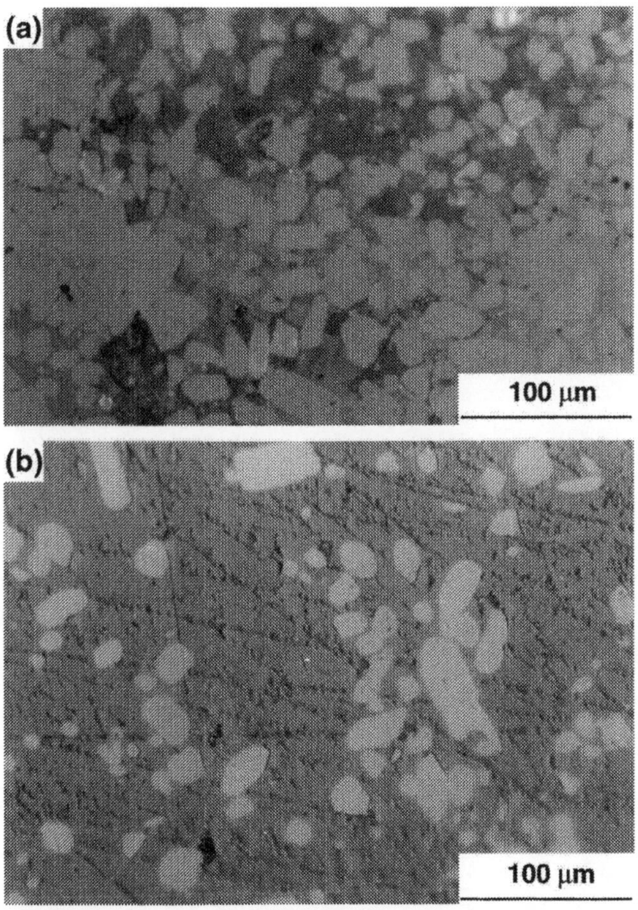

Fig. 5. Optical micrographs of the samples containing (a) 100 and (b) 20 mol% of the Nd-422 phase. A dense concentration of Nd-422 particles (white in contrast) and a nonuniform distribution of Ba-Cu-O liquid can be seen in (a).

Nd-422 content increases, resulting in an initial increase in the growth rate. Although ℓ continues to decrease as the concentration of Nd-422 increases further, the diffusion coefficient of Nd atoms from the Nd-422 particles into the liquid decreases because of an associated increase in the viscosity of the Ba-Cu-O liquid/Nd-422 phase mixture. The decrease in D_L subsequently results in a decrease in the growth rate. Beyond a certain limit, as the content of the Nd-422 inclusions in the liquid increases even further, the lack of Ba-Cu-O liquid supply to the Nd-123 growth front hinders severely the solidification process. Comparison of the observed dependence of growth rate on Nd-422 content with Eq. (1) suggests that $\Delta T \gg G\ell$ for the growth conditions used in the present study. This is consistent with the observed buildup in the Nd-422 particle concentration and clear discontinuity in the Ba-Cu-O liquid-phase content illustrated in the optical micrograph of the sample containing 100 mol% Nd-422 in Fig. 5(a). The corresponding micrograph for the sample containing 20 mol% Nd-422 is shown in Fig. 5(b) for purposes of comparison. It was not possible to grow a single grain for the former composition, even after holding the sample at the growth temperature for a long time. It may be concluded from the above observations that the addition of secondary phase inclusions can be optimized to improve the growth rate of the grain, in addition to yielding higher J_c samples.

IV. Conclusions

Large NdBCO single grains, consisting of Nd-422 phase particles embedded in a superconducting Nd-123 matrix, containing a wide range of Nd-422 phase inclusions, have been fabricated under 1% O_2 in N_2 by the top seeded melt growth technique. The growth rate of these grains along the major crystallographic axes has been measured as a function of both undercooling and Nd-422 phase content. The observed variation of growth rate with Nd-422 phase content, which exhibits a maximum for ≈10 mol%, is consistent with the planar solidification model of Cima *et al.* The growth rate observed along the *c*-axis is consistently higher than that along the *a*/*b* axes at all values of undercooling. The growth rate of NdBCO has been found to be significantly greater than that for YBCO along both growth directions, which is important in the development of a practical materials process.

References

[1]S. Jin, T. H. Tiefel, R. C. Sherwood, M. E. Davis, R. B. van Dover, S. W. Kammlott, R. A. Fastnacht, and H. D. Keith, "High Current Density in Bulk $YBa_2Cu_3O_{7-\delta}$ Superconductor," *Appl. Phys. Lett.*, **52,** 2074–76 (1988).

[2]M. Murakami, N. Sakai, H. Takaichi, and D. N. Matthews, "A New Process with the Promise of High J_c in Oxide Superconductor," *Jpn. J. Appl. Phys.*, **28,** 1189–94 (1989).

[3]W. Lo, D. A. Cardwell, C. D. Dewhurst, and S. L. Dung, "Fabrication of Large YBCO Grains Seeded Peritectic Solidification," *J. Mater. Res.*, **11,** 786–94 (1996).

[4]R. L. Meng, L. Gao, P. Gautier-Picard, D. Ramirez, Y. Y. Sun, and C. W. Chu, "Growth and Possible Size Limitation of Quality Single Grain $YBa_2Cu_3O_{7-\delta}$," *Physica C*, **232** [3–4] 337–46 (1994).

[5]S. I. Yoo, N. Sakai, H. Taguchi, and M. Murakami, "Melt Processing for Obtaining $NdBa_2Cu_3O_y$ Superconductors with High T_c and Large J_c," *Appl. Phys. Lett.*, **33,** L1000–L1003 (1994).

[6]N. Hari Babu, W. Lo, D. A. Cardwell, and A. M. Campbell, "The Irreversibility Behavior of NdBaCuO Fabricated by Top-Seeded Melt Processing," *Appl. Phys. Lett.*, **75,** 2891 (1999).

[7]Th. Wolf, A.-C. Bornarel, H. Kupfer, R. Meier-Hirmer and B. Obst, "High Irreversibility Fields and Current Densities in $NdBa_2Cu_3O_{y-\delta}$ Single Crystals and Melt-Textured Samples," *Phys. Rev. B*, **56** [10] 6308–19 (1997).

[8]M. J. Cima, M. C. Flemings, A. M. Firueredo, M. Nakade, H. Ishii, H. D. Brody, and J. S. Haggerty, "Semi-solidification of High Temperature Superconducting Oxides," *J. Appl Phys*, **72** [1] 179–90 (1992).

[9]Y. Nakamura, A. Endo, and Y. Shiohara, "Relation between the Undercooling and the Growth Rate in Y123 + 10% Y211," *J. Mater. Res.*, **11,** 1094 (1996).

[10]Ch. Krauns, M. Sumida, M. Tagami, Y. Yamada, and Y. Shiohara, "Solubility of RE Elements into Ba-Cu-O Melts and the Enthalpy of Dissolution," *Z. Phys. B*, **96,** 207–12 (1994).

[11]W. Lo, N. Hari Babu, D. A. Cardwell, Y. Shi, and D. M. Astill, "Fabrication and Characterization of Large NdBCO Grains Prepared under Low Oxygen Pressure," *J. Mater. Res.*, **15,** 33 (2000).

[12]W. Lo, D. A. Cardwell, and Y. H. Shi, "Spray Dried Pt-Doped Nd-Ba-Cu-O Precursor Powder for Seeded Peritectic Processing of Large Superconducting Grains," *Mater. Sci. Eng. B*, **B65,** 1 (1999).

[13]A. Endo, H. S. Chauhan, Y. Nakamure, and Y. Shiohara, "Relation between Growth Rate and Undercooling in Pt-Added $YBa_2Cu_3O_{7-\delta}$," *J. Mater. Res.*, **11** [5] 1114–19 (1996).

[14]K. Salama, A. S. Parikh, and L. Woolf, "High Rate Melt Texturing of $Nd_{1-x}Ba_{2+x}Cu_3O_{7-\delta}$ Type Superconductors," *Appl. Phys. Lett.*, **68** [14] 1993–95 (1996). □

journal

J. Am. Ceram. Soc., 83 [10] 2365–68 (2000)

Reactive Sintering and Retrograde Densification of Bulk Bismuth-Based Superconductors

Ryan L. Thayer,[*,†,‡] Sarah R. Schmidt,[*,†] Stephen E. Dorris,[†] Jeffrey W. Bullard,[*,§] and Michael T. Lanagan[*,†,‡]

Energy Technology Division, Argonne National Laboratory, Argonne, Illinois 60439–4838 and Department of Materials Science and Engineering University of Illinois at Urbana–Champaign, Urbana, Illinois 61801

Densification was observed when bulk ceramic superconductors with a composition of $Bi_{1.8}Pb_{0.4}Sr_2Ca_2Cu_3O_{10+\delta}$ were sintered for short times (<4 h) near the incongruent melting temperature. The initial shrinkage of the powder compacts was correlated with the existence of a transient liquid phase. Retrograde densification, which is a decrease in density, occurred for longer sintering times and was attributed to changes in pore structure. The results provide direction for future processing of bulk bismuth-based superconductors and powder-in-tube tapes for the fabrication of long-length wire.

I. Introduction

Bulk ceramic superconductors are candidates for a wide array of potential applications. Superconducting magnetic energy storage systems, fault-current limiters, and magnetic resonance imaging systems require current leads with a high current capacity and low thermal conductivity.

The reactive sintering of $[Bi,Pb]_{2+x}Sr_2Ca_2Cu_3O_{10+\delta}$, where $0 < x < 0.2$ (hereafter referenced as Bi-2223), has been used to achieve excellent phase purity and high critical current density (J_C).[1,2] The reactive sintering of a phase assemblage of $[Bi,Pb]_{2+x}Sr_2Ca_1Cu_2O_{8+\delta}$ (hereafter referenced as Bi-2212), Ca_2CuO_3, and CuO at high temperature results in the formation of a transient liquid phase.[1–6] The exact composition and volume fraction of the liquid, as a function of temperature and time, are difficult to characterize; however, a transient liquid phase is believed to fill several important roles during sintering. This liquid phase promotes reaction kinetics by decreasing the activation energy (thereby enhancing mass transport),[7] contributes to grain growth[8] and alignment,[9] and heals damage that is incurred during mechanical deformation.[10] Because the amount and distribution of lead in the system can influence liquid-phase formation, they also have important roles in Bi-2223 processing.[2,4]

Most of the research on Bi-2223 processing has been performed on powder-in-tube (PIT) tapes. PIT processing is a useful technology for the fabrication of long lengths of superconductors.[11] PIT tapes are fabricated by packing superconductor powder into a silver tube. The tube is swaged, drawn, rolled to a final thickness of ~200 μm, and heat-treated.

D. R. Clarke—contributing editor

Manuscript No. 189715. Received December 3, 1998; approved March 30, 2000.
Supported by the U.S. Department of Energy (DOE), Energy Efficiency and Renewable Energy, as part of a DOE program to develop electric power technology (under Contract No. W-31-109-Eng-38).

*Member, American Ceramic Society.
†Energy Technology Division, Argonne National Laboratory.
‡Current address: Materials Research Laboratory, Pennsylvania State University, University Park, PA 16802.
§Dept. of Materials Science and Engineering, University of Illinois at Urbana–Champaign.

Increasing the relative density and grain alignment while maintaining high phase purity is an important goal of processing both bulk and PIT tapes of Bi-2223. Usual pressureless sintering is frustrated by a phenomenon known as "retrograde densification," by which the relative density of bismuth-based superconductor tapes and bulk monoliths passes through a maximum and actually decreases significantly when annealed further. Retrograde densification of Bi-2223 is well documented[6,10,12–14] and poses an obstacle to the successful sintering of these materials; for example, a 40% decrease in density of bulk specimens was reported by Hatano *et al.*[12] Other researchers have found that PIT tapes expand 7% in thickness during processing.[13] For this reason, combined thermomechanical processes usually are developed empirically to obtain the most-desirable microstructures and highest J_C values. However, these processes, which typically involve the intermittent interruption of the sintering schedule to isostatically press the specimen, are time-consuming and inefficient. Therefore, it seems desirable that we should more clearly understand and, if possible, minimize the influences of retrograde densification.

The generally accepted mechanism of retrograde densification of bismuth-based superconductors is related to grain impingement. In an early work, Johnson and Rhodes[14] observed that grain growth in Bi-2212 is highly anisotropic and produces large platelike grains that grow more readily in the *a*/*b* plane than in the orthogonal direction. Based on observations of the grain morphology, they proposed that the mutual impingement of these grains during their growth pushes the microstructure apart, causing retrograde densification. To date, most studies of the microstructural development and properties of bismuth-based superconductors have focused on long annealing times (>50 h), well beyond the beginning of retrograde densification.

In this paper, we attempt to further elucidate the nature of retrograde densification by examining details of the microstructural development near the onset of this phenomenon during reactive sintering of bulk Bi-2223. The relative, bulk, and apparent densities are characterized as a function of time and are correlated with the evolution of the pore-space topology and microstructural morphology.

II. Experimental Procedure

A mixture of Bi-2212, Ca_2CuO_3, and CuO powders (from Superconductive Components, Inc., Columbus, OH), with a nominal Bi-2223 composition ($Bi_{1.8}Pb_{0.3}Sr_{2.05}Ca_{1.95}Cu_3O_x$), was pressed into pellets 2.54 cm in diameter and sintered at 860°C in air with dwell times up to 50 h. The sintering schedule was based on previous studies that focused on maximizing the J_C value and phase purity of the bulk bars.[6]

After sintering, the fracture surfaces of the pellets were examined via scanning electron microscopy (SEM) to determine the grain size. Then, the pellets were ground with a mortar and pestle and sieved through a 60-mesh screen. The resulting powder was characterized via X-ray diffractometry (XRD) to monitor the phase conversion to Bi-2223, differential thermal analysis (DTA) to observe the onset temperature of the liquid-phase formation, and helium pycnometry to determine the theoretical density. The post-sintering relative densities of the pellets were measured using the immersion method. Carbon content of the pellets was <0.02%, as measured using a carbon/sulfur determinator (Model C/S 300, Leco Corp., St. Joseph, MI).

For the immersion method, entire pellets were placed in isopropyl alcohol (IPA) on a hot plate near the boiling point of IPA for ~1 h until no air bubbles were observed emanating from the pellets, which signified that all the accessible pore space was filled. In addition, each pellet was fractured into three or four pieces, and the immersion method was repeated to compare the open porosity of the surface of the pellets with the open porosity of the exposed microstructure. The open porosity of the surface and the exposed microstructure were identical, within experimental error.

During sintering in air, changes in the linear dimension of the Bi-2223 specimens were measured continuously using a contact dilatometer (Model Theta Dilatronic, Theta Industries, Port Washington, NY). Specimens were 2.54 cm long and 0.381 cm in diameter; they were made by filling a rubber tube (inside diameter of 0.635 cm) with powder, then sealing and pressurizing the tube to 200 MPa in a cold isostatic press.

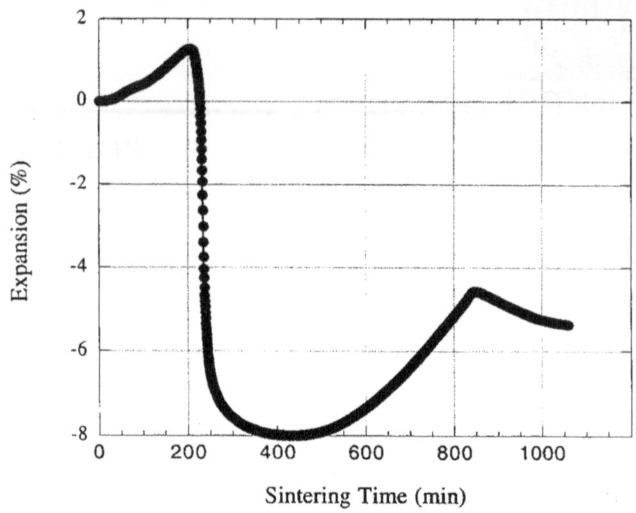

Fig. 2. Dilatometer data showing expansion as a function of sintering time; the sample was heated to the sintering temperature for the period of <200 min and cooled for the period of >850 min.

III. Results and Discussion

During sintering, the mixture of Bi-2212, Ca_2CuO_3, and CuO reacts to form Bi-2223. XRD shows that most of the Bi-2223 is formed within a period of 7.5–50 h, which is in agreement with results of Luo *et al.*[15] After 1000 min, XRD patterns (not shown) indicate a phase conversion of >90%. Additional heat treatment is necessary to remove the residual alkaline-earth cuprate phases.

Figure 1 shows the temperature that corresponds to the initial formation of a liquid phase as the reactive sintering of Bi-2223 progresses. A low-temperature endotherm with an onset temperature of 858°C is evident in the DTA pattern for sintering for 0 h; however, after sintering for 2.5 h, the first endotherm shrinks and the major endothermic event is the incongruent melting of the bismuth superconductor. This observation agrees with previous results.[2] The size of the first endotherm is dependent on the amount of reactive liquid; therefore, our results suggest that much, although not all, of the liquid is consumed very early in the sintering process.

Figure 2 shows that a maximum in linear shrinkage occurs at 3.67 h, after which the Bi-2223 expands. In the early sintering stages of single-component compacts, a liquid phase can contribute to densification by aiding in particle rearrangement.[16] In multicomponent systems, such as Bi-2223, the liquid phase can promote either densification or retrograde densification.[17] Factors that affect densification include the relative solubilities of the liquid and solid, as well as the degree of wetting of the solid by the liquid. Densification via liquid-phase sintering has been observed for highly anisometric microstructures other than Bi-2223, e.g., Si_3N_4.[18] If a large fraction of the liquid phase is consumed early in the sintering process (<4 h), an insufficient amount of liquid may remain for particle rearrangement at longer sintering times. DTA suggests that a significant portion of the liquid is, in fact, consumed during the first 3 h of heat treatment (Fig. 1).

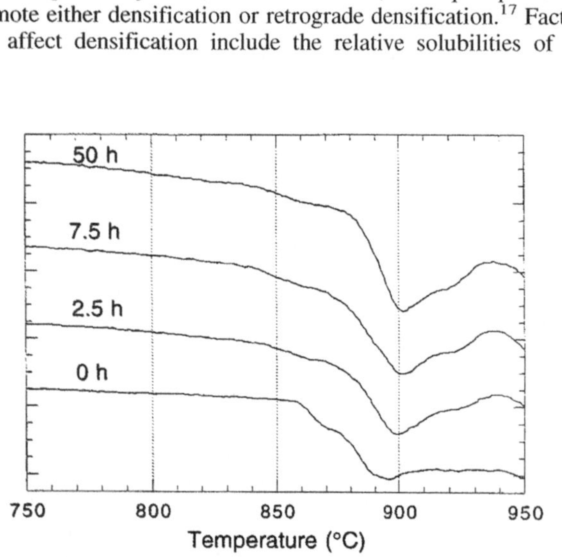

Fig. 1. DTA results showing the formation of a liquid phase during the sintering of Bi-2223.

Grain growth occurs throughout the sintering process (see Fig. 3), which agrees with earlier reports.[19] We found that grains grow in the *a*/*b* plane from a size of ~15 μm at 0.3 h to ~80 μm at 50 h, whereas grain growth in the *c*-direction is negligible. Although the amount of liquid probably is insufficient for particle rearrangement beyond 4 h, the residual liquid could be sufficient to enhance grain growth. The amount of liquid phase may be too small for DTA to detect the melting temperature for sintering times of >4 h. Acoustic-velocity measurements suggest the presence of a liquid phase at substantially longer times (20 h) for PIT tapes.[5]

Figure 3 also shows that significant grain growth and impingement occur during the first 3 h of sintering. During this time, the specimen density increases from ~50% to ~80% (see Fig. 4) and open porosity decreases from 50% to 0%. The data show grain growth and impingement before retrograde densification, which suggests that additional mechanisms contribute to this phenomenon.

Although grain pushing is a potential cause of retrograde densification in these materials, our experimental data reveal a close correlation between retrograde densification and a microstructural event. At the onset of retrograde densification, the pore volume that is accessible to IPA decreases to zero. The pore percolation threshold in this system occurs at a relative density of ~80%, which is fairly low for common porous media. However, systems such as sintered Bi-2223 with highly anisometric grains easily can exhibit percolation thresholds this low or even lower.[20]

It is known that vapor species that are insoluble in the solid phase can become trapped within isolated pores during the final stages of sintering, and that the back pressures that are generated by these trapped vapors can decrease or even completely offset the driving force for further densification.[21] This influence of trapped gases during final-stage sintering has been reported for oxygen that is enclosed in pores of $YBa_2Cu_3O_7$[22] and other common ceramics.[21] Trapped gases also are believed to be responsible for the end-point densities in alkali-halide powder compacts.[23]

The influence of trapped insoluble gases can explain why further densification may be retarded or stopped after pore closure; however, trapped gases by themselves cannot be responsible for decreases in the relative density. If, however, in addition to the vapor pressure of gas that is trapped from the atmosphere, the

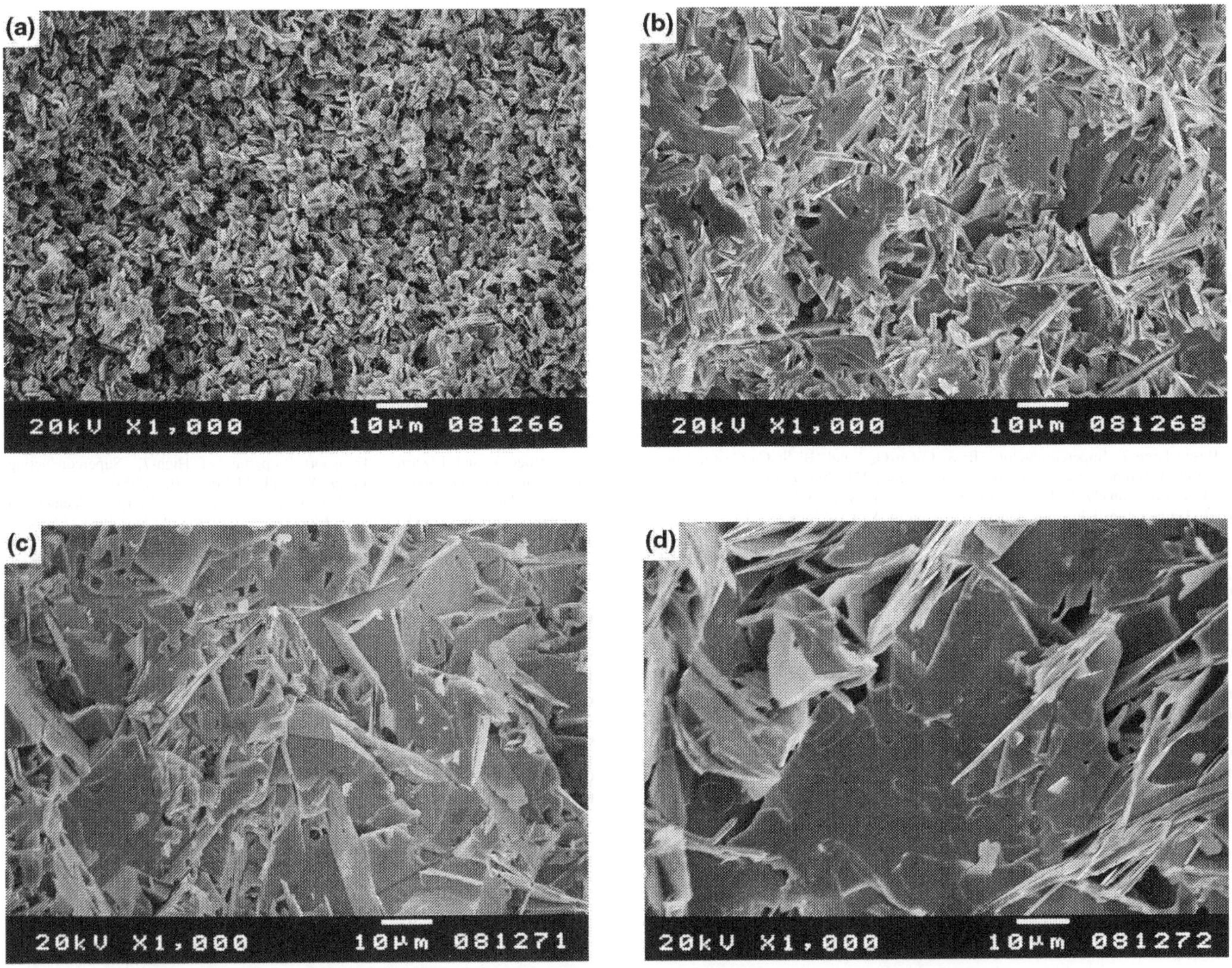

Fig. 3. SEM photomicrographs of fracture surfaces of Bi-2223 sintered at 860°C for (a) 0, (b) 2.5, (c) 7.5, and (d) 50 h.

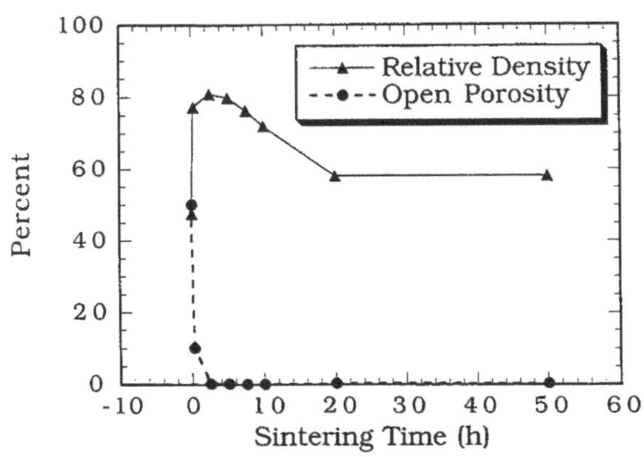

Fig. 4. Relative density and open porosity versus sintering time for Bi-2223.

vapor pressure is further increased by the decomposition or volatilization of some component of the solid, then the back pressure within isolated pores could exceed the driving force for densification and conceivably could result in a retrograde phenomenon. During the reactive sintering of Bi-2212 from 850°C to 901°C, others noted a 4% mass loss, which was attributed to the volatilization of adsorbed water and/or residual carbonates.[14] Retrograde densification also was observed in this same temperature range.[14]

As mentioned previously, the carbon content of the Bi-2223 used in this study was low (<0.02%); therefore, the vaporization of CO_2 should be negligible. However, the high equilibrium vapor pressure of lead, which was used here to promote the formation of a Bi-2223 phase, is known to result in a loss of lead from the solid. Therefore, the continued volatilization of lead beyond the pore percolation threshold may be responsible for the retrograde densification phenomena that have been observed in this and similar materials.[24]

IV. Summary and Conclusions

For the reactive sintering of bulk $[Bi,Pb]_{2+x}Sr_2Ca_2Cu_3O_{10+\delta}$, where $0 < x < 0.2$ (Bi-2223), retrograde densification is correlated closely to a transition in the pore topology from an interconnected network to individual closed cavities. The most-plausible casual relationship between this dramatic transition in densification behavior and the pore connectivity is that trapped gases within isolated pores, in combination with continually evolving vapor from the solid phase, attain pressures that are high enough to cause the pores to expand significantly.

In practice, to make current leads or powder-in-tube (PIT) tapes, intermittent pressing and heat-treatment processes are required to achieve high density and high critical current densities. The interruption of sintering after the onset of retrograde densification to cold isostatically press a Bi-2223 specimen, which is an often-adopted procedure in the manufacture of such material, may simply cause a fracture within the body that reconnects the pore space and allows densification to resume until the pore space again reaches percolation and becomes disconnected. Most PIT processes include a sintering period of 50 h, which is a period that has not been optimized. Our study suggests that sintering should be interrupted by a deformation process at the onset of retrograde densification, which can occur earlier than 50 h.

References

[1]D. Y. Kaufman, M. T. Lanagan, S. E. Dorris, J. T. Dawley, I. D. Bloom, M. C. Hash, N. Chen, M. R. DeGuire, and R. B. Poeppel, "Thermomechanical Processing of Reactively Sintered Ag-Clad $(Bi,Pb)_2Sr_2Ca_2Cu_3O_x$ Tapes," *Appl. Supercond.*, **1**, 81–91 (1993).

[2]S. E. Dorris, B. C. Prorok, M. T. Lanagan, S. Sinha, and R. B. Poeppel, "Synthesis of Highly Pure Bismuth-2223 by Two-Powder Process," *Physica C (Amsterdam)*, **212**, 66–74 (1993).

[3]R. H. Arendt, M. F. Garbauskas, and P. J. Bednarczyk, "An Alternate Preparation for $(Bi,Pb)_2Ca_2Sr_2Cu_3O_z$," *Physica C (Amsterdam)*, **176**, 126–30 (1991).

[4]P. Majewski, S. Kaesche, H.-L. Su, and F. Aldinger, "The Pb Solubility of the Bi-Based High-T_c Superconductors '$Bi_2Sr_2CaCu_2O_8$ ' and '$Bi_2Sr_2Ca_2Cu_3O_{10}$' as a Function of Temperature," *Physica C (Amsterdam)*, **221**, 295–98 (1994).

[5]M. T. Lanagan, D. S. Kupperman, G. A. Yaconi, S. H. Kilgore, and A. Saigal, "High-Temperature Ultrasonic Characterization of Ag-Clad Superconductor Tapes," *IEEE Trans. Appl. Supercond.*, **5** [2] 1475–78 (1995).

[6]M. P. Chudzik, B. J. Polzin, R. Thayer, J. J. Picciolo, B. L. Fisher, and M. T. Lanagan, "Fabrication and Characterization of (Bi,Pb)-Sr-Ca-Cu-O (2223) Bars," *IEEE Trans. Appl. Supercond.*, **7** [2] 2102–105 (1997).

[7]Y. S. Sung and E. E. Hellstrom, "Phase Regions and Kinetics of Formation of $(Bi,Pb)_2Sr_2Ca_2Cu_3O_x$ in Ag-Sheathed Tape," *J. Am. Ceram. Soc.*, **78** [8] 2003–2008 (1995).

[8]P. E. D. Morgan, J. D. Piché, and R. M. Housley, "Use of a Thermal Gradient to Study the Role of Liquid Phase During the Formation of Bismuth High Temperature Superconductors," *Physica C (Amsterdam)*, **191**, 179–84 (1992).

[9]J. S. Luo, S. E. Dorris, A. K. Fischer, J. S. LeBoy, V. A. Maroni, Y. Feng, and D. C. Larbalestier, "Mode of Lead Addition and Its Effects on Phase Formation and Microstructural Development in $Ag/(Bi,Pb)_2Sr_2Ca_2Cu_3O_x$ Composite Conductors," *Supercond. Sci. Technol.*, **9**, 412–21 (1996).

[10]M. Satou, Y. Yamada, S. Murase, T. Kitamura, and Y. Kamisada, "Densification Effect on the Microstructure and Critical Current Density in $(Bi,Pb)_2Sr_2Ca_2Cu_3O_x$Ag Sheathed Tape," *Appl. Phys. Lett.*, **64** [5] 640–42 (1994).

[11]G. N. Riley Jr. and W. L. Carter, "Commercialization of Ceramic Superconductor Composite Wires," *Am. Ceram. Soc. Bull.*, **72** [7] 91–97 (1993).

[12]T. Hatano, K. Aota, S. Ikeda, K. Nakamura, and K. Ogawa, "Growth of the 2223 Phase in Leaded Bi-Sr-Ca-Cu-O System," *Jpn. J. Appl. Phys.*, **27** [11] L2055–L2058 (1988).

[13]A. Oota, J. Iwaya, P. Songsak, T. Saigou, and M. Funakura, "Critical Current Density, Morphology and Microstructure of Ag-Sheathed $(Bi,Pb)_2Sr_2Ca_2Cu_3O_x$ Tapes," *Physica C (Amsterdam)*, **214**, 9–18 (1993).

[14]D. W. Johnson Jr. and W. W. Rhodes, "Retrograde Densification in $Bi_2Sr_2CaCu_2O_8$ Superconductors," *J. Am. Ceram. Soc.*, **72** [12] 2346–50 (1989).

[15]J. S. Luo, N. Merchant, V. A. Maroni, D. M. Gruen, B. S. Tani, W. L. Carter, and G. N. Riley Jr., "Kinetics and Mechanism of the $(Bi,Pb)_2Sr_2Ca_2Cu_3O_{10}$ Formation Reaction in Silver-Sheathed Wires," *Appl. Supercond.*, **1**, 101–107 (1993).

[16]W. D. Kingery, H. K. Bowen, and D. R. Uhlmann, *Introduction to Ceramics*, 2nd Ed.; p. 500. Wiley–Interscience, New York, 1976.

[17]R. M. German, *Liquid Phase Sintering*; p. 67. Plenum Press, New York, 1985.

[18]T. Ekstrom and M. Nygren, "SiAlON Ceramics," *J. Am. Ceram. Soc.*, **75** [2] 259–76 (1992).

[19]N. V. Vo, H. K. Liu, and S. X. Dou, "Effect of Sintering Periods on the Microstructure and Electrical Transport Properties of High-T_C Superconducting Bi-(Pb)-Sr-Ca-Cu-O Tapes," *J. Mater. Res.*, **11** [5] 1101–107 (1996).

[20]E. J. Garboczi, K. A. Snyder, J. F. Douglas, and M. F. Thorpe, "Geometrical Percolation Threshold of Overlapping Ellipsoids," *Phys. Rev. E: Stat. Phys., Plasmas, Fluids, Relat. Interdiscip. Top.*, **52**, 819–28 (1995).

[21]R. L. Coble and R. M. Cannon, "Current Paradigms in Powder Processing," *Mater. Sci. Res.*, **11**, 151 (1978).

[22]F. M. Costa, A. P. Goncalves, M. L. Almeida, and J. M. Vieira, "Grain Connectivity and Superconductivity of Bulk High Temperature Superconductors"; pp. 519–33 in *Physics and Materials Science of High Temperature Superconductors*, NATO Advanced Study Institute, Series E: Applied Sciences, Vol. 181. Edited by R. Kossowsky, S. Methfessel, and D. Wohlleben. Kluwer Academic Publishers, Dordrecht, The Netherlands, 1990.

[23]J. W. Bullard and A. W. Searcy, "Microstructural Development during Sintering of Lithium Fluoride," *J. Am. Ceram. Soc.*, **80** [9] 2395–400 (1997).

[24]T. Sata, K. Sakai, and S. Tashiro, "Vapor Pressures of Bismuth, Lead, and Copper Components in $Bi_2Sr_{1.7}CaCu_2O_y$ and $Bi_{1.7}Pb_{0.3}Sr_{1.7}Ca_2Cu_3O_y$ Superconductor Ceramics," *J. Am. Ceram. Soc.*, **75** [4] 805–808 (1992). □

journal

J. Am. Ceram. Soc., 84 [1] 183–87 (2001)

Phase Diagram Studies in the System Tl_2O_3–BaO–CaO–CuO–Ag

Peter J. Majewski, Arthur Jalowicki, Stefanie Kaesche, and Fritz Aldinger*

Max-Planck-Institut für Metallforschung, Pulvermetallurgisches Laboratorium, 70569 Stuttgart, Germany

The phase relations within the system Tl_2O_3–BaO–CaO–CuO including Ag have been studied with emphasis on the high-temperature superconducting phases $TlBa_2Ca_2Cu_3O_{8.5}$ (1223 phase), $Tl_2Ba_2Ca_2Cu_3O_{10}$ (2223 phase), $TlBa_2CaCu_2O_{6.5}$ (1212 phase), and $Tl_2Ba_2CaCu_2O_8$ (2212 phase) at 890°C in an oxygen atmosphere. 1223 has been found to be in equilibrium with a liquid phase that is Tl poor. 2223 and 2212 exhibit varying Tl/(Ba + Ca) ratios. The three-phase field 1223 + 2223 + 2212 has been identified. The results of this study emphasize that multiphase samples can be prepared which consist of three superconducting phases, each exhibiting a critical temperature of 100 K or above.

I. Introduction

The high-temperature superconducting phases in the system Tl_2O_3–BaO–CaO–CuO [$TlBa_2Ca_2Cu_3O_{8.5}$ (1223 phase, T_c = 110 K), $Tl_2Ba_2Ca_2Cu_3O_{10}$ (2223 phase, T_c = 125 K), and $Tl_2Ba_2CaCu_2O_8$ (2212 phase, T_c = 100 K)] are of great technological interest because of their high critical temperatures of 100 K or above.[1–4] For the preparation of defined single-phase or multiphase ceramics of these compounds, knowledge of the phase equilibria of the system Tl_2O_3–BaO–CaO–CuO is of great importance. To simulate the preparation conditions during the "oxide powder in tube" processing of high-temperature superconducting Ag-sheathed tapes (OPIT processing[5]) the phase equilibria were studied in the presence of about 80 vol% Ag.

The presentation of an entire five-component system represented by the system Tl–Ba–Ca–Cu–O at varying temperatures and constant pressure requires a five-dimensional space which is impossible to handle. Several reasonable simplifications have to be made to reduce the number of components to get a lower dimension representation of the system. A less reasonable simplification is to neglect one of the elements, because the high-temperature superconducting phases 1223, 2223, 2212, and 1212 exist within this five-dimensional system and not on one of its four-dimensional "surfaces." Therefore, the metal oxides Tl_2O_3, BaO, CaO, and CuO were considered as components instead of the elements, resulting in a reduction of the number of components from five to four. Nevertheless, four dimensions are still needed to describe this system. To figure the system by three dimensions the temperature has to be considered to be constant. However, it is emphasized that the phase diagrams presented for this system are only sections through a five-dimensional space.

The systems Tl–O,[6] BaO–CuO,[7,8] CaO–CuO,[9] and Tl_2O_3–CuO[10] have been studied in detail. The most significant aspect is the fact that Tl_2O_3 decomposes at about 780°C, forming Tl_2O and oxygen. In addition, it is emphasized that a liquid phase occurs in the system BaO–CuO at about 926°C in oxygen atmosphere (906°C in air)[8] that almost corresponds with the preparation temperature of 1223 ceramics.[11] The system Tl_2O_3–BaO–CuO has been studied in Ref. 10. The system contains the compound $Tl_2Ba_2CuO_6$ (2201 phase), which represents the $n = 1$ member of the homologous series $Tl_2Ba_2Ca_{n-1}Cu_nO_{4+2n}$ with $n = 1$ to 3. However, the analogous compound $TlBa_2CuO_{4.5}$, which represents the $n = 1$ member of the series $TlBa_2Ca_{n-1}Cu_nO_{2.5+2n}$ with $n = 1$ to 4, was not observed in Ref. 10. An isothermal section of the ternary system BaO–CaO–CuO has been studied partly in Ref. 12. The authors observed the ternary compounds $Ba_4CaCu_3O_8$ and $Ba_6CaCu_3O_{10}$. The quaternary system has been studied in Ref. 13 with emphasis on varying Tl_2O as well as oxygen partial pressure at 850°C and fixed BaO:CaO:CuO ratios of 2:1:2, 2:2:3, and 2:3:4. At 1 bar oxygen partial pressure 1223 samples contain 1223 and 2223 at $\log p_{Tl_2O} < 3$ and BaO:CaO:CuO = 2:2:3. Single-phase 1223 samples could not be prepared under the experimental conditions.

II. Experimental Procedure

Samples were prepared using Tl_2O_3, $BaCO_3$, $CaCO_3$, and CuO (purity > 99%), respectively. In the first step precursors were prepared by mixing $BaCO_3$, $CaCO_3$, and CuO and annealing them at 920°C in air for 48 h. After annealing the precursors were characterized by X-ray diffraction. The results of the analysis showed that $BaCO_3$ and $CaCO_3$ were completely reacted after this thermal treatment. The precursors were ground and mixed with appropriate amounts of Tl_2O_3 and Ag, and compacted by cold isostatic pressing (625 MPa). The pressed bodies were wrapped into Ag foil and sealed in a quartz glass tube containing pure oxygen atmosphere. The oxygen pressure in the tube was set to be 1 bar at 890°C. However, since reactions in the system can be expected to release oxygen to or absorb oxygen from the atmosphere, it cannot be excluded that the total pressure within the glass

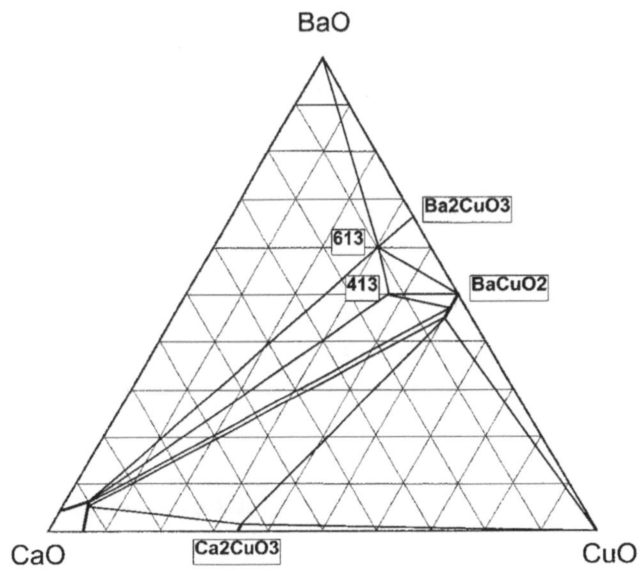

Fig. 1. The ternary system BaO–CaO–CuO at 890°–900°C at p_{O_2} = 1.

R. L. Snyder—contributing editor

Manuscript No. 189257. Received June 30, 1999; approved August 14, 2000.
Financial support was provided by the "Deutsche Forschungsgemeinschaft."
*Member, American Ceramic Society.

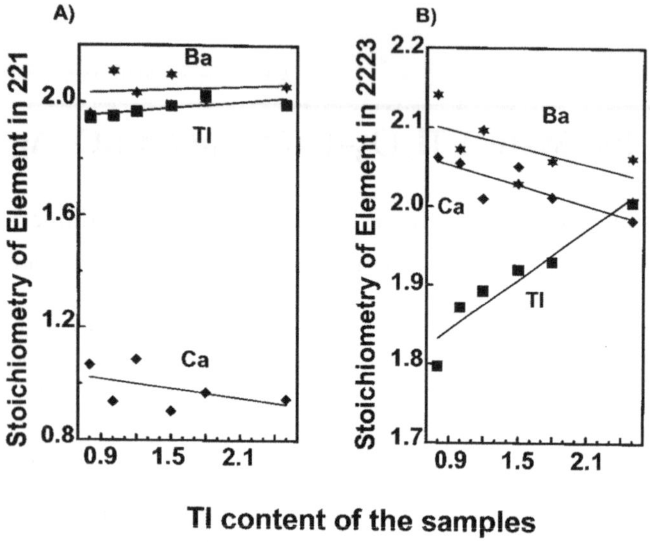

Fig. 2. (A) The stoichiometry of Tl, Ca, and Ba of 2212 vs the Tl content of the samples: (✱) Ba, (◆) Ca, (■) Tl. (B) The stoichiometry of Tl, Ca, and Ba of 2223 vs the Tl content of the samples: (✱) Ba, (◆) Ca, (■) Tl. Lines represent linear regression.

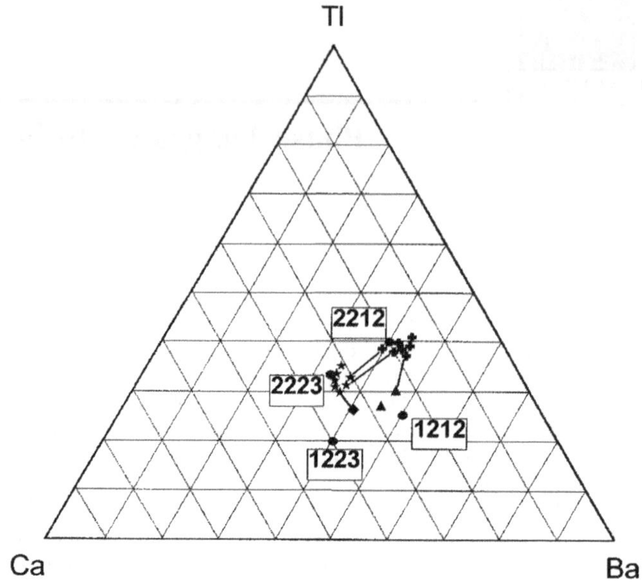

Fig. 3. Variation of the Tl, Ba, and Ca content of 2223, 2212, 1223, and 1212. The Cu content of the phases has been considered to be constant. The lines between marks indicate that phases with these compositions were found in one sample: (★) 2223, (✚) 2212, (◆) 1223, (▲) 1212.

tubes slightly varied during the sinter step. Nevertheless, because of the small amount of ceramic material compared with Ag (ceramic:Ag = 20:80 in volume percent) the oxygen partial pressure within the glass tubes could be expected to be almost 1. After annealing at 890°–900°C for 60 min the samples were cooled down to room temperature at 10°C/min. ICP-OES analysis of the samples showed that the processing caused a loss of about 40 wt% of Tl due to evaporation of Tl and reaction of Tl with the quartz glass tube that formed a light yellow layer on the inside of the tube.

The samples prepared to study the ternary system BaO–CaO–CuO were annealed at 890°–900°C for 48 h in flowing oxygen atmosphere without silver.

Phase identifications were performed using scanning electron microscopy with EDX (Zeiss DSM 982 Gemini) and X-ray diffraction (Siemens D-5000, Cu$K\alpha_1$ radiation, step size 0.02°, counting time 1.2 s). For the calculation of the stoichiometry of the compounds the Tl + Ba + Ca content of the phases was constrained to be 4 for the 1212 phase, 5 for the 2212 and 1223 phases, and 6 for the 2223 phase.

It is emphasized that because of the high content of Ag in the samples the diffraction patterns were characterized by strong reflections of Ag. To figure the reflections of the Tl–Ba–Ca–Cu–O phases the patterns were blown up, resulting in a high background of the diagram. Because of the high background, reflections of less intensity could not be detected.

III. Results and Discussion

(1) The Ternary System BaO–CaO–CuO

The ternary system BaO–CaO–CuO has been studied in detail, first because it is an important part of the quaternary system and second because several high-temperature superconducting phases exist in the system at high pressures.[14] For the investigation of the stability and phase relations of these compounds it is important to know the phase equilibria of the system at ambient pressure.

Two ternary phases were observed in the system (Fig. 1), the compositions of which were $Ba_4CaCu_3O_7$ (413 phase) and $Ba_6CaCu_3O_{10}$ (613 phase), confirming the results of Ref. 12.

A liquid phase was not observed at the annealing temperature of 890°–900°C, which correlates with the results of Ref. 8. Ca_2CuO_3

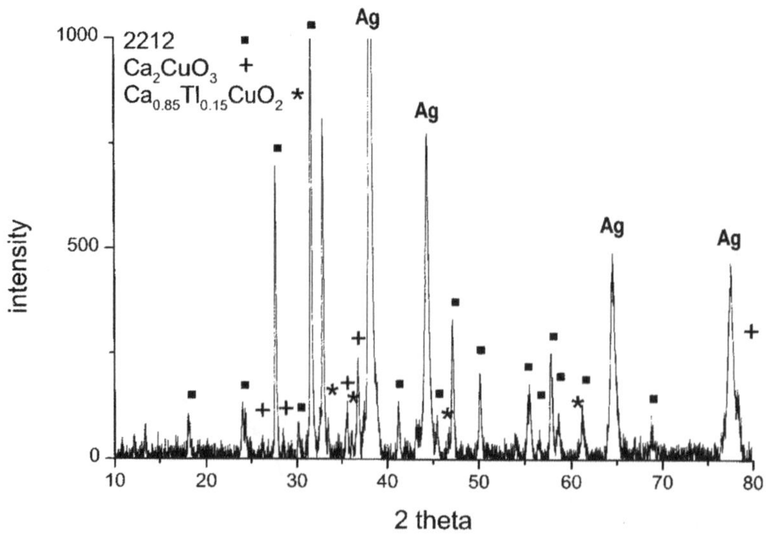

Fig. 4. XRD pattern of a sample consisting of 2212, Ca_2CuO_3, CuO, and the Ca–Tl–Cu–O compound ($Ca_{0.75}Tl_{0.15}CuO_{2+x}$).

Table I. Phases in Thermodynamic Equilibrium with 1223, 2223, 2212, and 1212

1223:	2223, 2212, 1212, $BaCuO_2$, $(Ca,Tl)CuO_{2+x}$, CuO, liquid
2223:	1223, 2212, $BaCuO_2$, Ba_2CuO_3, Ca_2CuO_3, $(Ca,Tl)CuO_{2+x}$, CuO
2212:	1223, 2223, 1212, $BaCuO_2$, Ba_2CuO_3, Ca_2CuO_3, $(Ca,Tl)CuO_{2+x}$, CuO
1212:	1223, 2212, $BaCuO_2$, Ba_2CuO_3, CuO, liquid

Table II. Observed Four-Phase Regions

(1) 1223 + $(Ca,Tl)CuO_{2+x}$ + $BaCuO_2$ + liquid
(2) 1223 + $(Ca,Tl)CuO_{2+x}$ + CuO + liquid
(3) 1223 + 2223 + $(Ca,Tl)CuO_{2+x}$ + $BaCuO_2$
(4) 1223 + 2223 + $(Ca,Tl)CuO_{2+x}$ + CuO
(5) 1223 + 2223 + 2212 + $BaCuO_2$
(6) 2223 + 2212 + $BaCuO_2$ + Ca_2CuO_3
(7) 2223 + 2212 + $(Ca,Tl)CuO_{2+x}$ + CuO
(8) 2223 + 2212 + $(Ca,Tl)CuO_{2+x}$ + Ca_2CuO_3
(9) 1223 + 1212 + $BaCuO_2$ + liquid

and $BaCuO_2$ exhibited a small solubility of Ba (2.9 at.%) and Ca (4.7 at.%), respectively. For CaO a Ba and Cu solid solubility of about 6 at.% was determined. The Ba and Ca solubility of CuO was less than 1 at.%.

(2) *The Quaternary System with Ag*

Even considering the loss of Tl by evaporation single-phase 1223 and 2223 samples could not be prepared. The 1223 samples contained mainly 2223 and 1223 besides $BaCuO_2$. This indicates that the nominal compositions of 1223 and 2223 lay outside the single-phase regions of these phases, which is comparable with the observations that were made for the (Bi,Pb)-2223 phase.[15]

Table III. Additional Four-Phase Regions Constructed from the Observed Four-Phase Regions

(1) 1223 + 2223 + 2212 + CuO
(2) 1223 + 2212 + 1212 + CuO
(3) 1223 + 1212 + CuO + liquid
(4) 1223 + 2212 + 1212 + $BaCuO_2$
(5) 2223 + $(Ca,Tl)CuO_{2+x}$ + Ca_2CuO_3 + $BaCuO_2$
(6) 2212 + $(Ca,Tl)CuO_{2+x}$ + Ca_2CuO_3 + CaO
(7) 2212 + 1212 + $BaCuO_2$ + CaO
(8) 2212 + $BaCuO_2$ + Ca_2CuO_3 + CaO
(9) $(Ca,Tl)CuO_{2+x}$ + Ca_2CuO_3 + CuO + liquid
(10) $(Ca,Tl)CuO_{2+x}$ + Ca_2CuO_3 + $BaCuO_2$ + liquid

Microprobe analysis indicated that with increasing Tl content of the samples the Tl content of the 2212 and 2223 phases increased. The Ba content appeared to be almost constant, whereas the Ca content slightly decreased (Figs. 2). The Ca contents of the phases were slightly in excess at low Tl contents and only insignificantly deficient at high Tl contents, indicating that Tl substituted Ca and vice versa. However, the significant decrease of the Tl content and the minor increase of the Ca content of the 2223 phase with decreasing Tl concentrations of the samples indicate that the Tl site of the 2223 crystal structure is partly vacant, whereas the analysis of the 2212 phase almost correlates with the ideal formula $Tl_2Ba_2CaCu_2O_x$.

EDX analysis of the 1223 phase of a sample consisting of 1223, 2223, $(Ca,Tl)CuO_2$, and $BaCuO_2$ showed that Tl substituted for Ca to some extent ($Tl_{1.43}Ba_2Ca_{1.56}Cu_3O_x$), whereas analysis of the 1212 phase of a sample consisting of 2212, 1212, and $BaCuO_2$ indicated that Tl obviously substituted for Ba ($Tl_{1.22}Ba_{1.83}Ca_{0.95}Cu_2O_x$) (Fig. 3). These observations suggest that superconducting compounds exhibit extended homogeneity regions with varying Tl:Ba:Ca ratios.

A Ca–Tl–Cu–O compound was found to be in equilibrium with the superconducting phases 1223, 2223, and 2212 at 900°C. The

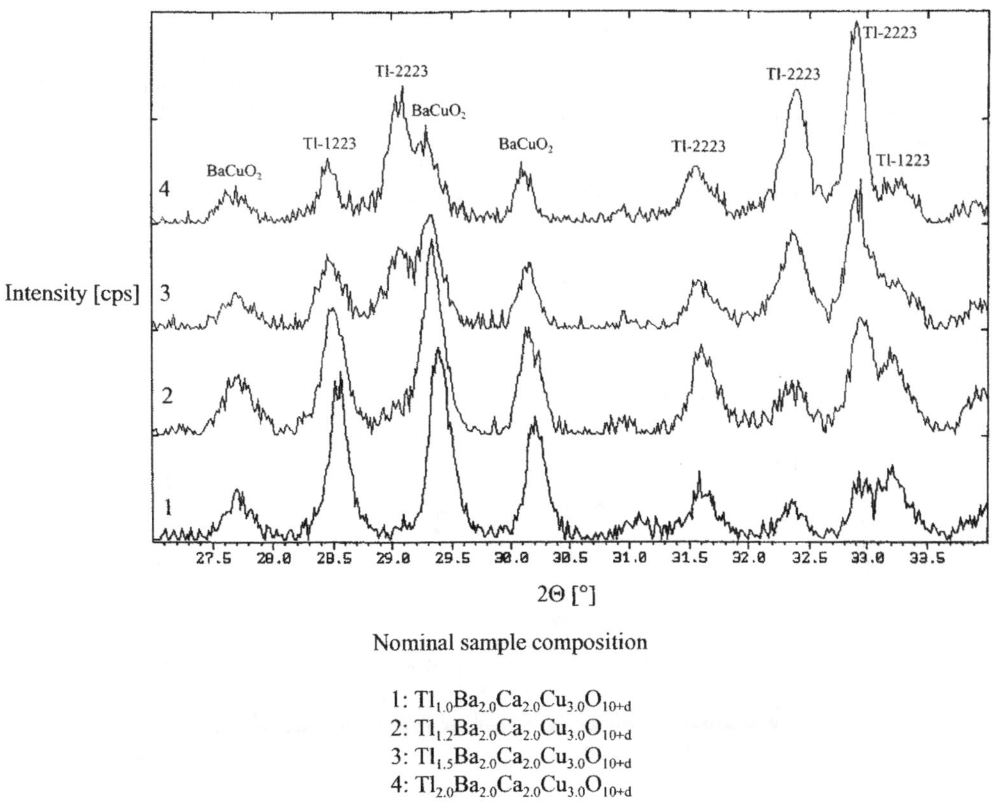

Fig. 5. XRD patterns of samples of the four-phase region 1223 + 2223 + $BaCuO_2$ + $(Ca,Tl)CuO_3$ with increasing Tl content.

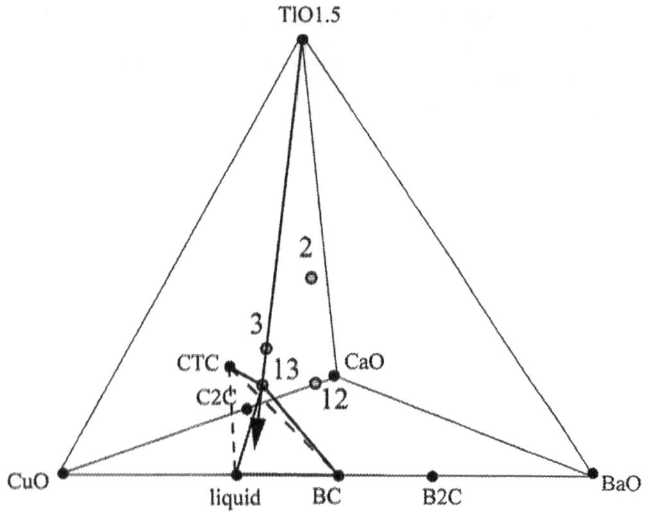

Fig. 6. Schematic representation of the occurrence of the liquid phase during TlO evaporation: (12) 1221, (13) 1223, (2) 2212, (3) 2223, (BC) $BaCuO_2$, (B2C) Ba_2CuO_3, (C2C) Ca_2CuO_3, (CTC) $(Ca,Tl)CuO_2$.

chemical composition was approximately $Ca_{0.75}Tl_{0.15}CuO_{2+x}$.[9] This compound could correspond to $Ca_{0.833}CuO_{2-x}$ ($CaCuO_2$ according to Ref. 16). However, $Ca_{0.833}CuO_{2-x}$ is stable only up to about 750°C.

In the system SrO–CaO–CuO a ternary phase with a stoichiometry of $Sr_{0.15}Ca_{0.85}CuO_2$ is stable at 900°C in oxygen atmosphere.[16–18] However, it has not been confirmed whether the Ca–Tl–Cu–O compound corresponds to $Sr_{0.15}Ca_{0.85}CuO_2$ with Sr substituted by Tl. The XRD pattern of a sample consisting of 2212, Ca_2CuO_3, CuO, and the Ca–Tl–Cu–O compound, as determined by EDX, is shown in Fig. 4. Two reflections seen in the XRD pattern could not account for 2212, Ca_2CuO_3, or CuO, but do correspond with the main reflections of $Ca_{0.85}Sr_{0.15}CuO_2$ after Ref. 12, indicating that the Ca–Tl–Cu–O compound correlates with the known $Ca_{0.85}Sr_{0.15}CuO_2$ structure.

The high-temperature superconducting phases 1212, 1223, 2212, and 2223 were found to be in equilibrium with the cuprates and CuO. 1223 and 1212 were in equilibrium with a liquid (Table I). Ten four-phase regions were observed within the system Tl_2O_3–BaO–CaO–CuO–Ag at 890°–900°C in oxygen atmosphere, which are listed in Table II. For the construction of the four-phase regions the ideal formulas of the phases have been considered. Taking the observed four-phase regions and the phase diagram of the system BaO–CaO–CuO into account, and considering the phase rules of heterogeneous phase equilibria, 10 additional four-phase regions can be constructed (Table III).

Although $Tl_2Ba_2CuO_6$ was known to be in equilibrium with CuO,[10] it was not found within the studied concentration region. As the melting temperature of $Tl_2Ba_2CuO_6$ is about 930°C,[10] it is believed that in the Ag-containing system the compound is already molten under the experimental conditions.

To test whether the prepared samples represent thermodynamic equilibrium, a series of samples were prepared with increasing Tl content located in the four-phase region 1223 + 2223 + $BaCuO_2$ + Ca_2CuO_3. It is expected that with increasing Tl concentration the content of 2223 increases and that of 1223 decreases. The XRD patterns of these samples support this and are shown in Fig. 5. This result gives evidence that the lever law of heterogeneous phase equilibria is fulfilled and the samples represent thermodynamic equilibria.

The 1223 phase was found to be in equilibrium with a liquid phase, which was indicated by a characteristic deformation of the samples during annealing. The composition of the liquid could not be determined, because a glass phase could not be found in those samples. However, considering that in the binary system BaO–CuO a liquid occurred at about 890°C in air and 920°C in pure oxygen atmosphere, it can be expected that the liquid phase in the quaternary system correlates with that liquid phase of the system BaO–CuO with small amounts of Ca and Tl. In this work liquid was detected at a lower temperature in an oxygen atmosphere as compared with Ref. 7, and no liquid was observed in the system BaO–CaO–CuO at 890°–900°C, suggesting that Tl and Ag reduce the melting temperature of the system significantly. Since the liquid phase of the system Bi_2O_3–SrO–CaO–CuO–Ag dissolves several percent Ag,[19] it can also be concluded that in the Tl system Ag is dissolved in the liquid.

For the processing of 1223 ceramics it is emphasized that the occurrence of a liquid phase at Tl-poor concentrations is important, since because of the high volatility of Tl_2O, local deficiency of Tl in the sample cannot be excluded. With decreasing Tl content the sample enters the four-phase region 1223 + $BaCuO_2$ + Ca–Tl–Cu–O + liquid. Further decrease of the Tl content by evaporation causes an increase of the amount of liquid and a decrease of the 1223 content (Fig. 6). The liquid forming at Tl-deficient concentration represents a thermodynamic stable phase during sintering which is different from the transient liquid phase that is observed during the preparation of the Bi superconductor $(Bi,Pb)_2Sr_2Ca_2Cu_3O_{10}$ (Bi-2223).[5] That liquid phase disappears during the preparation of Bi-2223, but the thermodynamic stable liquid phase that is in equilibrium with 1223 remains in the ceramic until the sample is cooled.

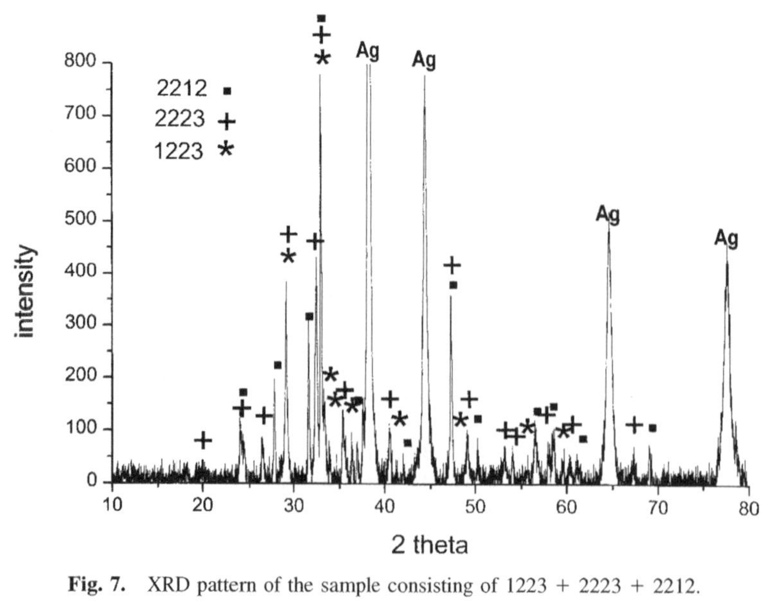

Fig. 7. XRD pattern of the sample consisting of 1223 + 2223 + 2212.

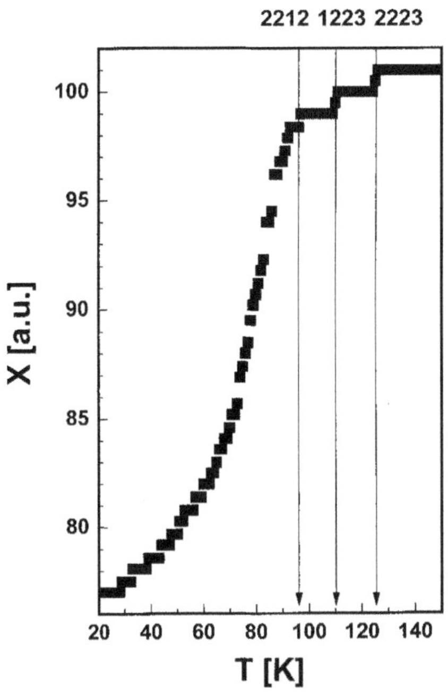

Fig. 8. Susceptibility versus temperature plot of a sample containing 1223 (T_c = 97 K), 2223 (T_c = 125 K), and 2212 (T_c = 110 K).

The occurrence of the liquid phase can be avoided by choosing a concentration slightly richer in Tl and Ca as compared with 1223, resulting in a sample consisting of 1223 + 2223 + 2212. It is emphasized that the existence of this three-phase region allows one to prepare ceramics that consist of three high-temperature superconducting phases with a critical temperature of ~100 K (Figs. 7 and 8).

IV. Conclusions

This study shows that during the preparation of 1223 in Ag-sheathed tapes a thermodynamically stable liquid occurs besides 1223 at Tl-deficient compositions which cannot be excluded because of the evaporation of TlO. However, by choosing a composition of a sample within the three-phase region 1223 + 2223 + 2212 the occurrence of the liquid phase can be avoided. Furthermore, because of the existence of the 1223 + 2223 + 2212 three-phase region it is possible to prepare multiphase samples consisting of three high-temperature superconducting phases where each of them exhibits a critical temperature of 100 K or above.

References

[1]S. S. P. Parkin, V. Y. Lee, A. I. Nazzal, R. Savoy, R. Beyers, and S. J. La Placa, "$Tl_1Ca_{n-1}Ba_2Cu_nO_{2n+3}$ (n = 1, 2, 3): A New Class of Crystal Structures Exhibiting Volume Superconductivity at up to ~110 K," *Phys. Rev.*, **61**, 750–53 (1988).

[2]S. S. P. Parkin, V. Y. Lee, E. M. Engler, A. I. Nazzal, T. C. Huang, G. Gorman, R. Savoy, and R. Beyers, "Bulk Superconductivity at 125 K in $Tl_2Ca_2BaCu_3O_x$," *Phys. Rev. Lett.*, **60**, 2539–42 (1988).

[3]D. P. Matheis and R. L. Snyder, "The Crystal Structures and Powder Diffraction Patterns of the Bismuth and Thallium Ruddlesden–Popper Oxide Superconductors," *Powder Diffr.*, **5**, 8–25 (1990).

[4]R. N. Bhattacharya, P. A. Parilla, and R. D. Blaugher, "Processing of Thick Film Electrodeposited Thallium Oxide Superconductors," *Physica*, **C211**, 475–87 (1993).

[5]S. X. Dou and H. K. Liu, "Ag-Shealthed Bi(Pb)SrCaCuO Superconducting Tapes," *Supercond. Sci. Technol.*, **6**, 297–314 (1993).

[6]W. L. Holstein, "Thermodynamics of the Volatilization of Thallium(I) Oxide from Tl_2O, Tl_4O_3, and Tl_2O_3," *J. Phys. Chem.*, **97**, 4224–30 (1993).

[7]W. Zhang, K. Osamura, and S. Ochiai, "Phase Diagram of the BaO–CuO Binary System," *J. Am. Ceram. Soc.*, **73**, 1958–64 (1990).

[8]B. J. Lee and D. N. Lee, "Thermodynamic Evaluation for the Y_2O_3–BaO–CuO_x System," *J. Am. Ceram. Soc.*, **74**, 78–84 (1991).

[9]D. Risold, B. Hallstedt, and L. J. Gauckler, "Thermodynamic Assessment of the Ca–Cu–O System," *J. Am. Ceram. Soc.*, **78**, 2655 (1995).

[10]T. K. Jondo, C. Opagiste, J. L. Jorda, M. Th. Cohen-Adad, F. Sibieude, M. Couach, and A. F. Khoder, "Phase, Equilibria in the Binary Thallium Oxide ($TlO_{1.5}$)–Cupric Oxide, $TlO_{1.5}$–Barium Oxide and Ternary $TlO_{1.5}$–BaO–CuO Systems," *J. Alloys Compd.*, **195**, 53–56 (1993).

[11]K. C. Goretta, N. Chen, M. T. Lanagan, S. E. Dorris, J. Hu, C.-T. Wu, and R. B. Poeppel, "Processing $TlBa_2Ca_2Cu_3O_x$ Powders," *Supercond. Sci. Technol.*, **5**, 534–37 (1992).

[12]K. A. Kubat-Martin, E. Garcia, and D. E. Peterson, "$Ba_4CaCu_3O_x$ a New Phase in the System Ba–Ca–Cu–O," *Physica C*, **172**, 75–80 (1990).

[13]T. L. Aselage, E. L. Venturini, S. B. VanDeusen, T. J. Haedley, M. O. Eatough, and J. A. Voigt, "Phase, Stability and Properties of Near-Equilibrium Tl–Ca–Ba–Cu–O Superconductors," *Physica C*, **203**, 25–36 (1992).

[14]M. A. Alario-Franco, C. Chaillout, J. J. Capponi, J. L. Thollence, and B. Souletie, "A New HTSC, Family: The Copper Analogues of the Single-Layer Hg or Tl Copper Oxide Superconductors," *Physica C*, **222**, 52–56 (1994).

[15]P. Majewski, S. Kaesche, and F. Aldinger, "Fundamental Material Aspects Underlying the Preparation of High-Temperature Superconducting $(Bi,Pb)_{2+x}Sr_2Ca_2Cu_3O_{10+d}$ Ceramics," *J. Am. Ceram. Soc.*, **80**, 1174–80 (1997).

[16]R. S. Roth, C. J. Rawn, J. J. Ritter, and B. P. Burton, "Phase Equilibria of the System SrO–CaO–CuO," *J. Am. Ceram. Soc.*, **72**, 1545 (1989).

[17]K. Schulze, P. Majewski, B. Hettich, and G. Petzow, "Phase Equilibria in the System Bi_2O_3–SrO–CaO–CuO with Emphasis on the High-T_c Superconducting Compounds," *Z. Metallkd.*, **81**, 836–42 (1991).

[18]D. Risold, B. Hallstedt, and L. J. Gauckler, "Thermodynamic Modeling and Calculation of Phase Equilibria in the Strontium–Calcium–Copper–Oxygen System at Ambient Pressure," *J. Am. Ceram. Soc.*, **80**, 537–50 (1997).

[19]P. Majewski, A. Sotelo, H. Szillat, S. Kaesche, and F. Aldinger, "Phase Diagram Studies in the Quasi-Binary System Ag-'$Bi_2Sr_2CaCu_2O_8$'," *Physica*, **C275**, 47–51 (1997). □

J. Am. Ceram. Soc., 85 [5] 1190–96 (2002)

Superplastic Deformation in Fine-Grained $YBa_2Cu_3O_{7-x}$

Jondo Yun

Division of Advanced Materials Engineering, Kyungnam University, Masan, 631-701, Korea

Ye T. Chou* and Martin P. Harmer*

Department of Materials Science and Engineering and Materials Research Center, Lehigh University, Bethlehem, Pennsylvania 18015

The superplastic behavior of $YBa_2Cu_3O_{7-x}$ ceramic superconductors was studied. Large compressive deformation over 100% strain was measured in the temperature range of 775°–875°C, with a strain rate of 1×10^{-5} to 1×10^{-3}/s, and a grain size of 0.5–1.4 μm. The nature of the deformation was investigated in terms of three deformation parameters: the stress exponent (n), the grain size exponent (p), and the activation energy (Q). The measured values of these parameters were $n = 2 \pm 0.3$, $p = 2.7 \pm 0.7$, and $Q = 745 \pm 100$ kJ/mol. With the aid of the deformation map, the deformation mechanism was identified as grain boundary sliding accommodated by grain boundary diffusion. The conclusion is consistent with the microstructural observations made by SEM and TEM: the invariance of equiaxed grain shape, the absence of significant dislocation activity, no grain boundary second phases, and no significant texture development.

I. Introduction

Since the discovery of superplasticity in α + β brass,[1] many metallic systems have been found to exhibit superplastic flow.[2] During the last three decades, superplasticity has also been observed in ceramic materials at high temperatures.[3,4] Although a number of theories have been proposed to explain the phenomenon,[5–9] the basic mechanism is still not well understood.

One of the promising high-T_c oxide superconductors is $YBa_2Cu_3O_{7-x}$, which has a T_c of about 95 K. Along with other physical properties that have been extensively studied, the plastic deformation has also been investigated, and extended strains have been observed both in compression and in tension.[10–16] These observations are very encouraging and may lead to a potential manufacturing process to produce ceramic superconductors in useful shapes, such as current leads, magnetic shields, or even wires. However, there have been no detailed discussions of these observations, especially the understanding of the deformation mechanisms.

As discussed in an early report,[17] the deformation mechanism of coarse-grained $YBa_2Cu_3O_{7-x}$ studied by previous investigators was diffusional creep (DC) with a stress exponent of 1 ($n = 1$), except for the case of very high flow stress (>1 GPa). However, our preliminary work[15,16] of fine-grained $YBa_2Cu_3O_{7-x}$ samples showed that the stress exponent is 2, suggesting that a different mechanism could be active during the deformation. The purpose of this study is to determine the deformation parameters in a more precise manner, and to identify the controlling mechanism for the superplastic deformation in fine-grained $YBa_2Cu_3O_{7-x}$.

II. Experimental Procedure

(1) Sample Preparation

Commercial $YBa_2Cu_3O_{7-x}$ powders (Rhone Poulenc, Inc.) were ground in an agate mortar and pressed uniaxially at 69 MPa and then isostatically at 345 MPa into rectangular compacts of 6 mm × 10 mm × 25 mm. The compacts were dried at 100°C for 1 h and homogenized at 800°C for 4 h. They were subsequently sintered at 875°C in air or at 910°C in oxygen for various periods of time to prepare different sets of samples with varying grain sizes. Table I lists the grain size and bulk density of each set of prepared samples, which are named 00FF, 00F, 00L, and 00LL and had grain sizes of 0.54, 0.78, 1.02, and 1.36 μm. After sintering, all samples were annealed in flowing oxygen at 500°C for 24 h. The annealed samples were cut and ground into rectangular bars with two sets of dimensions: 3.0 mm × 3.0 mm × 6.1 mm, and 3.8 mm × 3.8 mm × 7.6 mm.

(2) High-Temperature Compression Test

The ground samples were tested in a split-type furnace attached to a hydraulic-type Instron testing unit controlled by a computer. Each sample was placed between the lower loading block and the upper round pedestal, which was made of silicon carbide and used for the alignment. The samples were preheated at a rate of 10°C/min and held for 1.5 h at the assigned test temperature before loading. They were then deformed under compression at temperatures of 750°, 775°, 800°, 825°, 850°, and 875°C at a constant strain rate ranging from 1.0×10^{-5}/s to 1.0×10^{-3}/s. The strain rate was changed during the test when the stress exponent was determined. A calibration sample was placed near the test sample during the test for comparison. The calibration sample was not loaded, but was kept under the same heating and cooling cycle.

(3) Characterization

X-ray diffraction (XRD) analysis was conducted for samples before and after 50% deformation to examine the stability of the $YBa_2Cu_3O_{7-x}$ phase and the change in the degree of preferred

Table I. Conditions for the Preparation of $YBa_2Cu_3O_{7-x}$ Samples with Various Grain Sizes

	Sintering conditions				
Sample	Temperature (°C)	Time (h)	Atmosphere	GS (μm)[†]	D_b (%)[‡]
00FF	875	6	Air	0.54	70
00F	910	1	Oxygen	0.78	76
00L	910	3	Oxygen	1.02	82
00LL	910	6	Oxygen	1.36	86

[†]Grain size. [‡]Relative bulk density.

D. J. Green—contributing editor

Manuscript No. 187719. Received June 12, 2001; approved February 11, 2002.
*Member, American Ceramic Society.

orientations. The samples were cut in two directions, perpendicular and parallel to the axis of compression. The cross sections were exposed to Cu$K\alpha$ X-rays to obtain the XRD pattern. For both as-sintered and as-deformed samples, the density was measured by the Archimedes method. For the ground sample, the density was determined by measuring the sample dimensions. Both methods gave consistent values within 1% or 2% of error.

The as-sintered, deformed, and calibration samples (all oxygen-annealed) were cut, ground, and polished. The mirror-polished sample was etched with 0.3% nitric acid for 2–4 s.[18] The microstructures of the etched samples were examined using a Zeiss light optical microscope and a JSM-6300F scanning electron microscope (SEM). The grain size of all samples was measured by the linear intercept method. A conversion factor of 1.126, as specified by the ASTM,[19] was adapted to calculate the grain size. TEM analysis was conducted for selective samples before and after 50% deformation. Philips EM400 and EM430 transmission electron microscopes were used to observe the fine structures, including the grain boundaries, dislocations, and other types of defects.

(4) *Data Analysis*

The applied stress was measured at a true strain of 4% for each test. The stress exponent, n, was determined from a log–log plot of applied stress versus strain rate using the equation $n = 1/m$, where m is the strain rate sensitivity and the slope of the curve. The activation energy, Q, was determined from a least-squares fit to a plot of log (applied stress) vs $1/T$. The grain size exponent, p, was determined from a least-squares fit of log (effective stress) vs log (grain size). Because of the presence of porosities, the actual stress, which is referred to as the effective stress (σ_{eff}) in the sample, was higher than the applied stress (σ_{a}). The effective stress is a function of the porosity and is estimated by the equation[20,21]

$$\sigma_{\mathrm{eff}} = e^{3.7P}\sigma_{\mathrm{a}} \tag{1}$$

where P is the porosity. It is noted that in the determination of n or Q, samples of the same porosities were used for testing. Thus, in the calculations, the porosity was a constant, and either the applied or effective stress could be used.

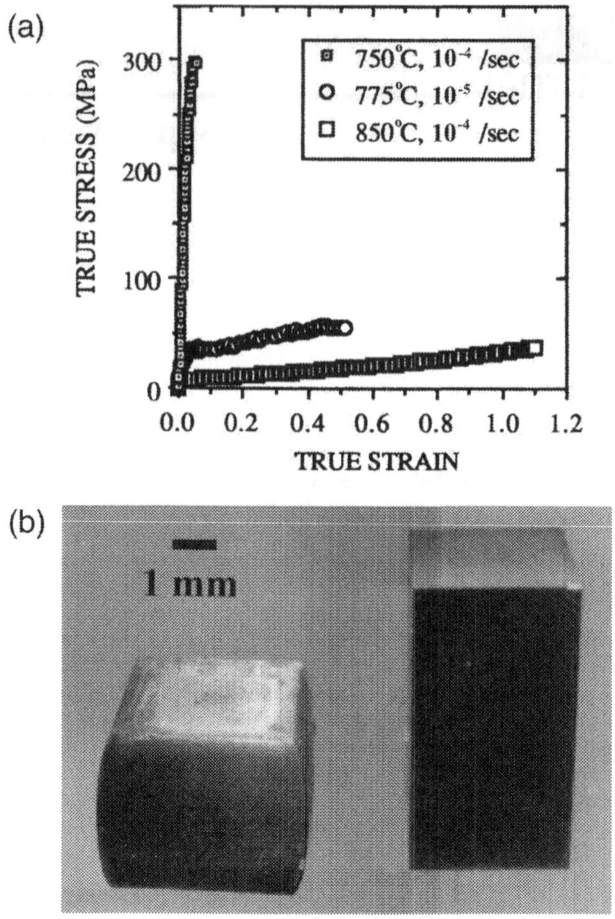

Fig. 1. (a) Stress–strain curves of the $YBa_2Cu_3O_{7-x}$ samples under compression, and (b) the shapes of the $YBa_2Cu_3O_{7-x}$ samples before and after 60% deformation at 850°C and a strain rate of 2×10^{-4}/s.

III. Results

(1) *Deformation Behavior*

All tested samples, except for those deformed at 750°C, showed a large plastic deformation (Fig. 1). The flow stress increased slowly with increasing strain due to strain hardening, and no yield drop was observed. The sample had a barrel shape after deformation due to the stress gradient along the end surface, resulting from the friction between the sample and the pedestal. The analysis for the stress at the end surface showed that the stress variation was not large (less than 9% of the average stress) at 4% strain where the flow stress was measured.[21] Note that the measured stress was an average value. The details of deformation behavior have been reported elsewhere.[15,21]

When the fine-grained samples showed stable large plastic deformation, coarse-grained samples (>1.0 μm) showed shear marks (X-shape) on their side surfaces after deformation under an applied stress of 80 MPa or higher. These marks would give a measure of the concentration of stress and strain along the diagonal axis. The associated stress–strain curve of these samples showed a yield drop without exception, where the applied stress decreased after reaching a maximum. This suggested that the deformation process was unstable under such conditions. The generation of shear marks was sensitive to the applied stress and grain size, but insensitive to the strain rate, as shown in Fig. 2.

(2) *Determination of Deformation Parameters*

The stress exponent (n), the grain size exponent (p), and the activation energy for deformation (Q) were determined. In the temperature range of 800°–850°C (Fig. 3), and in the grain size range of 0.5–1.4 μm (Fig. 4), the stress exponent was 2.0 ± 0.3 at low stresses (<50 MPa). However, it decreased to 1.0 at higher stresses (>50 MPa) in the 00L and 00LL samples deformed at 850°C. Also at 800°C, similar values for the stress exponent, 1.1 ± 0.1, were observed at high stresses in the 00L and 00LL samples, as shown in Fig. 5. The results show that there were two regimes, low- and high-stress regimes, for which different mechanisms were responsible. The grain size exponent was determined in the low-stress regime at a strain rate of 5×10^{-5}/s at 850°C. It was 2.7 ± 0.7, after taking consideration of porosity (Fig. 6). The activation energy for deformation was 745 ± 100 kJ/mol in the temperature range of 775°–850°C, as shown in Fig. 7.

Chen *et al.*[22] studied the lattice self-diffusivity of cations in $YBa_2Cu_3O_{7-x}$, and determined the activation energies of Y and Ba to be 1000 and 890 ± 80 kJ/mol, respectively. The activation energy for lattice self-diffusion of Cu was determined to be ~260 kJ/mol,[23] and, for oxygen, ~100 kJ/mol.[24,25] The diffusion rate of Y is the slowest in the compound as indicated by the highest activation energy. Since the slowest ion controls the overall process, the Y-ion diffusion would control the overall diffusion in $YBa_2Cu_3O_{7-x}$, and the activation energy for lattice diffusion in $YBa_2Cu_3O_{7-x}$ is then taken to be 1000 kJ/mol. Then the measured value of activation energy for the deformation, 745 ± 100 kJ/mol, is much lower than the activation energy for lattice diffusion of $YBa_2Cu_3O_{7-x}$, suggesting that it would be from the grain boundary diffusion. It is noted that the activation energy obtained in the present study is in good agreement with that of plastic deformation in $YBa_2Cu_3O_{7-x}$ in the previous reports.[17,26]

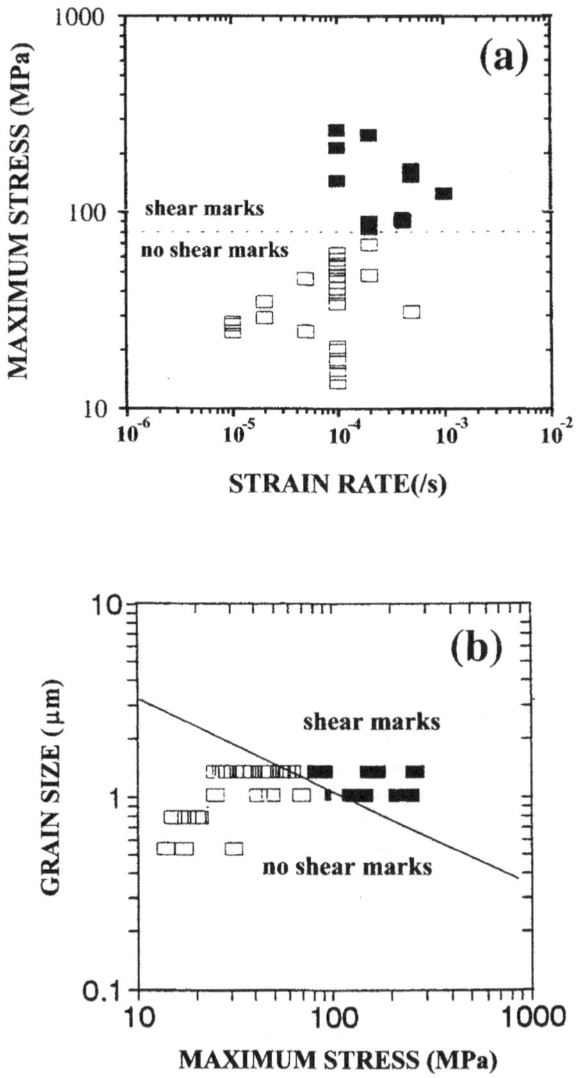

Fig. 2. Conditions for the occurrence of shear marks: (a) the maximum applied stress versus strain rate plot showing its sensitivity to the applied stress, (b) the grain size versus maximum applied stress plot showing the boundary between two different deformation behaviors.

Based on all deformation parameters determined above, a constitutive equation can be written as

$$\frac{d\varepsilon}{dt} = K\left(\frac{\sigma_{eff}^2}{d^{2.7}}\right)e^{-745000/RT} \tag{2}$$

where K is a constant.

(3) *Microstructural Change*

When the test temperature was 775°C, the as-sintered, deformed, and calibration samples (00FF) all had the same grain size, about 0.5 μm, indicating that there was neither static nor dynamic grain growth during the deformation process. However, at 850°C, the grain size changed from 0.5 to 0.7 μm in the calibration sample and to 0.8 μm in the deformed sample. On the other hand, the grain shape remained unchanged during the deformation. Figure 8 illustrates the microstructures of 00FF samples deformed to 50% at 850°C and 1.0×10^{-4}/s. These microstructures were taken from cross sections parallel and perpendicular to the stress axis, respectively.

The presence of grain boundary phase was examined in the 00FF as-sintered sample by high-resolution TEM. The lattice images from two adjacent grains separated by a grain boundary are

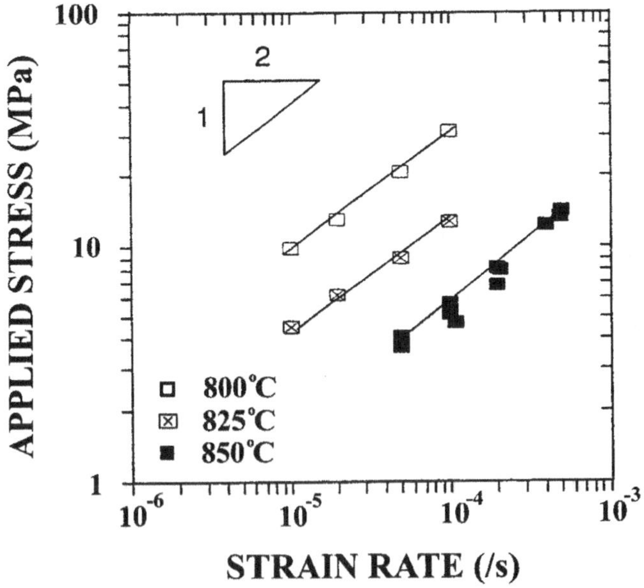

Fig. 3. Plot of applied stress versus strain rate in samples tested at 800°, 825°, and 850°C.

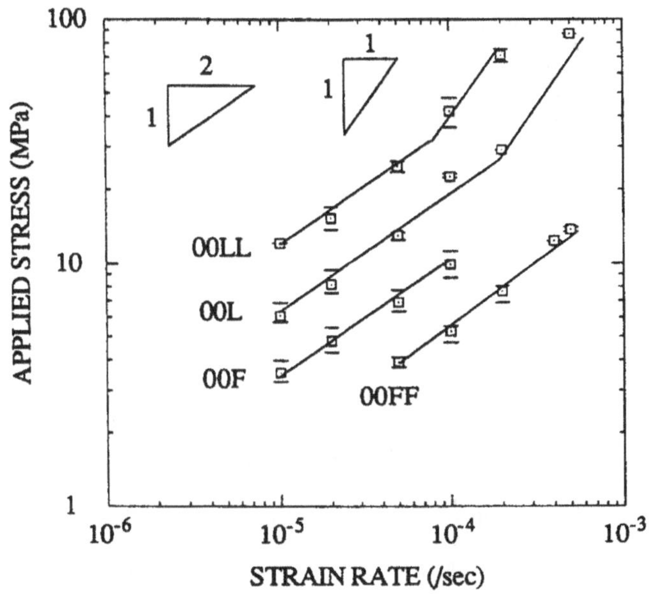

Fig. 4. Plot of applied stress versus strain rate for samples of various grain sizes, deformed at 850°C. Slopes of $n = 2$ and $n = 1$ are marked for comparison.

shown in Fig. 9. Also shown in the insert of the figure is the selected area diffraction pattern (SADP) from the grain boundary. No grain boundary phase thicker than 1 nm was observed. Figure 10 shows the TEM microstructures of the 00FF samples before and after 50% deformation at 850°C and 1.0×10^{-4}/s. A number of twin boundaries were observed in the sample before and after deformation. Since the equilibrium phase at the deformation temperature is tetragonal, twin boundaries could not be formed during the deformation. The observed twins must have been generated by oxygenation during the cooling and annealing stages.[27] There were no intragranular dislocations observed in the samples before and after deformation, except for the Frank partial dislocations at the edges of the stacking faults. Since the flow stress level during the superplastic deformation was relatively low (8–50 MPa), the formation of intragranular dislocations would be unlikely. Note that the previous studies on plastic deformation also

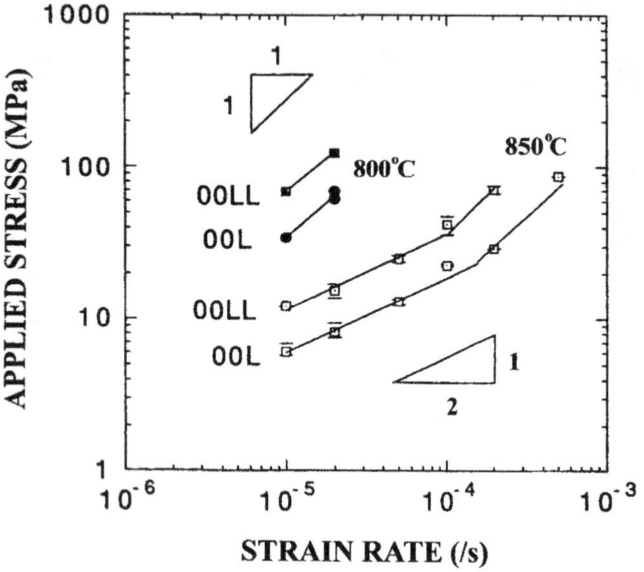

Fig. 5. Plot of applied stress versus strain rate in samples (00LL and 00L with grain sizes of 1.4 and 1.0 μm, respectively) tested at 800° and 850°C. Note $n \cong 1$ at high stresses.

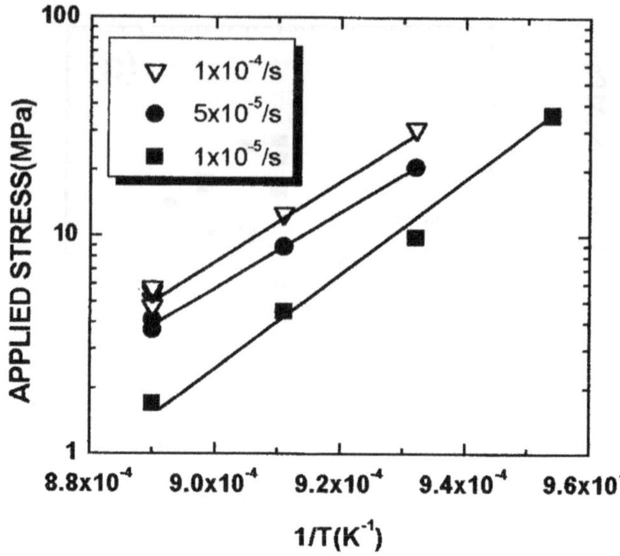

Fig. 7. Plot of applied stress versus temperature. The slope determines the activation energy.

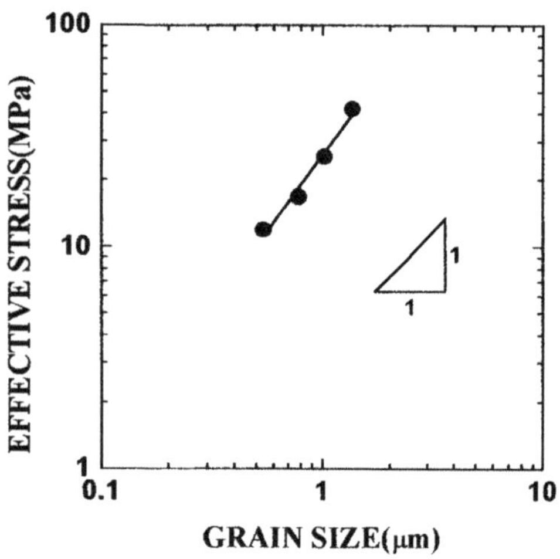

Fig. 6. Plot of effective stress versus grain size of the samples at a constant strain rate of 5×10^{-5}/s and 850°C. The slope for $p = 2$ is marked for comparison. Note that p is a multiplication of slope and stress exponent ($n = 2$).

showed the absence of intragranular dislocation activity,[28] except at very high stress level (~1 GPa)[29] or at room temperature.[30,31]

IV. Discussion

The above experimental results show that the mechanism for superplastic deformation is different from that for DC.[28] At a flow stress of about 50 MPa, there was an indication of a transition in mechanisms from the one with $n = 2$ at low stress to another with $n = 1$ at high stress. The activation energy obtained was much lower than that for the lattice self-diffusion. The grain size exponent, however, was nearly the same as that for DC. For the present case, the following three mechanisms could be considered for the observed superplastic deformation: (1) liquid phase related superplasticity, (2) grain boundary sliding (GBS) with accommodation by dislocations, and (3) GBS with accommodation by diffusion.

Chen and Hwang[32] observed that the stress exponent for the superplastic deformation of SiAlON decreased at a characteristic stress, similar to the present case. They found thin films of liquid at the grain boundary and explained that this phenomenon was due to the presence of a repulsive force between the initially wetted grains, which made the grain boundary rigid and dry at the characteristic stress. Their theory seems partially consistent with the present observation that the stress exponent suddenly decreased at a characteristic stress. However, in the absence of the grain boundary liquid phase, as supported by the HREM results, this theory cannot explain the present findings.

The theoretical model of GBS with accommodation by dislocations predicts $n = 2$, $p = 2$, and $Q = Q_{gb}$, where Q_{gb} is the activation energy of grain boundary diffusion. The values are close to the results obtained in the present study, suggesting that the prevailing deformation mechanism would be related to dislocation activities. However, there is no evidence of dislocation reaction in the observed microstructures. Moreover, the model was found invalid in terms of its dependence on the other deformation processes.

When multiple mechanisms are activated simultaneously and their rates are comparable, the overall rate is a function of the deformation rates of individual processes. The function is given as either a direct or an inverse summation, depending on the relationship of deformation processes that take place. When the processes operate independently, the faster process dominates the overall rate. When the processes are mutually dependent and occur sequentially, the slower process dominates the overall process. In the case of an independent relationship, the stress exponent increases with increasing stress, because the process with a higher stress exponent is the faster one. On the other hand, in the case of a dependent relationship, the stress exponent decreases with increasing stress.

The observation of decreasing stress exponent at a transition from $n = 2$ to $n = 1$ with increasing stress indicates that the observed superplastic deformation mechanism with $n = 2$ is dependent on the observed deformation process with $n = 1$. The deformation process with $n = 1$ exhibits Newtonian flow behavior and could be DC, or GBS with accommodation by diffusion.[7,8,33] Then the mechanism for superplastic deformation (SP) should be dependent on DC or GBS with accommodation by diffusion. In this context, GBS with accommodation by dislocations cannot respond to the observed deformation, because GBS with accommodation by the dislocation process is independent of DC or GBS with accommodation by diffusion. If it occurs, then it should occur at a higher stress level than DC with $n = 1$, contrary to the present observation.

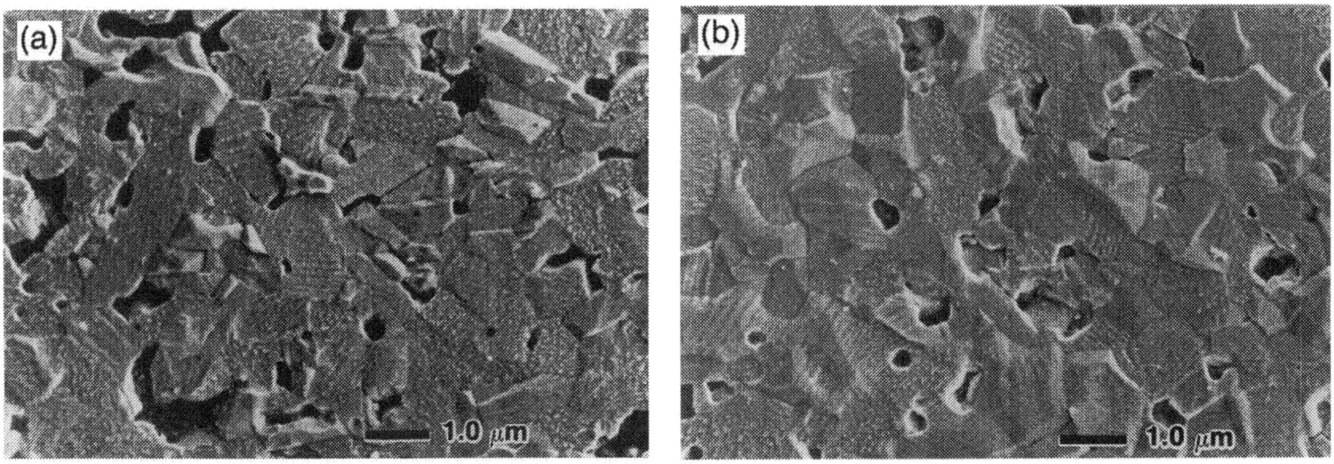

Fig. 8. Microstructures of a $YBa_2Cu_3O_{7-x}$ sample after deformation to 50% at 850°C: (a) taken from cross sections parallel to the stress axis and (b) perpendicular.

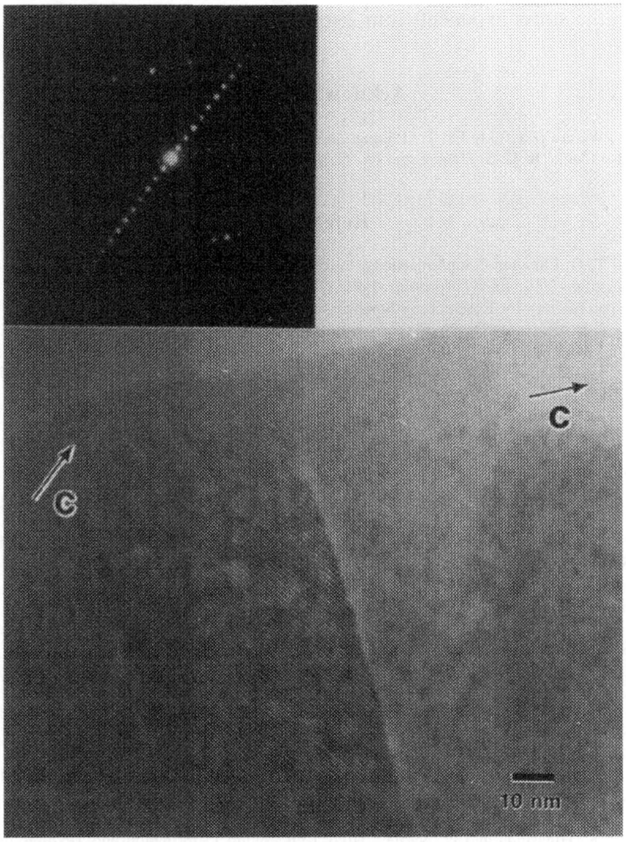

Fig. 9. TEM micrograph showing a grain boundary and lattice images in two pertinent grains. The SADP in the insert is taken at the grain boundary. Note that the grain boundary is free of second phase.

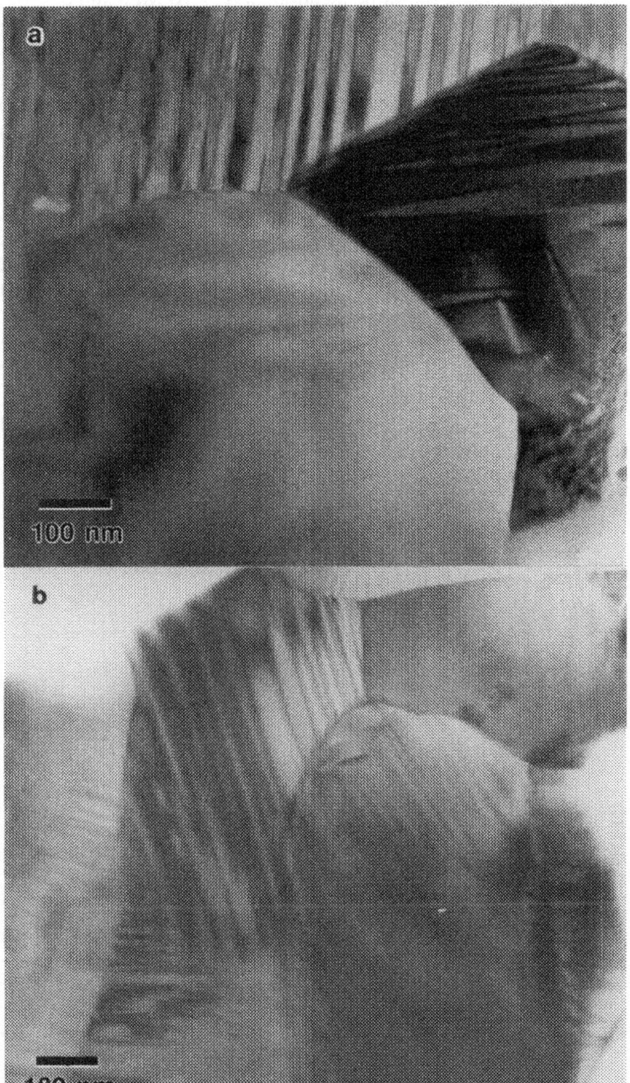

Fig. 10. TEM micrographs of $YBa_2Cu_3O_{7-x}$ samples (a) before and (b) after 50% deformation at 850°C and 1.0×10^{-4}/s.

A process that is dependent on diffusion would be the interface reaction.[34–37] Ashby[34] pointed out that when the grain boundary is not a perfect sink or source for vacancies, the reaction rate at the boundary may be slower than the diffusion rate, and hence the overall rate is controlled by the reaction rate at the boundary, which is again a function of the mobility and density of grain boundary dislocations. When the interface reaction controls the process, the rate equation is a function of σ^2/d^p, where p is either 1[8,34,36] or 2.[37] It is thus suggested that the GBS with accommodation by diffusion controlled by interface reaction is the dominant mechanism for the observed superplastic deformation. The suggested mechanism is consistent with the microstructural observations, such as the maintenance of grain morphology and the absence of texture, dislocations, and grain boundary phases.

The suggested mechanism was inserted in the deformation map of our previous study[17,38] as shown in Fig. 11. The map shows that the boundary between the interface reaction control (IRC) and the diffusion control (DC) process passes through the top right corner of the solid rectangle that covers the present data. This agrees with the observation that a transition occurred at a stress of 50 MPa and grain size of 1 μm, and that the stress exponent changed from 2 to 1. Similar transitions were observed in other ceramics as well. Cannon, Rhodes, and Heuer[39] observed the transition from DC to IRC in creep of alumina at stresses between 10 and 100 MPa with $d = 1.2$ μm. Wakai and Nagano[40] observed the interface reaction control process on superplastic ZrO_2 at stresses <100 MPa for $d =$ 1.4 μm. Nauer and Carry[41] observed the transition at a stress of 60 MPa when d is 0.8 μm. Furushiro *et al.*[42] observed a transition from the $n = 2$ process at low stress to the $n = 1$ process at high stress (>20 MPa) in a superplastic Al_2O_3–20 wt% ZrO_2 composite with submicrometer grain sizes (0.4 μm). Okamoto *et al.*[43] also observed a similar transition at about 100 MPa in fine-grained (0.5 μm) Al_2O_3 and suggested that the deformation mechanism would be interface reaction controlled GBS. Schneibel and Hazzledine[37] predicted in their theoretical work that the boundary between IRC and DC would lie at the same location, at $\sigma = 10$ MPa and $d = 0.5$ μm for any material. These results indicate that the transition from DC to IRC would occur at a stress of 10–100 MPa and a grain size of about 1 μm. It is very consistent with the deformation map constructed here; that is, the boundary is located at a stress of 50 MPa and a grain size of 1 μm. Jimenez-Melendo *et al.*[44] also reported a boundary at 1 μm grain size in their study of the deformation map of $YBa_2Cu_3O_{7-x}$ superconductors.

It should be noted that in Fig. 2(b) the boundary that separates the conditions for the presence of shear marks is located at a stress of 80 MPa and a grain size of 1 μm. Since the stress–strain curves of the tests in which the shear marks were observed show a yield drop, the presence of shear marks may indicate the loss of plasticity. If the generation of shear marks reflects the end of superplasticity, the observed boundary may refer to the location of the boundary between the GBS and Coble creep (CC) domains. Since the observed boundary in the deformation map is located close to the boundary between the IRC and DC, the superplastic deformation with a mechanism of diffusion-controlled GBS would be observed in a very short range of stress and grain size.

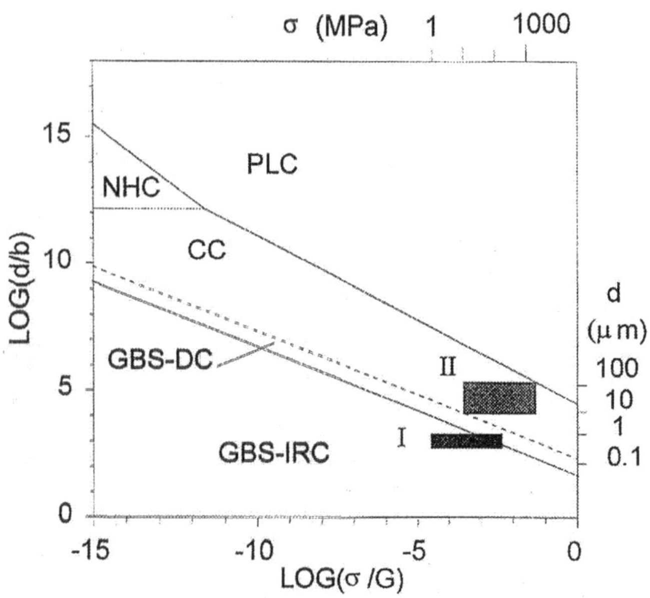

Fig. 11. Langdon–Mohamed type deformation map for the superplastic deformation of $YBa_2Cu_3O_{7-x}$ at 850°C including the grain boundary sliding mechanism: (I) present study of superplasticity, (II) previous studies of creep; (PLC) power law creep, (NHC) Nabarro–Herring creep, (CC) Coble creep, (GBS) grain boundary sliding, (DC) diffusion controlled, (IRC) interface reaction controlled.

V. Conclusions

(1) The superplastic deformation of $YBa_2Cu_3O_{7-x}$ superconductors with fine grain sizes of 0.5–1.4 μm was investigated in the temperature range of 775° to 850°C and at strain rates of 1.0×10^{-5} to 5×10^{-4}/s in a compression mode. The stress–strain curve showed the stable deformation behavior up to or over 100% strain.

(2) The deformation parameters, the stress exponent (n), the grain size exponent (p), and the activation energy (Q) were determined. Their values were $n = 2.0 \pm 0.3$, $p = 2.7 \pm 0.7$, and $Q = 745 \pm 100$ kJ/mol. The microstructures showed invariance of equiaxed grain shape, no dislocation activity, no grain boundary phase, and no significant texture development after a large deformation.

(3) Based on the quantitative data and the observed microstructures, it was suggested that the mechanism for superplastic deformation is grain boundary sliding (GBS) with accommodation by diffusion, controlled by interface reaction.

(4) The transition of deformation mechanism, from interface-reaction-controlled (IRC) to diffusion-controlled (DC) GBS, was observed at a flow stress of 50 MPa, characterized by a decrease in the stress exponent from 2 to 1 as the stress increased.

Acknowledgments

We are grateful to Dr. T. G. Langdon for helpful discussions and suggestions. We also thank Dr. C.-S. Hwang and Dr. C. W. Yang for technical assistance.

References

[1]T. G. Langdon, "Superplasticity: An Historical Perspective"; pp. 3–12 in *Superplasticity in Advanced Materials.* Edited by S. Hori, M. Tokizane, and N. Furushiro. Japan Society for Research on Superplasticity, Osaka, Japan, 1991.

[2]J. W. Edington, K. N. Melton, and C. P. Cutler, "Superplasticity," *Prog. Mater. Sci.*, **21** [2] 61–158 (1976).

[3]F. Wakai, S. Sakaguchi, and Y. Matsuno, "Superplasticity of Yttria-Stabilized Tetragonal ZrO_2 Polycrystals," *Adv. Ceram. Mater.*, **1** [3] 259–63 (1986).

[4]Y. Maehara and T. G. Langdon, "Review—Superplasticity in Ceramics," *J. Mater. Sci.*, **25**, 2275–86 (1990).

[5]A. Ball and M. M. Huchison, "Superplasticity in the Aluminum–Zinc Eutectoid," *Met. Sci. J.*, **3**, 1–7 (1969).

[6]T. G. Langdon, "Grain Boundary Sliding as a Deformation Mechanism during Creep," *Philos. Mag.*, **22**, 689–700 (1970).

[7]R. C. Gifkins, "Grain-Boundary Sliding and Its Accommodation during Creep and Superplasticity," *Metall. Trans. A*, **7A**, 1225–32 (1976).

[8]M. F. Ashby and R. A. Verrall, "Diffusion-Accommodated Flow and Superplasticity," *Acta Metall.*, **21**, 149–63 (1973).

[9]O. A. Kaibyshev, *Superplasticity of Alloys, Intermetallics, and Ceramics.* Springer-Verlag, Berlin, Germany, 1992.

[10]O. A. Kaibyshev, R. M. Imaev, and M. F. Imaev, "Superplasticity of the Ceramic Compound $YBa_2Cu_3O_{7-x}$," *Sov. Phys.—Dokl. (Engl. Transl.)*, **34** [4] 375–76 (1989).

[11]G. Bussod, A. Pechenik, C.-T. Chu, and B. Dunn, "Effect of Temperature and Strain Rate on the Plastic Deformation of Fully Dense Polycrystalline $YBa_2Cu_3O_{7-x}$ Superconductor," *J. Am. Ceram. Soc.*, **72** [1] 137–39 (1989).

[12]Y. Kodama and F. Wakai, "Hot Deformation and Superconductivity of $YBa_2Cu_3O_{7-x}$ Ceramics"; pp. 113–16 in *Advances in Superconductivity II.* Edited by T. Ishiguro and K. Kajimura. Springer-Verlag, Tokyo, Japan, 1990.

[13]K. Higashida and N. Narita, "High Temperature Deformation and Textures in Oxide Superconductors"; pp. 805–808 in *Advances in Superconductivity III.* Edited by K. Kajimura and H. Hayakawa. Springer-Verlag, Tokyo, Japan, 1991.

[14]W.-J. Kang, S. Hanada, Y. Wadayama, and A. Nagata, "Microstructural Control of $YBa_2Cu_3O_{7-x}$ by High Temperature Compression," *Mater. Sci. Eng.*, **B13** 125–32 (1992).

[15]J. Yun, M. P. Harmer, and Y. T. Chou, "Superplastic Flow in Fine-Grained $YBa_2Cu_3O_{7-x}$ High T_c Superconductors," *Scr. Metall. Mater.*, **29**, 267–71 (1993).

[16]J. Albuquerque, M. P. Harmer, and Y. T. Chou, "Tensile Superplastic Deformation of $YBa_2Cu_3O_{7-x}$ High T_c Superconductors," *Acta Mater.*, in press.

[17]J. Yun, M. P. Harmer, and Y. T. Chou, "Deformation Mechanism Map for Creep in $YBa_2Cu_3O_{7-x}$," *J. Mater. Sci.*, **30**, 4906–11 (1995).

[18]J. Yun, G. Güven, and Y. T. Chou, "Chemical Etching of $YBa_2Cu_3O_{7-x}$," *J. Mater. Sci. Lett.*, **12**, 1080–82 (1993).

[19]ASTM Designation E112-85, "Standard Methods for Determining the Average Grain Size," *Annual Book of ASTM Standards.* American Society for Testing and Materials, Philadelphia, PA, 1986.

[20]J. Yun and Y. T. Chou, "The Effective Stress in Deformation of Porous Materials," unpublished work.

[21]J. Yun, "Superplastic Deformation of High T_c Superconductors: $YBa_2Cu_3O_{7-x}$ and $YBa_2Cu_3O_{7-x}$/Ag"; Ph.D. Dissertation. Lehigh University, Bethlehem, PA, 1994.

[22]N. Chen, S. J. Rothman, and J. L. Routbort, "Tracer Diffusion of Ba and Y in $YBa_2Cu_3O_x$," *J. Mater. Res.*, **7** [9] 2308–16 (1992).

[23]J. L. Routbort, S. J. Rothman, N. Chen, J. N. Mundy, and J. E. Baker, "Site Selectivity and Cation Diffusion in $YBa_2Cu_3O_{7-x}$," *Phys. Rev. B.*, **43** [7] 5489–97 (1991).

[24]K. N. Tu, N. C. Yeh, S. I. Park, and C. C. Tsuei, "Diffusion of Oxygen in Superconducting $YBa_2Cu_3O_{7-x}$ Ceramic Oxides," *Phys. Rev. B*, **39** [1] 304–14 (1989).

[25]J. L. Tallon and M. P. Staines, "Oxygen Self-Diffusion Coefficient in the Superconductor $YBa_2Cu_3O_{7-x}$ from Internal Friction Measurement," *J. Appl. Phys.*, **68** [8] 3998–4001 (1990).

[26]M. Jimenez-Melendo, A. R. De Arellano-Lopez, A. Dominguez-Rodriguez, and K. C. Goretta, and J. L. Routbort, "Diffusion-Controlled Plastic Deformation of $YBa_2Cu_3O_x$," *Acta Met. Mater.*, **43** [6] 2429–34 (1995).

[27]O. R. Monteiro, J. W. Evans, and S. M. Johnson, "*In Situ* Observation of the Orthorhombic–Tetragonal Phase Transformation in $YBa_2Cu_3O_{7-x}$," *J. Appl. Phys.*, **69** [4] 2414–17 (1991).

[28]K. C. Goretta, J. L. Routbort, A. C. Biondo, Y. Gao, A. R. Arellano-Lopez, and A. Dominguez-Rodriguez, "Compressive Creep of $YBa_2Cu_3O_{7-x}$," *J. Mater. Res.*, **5** [12] 2766–70 (1990).

[29]M. J. Kramer, L. S. Chumbley, and R. W. McCallum, "Analysis of Deformed $YBa_2Cu_3O_{7-x}$," *J. Mater. Sci.*, **25**, 1978–86 (1990).

[30]J. Rabier and M. F. Denanot, "Plastic Deformation of $YBa_2Cu_3O_{7-x}$ and Related Structural Defects," *Rev. Phys. Appl.*, **25**, 55–59 (1990).

[31]J. Rabier, P. D. Tall, and M. F. Denanot, "On the Dissociation of Dislocations in $YBa_2Cu_3O_{7-x}$," *Philos. Mag.*, **A67**, 1021–35 (1993).

[32]I-W. Chen and S.-L. Hwang, "Shear Thickening Creep in Superplastic Silicon Nitride," *J. Am. Ceram. Soc.*, **75** [5] 1073–79 (1992).

[33]J. R. Spingarn and W. D. Nix, "Diffusional Creep and Diffusionally Accommodated Grain Rearrangement," *Acta Metall.*, **26**, 1389–98 (1978).

[34]M. F. Ashby, "On Interface-Reaction Control of Nabarro–Herring Creep and Sintering," *Scr. Metall.*, **3**, 837–42 (1969).

[35]G. W. Greenwood, "The Possible Effects on Diffusion Creep of Some Limitation of Grain Boundaries as Vacancy Sources or Sinks," *Scr. Metall.*, **4**, 171–74 (1970).

[36]B. Burton, "Interface Reaction Controlled Diffusional Creep: A Consideration of Grain Boundary Dislocation Climb Sources," *Mater. Sci. Eng.*, **10**, 9–14 (1972).

[37]J. H. Schneibel and P. M. Hazzledine, "The Role of Coble Creep and Interface Control in Superplastic Sn–Pb Alloys," *J. Mater. Sci.*, **18**, 562–70 (1983).

[38]J. Yun and Y. T. Chou, "Deformation Mechanism Map for Creep and Superplastic Deformation in $YBa_2Cu_3O_{7-x}$ Ceramic Superconductors," *J. Kor. Ceram. Soc.*, **33** [6] 718–24 (1996).

[39]R. M. Cannon, W. H. Rhodes, and A. H. Heuer, "Plastic Deformation of Fine-Grained Alumina (Al_2O_3): I, Interface-Controlled Diffusional Creep," *J. Am. Ceram. Soc.*, **63** [1–2] 46–53 (1980).

[40]F. Wakai and T. Nagano, "The Role of Interface-Controlled Diffusion Creep on Superplasticity of Yttria-Stabilized Tetragonal ZrO_2 Polycrystals," *J. Mater. Sci. Lett.*, **7**, 607–609 (1988).

[41]M. Nauer and C. Carry, "Creep Parameters of Yttria Doped Zirconia Materials and Superplastic Deformation Mechanisms," *Scr. Met. Mater.*, **24**, 1459–63 (1990).

[42]N. Furushiro, T. Tanizawa, K. Akashiro, and S. Takeshita, "Grain Boundary Sliding during Superplastic Deformation in Ceramic Materials"; pp. 245–50 in *Superplasticity in Advanced Materials.* Edited by S. Hori, M. Tokizane, and N. Furushiro. Japan Society for Research on Superplasticity, Osaka, Japan, 1991.

[43]Y. Okamoto, M. Inoue, T. Shiono, K. Hayashi, and T. Nishikawa, "Diffusion Controlled and Interface Reaction Controlled Deformation in Fine-Grained Alumina and Y-TZP"; pp. 269–74 in *Superplasticity in Advanced Materials.* Edited by S. Hori, M. Tokizane, and N. Furushiro. Japan Society for Research on Superplasticity, Osaka, Japan, 1991.

[44]M. Jimenez-Melendo, A. Dominguez-Rodriguez, and J. L. Routbort, "Deformation Maps of $YBa_2Cu_3O_{7-x}$ Superconductors," *Scr. Metall. Mater.*, **32** [4] 621–26 (1995). □

J. Am. Ceram. Soc., **86** [2] 257–65 (2003)

Reel-to-Reel Continuous Chemical Solution Deposition of Epitaxial Gd_2O_3 Buffer Layers on Biaxially Textured Metal Tapes for the Fabrication of $YBa_2Cu_3O_{7-\delta}$ Coated Conductors

Tolga Aytug,[†,‡] Mariappan P. Paranthaman,[*,†] Byeongwon W. Kang,[§] David B. Beach,[†] Srivatsan Sathyamurthy,[†] Eliot D. Specht,[§] Dominic F. Lee,[*,§] Roeland Feenstra,[‡] Amit Goyal,[*,§] Donald M. Kroeger,[§] Keith J. Leonard,[§] Patrick M. Martin,[§] and David K. Christen[*,‡]

Chemical Sciences Division, Solid State Division, and Metals and Ceramics Divison, Oak Ridge National Laboratory, Oak Ridge, Tennessee 37831

A nonvacuum reel-to-reel dip-coating system has been used to continuously fabricate epitaxial Gd_2O_3 buffer layers on mechanically strengthened, biaxially textured Ni–(3 at.% W–1.7 at.% Fe), defined as Ni-alloy, metal tapes. Because of its significance as a seed layer, optimum processing conditions (postannealing speed and temperature) of Gd_2O_3 buffer layers have been studied. Highly textured films were obtained under reducing (96% Ar + 4% H_2) atmosphere at temperatures between 1100° and 1150°C; postannealing speed did not significantly affect the crystalline quality of the Gd_2O_3. Scanning electron microscopy revealed a continuous, dense, and crack-free surface morphology for these dip-coated buffers. The Gd_2O_3 layer thickness led to pronounced differences in the growth characteristics of the subsequent YSZ and CeO_2 layers deposited by rf-magnetron sputtering. Epitaxial YBCO films grown by pulsed laser deposition on the short prototype CeO_2/YSZ/Gd_2O_3/Ni–(3 at.% W–1.7 at.% Fe) conductors yielded self-field critical current densities (J_c) as high as 1.2 × 10^6 A/cm^2 at 77 K. Pure Ni tapes were used to assess the viability of dip-coated buffers for long length coated conductor fabrication. The YBCO films, grown on 80 cm long and 1 cm wide CeO_2/YSZ/Gd_2O_3 buffered Ni tapes by the industrially scalable *ex situ* BaF_2 precursor process, exhibited end-to-end self-field J_c values of 6.25 × 10^5 A/cm^2 at 77 K. These results demonstrate the reproducible epitaxy of solution-derived seed layers on pure Ni and Ni-alloy tapes as well as underscore the viability of solution approaches for the production of long length YBCO-based coated conductors.

I. Introduction

Recently, scalable techniques for achieving long lengths of biaxially textured buffer layers and superconductors on Ni or Ni-alloy substrates have been proposed. One of the approaches uses the concept of rolling-assisted biaxially textured substrates (RABiTS).[1,2] This technique employs thermomechanical processing to obtain sharp biaxially textured metal tapes, which can be produced over long lengths. The primary issue for the successful development of coated conductors for high-temperature superconducting (HTS) wire applications is to chemically isolate the HTS film from the metal substrate, preventing diffusion of Ni into the superconductor during high-temperature processing, and also provide an epitaxial template for growth of the HTS coating. Various combinations of oxide buffer layers have been tried on metallic substrates for the development of HTS-coated conductors.[1–14] The most successful tape architecture has been the combination of CeO_2 and yttria-stabilized zirconia (YSZ), yielding $YBa_2Cu_3O_{7-\delta}$ (YBCO) films with high critical current densities (J_c) exceeding 10^6 A/cm^2 (77 K, 0 T).[2,8,10–13] Fabrication of these buffer layers is generally conducted by vapor deposition processes such as electron beam evaporation (e-beam), pulsed laser deposition (PLD), or magnetron sputtering.[1–13] A major challenge in using any of these buffer-deposition techniques is prevention of unfavorably oriented NiO on the Ni surface.[2,10–13,15] This requirement is difficult in principle since the growth of oxide films generally requires elevated temperatures and oxidizing environment. Another issue of concern has been the formation of cracks in the base CeO_2 layer at thicknesses greater than a few hundred angstroms;[10,16,17] it is well understood that integrity of a seed buffer layer is crucial for reproducible epitaxial growth of subsequent oxide buffer and HTS layers.

For the commercial development of a practical superconducting wire technology, more adaptive and easily scalable processing techniques are highly required. Nonvacuum chemical solution based processes,[18–28] such as sol–gel and metal-organic decomposition (MOD), offer many desirable aspects, such as precise control of metal oxide precursor stoichiometry and composition, mechanical simplicity, and low cost. Recently, it has been reported that various RE_2O_3 (rare-earth (RE) = Gd, Yb, and Eu) and $RE_2Zr_2O_7$ (RE = La and Nd) oxide films can grow epitaxially on textured Ni substrates by using solution-based methods.[23–25,28] While short segments of sol–gel seeded Ni substrates with sputtered CeO_2/YSZ overlayers have yielded YBCO with J_c in the range of 0.4 × 10^6–1.1 × 10^6 A/cm^2 at 77 K,[23–25,28] long lengths of high-J_c YBCO coatings on solution seed buffer layers have not yet been produced.

Much of the work reported using the RABiTS technique has used high-purity Ni (99.99%) as the texture material. However, the low mechanical strength and ferromagnetism of pure Ni present significant challenges in long length manufacturing and hinders its usage in applications where AC losses are an issue. Therefore, it is desirable to work with textured alloy substrates that are much stronger and nonmagnetic, or have reduced magnetism. Recently, the successful texturing of Ni strengthened with small additions (3–5 at.%) of W and (1.7 at.%) Fe has been reported.[29,30] These substrates, typically 50 μm thick, accommodate similar annealing

B. A. Tuttle—contributing editor

Manuscript No. 187380. Received October 23, 2001; approved November 26, 2002.

This work was supported by the U.S. Department of Energy, Division of Materials Sciences, Office of Science, Office of Power Technologies—Superconductivity Program, Office of Energy Efficiency and Renewable Energy. The research was performed at the Oak Ridge National Laboratory, managed by U.T.-Battelle, LLC, for the U.S. Department of Energy under Contract No. DE-AC05-00OR22725. Acknowledgments also extended to Oak Ridge Associated Universities for their support.

*Member, American Ceramic Society.
†Chemical Sciences Division.
‡Solid State Division.
§Metals and Ceramics Division.

conditions and show comparable grain sizes to those of pure Ni. After recrystallization, however, they exhibit yield strengths (YS) of 143 and 154 MPa at 0.02% and 0.2% strain, respectively, which are much higher than those of pure Ni (YS = 40 and 58 MPa, respectively).[29] In addition, the saturation magnetization, M, of a Ni–(3 at.% W–1.7 at.% Fe), [Ni-alloy], (M = 40 G·cm^3/g) was found to be significantly lower than that of pure Ni (M = 57.5 G·cm^3/g at T = 0 K). Development of oxide buffer layers and HTS coatings by both vacuum and nonvacuum processes on these alloys has just begun.[28,30,31] Recently, high-quality YBCO films with J_c (self-field, 77 K) values of 1.9×10^6 A/cm^2 have been demonstrated on such substrates where the oxide buffer layers have been grown by all-vacuum processes.[31] In addition to vacuum-based techniques, oxide buffer layers have also been grown directly on Ni-alloy tapes by nonvacuum solution-based methods. However, despite a good crystalline structure, to date there have been no reports on the growth of high-J_c YBCO films on these sol–gel buffered alloy tapes.

In this paper, first we report the epitaxial growth and property characterizations of sol–gel processed, reel-to-reel dip-coated Gd_2O_3 seed buffer layers on textured Ni-alloy substrates. Effects of processing parameters on seed-layer quality and carbon retention were systematically explored. Second, as a proof of principle, we present the growth of high- quality YBCO coatings on short segments of these Gd_2O_3/Ni–(3 at.% W–1.7 at.% Fe) tapes with sputtered CeO_2/YSZ cap buffer layers. Third, we extend short sample studies to demonstrate for the first time the continuous fabrication of high-J_c HTS coatings in long lengths on pure Ni tapes that have solution processed seed layers. The electrical transport properties, measured over lengths (100 cm) of YBCO films grown using the *ex situ* BaF_2 precursor process, are also reported.

II. Experimental Procedure

(1) Preparation of Biaxially Textured Ni and Ni-Alloy Substrates

Biaxially textured Ni (99.99%) and Ni–(3 at.% W–1.7 at.% Fe) substrates were obtained by progressive cold rolling of polycrystalline, randomly oriented Ni and Ni-alloy bars to a total deformation of 95%–98%. After mechanical deformation, 50 μm thick and 1 cm wide substrates were first cleaned with isopropyl alcohol in a reel-to-reel ultrasonic cleaning unit, and then annealed at 1100°C (for pure Ni) and 1250°C (for Ni-alloy) for 1 h in a reducing, forming gas (96% Ar + 4% H_2) atmosphere. The procedure yields the desired {100}⟨100⟩ cube texture to typical levels of 6–8° FWHM.

(2) Preparation of Gd_2O_3 Solution and Coating Procedure

A 2-methoxyethanol solution of gadolinium methoxyethoxide/acetate, prepared by an alkoxide sol–gel synthesis route, was used to coat the Ni and Ni-alloy tapes. The solution preparation was conducted under Ar gas atmosphere using a Schlenck-type apparatus. To avoid premature hydrolysis, which can induce precipitation, the as-received gadolinium acetate powder was first purified by dissolving in a mixture of water and acetic acid at a ratio of 3:1, filtered, and then evaporated to dryness in an oven maintained at 150°C for overnight to remove absorbed moisture. Then the purified gadolinium acetate of about 3.34 g was dissolved in 40 mL of 2-methoxyethanol in a 250 mL round-bottomed flask. The flask was refluxed for 1 h with excess 2-methoxyethanol to exchange the acetic acid ligand with the methoxyethoxide ligand. The final volume was adjusted to 40 mL to produce a 0.5M Gd_2O_3 precursor solution.

In a reel-to-reel dip-coating unit, tapes with dimensions of 10 cm in length and 1 cm in width were coated to determine the optimum processing parameters of Gd_2O_3 seed layers on Ni-alloy. The textured tapes were continuously pulled through the Gd_2O_3 precursor solution at a constant speed in the range of 20–300 cm/h. These double-sided, dip-coated tapes were then transported, in a flowing forming gas atmosphere, through a preheated furnace at various temperatures (T_{an}) ranging from 1050° to 1250°C. For this processing, different tape transport speeds (S_{an}) in the range of 5 to 30 cm/h were investigated, resulting in heating rates of 6° to 40°C/min, respectively. The rate is limited by the length of the uniform hot zone for the present furnace, which spans ~10 cm. After the growth conditions of Gd_2O_3 on Ni-alloys were optimized based on structural and morphologic characterizations, the same parameters were used for the growth on Ni tapes. In fact, from detailed studies (not shown here) the growth conditions of Gd_2O_3 on both substrates are found to be similar.

(3) Preparation of YBCO Test Structures

Sputtering by the rf-magnetron technique was used to deposit both YSZ and CeO_2 overlayers on short segments of the dip-coated substrates, yielding the buffer-layer sequence CeO_2/YSZ/Gd_2O_3/Ni–(3 at.% W–1.7 at.% Fe). Oxide sputtering targets were 2 in. in diameter and the system base pressure was 1×10^{-6} torr. Samples were attached onto a heated substrate block assembly by clipping at both ends, in an on-axis sputtering geometry. Sputtering conditions for both oxides consist of a mixture of 10 mtorr of forming gas and 2×10^{-6} torr of H_2O, and a substrate temperature of 780°C. Typical film thicknesses for these YSZ and CeO_2 layers were around 200 and 20 nm, respectively.

To assess the quality of the CeO_2/YSZ/Gd_2O_3/Ni-alloy multilayer structure, YBCO films were deposited by pulsed laser deposition (PLD), using a XeCl excimer laser system, operated with energy density of ≈4 J/cm^2. Details of the experimental conditions were reported elsewhere.[7–9] Typical YBCO film thicknesses were 200 nm.

(4) Characterization Techniques

X-ray diffraction analysis was used to analyze phase, structure, and texture (i.e., the orientation relationship of the grains relative to the substrate material geometry) of the buffer layers as well as the YBCO coatings deposited on short samples. A Philips Model XRG3100 diffractometer with Cu$K\alpha$ radiation was used to record powder diffraction patterns. Texture analysis was achieved using a Rigaku rotating anode X-ray generator with a graphite incident-beam monochromator selecting Cu$K\alpha$ with slits defining 2 mm × 2 mm beam. Pole figures were collected with a 4-circle diffractometer. The out-of-plane (ω-rocking curve scans) alignment of the films was measured by rocking the desired (00l) planes about an axis perpendicular to the scattering plane (ω-axis) and the in-plane (ϕ-scan) alignments were measured by rotating the desired (hkl) planes around the normal to the sample plane. Peak width FWHM values were determined by least-squares fitting to Gaussian line shapes. The texture of long-length samples was measured with a larger diffractometer (Model 4-circle, Huber) to accommodate a reel-to-reel translation stage. Data were obtained as a function of position by continuously measuring out-of-plane (ω-rocking curve scan) and in-plane (ϕ-scan) alignments, while translating the sample. Microstructural investigations were conducted using a Hitachi S-4100 high-resolution field emission type scanning electron microscope (HRSEM). The thickness of the buffer and superconducting layers was determined by Rutherford backscattering spectroscopy (RBS). In addition, ion spectroscopy techniques were used to determine the carbon content of the dip-coated buffers, by using a resonance for the 1.73 MeV H^+ ions at near normal incidence. A standard four-point contact technique was applied to measure the superconducting critical temperature (T_c) and J_c of the YBCO films. For short samples, the distance between voltage contacts was 4 mm and the values of J_c were assigned at a 1 μV/cm criterion. Electrical contacts of silver were deposited onto the samples using dc-magnetron sputtering followed by an O_2 annealing in 1 atm for 30 min at 500°C.

III. Results and Discussion

(1) Growth of Gd_2O_3 Seed Layers on Biaxially Textured Ni-Alloy Substrates

Figure 1 shows the XRD θ–2θ spectrum for a series of dip-coated Gd_2O_3 seed layers processed at various temperatures

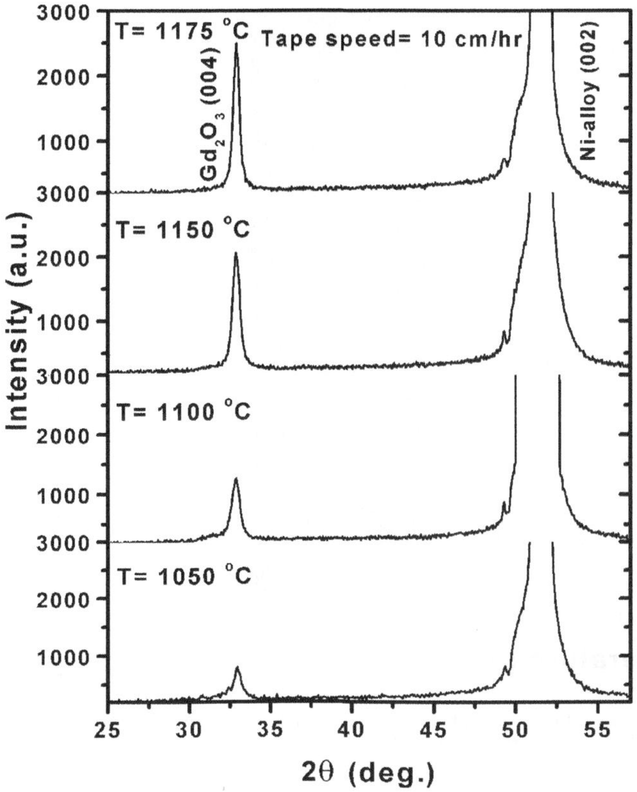

Fig. 1. XRD θ–2θ patterns of dip-coated Gd_2O_3 films grown on biaxially textured Ni–(3 at.% W–1.7 at.% Fe) substrates at various annealing temperatures.

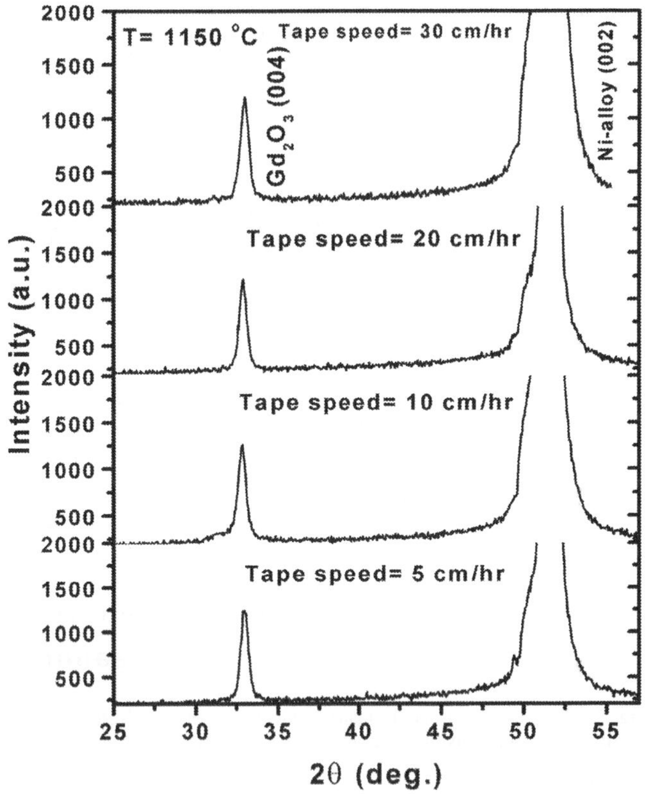

Fig. 2. XRD θ–2θ patterns of dip-coated Gd_2O_3 films grown on biaxially textured Ni–(3 at.% W–1.7 at.% Fe) substrates at 1150°C, for various annealing speeds.

ranging from 1050° to 1170°C at a fixed S_{an} of 10 cm/h. At $T_{dep} \leq 1100$°C, broad and relatively low intensity (00*l*) Gd_2O_3 reflections are observed as evidence of inhomogeneity or high structural defect density. For 1100°C $< T_{dep} <$ 1175°C, the (00*l*) peak intensities increase and the reflections sharpen, indicating substantial improvement in the crystalline quality of the films. Since dip-coated film thickness is directly proportional to the coating speed, which was kept constant for each individual sample, comparison of the relative Gd_2O_3 (00*l*) peak intensities reflects the crystalline quality of the films. At temperatures above 1200°C, however, the films showed extra Gd_2O_3 peaks reflections, which are indexed as belonging to Gd_2O_3 monoclinic phase components.[32,33] To investigate the influence of annealing speed on the crystalline quality of Gd_2O_3, a series of films were annealed at T_{an} = 1150°C in different S_{an} in the range of 5–30 cm/h. From the comparable (00*l*) θ–2θ peak intensities, shown in Fig. 2, it is apparent that S_{an} does not significantly affect the crystalline quality of the Gd_2O_3. The effect of T_{an} on the in-plane and the out-of-plane texture of Gd_2O_3 layers was also studied at a fixed S_{an} = 10 cm/h, and the results are shown in Figs. 3(a) and (b), respectively. Note that biaxially textured Ni-alloy substrates generally have slightly different in-plane and out-of-plane textures depending on the specific rolling and crystallization conditions. Thus, to make a direct and more quantitative comparison between the processing parameters and the crystalline quality of the grown films, the FWHM peak-width ratio relative to the underlying Ni-alloy is plotted as a function of T_{an}. Substantial improvement in the in-plane and out-of-plane alignment of the Gd_2O_3 layers is observed with an increase in temperature above 1050°C. Depositions at $T_{an} >$ 1100°C, however, did not significantly affect the epitaxial quality. Similarly the effect of S_{an} on the epitaxial growth of the films (for the same samples studied in Fig. 2) can be observed from the results of ω- and ϕ-scan peak-width ratios, as shown in Fig. 4, for a fixed T_{an} = 1150°C. While S_{an} does not have a remarkable influence on the in-plane alignment of the films, an increase in S_{an} over 20 cm/h has a negative effect on the out-of-plane texture, as evidenced by the gradual increase in the ω-scan FWHM ratios. Since the uniform hot-zone length is ~10 cm for the present furnace, the latter effect for higher S_{an} can be correlated with the decreased residence time in the effective hot-zone region. On the other hand, the in-plane alignment is likely governed by the crystallographic texture of the underlying substrate. Hence, the combined observations of Figs. 1–4 suggest that the optimal range for epitaxial growth of Gd_2O_3 on Ni–(3 at.% W–1.7 at.% Fe) should be a T_{an} of 1100° to 1175°C and $S_{an} <$ 30 cm/h.

The epitaxial quality of the Gd_2O_3 films was examined by XRD pole figure analysis. The logarithmic-scale (222) pole figures for two films processed at T_{an} = 1150°C and S_{an} = 20 cm/h, having thin (25 nm) and thick (73 nm) Gd_2O_3 layers, are shown in Figs. 5(a) and (b), respectively. The data for both films indicate single-domain epitaxy, with the Gd_2O_3[001]//substrate[001] and Gd_2O_3[110]//substrate[100]. This result is important, since any other orientational domains could lead to the potential disruption of current flow by high-angle grain boundaries in the HTS layer.

For implementation of nonvacuum dip-coated Gd_2O_3 seed layers into long-length wire manufacturing, it is important to grow dense and crack-free layers with minimum carbon contamination. Resonant RBS has been used to analyze the carbon levels (number of carbon atoms/number of gadolinium atoms) in several Gd_2O_3 samples, with an interest on the correlation with respect to T_{an} and S_{an}. The results are summarized in Table I. The slight variations in thickness for the same coating speed are due to differences in the solution viscosity. While we observe no strong dependence of carbon content on either T_{an} or S_{an}, the overall carbon levels are significant (≈0.4–0.6 carbon/gadolinium atom). These levels, however, do not adversely affect the surface morphology of the thinner buffers, as shown in the following. Plan-view HRSEM micrographs of the surface morphology for 25 and 73 nm thick Gd_2O_3 films fabricated on biaxially textured Ni-alloy substrates are presented in Figs. 6(a) and (b), respectively. While each sample exhibits a continuous, crack-free, and dense surface morphology, significant surface roughening develops for the thicker coatings (Fig.

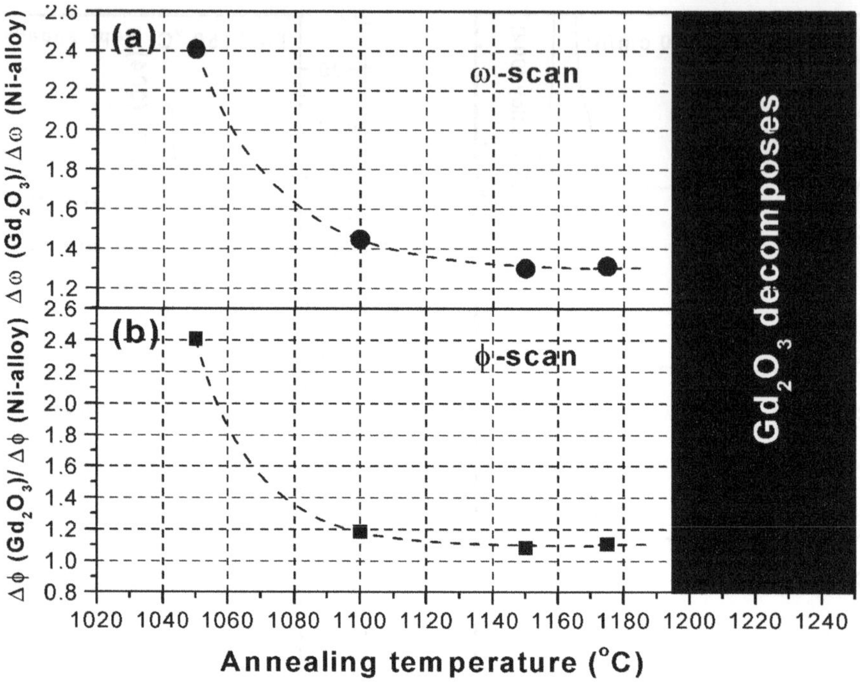

Fig. 3. Dependence of the normalized XRD FWHM peak widths, quantifying the crystallinity of Gd_2O_3 films as a function of postannealing temperature: (a) the out-of-plane distribution width $\Delta\omega$; (b) the in-plane distribution width $\Delta\phi$. The lines serve as a visual guide.

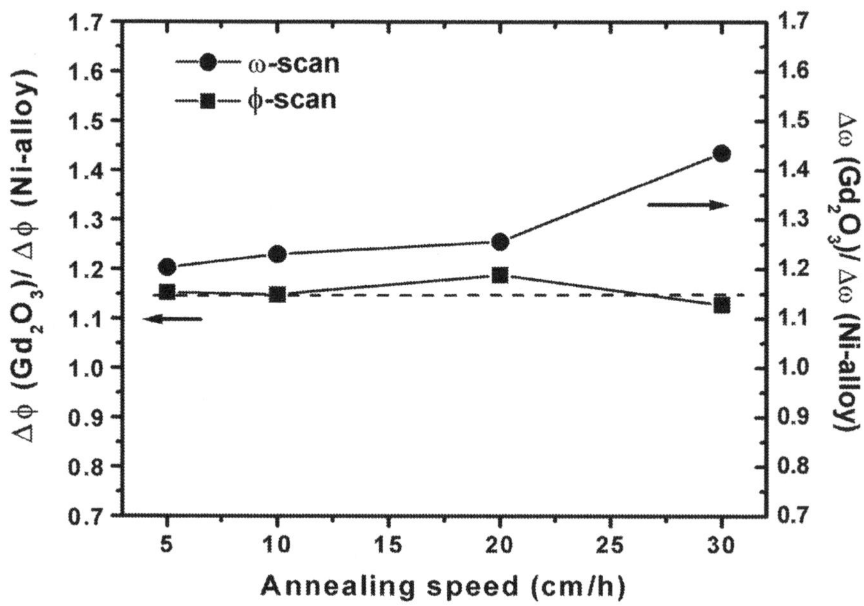

Fig. 4. Effect of postannealing speed on the FWHM ratios of ω-rocking curves and ϕ-scans for Gd_2O_3 films on Ni–(3 at.% W–1.7 at.% Fe) tapes, fabricated at 1150°C. The dashed line serves as a visual guide.

6(b)). On the other hand, thinner coatings (25 nm) exhibit extremely smooth and featureless surface microstructure (Fig. 6(a)), with an rms roughness of ~3 nm as determined by atomic force microscopy measurements. Such surface morphology should best serve the goal of obtaining high-quality YBCO coatings, despite the internal carbon content of the Gd_2O_3. The effects of Gd_2O_3 seed layer thickness on the growth characteristics of subsequent oxide buffer layers are discussed in the following section.

(2) Structural and Superconducting Property Characterizations of YBCO/CeO_2/YSZ/Gd_2O_3/Ni-Alloy Multilayer Structure

After optimizing the fabrication process for dip-coated Gd_2O_3 films on Ni-alloys, we have first deposited YSZ and CeO_2 overlayers by rf-magnetron sputtering, and then investigated the quality of this solution seeded CeO_2/YSZ/Gd_2O_3 buffer layer structure by depositing YBCO films. In this architecture the YSZ buffer layer acts as a Ni diffusion barrier and the CeO_2 layer ensures a better lattice match with YBCO. Figures 7(a) and (b) show the results of XRD θ–2θ scans for the CeO_2/YSZ bilayers deposited on Gd_2O_3/Ni-alloy substrates, where the Gd_2O_3 layer thickness is 25 and 73 nm, respectively. The data show that CeO_2/YSZ bilayers grown on thicker Gd_2O_3 layers have a polycrystalline nature (Fig. 7(b)), whereas films deposited on thinner seed layers exhibit excellent *c*-axis oriented growth (Fig. 7(a)). This growth behavior seems overwhelmingly governed by the surface morphology of the Gd_2O_3 films. It is known that surface roughness exacerbates adverse interfacial reaction between the

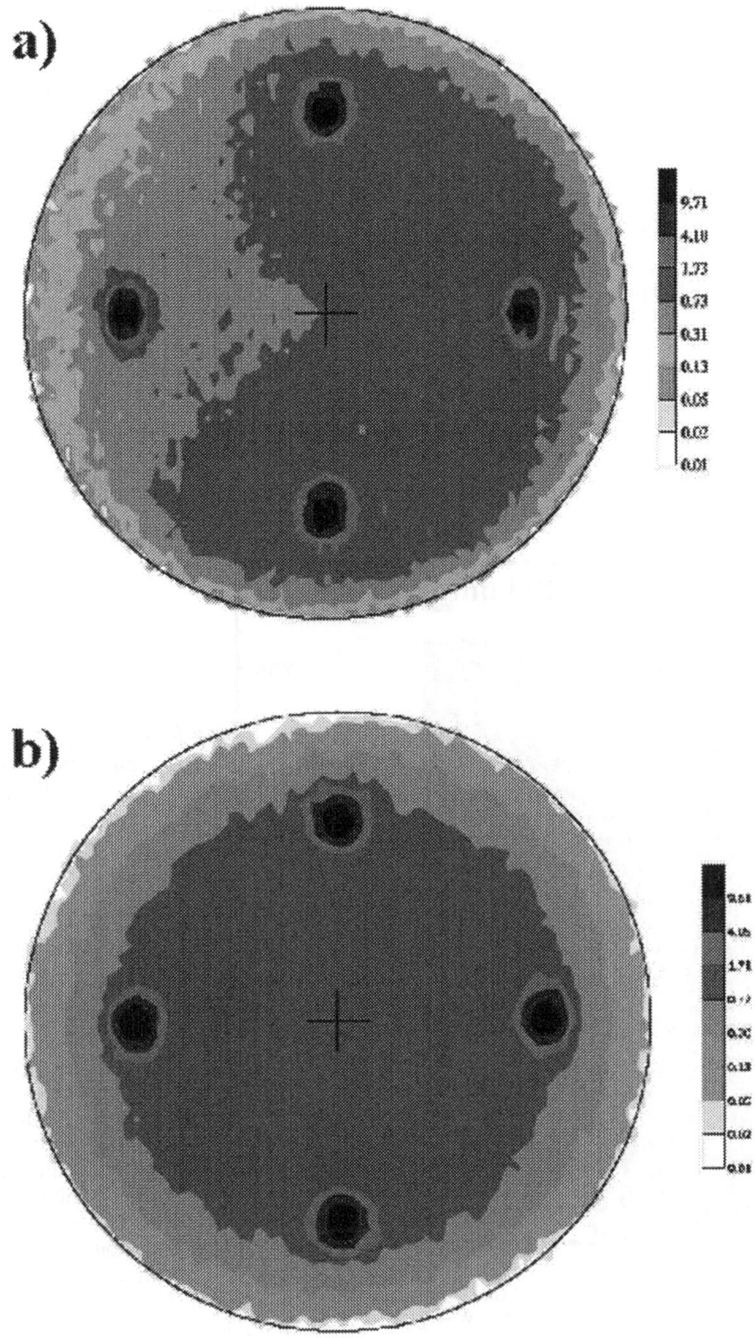

Fig. 5. The (222) logarithmic pole figure of (a) 25 nm and (b) 73 nm thick dip-coated Gd_2O_3 film processed on biaxially textured Ni–(3 at.% W–1.7 at.% Fe) at 1150°C with an annealing speed of 20 cm/h.

film and the substrate. Therefore, the relatively coarse surface morphology of thicker Gd_2O_3 films could contribute to polycrystalline growth behavior of the CeO_2/YSZ overlayers.

Following deposition of CeO_2/YSZ overlayers, suitability of the CeO_2/YSZ/Gd_2O_3/Ni-alloy structure was tested with the growth of YBCO films. A typical XRD θ–2θ scan is shown in Fig. 7(c), where the pattern indicates only (00*l*) reflections for the YBCO, demonstrating *c*-axis perpendicular to the substrate surface. After YBCO deposition, weak NiO impurity peaks are observed at 37° and 44°, implying minor oxidation of the metal substrate interface during the HTS processing step. However, this occurrence seems to have no adverse effect on the superconducting properties. The relative in-plane and out-of-plane textures between individual layers are documented in Figs. 8(a) and (b), respectively. The ω-rocking curve scans (Fig. 8(a)) obtained through the (002) peak reflections for Ni, YSZ, and CeO_2 (004), reflection for Gd_2O_3, and

Table I. Processing Conditions and the Resulting Carbon Contents of Several Dip-Coated Gd_2O_3 Films for a Series of Annealing Temperatures, Annealing Speeds, and Coating Speeds

Temperature (°C)	Coating speed (cm/h)	Annealing speed (cm/h)	Gd_2O_3 thickness (Å)	No. of carbon at. / No. of Gd at.
1050	300	10	730	0.59
1100	300	10	670	0.58
1150	300	10	710	0.37
1175	300	10	730	0.39
1150	20	5	185	0.55
1150	20	20	130	0.49
1150	20	30	250	0.41

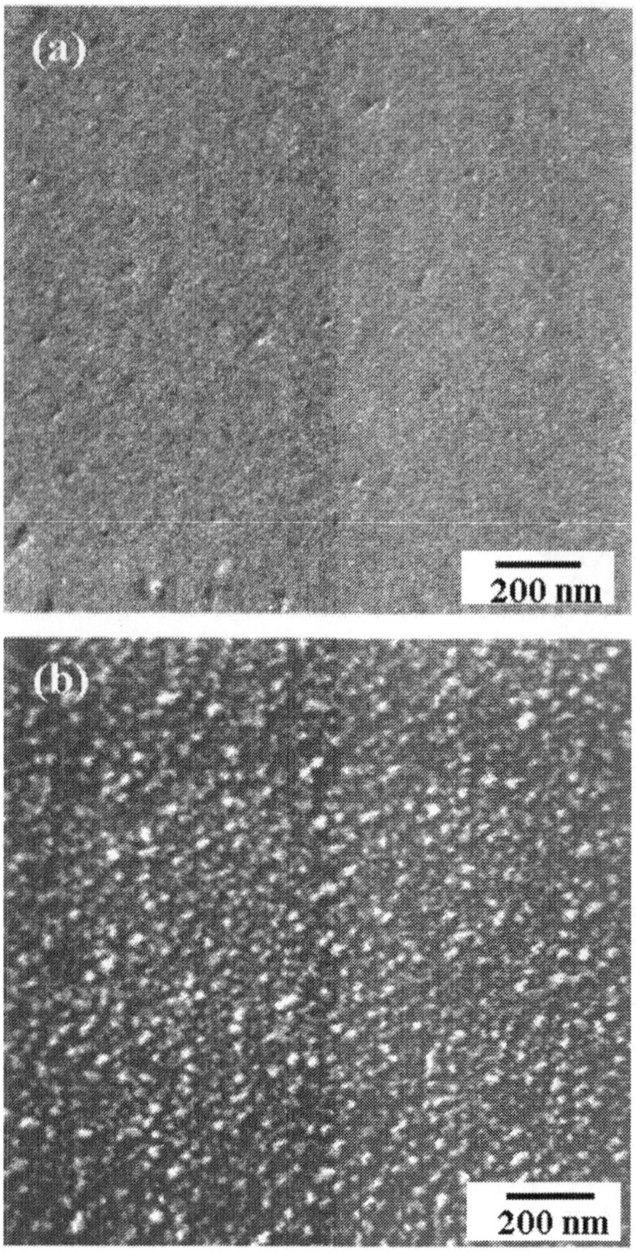

Fig. 6. High-resolution SEM micrographs for (a) 25 nm and (b) 73 nm sol–gel dip-coated Gd_2O_3 seed layers on (100) textured Ni–(3 at.% W–1.7 at.% Fe) substrates.

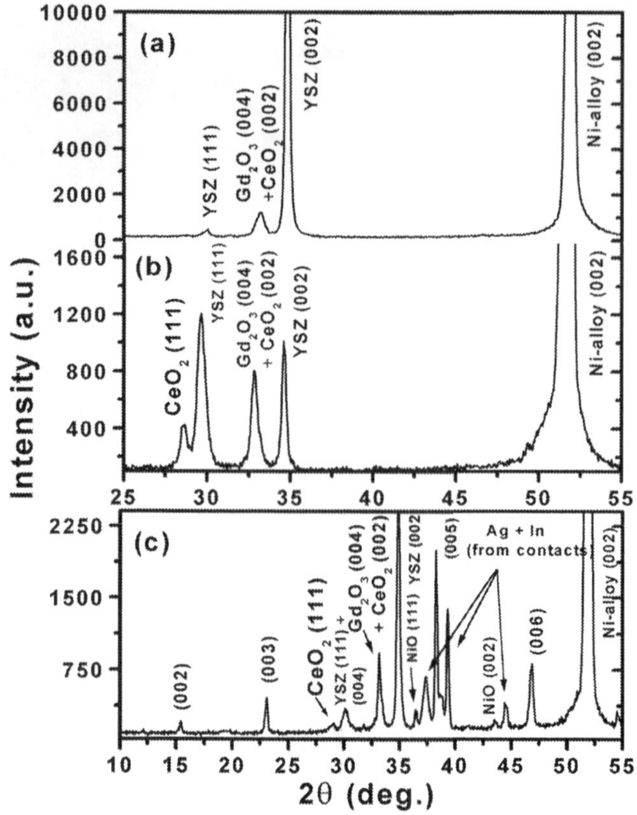

Fig. 7. XRD θ–2θ scans, of the CeO_2/YSZ bilayers on Gd_2O_3/Ni–(3 at.% W–1.7 at.% Fe) substrates, where the Gd_2O_3 layer thickness is about (a) 25 and (b) 73 nm. (c) A typical XRD θ–2θ spectrum for a YBCO film on the CeO_2/YSZ/Gd_2O_3(20 nm)/Ni–(3 at.% W–1.7 at.% Fe) multilayer structure. YBCO peaks are labeled by (00*l*) indexes.

the (006) reflection for YBCO, yield peak-width FWHM values ($\Delta\omega$) of 5.2°, 7.2°, 7.7°, 6.4°, and 5.8°, respectively, indicating good out-of-plane crystallographic alignment among the layers. In-plane textures of the individual layers for the same sample are illustrated in Fig. 8(b), where ϕ-scan data through the YBCO(103), CeO_2(111), YSZ(111), Gd_2O_3(222), and Ni(111) reflections reveal four equally spaced peaks, with similar in-plane FWHM values ($\Delta\phi$) of about 8–9°. These in-plane epitaxial orientations correspond to CeO_2, YSZ, and Gd_2O_3 basal planes that are rotated 45° relative to in-plane Ni-alloy substrate (100) axis, whereas the *a*-axis and *b*-axis of the YBCO are rotated 45° with respect to the top CeO_2 axes. The observed 45° rotations are anticipated according to the lattice match considerations between subsequent layers.

Since the qualifying test for any coated-conductor buffer-layer structure is compatibility with the deposition of high-quality, high-J_c YBCO coatings, electrical transport and superconducting property measurements were conducted on the same sample. Figure 9 compares the magnetic field (*H*//*c*-axis) dependence of J_c at 77 K with those of YBCO films grown on solution derived CeO_2/YSZ/Gd_2O_3(dip-coated)/Ni and on standard, all vacuum deposited, buffer layer RABiTS architecture of CeO_2/YSZ/CeO_2/Ni. The inset of this figure shows the superconducting transition region of the temperature-dependent resistivity, yielding a zero-resistance T_c = 90.5 K. The zero-field transport J_c of the YBCO/CeO_2/YSZ/Gd_2O_3/Ni-alloy sample is 1.2×10^6 A/cm^2 and it exhibits a high irreversibility field (H_{irr}) of 7 T. This J_c performance is comparable to that of epitaxial YBCO films on CeO_2/YSZ/Gd_2O_3(dip-coated) and all-vacuum CeO_2/YSZ/CeO_2 buffered biaxially textured Ni substrates. Thus, the microstructural requirements for high-J_c are imparted in YBCO through the CeO_2/YSZ/Gd_2O_3, including a near absence of weak-linked high-angle grain boundaries and an overall high degree of in- and out-of-plane alignment. The slight differences in high-field J_c behavior of the different samples arise from subtle variations in flux pinning properties, which can be associated with small changes in YBCO deposition conditions, composition, and substrate surface properties.

(3) Long-Length Fabrication of RABiTS Architecture Based on Dip-Coated Gd_2O_3/Ni Tapes

Having realized excellent structure and superconducting properties of YBCO films on short segments of CeO_2/YSZ capped, sol–gel synthesized Gd_2O_3 seed layers, we devoted our efforts to long-length implementation of the same architecture on Ni tapes. Figure 10 plots the XRD ω-rocking curve intensity distributions for the CeO_2(002) + Gd_2O_3(004) and YSZ(002) peak reflections taken along a 1 m length of CeO_2/YSZ/Gd_2O_3(dip-coated) buffered Ni tape. The tape was drawn through the solution at a rate of 80 cm/h, and the film was annealed in a preheated furnace at a speed of 20 cm/h at 1130°C. It should be mentioned that by reducing the concentration of Gd_2O_3 precursor solution from 0.5*M* to 0.25*M*, 80 cm/h coating speed results in a layer thickness of

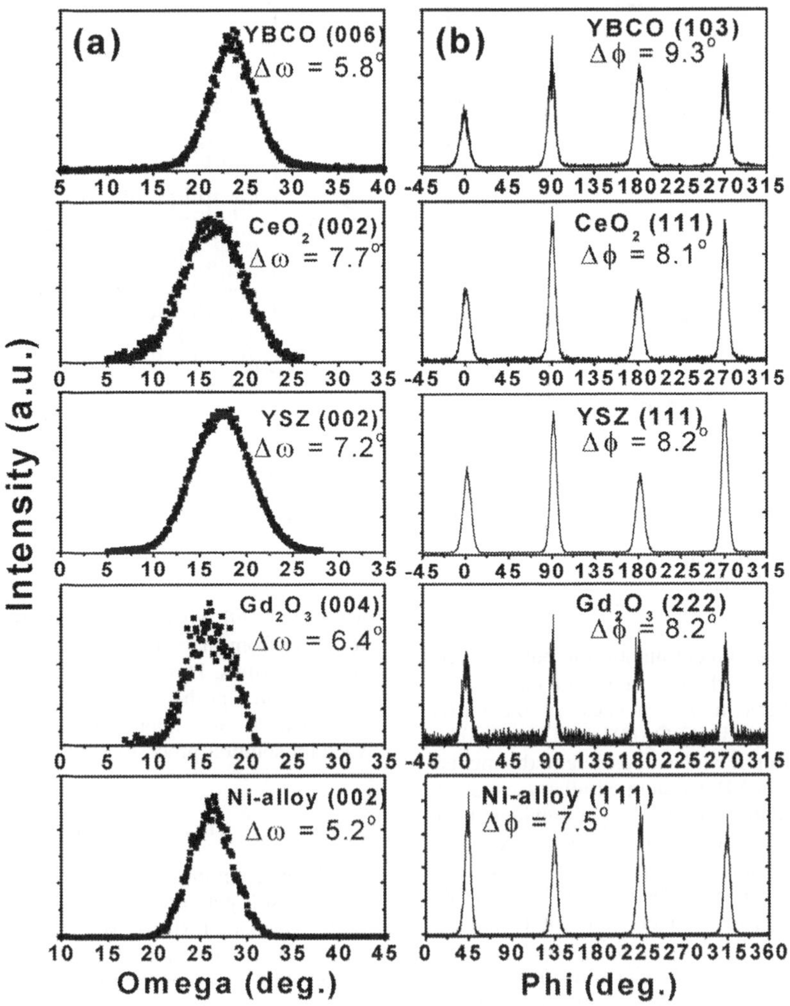

Fig. 8. XRD (a) out-of-plane (ω-scans) and (b) in-plane (ϕ-scans) texture for an epitaxial YBCO film (200 nm thick) on a $CeO_2/YSZ/Gd_2O_3$/Ni–(3 at.% W–1.7 at.% Fe) multilayer structure. The FWHM (Δω and Δϕ) for each scan are indicated.

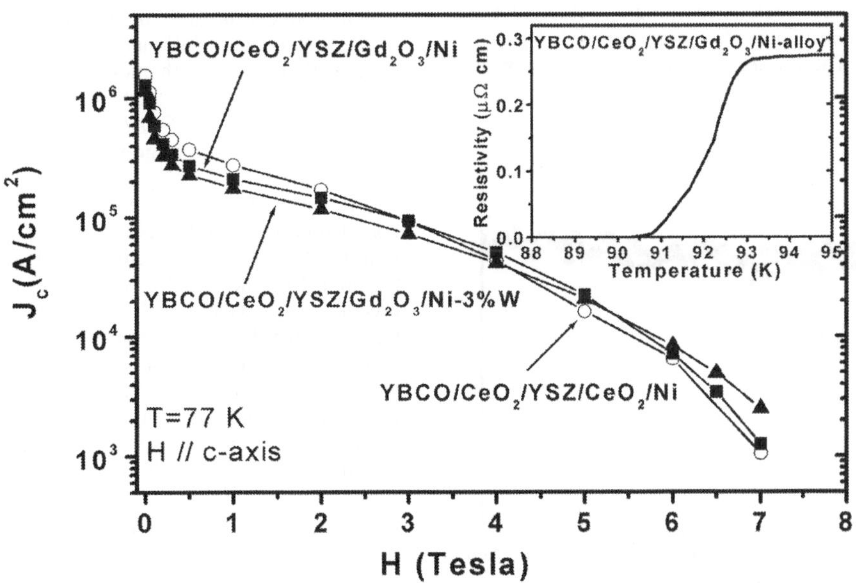

Fig. 9. Magnetic field dependence of J_c, measured at 77 K, for a YBCO film on the $CeO_2/YSZ/Gd_2O_3$/Ni–(3 at.% W–1.7 at.% Fe) multilayer structure. Also shown for comparison are J_c (*H*, 77 K) for the YBCO/$CeO_2/YSZ/Gd_2O_3$/Ni and YBCO/CeO_2/YSZ/CeO_2/Ni. The inset shows the resistive superconducting transition region of the present sample.

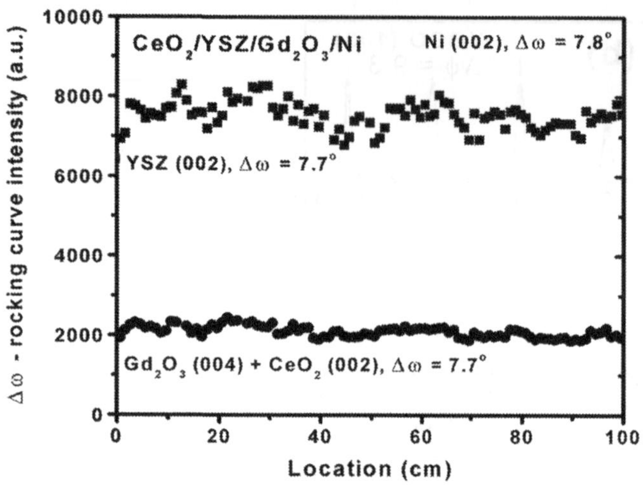

Fig. 10. XRD ω-rocking curve intensity distributions for the $CeO_2(002)$ + $Gd_2O_3(004)$ and the YSZ(002) peak reflections as a function of the length over a 1 m vacuum and nonvacuum processed CeO_2/YSZ/Gd_2O_3 buffer layers on biaxially textured Ni tape.

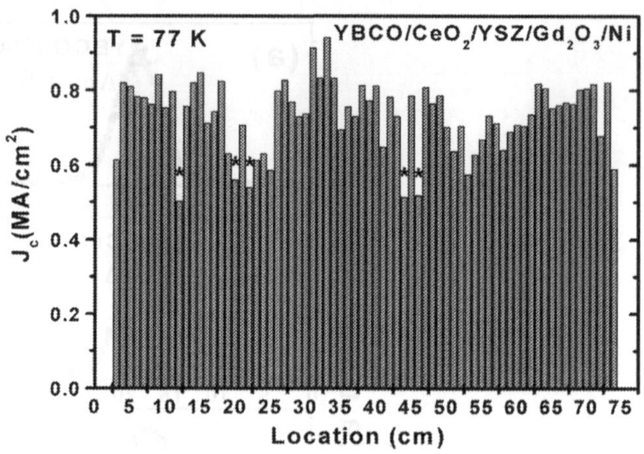

Fig. 12. Sectional J_c (77 K, self-field) values for the YBCO film obtained on 1 m long sol–gel seeded CeO_2/YSZ/Gd_2O_3/Ni tape. Visible line features marked by ★.

~20–25 nm. Uniform *c*-axis alignment among the buffer layers can be seen from this figure. Slight changes in the ω-rocking curve intensities are due to variations in the thickness of the buffer layers. Since the Bragg angles for the $CeO_2(002)$ and $Gd_2O_3(004)$ reflections are nearly coincident, it was not possible to distinguish them within the resolution of our XRD system. Similarly, excellent uniformity for the in-plane texture of the YSZ layer is evident from the plot (Fig. 11) of XRD ϕ-scan intensity distribution recorded from the YSZ(111) reflection along the tape length. The ω- and ϕ-scan FWHM values of 7.7° and 8.7° obtained for CeO_2–YSZ–Gd_2O_3 buffers and YSZ layer, respectively, are comparable to those of the Ni substrate ($\Delta\omega = 7.8°$, $\Delta\phi = 8.5°$). The quality of this tape was examined by depositing YBCO films using the *ex situ* BaF_2 precursor process in a reel-to-reel system. Details of this experimental procedure were reported earlier.[34,35] The results of electrical transport J_c measurements at 77 K and zero applied field, conducted over a 80 cm long section of the same YBCO/CeO_2/YSZ/Gd_2O_3/Ni tape, are shown in Fig. 12. The sectional J_c values were taken at each 1-cm increment along the tape length. The tape width is 1 cm and the thickness of the YBCO coating is 300 nm. It is clear from the figure that the sample exhibits an average J_c value of 7.5×10^5 A/cm^2 and it supports an end-to-end J_c of 6.25×10^5 A/cm^2, with some regions reaching J_c values of 1×10^6 A/cm^2. The latter value represents the best-recorded performance on short samples. The low J_c sections marked with ★ are associated with visible line features along the YBCO thickness, resulting from the variations in the composition of YBCO(BaF_2) precursor films. Despite the existence of these flaws, the present results represent the first reported achievement of high-J_c YBCO performance on a long length RABiTS structure having a non-vacuum solution processed seed layer.

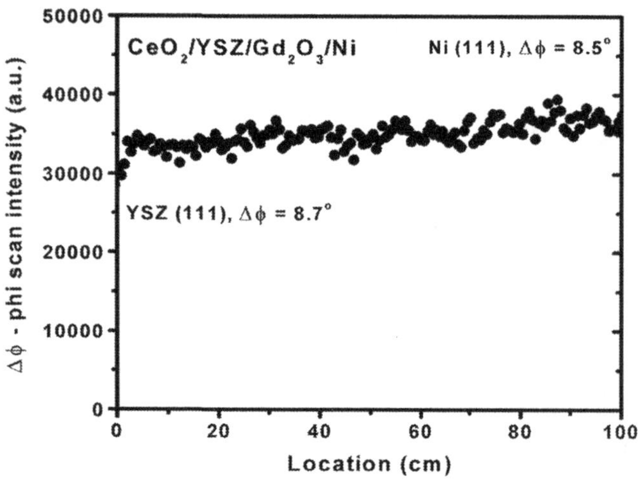

Fig. 11. Longitudinal distribution of XRD ϕ-scan intensity of YSZ(111) peak reflection for the same 1 m tape, shown in Fig. 10.

IV. Summary and Conclusions

We have demonstrated the fabrication of high-J_c YBCO coatings on sol–gel processed, reel-to-reel dip-coated Gd_2O_3 seed buffer layers on textured, strengthened Ni and Ni-alloy substrates, with sputtered CeO_2/YSZ overlayers. High-quality Gd_2O_3 layers were achieved at temperatures between 1100° and 1150°C under reducing forming gas (96% Ar + 4% H_2) atmosphere. It has been found that the postannealing throughput speed (S_{an}) does not have a significant affect on the in-plane alignment of Gd_2O_3. However, the out-of-plane alignment has a tendency to degrade for S_{an} exceeding 20 cm/h. Whether these findings will limit processing speeds requires further study of the independent affects of ramp rate and residence time in the furnace. The surface roughness of Gd_2O_3 films was found to depend strongly on the thickness of the layers by the combination of coating speed and solution viscosity, which produced notable differences in the growth characteristics of the subsequent YSZ and CeO_2 barrier and cap layers. The combinations of smooth morphology and crystalline cap buffer layers on thin Gd_2O_3 yielded YBCO films with self-field J_c values as high as 1.2×10^6 A/cm^2 at 77 K. The magnetic field performance of J_c was as good as that obtained for YBCO films on standard CeO_2/YSZ/CeO_2 buffered Ni substrates.

After establishing this proof of principle on short prototype conductors, we have presented our results on a length demonstration (100 cm) of dip-coated Gd_2O_3 seed layers on biaxially textured Ni. The YBCO films grown using the *ex situ* BaF_2 precursor approach on this nonvacuum and vacuum derived CeO_2(sputtered)/YSZ(sputtered)/Gd_2O_3(dip-coated)/Ni multilayer structure supported a mean self-field J_c value of 7.5×10^5 A/cm^2 at 77 K with some regions reaching 1×10^6 A/cm^2. These observations underscore the potential viability of nonvacuum chemical solution-based synthesis routes for the development and manufacturing of RABiTS-based coated conductors.

References

[1]A. Goyal, D. P. Norton, J. D. Budai, M. Paranthaman, E. D. Specht, D. M. Kroeger, D. K. Christen, Q. He, B. Saffian, F. A. List, D. F. Lee, P. M. Martin, C. E. Klabunde, E. Hatfield, and V. K. Sikka, "High Critical Current Density Superconducting Tapes by Epitaxial Deposition of $YBa_2Cu_3O_x$ Thick Films on Biaxially Textured Metals," *Appl. Phys. Lett.*, **69**, 1795 (1996).

[2]D. P. Norton, A. Goyal, J. D. Budai, D. K. Christen, D. M. Kroeger, E. D. Specht, Q. He, B. Saffian, M. Paranthaman, C. E. Klabunde, D. F. Lee, B. C. Sales, and F. A. List, "Epitaxial $YBa_2Cu_3O_7$ on Biaxially Textured Nickel (001): An Approach to Superconducting Tapes with High Critical Current Density," *Science (Washington D.C.)*., **274**, 755 (1996).

[3]C. M. Carlson, J. C. Price, P. A. Parilla, D. S. Ginley, D. Niles, R. D. Blaugher, A Goyal, M. Paranthaman, D. M. Kroeger, and D. K. Christen, "Laser-Ablated Epitaxial $LaAlO_3$ Buffer Layers on Biaxially Textured Ni Substrates for Superconducting Tapes," *Physica C*, **304**, 82 (1998).

[4]M. Paranthaman, D. F. Lee, A. Goyal, E. D. Specht, P. M. Martin, X. Cui, J. E. Mathis, R. Feenstra, D. K. Christen, and D. M. Kroeger, "Growth of Biaxially Textured RE_2O_3 Buffer Layers on Rolled-Ni Substrates Using Reactive Evaporation for HTS-Coated Conductors," *Supercond. Sci. Technol.*, **12**, 319 (1999).

[5]C. Cantoni, T. Aytug, D. T. Verebelyi, M. Paranthaman, E. D. Specht, D. P. Norton, and D. K. Christen, "Conductive Buffer Layers and Overlayers for the Electrical Stability of Coated Conductors," *IEEE Trans. Appl. Supercond.*, **11**, 3309 (2001).

[6]Q. He, D. K. Christen, R. Feenstra, D. P. Norton, M. Paranthaman, E. D. Specht, D. F. Lee, A. Goyal, and D. M. Kroeger, "Growth of Biaxially Oriented Conductive $LaNiO_3$ Buffer Layers on Textured Ni Tapes for High-T_c-Coated Conductors," *Physica C*, **314**, 105 (1999).

[7]T. Aytug, J. Z. Wu, C. Cantoni, D. T. Verebelyi, E. D. Specht, M. Paranthaman, D. P. Norton, D. K. Christen, R. E. Ericson, and C. L. Thomas, "Growth and Superconducting Properties of $YBa_2Cu_3O_{7-\delta}$ Films on Conductive $SrRuO_3$ and $LaNiO_3$ Multilayers for Coated Conductor Applications," *Appl. Phys. Lett.*, **76**, 760 (2000).

[8]T. Aytug, J. Z. Wu, B. W. Kang, D. T. Verebelyi, C. Cantoni, E. D. Specht, A. Goyal, and M. Paranthaman, "An All-Sputtered Buffer Layer Architecture for High-J_c $YBa_2Cu_3O_{7-\delta}$ Coated Conductors," *Physica C*, **340**, 33 (2000).

[9]T. Aytug, B. W. Kang, C. Cantoni, E. D. Specht, M. Paranthaman, A. Goyal, D. K. Christen, D. T. Verebelyi, J. Z. Wu, R. E. Ericson, C. L. Thomas, C.-Y. Yang and S. E. Babcock, "Growth and Characterization of Conductive $SrRuO_3$ and $LaNiO_3$ Multilayers on Textured Ni Tapes for High-J_c $YBa_2Cu_3O_{7-\delta}$ Coated Conductors," *J. Mater. Res.*, **16**, 2661 (2001).

[10]M. Paranthaman, A. Goyal, F. A. List, E. D. Specht, D. F. Lee, P. M. Martin, Q. He, D. K. Christen, D. P. Norton, J. D. Budai, and D. M. Kroeger, "Growth of Biaxially Textured Buffer Layers on Rolled-Ni Substrates by Electron Beam Evaporation," *Physica C*, **275**, 266 (1997).

[11]J. E. Mathis, A. Goyal, D. F. Lee, F. A. List, M. Paranthaman, D. K. Christen, E. D. Specht, D. M. Kroeger, and P. M. Martin, "Biaxially Textured Y-123 Conductors on RABiTS with Critical Current Densities of 2–3 MA/cm^2," *Jpn. J. Appl. Phys.*, **37**, L1379 (1998).

[14]C. Park, D. P. Norton, D. K. Christen, D. T. Verebelyi, R. Feenstra, J. D. Budai, A. Goyal, D. F. Lee, E. D. Specht, D. M. Kroeger, and M. Paranthaman, "Long Length Fabrication of YBCO on Rolling Assisted Biaxially Textured Substrates (RABiTS) Using Pulsed Laser Deposition," *IEEE Trans. Appl. Supercond.*, **9**, 2276 (1999).

[13]F. A. List, A. Goyal, M. Paranthaman, D. P. Norton, E. D. Specht, D. F. Lee, and D. M. Kroeger, "High J_c YBCO Films on Biaxially Textured Ni with Oxide Buffer Layers Deposited Using Electron Beam Evaporation and Sputtering," *Physica C*, **302**, 87 (1998).

[14]V. Boffa, C. Annino, D. Bettinelli, S. Ceresara, L. Ciontea, F. Fabbri, V. Galluzzi, U. Gambardella, G. Celentano, G. Grimaldi, A. Mancini, T. Petrisor, and P. Scardi, "Epitaxial growth of Heterostructures on Biaxially Textured Metallic Substrates for $YBa_2Cu_3O_{7-x}$ Tape Fabrication," *Philos. Mag. B*, **80**, 979 (2000).

[15]T. J. Jackson, B. A. Glowacki, and G. J. Schmitz, "Oxidation Thermodynamics of Metal Substrates during the Deposition of Buffer Layer Oxides," *Physica C*, **296**, 215 (1998).

[16]S. Oh, J. Yoo, K. Lee, J. Kim, and D. Youm, "Comparative Study on the Crack Formations in the CeO_2 Buffer Layers for YBCO Films on Textured Ni Tapes and Pt Tapes," *Physica C*, **308**, 91 (1998).

[17]X. Cui, F. A. List, D. M. Kroeger, A. Goyal, D. F. Lee, J. E. Mathis, E. D. Specht, P. M. Martin, R. Feenstra, D. T. Verebelyi, D. K. Christen, and M. Paranthaman, "Continuous Growth of Epitaxial CeO_2 Buffer Layers on Rolled Ni Tapes by Electron Beam Evaporation," *Physica C*, **316**, 27 (1999).

[18]E. Celik, I. H. Mutlu, and Y. S. Hascicek, "Sol–Gel Processing of Buffer Layers on Ni Tape for YBCO Surface Coated Conductors," *IEEE Trans. Appl. Supercond.*, **10**, 1162 (2000).

[19]S. Sathyamurthy and K. Salama, "Chemical Solution Deposition of Highly Oriented Strontium Titanate Buffer Layers for Coated Conductors," *Supercond. Sci. Technol.*, **13**, L1 (2000).

[20]J. C. Schmidt, A. Tiggers, and G. J. Schmitz, "Synthesis of Polycrystalline $BaZrO_3$ Coatings," *Mater. Sci. Eng., B*, **53**, 115 (1998).

[21]T. A. Gladstone, J. C. Moore, A. J. Wilkinson, and C. M. R. Grovenor, "Fabrication of Biaxially Textured Ni Substrates and $LaNiO_3$ Buffer Layers for Tl-1223 Thick Films," *IEEE Trans. Appl. Supercond.*, **9**, 2252 (1999).

[22]J. S. Morrell, Z. B. Xue, E. D. Specht, A. Goyal, P. M. Martin, D. F. Lee, R. Feenstra, D. T. Verebelyi, D. K. Christen, T. G. Chirayil, M. Paranthaman, C. E. Valet, and D. B. Beach, "Epitaxial Growth of Gadolinium Oxide on Roll-Textured Nickel Using a Solution Growth Technique," *J. Mater. Res.*, **15**, 621 (2000).

[23]M. Paranthaman, T. G. Chirayil, F. List, X. Cui, A. Goyal, D. F. Lee, E. D. Specht, P. M. Martin, R. K. Williams, D. M. Kroeger, J. S. Morrell, D. B. Beach, R. Feenstra, and D. K. Christen, "Fabrication of Long Lengths of Epitaxial Buffer Layers on Biaxially Textured Nickel Substrates Using a Continuous Reel-to-Reel Dip-Coating Unit," *J. Am. Ceram. Soc.*, **84**, 273 (2001).

[24]T. G. Chirayil, M. Paranthaman, D. B. Beach, D. F. Lee, A. Goyal, R. K. Williams, X. Cui, D. M. Kroeger, D. T. Verebelyi, and D. K. Christen, "Epitaxial Growth of $La_2Zr_2O_7$ Thin Films on Rolled-Ni Substrates by Sol–Gel Process for High T_c Superconducting Tapes," *Physica C*, **336**, 63 (2000).

[25]D. B. Beach, C. E. Valet, M. Paranthaman, E. D. Specht, J. S. Morrell, and Z. B. Xue, "Sol–Gel Synthesis of Rare Earth Aluminate Films as Buffer Layers for High T_c Superconducting Films," *Mater. Res. Soc. Symp. Proc.*, **495**, 263 (1998).

[26]M. W. Rupich, W. Palm, W. Zhang, E. Siegal, S. Annavarapu, L. Fritzemeier, M. D. Teplitsky, C. Thieme, and M. Paranthaman, "Growth and Characterization of Oxide Buffer Layers for YBCO Coated Conductors," *IEEE Trans. Appl. Supercond.*, **9**, 1527 (1999).

[27]T. G. Chirayil, M. Paranthaman, D. B. Beach, J. S. Morrell, E. Y. Sun, A. Goyal, R. K. Williams, D. F. Lee, P. M. Martin, D. M. Kroeger, R. Feenstra, D. T. Verebelyi, and D. K. Christen, "Epitaxial Growth of Yb_2O_3 Buffer Layers on Biaxially Textured-Ni (100) Substrates by Sol–Gel Process," *Mater. Res. Soc. Symp. Proc.*, **574**, 51 (1999).

[28]M. Paranthaman, T. G. Chirayil, S. Sathyamurthy, D. B. Beach, A. Goyal, F. A. List, D. F. Lee, X. Cui, S. W. Lu, B. W. Kang, E. D. Specht, P. M. Martin, D. M. Kroeger, R. Feenstra, C. Cantoni, and D. K. Christen, "Fabrication of Long Lengths of YBCO Coated Conductors Using a Continuous Reel-to-Reel Dip-Coating Unit," *IEEE Trans. Appl. Supercond.*, **11**, 3146 (2000).

[29]A. Goyal, R. Feenstra, M. Paranthaman, J. R. Thompson, B. W. Kang, C. Cantoni, D. F. Le, F. A. List, P. M. Martin, E. Lara-curio, C. Stevens, D. M. Kroeger, M. Kowalewski, E. D. Specht, T. Aytug, S. Sathyamurthy, R. K. Williams, and R. E. Ericson, "Strengthened, Biaxially Textured Ni Substrate with Small Alloying Additions for Coated Conductor Applications," *Physica C*, **382**, 251 (2002).

[30]W. Prusseit, R. Nemetschek, R. Semerad, B. Holzaphel, and J. Erickemeyer, "YBCO and Buffer Layer Deposition on Long Length Metal Tape Substrates by Thermal Evaporation"; p. 87 in Program and Extended Abstracts of the 5th International Workshop on Superconductivity Co-sponsored by ISTEC and MRS (Honolulu, HI, 2001). MRS, Boston, MA, 2001.

[31]M. Paranthaman, S. Sathyamurthy, T. Aytug, D. B. Beach, A. Goyal, B. W. Kang, R. Feenstra, D. K. Christen, D. F. Lee, P. M. Martin, L. Heatherly Jr., and D. M. Kroeger, "High Current Density YBCO Coated Conductors on Strengthened Biaxially Textured Ni–W Substrates"; p. 87 in Program and Extended Abstracts of the 5th International Workshop on Superconductivity Co-sponsored by ISTEC and MRS (Honolulu, HI, 2001). MRS, Boston, MA, 2001.

[32]H. R. Hoekstra, "Phase Relationships in the Rare Earth Sesquioxides at High Pressure," *Inorg. Chem.*, **5**, 754 (1966).

[33]M. Paranthaman, R. Feenstra, D. F. Lee, D. B. Beach, J. S. Morrell, T. G. Chirayil, A. Goyal, X. Cui, D. T. Verebelyi, J. E. Mathis, P. M. Martin, D. P. Norton, E. D. Specht, D. K. Christen, and D. M. Kroeger, "Demonstration of High Current Density YBCO Coated Conductors on RE_2O_3-Buffered Ni Substrates with Two New Alternative Architectures," *Adv. Cryog. Eng. Mater.*, **46**, 879 (2000).

[34]D. F. Lee, F. A. List, D. M. Kroeger, K. W. Childs, D. O'Neill, and W. B. Robbins, "Long-Length Conversion of Barium Fluoride Precursor Films on RABiTS"; p. 73 in Program and Extended Abstracts of the 5th International Workshop on Superconductivity Co-sponsored by ISTEC and MRS (Honolulu, HI, 2001). MRS, Boston, MA, 2001.

[35]S. W. Lu, F. A. List, D. F. Lee, X. Cui, M. Paranthaman, B. W. Kang, D. M. Kroeger, A. Goyal, P. M. Martin, and R. E. Ericson, "Electron Beam Co-evaporation of $Y–BaF_2–Cu$ Precursor Films for $YBa_2Cu_3O_{7-\delta}$ Coated Conductors," *Supercond. Sci. Technol.*, **14**, 218 (2001). □

journal

J. Am. Ceram. Soc., 86 [10] 1800–802 (2003)

Preparation of Dense MgB_2 Bulk Superconductors by Spark Plasma Sintering

Soo-Yong Lee, Sang-Im Yoo, Young-Woon Kim, Nong M. Hwang, and Doh-Yeon Kim*

School of Materials Science and Engineering and Center for Microstructure Science of Materials, Seoul National University, Seoul 151-744, Korea

Fully dense MgB_2 bulk specimens (~higher than 99% dense) were prepared using spark plasma sintering (SPS) at 1250°C for 15 min. Microstructure analyses revealed that faceted MgO particles of ~8% volume fraction were dispersed in the MgB_2 matrix. A sharp superconducting transition with an onset temperature of 38.5 K was confirmed by both magnetization and resistivity measurements.

I. Introduction

SINCE the recent discovery of superconductivity at 39 K in MgB_2,[1] this material has been investigated by many researchers. MgB_2 bulk specimens were usually prepared from a mixture of Mg and B in a sealed Ta tube with excess Mg, but these specimens were very porous and mechanically weak.[2,3] The sintered MgB_2 specimens also showed a porous spongelike microstructure due to the high volatility of Mg at elevated temperatures.[4–6] Thus, a processing method to achieve a fully dense MgB_2 bulk specimen has been sought for various purposes. Until now, sintering under an extremely high pressure such as several gigapascals[3,5,7] or hot isostatic pressing at 100 MPa[8] has been effective for the densification of MgB_2 bulk specimens. Otherwise, as reported by Zhao *et al.*,[9] it is necessary to use a sintering additive such as Ti for the densification of the MgB_2 bulk at ambient pressure. Except for an MgB_2 wire reported by Canfield *et al.*,[10] a fully dense bulk specimen has rarely been achieved.

On the other hand, spark plasma sintering (SPS) has been introduced recently and widely used to make fully dense and fine-grained ceramic materials.[11–14] Although the mechanism of enhanced densification has not yet been clearly understood, the process itself can be described simply; during sintering of the powder compact under a relatively low pressure (around 30 MPa), a pulsed direct current, which is expected to generate a spark discharge at the voids between the particles, is applied to the compact. Through the SPS process, a relatively low sintering temperature and a short holding time are generally sufficient to achieve a full densification. Therefore, compared with the normal sintering process at relatively high temperature, which results in a serious Mg evaporation, the SPS process is expected to be advantageous for the densification of MgB_2. We therefore tried the SPS process to prepare the highly dense MgB_2 bulk specimens. Microstructure and superconducting transition behavior of the SPS-processed MgB_2 specimen were also checked.

S. M. Mukhopadhyay—contributing editor

Manuscript No. 186761. Received August 6, 2002; approved April 28, 2003.
This work was supported by the Korean Science and Engineering Foundation (KOSEF) through the Center for Strongly Correlated Materials Research (CSCMR) (2002) and by the Korean Ministry of Science and Technology through the Creative Research Initiatives.

*Member, American Ceramic Society.

II. Experimental Procedure

The starting material was a commercial MgB_2 powder (Testbourne, Ltd., Hampshire, U.K.) for which the manufacturer claimed 99.5% purity and −100 mesh in particle size. The main impurities determined by spectroscopy were reported to be C (~3000 ppm) and Fe (<1000 ppm). The powders were placed in a graphite die (20 mm in diameter) and heated to 1250°C under a vacuum of 2.7 Pa. Then, uniaxial pressure of 30 MPa was applied. The applied electric current during sintering was 1200 A. After sintering at 1060°C for 15 min, the electric current flow was stopped and the pressure was released. This sintering condition was determined to be most effective from the preliminary experiments.

After polishing the MgB_2 specimens obtained, the porosity level was checked by the point counting method. Phase identification of the specimen was conducted with powder X-ray diffraction (XRD). The fracture surface of the specimen was observed through scanning electron microscopy (SEM), and transmission electron microscopy (TEM) was also used for microstructure analysis. Magnetization measurements were performed with a Quantum Design MPMS SQUID magnetometer. The dc resistivity as a function of temperature was measured with the standard four-probe method. For electrodes, a silver epoxy was pasted on the specimen and hardened at 150°C in air.

III. Results and Discussion

Figure 1 shows XRD patterns obtained from the MgB_2 starting powders and the SPS-processed specimen. As can be observed, major peaks in both cases were from the MgB_2 phase. However, minor peaks due to the MgO phase were also observed, which is likely to be due to the surface layer of the MgB_2 particles. Compared with the initial powders, it appears that the amount of MgO in the sintered specimen has slightly increased. On the other hand, a full densification of MgB_2 was achieved through the SPS. The porosity level of the sintered specimen was less than 0.5%. Figures 2(a) and (b) show the fractured and polished surface of the specimen, respectively. Except for a few residual spherical pores trapped inside the MgB_2 grain (Fig. 2(a)) and intergranular pores (Fig. 2(b)), the pores were rarely observed throughout the entire microstructure.

TEM observation has shown that MgO particles are present within the MgB_2 grains and also at the grain boundaries (Fig. 3). The MgO crystal phase was confirmed by using both energy dispersive X-ray spectroscopy and selected area electron diffraction. MgO crystals were highly faceted along (100) crystallographic planes. The average size of MgO particles and that of MgB_2 grains were determined to be 400 ± 180 and 900 ± 440 nm, respectively. Throughout the specimen, MgB_2 grain boundaries were clean and other second phases such as MgB_4 or Mg were not observed. Note that these second phases are usually present in the MgB_2 bulk specimens prepared by other fabrication processes.[8,15]

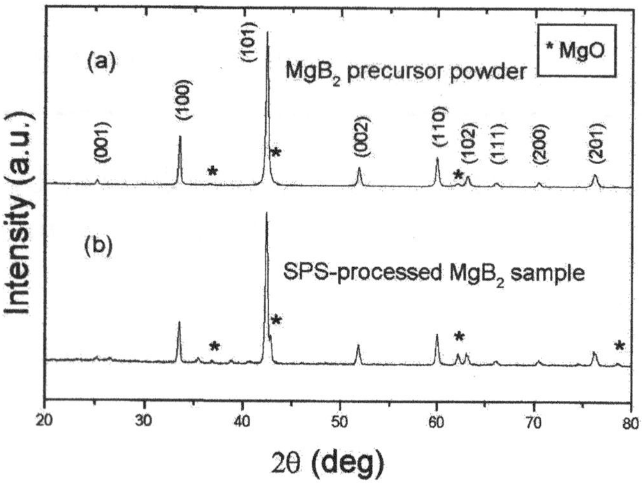

Fig. 1. XRD patterns of (a) the iniital MgB_2 powder and (b) the SPS-processed MgB_2 specimen.

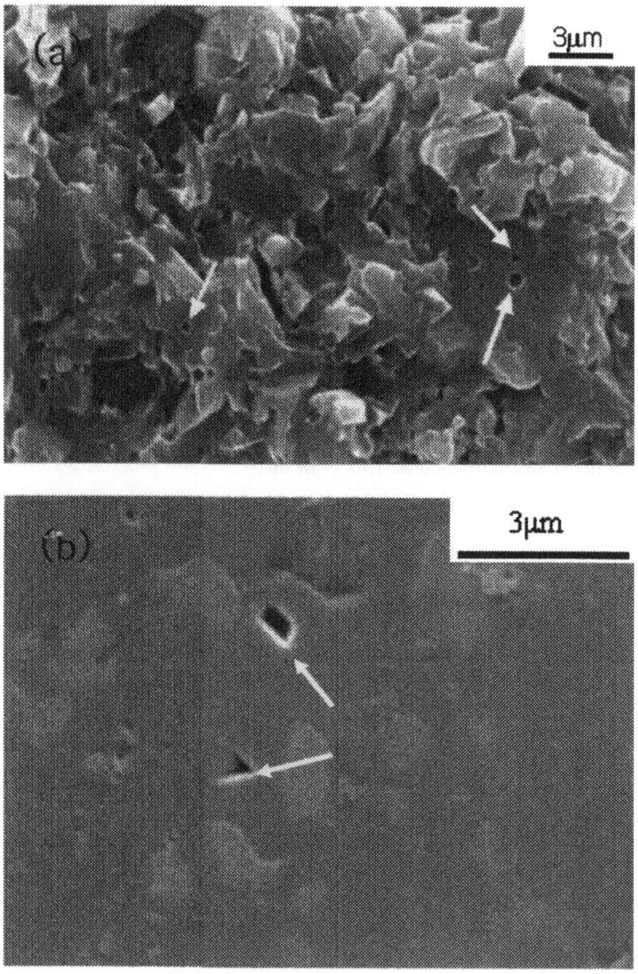

Fig. 2. Micrograph of (a) the polished and (b) the fractured surface of the MgB_2 specimen. The arrows indicate the pores.

The superconducting transition behaviors of the specimen are shown in Fig. 4. A rectangular MgB_2 specimen with the dimensions of $1.10 \times 1.25 \times 4.90$ mm^3 was used. The magnetization curves in Fig. 4(a) were measured following zero-field-cooling (ZFC) and field-cooling (FC) procedures. A field of 20 Oe was applied parallel to the long axis of the specimen. A sharp superconducting transition with an onset T_c of 38.5 K and a

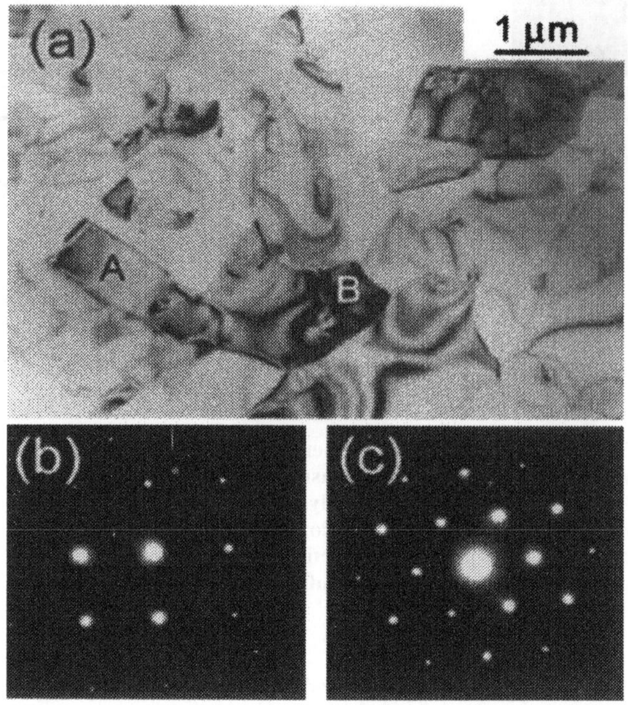

Fig. 3. (a) Bright-field image of the SPS-processed MgB_2 specimen. Areas marked as "A" and "B" are MgO and MgB_2, respectively. (b) and (c) are selected area diffraction patterns of areas A and B, respectively.

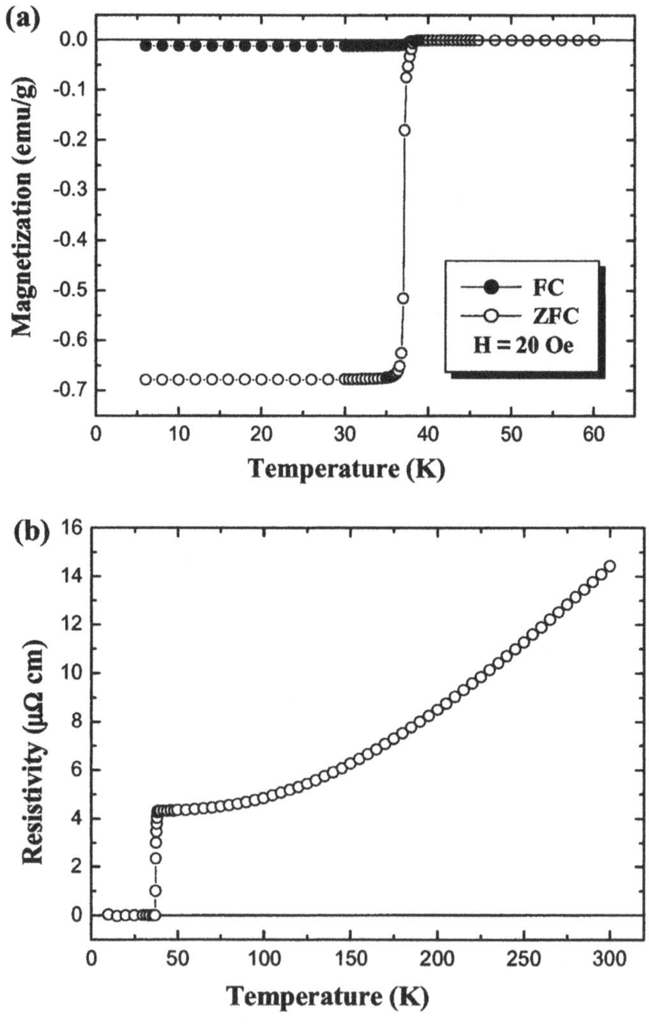

Fig. 4. (a) Magnetization curve and (b) resistivity curve of the SPS-processed MgB_2 specimen.

transition width of 0.5 K from a 10% to 90% drop can be noted. A slight depression of the onset T_c is likely to be due to the small amount of impurities included in the initial powder. Compared with the ZFC curve, a much smaller diamagnetic signal (~2% of the ZFC signal at 10 K) in the FC curve represents a strong flux pinning. Fine MgO particles and MgB_2 grain boundaries shown in Fig. 4 are believed to act as flux pinning sites. The enhancement of flux pinning is anticipated through the optimization of the process.

The variation of resistivity of the MgB_2 specimen as a function of temperature is displayed in Fig. 4(b). For this measurement, the electric current of 30 mA, corresponding to the current density of ~2.2 A/cm^2, was applied through the specimen. As shown in the figure, onset T_c and zero resistivity occur at 38.5 K and 37 K, respectively. A slightly broader transition width of 1 K compared with Fig. 4(a) was observed. This is probably due to a somewhat higher applied current density. The resistivity values of the specimen at 40 and 300 K were 4.3 and 14.4 μΩ·cm, respectively, leading to a residual resistivity (RR) ratio of ~3.4. At the normal state from 300 K to T_c, it is noted that the resistivity values are lower compared with those previously reported.[2–5,16–18] These low-resistivity values are expected to be the result of a fully dense microstructure with clean grain boundaries.

IV. Summary

Fully dense MgB_2 bulk specimens can be obtained by the SPS process. The specimen prepared showed a sharp superconducting transition with an onset T_c of 38.5 K. Fully dense specimens showed relatively low resistivity values at the normal state.

References

[1]J. Nagamatsu, N. Nakagawa, T. Muranaka, Y. Zenitani, and J. Akimitsu, "Superconductivity at 39 K in Magnesium Diboride," *Nature (London)*, **410**, 63–64 (2001).

[2]S. L. Bud'ko, G. Lapertot, C. Petrovic, C. E. Cunningham, N. Anderson, and P. C. Canfield, "Boron Isotope Effect in Superconducting MgB_2," *Phys. Rev. Lett.*, **86** [9] 1877–80 (2001).

[3]C. U. Jung, M.-S. Park, W. N. Kang, M.-S. Kim, K. H. P. Kim, S. Y. Lee, and S.-I. Lee, "Effect of Sintering Temperature under High Pressure on the Superconductivity of MgB_2," *Appl. Phys. Lett.*, **78** [26] 4157–59 (2001).

[4]A. Handstein, D. Hinz, G. Fuchs, K.-H Muller, K. Nenkov, O. Gutfleisch, and V. N. Narozhnyi, "Fully Dense MgB_2 Superconductor Textured by Hot Deformation," *J. Alloys Compds.*, **329**, 285–89 (2001).

[5]Y. Takano, H. Takeya, H. Fujii, H. Kumakura, T. Hatano, K. Togano, H. Kito, and H. Ihara, "Superconducting Properties of MgB_2 Bulk Materials Prepared by High-Pressure Sintering," *Appl. Phys. Lett.*, **78** [19] 2914–16 (2001).

[6]A. Serquis, Y. T. Zhu, E. J. Peterson, J. Y. Coulter, D. E. Peterson, and F. M. Mueller, "Effect of Lattice Strain and Defects on the Superconductivity of MgB_2," *Appl. Phys. Lett.*, **79** [26] 4399–401 (2001).

[7]D. C. Larbalestier, L. D. Cooley, M. O. Rikel. A. A. Polyanskii, J. Jiang, S. Patnaik, X. Y. Cai, D. M. Feldmann, A. Gurevich, A. A. Squitieri, M. T. Naus, C. B. Eom, E. E. Hellstrom, R. J. Cava, K. A. Regan, N. Rogado, M. A. Hayward, T. He, J. S. Slusky, P. Khalifah, K. Inumaru, and M. Haas, "Strongly Linked Current Flow in Polycrystalline Forms of the Superconductor MgB_2," *Nature (London)*, **410**, 186–89 (2001).

[8]T. C. Shields, K. Kawano, D. Holdom, and J. S. Abell, "Microstructure and Superconducting Properties of Hot Isostatically Pressed MgB_2," *Supercond. Sci. Technol.*, **15**, 202–205 (2002).

[9]Y. Zhao, Y. Feng, C. H. Cheng, L. Zhou, Y. Wu, T. Machi, Y. Fudamoto, N. Koshizuka, and M. Murakami, "High Critical Current Density of MgB_2 Bulk Superconductor Doped with Ti and Sintered at Ambient Pressure," *Appl. Phys. Lett.*, **79** [8] 1154–56 (2001).

[10]P. C. Canfield, D. K. Finnemore, S. L. Bud'ko, J. E. Ostenson, G. Lapertot, C. E. Cunningham, and C. Petrovic, "Superconductivity in Dense MgB_2 Wires," *Phys. Rev. Lett.*, **86** [11] 2423–26 (2001).

[11]J.-K. Park, U.-I. Chung, N. M. Hwang, and D.-Y. Kim, "Preparation of Dense PMN-PT Ceramics by Spark Plasma Sintering," *J. Am. Ceram. Soc.*, **84** [12] 3057–59 (2001).

[12]D. S. Perera, M. Tokita, and S. Morricca, "Comparative Study of Fabrication of Si_3N_4/SiC Composites by Spark Plasma Sintering and Hot Isostatic Pressing," *J. Eur. Ceram. Soc.*, **18**, 401–404 (1998).

[13]E. M. Carrillo-Heian, C. Unuvar, J. C. Gibeling, G. H. Paulino, and Z. A. Munir, "Simultaneous Synthesis and Densification of Niobium Silicide/Niobium Composites," *Scr. Mater.*, **45**, 405–12 (2001).

[14]J. Hong, L. Gao, S. D. D. L. Torre, H. Miyamoto, and K. Miyamoto, "Spark Plasma Sintering and Mechanical Properties of $ZrO_2(Y_2O_3)$–Al_2O_3 Composites," *Mater. Lett.*, **43**, 27–31 (2000).

[15]N. Rogado, M. A. Hayward, K. A. Regan, Y. Wang, N. P. Ong, J. M. Rowell, and R. J. Cava, "Low Temperature Synthesis of MgB_2," *J. Appl. Phys.*, **91** [1] 274–77 (2001).

[16]C. U. Jung, M.-S. Park, W. N. Kang, M.-S. Kim, S. Y. Lee, and S.-I. Lee, "Temperature- and Magnetic-Field-Dependent Resistivity of MgB_2 Sintered at High-Temperature and High-Pressure Condition," *Physica C*, **353**, 162–66 (2001).

[17]M. Dhalle, P. Toulemonde, C. Beneduce, N. Musolino, M. Decroux, and R. Flukiger, "Transport and Inductive Critical Current Densities in Superconducting MgB_2," *Physica C*, **363**, 155–65 (2001).

[18]C. E. Cunningham, C. Petrovic, G. Lapertot, S. L. Bud'ko, F. Labbs, W. Straszheim, D. K. Finnemore, and P. C. Canfield, "Synthesis and Processing of MgB_2 Powders and Wires," *Physica C*, **353**, 5–10 (2001). □

Ceramic Transactions

SURFACE-OXIDATION EPITAXY METHOD FOR CRITICAL CURRENT CONTROL OF $YBa_2Cu_3O_{7-\delta}$ COATED CONDUCTORS

Kaname Matsumoto
Department of Materials Science and Engineering, Kyoto University
Yoshida-honmachi, Sakyo-ku
Kyoto 606-8501, Japan

Izumi Hirabayashi
Superconductivity Research Laboratory, Div. V, ISTEC
2-4-3, Mutsuno, Atsuta-ku
Nagoya 456-8587, Japan

Kozo Osamura
Department of Materials Science and Engineering, Kyoto University
Yoshida-honmachi, Sakyo-ku
Kyoto 606-8501, Japan

ABSTRACT

The surface quality of NiO buffer, produced by surface-oxidation epitaxy (SOE) method, was improved by a polishing and a cap layer coating techniques. Critical current density (J_c) of the $YBa_2Cu_3O_x$ (YBCO) film directly deposited on the polished SOE-NiO buffer by pulsed laser deposition (PLD) method was increased up to 0.17MA/cm^2 (77K, 0T) from 0.03-0.05MA/cm^2 (77K, 0T, on the non-polished buffer).

In addition, a cap layer of perovskite oxide ($BaSnO_3$) was prepared for promoting the epitaxial growth of YBCO films on SOE-NiO buffer. As a result, J_c of the YBCO film formed on $BaSnO_3$/NiO/Ni substrate reached 0.45MA/cm^2 (77K, 0T). This is because the number of superconductive weak coupling in the YBCO film was reduced, by the flattening of the NiO surface and the coating of the perovskite oxide cap layer.

INTRODUCTION

Simple and low cost production methods to form biaxially textured $YBa_2Cu_3O_{7-\delta}$ (YBCO) films on long-length metallic substrates are necessary; in order to obtain high critical current density (J_c) YBCO coated conductors. We have proposed a method to produce YBCO coated conductors by using surface oxidized NiO layer grown on long-length nickel tapes. The biaxial orientation of NiO can be realized by considering the epitaxial growth of NiO on the nickel surface. We call this technique surface-oxidation epitaxy (SOE) method.[1, 2] The formation of NiO on cube-textured Ni, or Ni-based alloys has been widely investigated recently.[3,4,5]

Biaxially textured NiO layer functions as a barrier layer of chemical reaction between YBCO and nickel, and as a template for the epitaxial growth of YBCO film. However, according to the research so far, the J_c of YBCO film directly deposited on NiO was only 0.03-0.05MA/cm^2 (77K, 0T).[6] Although the NiO crystal made by SOE is highly in-plane oriented, the surface roughness induced by grooves in the NiO grain boundaries remains large. Such a rough surface may generate the tilt grain boundaries in YBCO films deposited on NiO, and bring about the J_c decrease when the tilt angle is high. Furthermore, a Ni contamination may also bring about the degradation of critical temperature (T_c) of YBCO, since Ni atom from the underlying Ni substrate is easy to diffuse in the grain boundary.

An insertion of MgO cap layer between NiO and YBCO film was very effective in order to solve these problems, and J_c of 0.3 MA/cm^2 (77K, 0T) was achieved on SOE-NiO.[2] Nevertheless, the possibility that the local degradation of J_c happens, in the rough surface area, has still remained because the roughness of the NiO surface has not been essentially improved. We have tried to polish SOE-NiO surface mechanically, and also insert a perovskite oxide cap layer between NiO and YBCO film for promoting the epitaxial growth of YBCO films on SOE-NiO buffer. Consequently, high J_c value of 0.45 MA/cm^2 (77K, 0T) was successfully achieved on SOE-NiO.[7] In this paper, we describe J_c control of the YBCO films on the Ni tapes through SOE process.

ORIENTATION CONTROL OF NiO LAYER BY SOE METHOD

An oxide of underlying metal is formed on the metal surface by a thermal oxidation. In an initial stage of the formation of the oxide, the fixed relation exists between crystal orientation of the grown oxide and that of the underlying metal surface. The typical preferred orientations of NiO on Ni(100) face are NiO(100) and NiO(111).[8] Crystal orientation relationships between thermally oxidized NiO and Ni surfaces, which were observed at 500°C in oxygen atmosphere, are summarized in Table I.

Table I. Orientation relationships between oxidized NiO on several Ni surfaces

Metal	Oxide	Oxidation condition	Crystal plane Ni NiO	Crystal orientation Ni NiO
Ni (fcc) a=3.53 Å	NiO (cubic) a=4.17 Å	1atm, O_2 500℃ 10~120min	{100}//{111}	[110]//[110]
			{100}//{100}	[110]//[110]
			{110}//{110}	[110]//[110]
			{110}//{114}	[110]//[110]
			{111}//{111}	[110]//[110]
			{113}//{110}	[110]//[110]

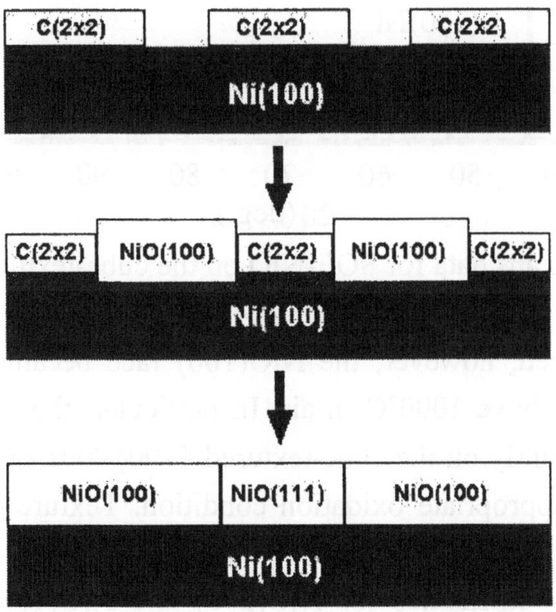

Fig.1. Schematic view of the initial stages of oxidation of Ni(100) surface.

The oxidation of metal begins by adsorbing oxygen molecules on the metallic surface. In the oxidation of (100) face of nickel single crystal, an adsorption of oxygen atoms by a two-dimensional arrangement is generated. A schematic view of the initial stages of Ni oxidation is presented in Fig.1. The layer of the simple cubic lattice in the period equal to nickel is gradually formed, when the adsorption advances, and finally, the isolated NiO nuclei is formed.[9] The NiO nuclei grow and coalesce, and the Ni surface would be covered with NiO. Then, a fixed crystal orientation relation occurs between NiO and Ni by the regulation of crystal orientation on the surface of underlying metal. Generally, (100)- and (111)-oriented NiO species coexists on the Ni(100) face because lattice misfit between NiO and Ni is large (misfit=18.1%).

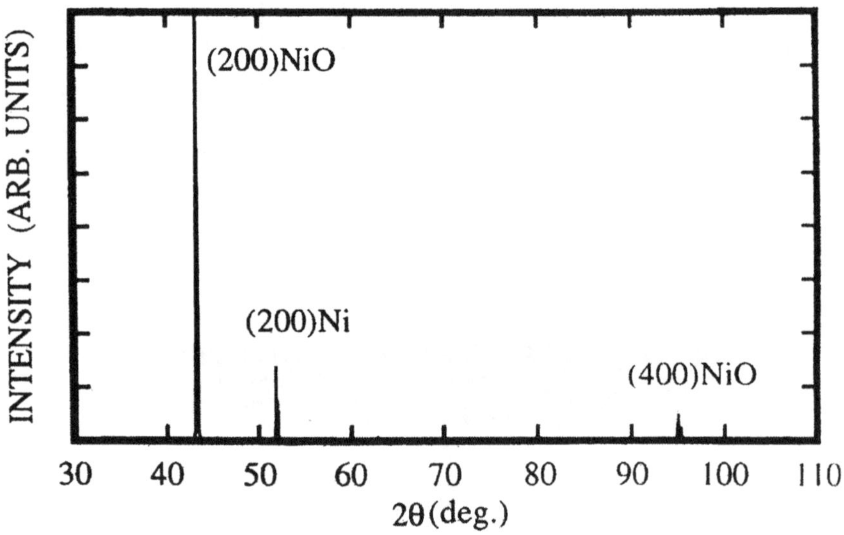

Fig.2. X-ray θ–2θ scans data for SOE-NiO on the cube-textured Ni substrate.

We have discovered, however, the NiO(100) face became dominant at the elevated temperature above 1000°C in air. In particular, the (100)-oriented NiO grains grew predominantly on the cube-textured {100}<001> nickel tape when it was oxidized under appropriate oxidation condition. Texture degree defined by I(200)/{I(200) +I(111)} have reached 0.99 at 1200°C, where I(200) and I(111) mean the X-ray θ–2θ intensities for NiO(200) and NiO(111) peaks. Figure 2 shows the typical X-ray θ–2θ scan data for the SOE-treated NiO/Ni substrate.

Strong NiO(200) and (400) peaks are observed. This indicates the outermost layer of NiO is almost composed of the (100)-oriented NiO grains. SOE method utilizes these phenomena.

Cube-textured Ni tapes, prepared by the combination of rolling and heat treatment, are used for the SOE processing. The recrystallization heat treatment at 700-800°C is conducted in order to achieve both the flatness and the in-plane orientation of the Ni surface. The SOE heat treatment is carried out subsequently at around 1200°C to form the biaxially textured NiO layer on the Ni surface. Then, crystal orientation relation between NiO and Ni becomes NiO{001}//Ni{001} and NiO<110>//Ni<110>. Highly biaxial texture of SOE-grown NiO layer has been confirmed by the X-ray pole figure measurement. A typical FWHM of in-plane texture, $\Delta\phi$, of NiO(111) peak is 10-14 degrees and FWHM of out of plane texture, $\Delta\theta$, of NiO(200) peak become 6-8 degrees.

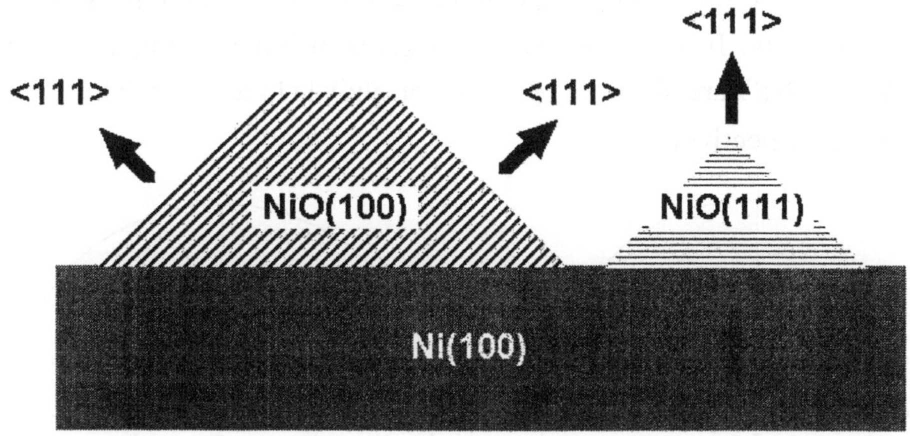

Fig.3. Orientation mechanism of biaxially textured SOE-NiO on Ni substrate.

At the initial stage of SOE process, (100)- and (111)-oriented NiO grains coexist. Why does NiO(100) face dominates on the cube-textured Ni tape after high temperature thermal oxidation? The reason would be speculated as follows: the growth rate of (111) face of cube crystal is larger than that of (100) face because (111) face is the non-singular face. Therefore, the NiO(111) grains grow only in the perpendicular direction, while it is permitted that the NiO(100) grains grow perpendicularly and horizontally, as illustrated in Fig.3. This means that the

NiO(100) grains laterally and quickly grow faster than the NiO(111) grains.

The (100)-oriented NiO grains coalesce and bury (111)-oriented NiO grains, and eventually covers all NiO surfaces. The NiO(100) grains grow by keeping epitaxial relation with Ni(100) surface, so that the biaxial orientation of NiO layer realizes. However, NiO surface becomes rough due to the coalescence of NiO(100) grains. The typical surface roughness of the SOE-NiO is approximately 30-150nm. The defects on the oxide buffer will induce the degradation of J_c of the YBCO films; hence the improvement of surface quality of SOE-NiO is strongly desired.

IMPROVEMENT OF J_c IN YBCO FILM ON SOE-NiO

YBCO films, prepared on SOE-NiO by pulsed laser deposition (PLD) with KrF excimer laser, show cube-on-cube epitaxy. A typical film formation condition is as follows: 700°C-substrate temperature, 200mTorr oxygen pressure and 2 J/cm^2 energy density. In Fig.4, X-ray pole figures of NiO(111) and YBCO(103) peaks for YBCO/NiO are shown. $\Delta\phi$ values of NiO(111) and YBCO(103) are 12 and 11 degrees, respectively.

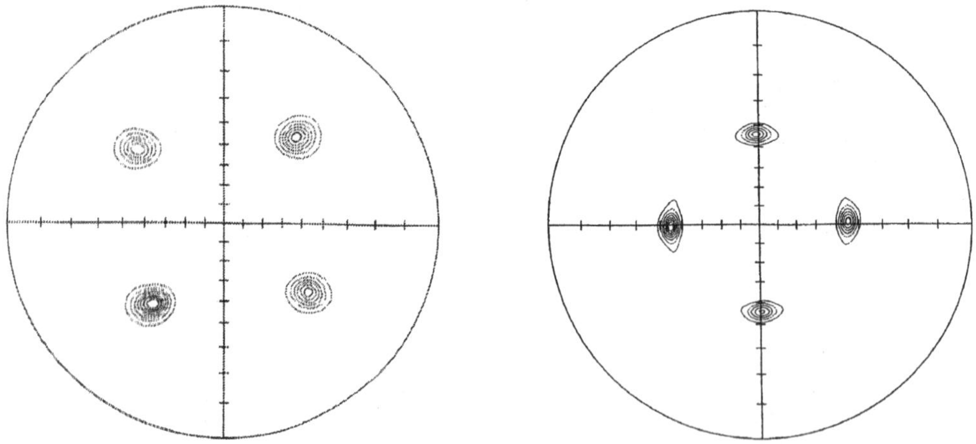

(a) NiO(111) (b) YBCO(103)

Fig.4. X-ray pole figures of NiO(111) and YBCO(103) peaks for YBCO/NiO.

However J_c of YBCO films deposited on the bare SOE-NiO was only 0.03-0.05 MA/cm^2 (77K, 0T). This is one order of magnitude lower than that expected from in-plane texture of YBCO. Recently, the YBCO films with high J_c of 2-3MA/cm^2 (77K, 0T) have been successfully deposited on NiO(100) single crystal by PLD.[10] Considering the potentiality of NiO(100) surface as a template for epitaxial growth of YBCO, surface roughness of NiO and/or contamination of YBCO film by Ni element through NiO grain boundaries seem to be causes of J_c degradation in YBCO/SOE-NiO system.

Fig.5. TEM cross-section of the interface between MgO layer and SOE-NiO.

It is effective to form a thin oxide cap layer on NiO in order to solve this problem. Improvement of surface quality of NiO has been attained by using thin MgO cap layer, and J_c of 0.3MA/cm^2 (77K, 0T) was obtained. The surface roughness of the buffer might be reduced with an excellent wetting property of MgO on NiO. TEM photograph of the interface between NiO and MgO is shown in Fig.5, where MgO was epitaxially and continuously grown on the SOE-NiO surface. MgO and NiO have a same crystal structure (rock-salt type) and they are completely soluble each other in solid state at all compositions. As a result the rough surface of NiO was nicely masked with MgO cap layer. MgO was also very effective in diffusion prevention of the Ni element. According to local EDX analysis, Ni did not diffuse in the MgO cap layer so that the Ni contamination was perfectly suppressed.

Even though MgO grows epitaxially on NiO, a careful control of the exact YBCO film deposition condition is necessary in order to form high J_c YBCO films on MgO. This is because the YBCO grains with cube-on-cube relation and with 45-degrees rotated orientation are easy to coexist on MgO.[11] Thus, the perovskite-type materials, which have better chemical compatibility to YBCO than that of MgO, were examined as a new cap layer in the following work.

MECHANICAL POLISHING OF SOE-NiO SURFACE

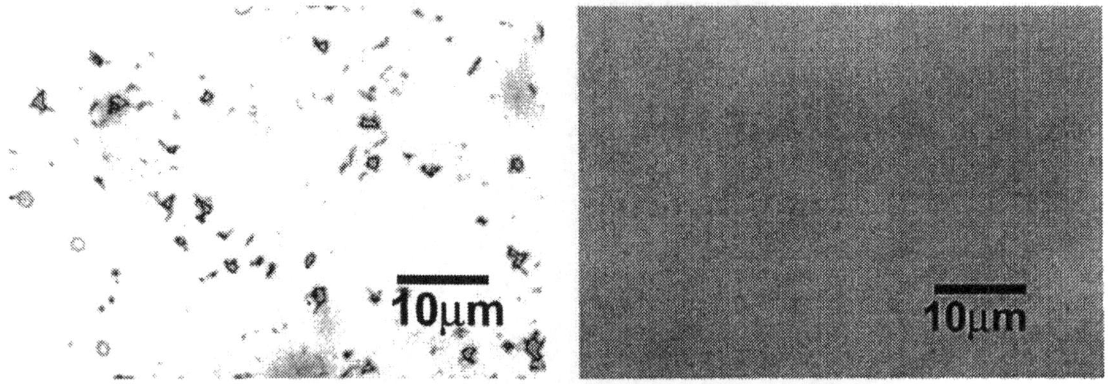

(a) Non-polished (b) Polished

Fig.6. Photographs of Non-polished and polished SOE-NiO surfaces

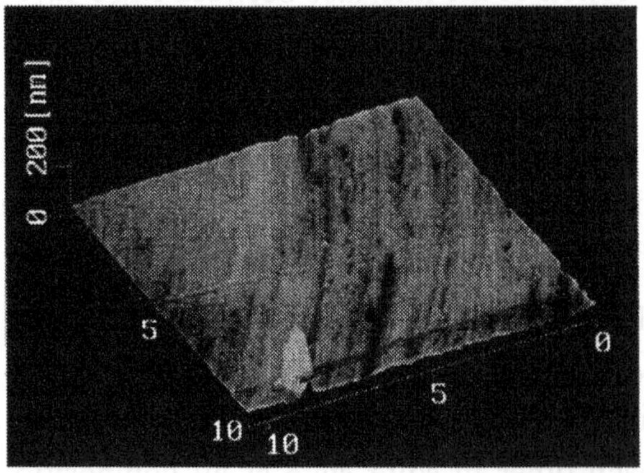

Fig.7. AFM picture of the polished SOE-NiO surface

In order to solve the rough-surface problem, the polishing of the SOE-NiO was investigated by using both 1 μm and 0.25 μm diamond pastes. Photographs of the non-polished and polished NiO surfaces are shown in Fig.6. The highly smooth and dense NiO surface was realized by this technique. The thickness of polished NiO layer was about 3 μm. No cracking in NiO occurred during polishing. We observed the surface morphology of polished NiO by AFM. Fig.7 shows the AFM image of the surface morphology of smooth NiO. Scan area of NiO was set to 10×10 μm^2. From the measurement of AFM image, the average surface roughness was 1.1 nm for smooth area and 5.5 nm for whole area.

The YBCO film deposited on the polished NiO surface by PLD showed strong *c*-axis orientation and cube-on-cube epitaxy. In-plane texture $\Delta\phi$ that determined from YBCO(103) peak was also 11 degrees. This value was similar to those of the other texturing technique. We have measured critical current (I_c) of the film at 77K and 0T. The bridge width and length formed in the YBCO film were 1.8 mm and 2.0 mm, and also, YBCO film thickness on NiO was 0.5 μm. Four terminals were attached by silver paste to the YBCO and the specimen was immersed into the liquid nitrogen. I_c of the YBCO film reached 1.52 A; namely, this corresponded to J_c of 0.17 MA/cm^2 (77K, 0T). These results indicate that the smooth NiO surface is crucial in order to produce the high J_c YBCO films on SOE-NiO.

PEROVSKITE CAP LAYER COATING

On the polished SOE-NiO, the perovskite oxide cap layer and YBCO films were grown by PLD. The perovskite materials, such as $BaTiO_3$, $BaSnO_3$ and $BaZrO_3$ were considered as a candidate since they included BaO layer and the excellent chemical compatibility with YBCO film was expected. As a first step, we chose $BaSnO_3$ material as a new cap layer because the crystal lattice constant is 0.411 nm.[12] The value is approximate for the lattice constant of NiO (0.416 nm) and misfit between them is -1.2%. Figure 8 shows the new configuration of YBCO /cap layer/SOE-NiO.

Deposition of $BaSnO_3$ and YBCO on the polished SOE-NiO was carried out by PLD at an energy density of $2J/cm^2$ and repetition rate of 10-25 Hz. The substrate temperature was varied in the temperature range of 700-780°C and oxygen pressure was changed from 30 to 200mTorr to determine the optimal deposition condition. The bridges of 1.0 mm width and 2.0 mm length were patterned in YBCO films for I_c measurement.

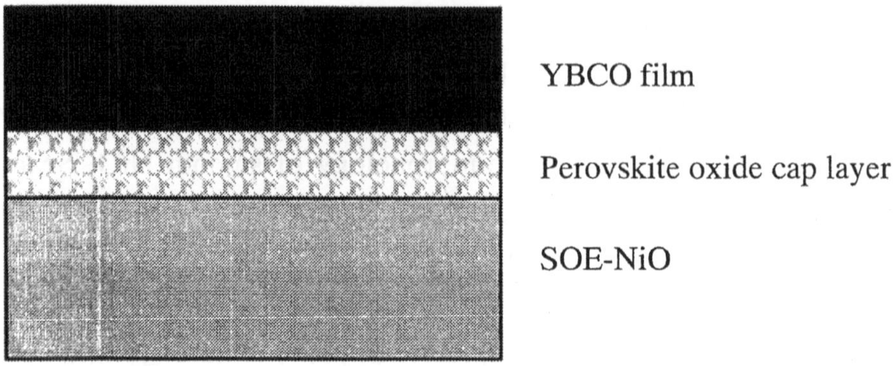

Fig.8. Multi-layer structure of YBCO/cap layer/SOE-NiO.

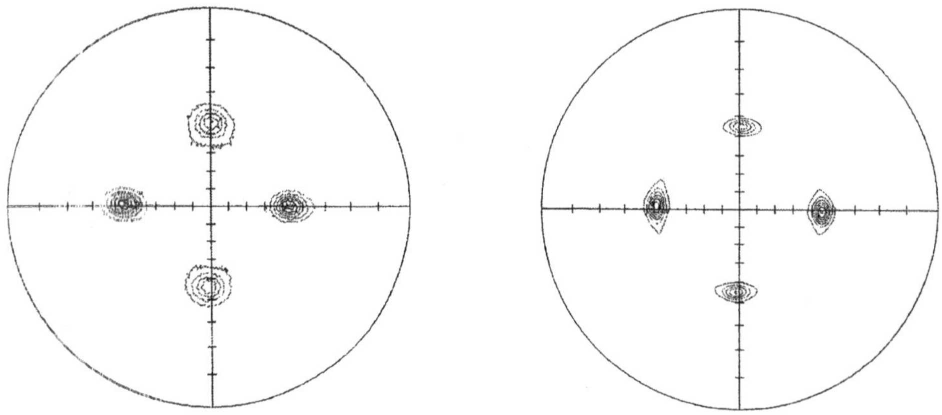

(a) $BaSnO_3$(110) (b) YBCO(103)

Fig.9. X-ray pole figures of $BaSnO_3$(110) and YBCO(103) peaks forYBCO/$BaSnO_3$/SOE-NiO configuration.

The $BaSnO_3$ films deposited on the NiO were *c*-axis oriented and in-plane textured and their thickness were 0.3-1.0 μm. Subsequently, YBCO films were deposited on the $BaSnO_3$ layer by PLD. The thickness of YBCO films on $BaSnO_3$/NiO was arranged to 0.5 μm. Figure 9 shows the pole figures of $BaSnO_3$(110) and YBCO(103) planes for YBCO/$BaSnO_3$/SOE-NiO configuration.

It is proven that $BaSnO_3$ and YBCO films epitaxially grow on the polished NiO(100) surface with cube on cube relation. In-plane texture, $\Delta\phi$ of YBCO(103) X-ray peak, was 10.4 degrees and out of plane texture, $\Delta\theta$, for YBCO(005) X-ray peak was 2.5 degrees. This means that YBCO film on the $BaSnO_3$/NiO substrate had the advanced biaxial texture.

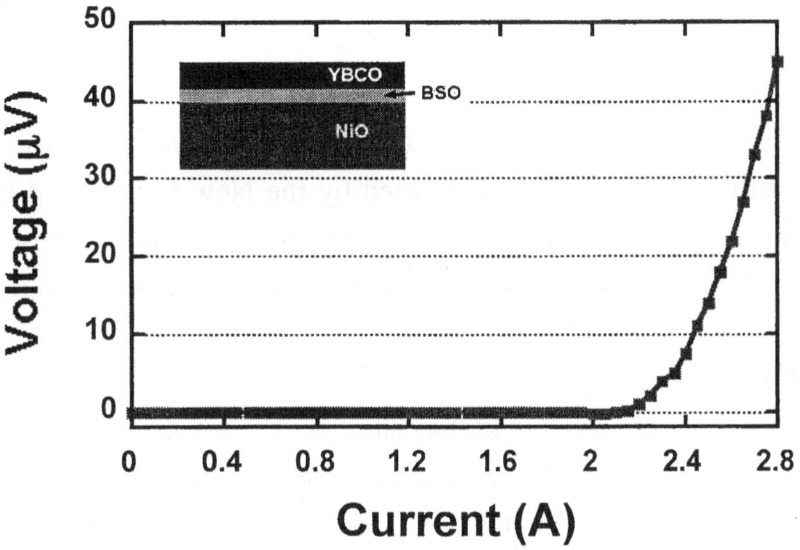

Fig.10. Current-voltage characteristics for YBCO/$BaSnO_3$/SOE-NiO system.

Current-voltage curve at 77K and 0T for YBCO/$BaSnO_3$/SOE-NiO specimen is presented in Fig.10. I_c of the patterned specimen was 2.25 A, which corresponded to J_c of 0.45 MA/cm^2 (77K, 0T). The improvement of J_c is due to the flattening of the NiO surface by the polishing and the good chemical compatibility between YBCO and $BaSnO_3$. It is considered that the perovskite material, such as $BaSnO_3$, is one of the candidates of cap layer on the NiO surface to obtain high J_c properties in the YBCO coated conductors by SOE method.

CONCLUSIONS

The surface roughness of NiO buffer by surface-oxidation epitaxy was improved by polishing technique. The surface roughness of NiO was reduced in this manner to 1-5nm, and J_c of 0.17MA/cm^2 (77K, 0T) was achieved for even YBCO films deposited on the bare NiO surface. In addition, the coating of the perovskite oxide cap layer was also examined instead of the MgO cap layer in order to promote the epitaxial growth of YBCO. We have obtained J_c of $0.45 AM/cm^2$ (77K, 0T) in $YBCO/BaSnO_3/NiO/Ni$ system. The improvement of J_c was due to both the flattening of NiO surface and good epitaxy between YBCO and $BaSnO_3$. It is considered that the perovskite material, such as $BaSnO_3$, is one of the candidates of cap layer on the NiO surface to obtain high J_c properties in YBCO coated conductors by SOE method.

ACKNOWLEDGMENTS

The author would like to thank T. Watanabe of Furukawa Electric Co., for valuable discussion. This work was supported by the New Energy and Industrial Technology Development Organization (NEDO) as Collaborative Research and Development of Fundamental Technologies for Superconductivity Applications.

REFERENCES

[1]K. Matsumoto, Y. Niiori, I. Hirabayashi, N. Koshizuka, T. Watanabe, Y. Tanaka, and M. Ikeda, "New Fabrication Method of High-J_c $YBa_2Cu_3O_7$ Superconducting Films on Flexible Metallic Substrates", pp.611 in *Adv. Superconductivity X*, Springer, Tokyo, 1998.

[2]K. Matsumoto, S. B. Kim, I. Hirabayashi, T. Watanabe, N. Uno, and M. Ikeda, "High Critical Current Density $YBa_2Cu_3O_7$ Tapes Prepared by the Surface-Oxidation Epitaxy Method", *Physica C***330** 150(2000).

[3]T. Watanabe, K. Matsumoto, T. Tanigawa, T. Maeda, and I. Hirabayashi, "Surface-Oxidation epitaxy of Ni-Clad Ni-20wt%Cr and Ni-Clad Austenitic Stainless Steel tapes for Y-Ba-Cu-O Coated Conductors", *IEEE Trans. Appl. Supercond.*, **11** 3134(2001).

[4]Z.Lockman, X. Qi, A. Berenov, R. Nast, W. Goldacker, J. MacManus-Doriscoll, "Study of Thermal oxidation of NiO buffers on Ni-based Tapes for Superconducting Substrates", *PhysicaC***351** 34(2001).

[5]T. Petrisor, V. Boffa, G. Celentarno, L. Ciontea, F. Fabbri, V. Galluzzi, U. Gambardella, A. Mancini, A. Rufoloni, and E. Varesi, "Epitaxial Oxidation of Ni-V Biaxially Textured Tapes", *PhysicaC* (2002), in press.

[6]K. Matsumoto, S. B. Kim, J. G. Wen, I. Hirabayashi, T. Watanabe, N. Uno, and M. Ikeda, "Fabrication of In-Plane Aligned YBCO Tapes Using NiO Buffer

Layers Made by Surface-Oxidation Epitaxy", *IEEE Trans. Appl. Supercond.* **9** 1539(1999).

[7]K. Matsumoto, I. Hirabayashi, and K. Osamura, "Surface-Oxidation Epitaxy Method to Control Critical Current of $YBa_2Cu_3O_7$ Coated Conductors", *Physica* C (2002), in press.

[8]N.N. Khoi, W. W. Smeltzer, and J. D. Embury, "Growth and Structure of Nickel Oxide on Nickel Crystal Faces", *J. of Electrochem. Soc.*, **122** 1495(1975).

[9]R.S. Sakai, A. P. Kaduwela, m. Sagurton, J. Osterwalder, D. J. Friedman, and C. S. Fadley, "X-ray Photoelectron Diffraction and Low-Energy Electron Diffraction Study of the Interaction of Oxygen with the Ni(100) Surface: c(2x2) to Saturated Oxide", *Surface Science* **282** 33(1993).

[10]T. Maeda, H. Iwai, T. Watanabe, and I. Hirabayashi, unpublished.

[11]D. M. Hwang, T. S. Ravi, R. Ramesh, S. W. Chan, C. Y. Chen, L. Nazar, "Application of a Near coincident Site Lattice Theory to the Orientations of $YBa_2Cu_3O_7$ Grains on (001)MgO Substrates", *Appl. Phys. Lett.* **57** 1690(1990).

[12]S. Miyazawa and M. Mukaida, "Transmission electron Microscope Observation of Interfaces in YBa2Cu3Ox/BaSno3/YBa2Cu3Ox Trilayers", *Jpn. J. Appl. Phys.*, **37** L949(1998).

CONTROL OF THE SULFUR *C* (2 × 2) SUPERSTRUCTURE ON {100}<100>-TEXTURED METALS FOR RABITS APPLICATIONS.

C. Cantoni, D. K. Christen, L. Heatherly, F. A. List, A. Goyal, G. W. Ownby, and D. M. Zehner

Oak Ridge National Laboratory, 1 Bethel Valley Rd., Oak Ridge, TN 37831

ABSTRACT

We investigate the influence of a chemisorbed S template with centered (2 × 2) structure on the epitaxial growth of commonly used oxide buffer layers on {100}<100> metals for RABiTS fabrication. Our study involves growth of CeO_2, Y-stabilized Zr_2O_3, Gd_2O_3, $LaMnO_3$ and $SrTiO_3$ seed layers on biaxially textured Ni and Ni-alloys. We also discuss the effect of an incomplete c (2 × 2) surface coverage on the seed layer texture and the use of H_2S in a pre-deposition anneal as a mean to control the superstructure coverage and optimize the seed layer texture.

INTRODUCTION

Recently we found that sulfur impurity atoms present on the Ni (001) surface are determinant for the epitaxial growth of certain oxide films used as seed buffer layers in RABiTS conductors (e.g.: CeO_2 or Y_2O_3-stabilized ZrO_2) [1,2]. Our results showed that chemical and structural properties of the textured metal surface have to be considered in order to obtain the desired texture for the buffer layers. Using reflection high-energy electron spectroscopy (RHEED), and Auger electron spectroscopy (AES), our experiments revealed the existence of a sulfur superstructure on the textured Ni surface that forms after surface segregation of sulfur contained as a common impurity in the metal bulk. During the high-temperature texturing anneal S atoms diffuse to the surface of the tape and arrange in a centered (2×2) superstructure. However, depending on the initial S concentration and/or specific annealing conditions, the superstructure layer can exhibit different coverage.

Here we analyze the relationship between S surface concentration and seed layer texture, and present a method that controls and optimizes the superstructure coverage on the metal, providing reproducible quality of the seed layer texture. In addition, we extend our study of the influence of the S superstructure to the case of biaxially textured Ni alloys, such as Ni-3%W, and Ni-13%Cr.

SEED LAYERS ON {100}<100> Ni BY PULSED LASER DEPOSITION

CeO_2 seed layers are successfully deposited by vapor deposition techniques like e-beam evaporation, reactive sputtering, and pulsed laser deposition (PLD). All these techniques use a background H_2O pressure of about 1×10^{-5} Torr and a substrate temperature in the range of 600 to 750 °C. Water dissociation provides enough oxygen in the background to form and stabilize the CeO_2 phase without oxidizing the Ni [3]. The stability of the Ni surface in the presence of water with a partial pressure in the range of 1×10^{-5} Torr was investigated by *in situ* RHEED performed inside the laser molecular beam epitaxy (MBE) chamber used to fabricate films for this study. It was found that such H_2O partial pressure does not cause any structural modification of the sample surface that can be related to growth of NiO islands or adsorbed O. The documented first stage of Ni oxidation consists of a chemisorbed O layer with $p\ (2 \times 2)$ structure up to a coverage of 0.25 ML, and with subsequent $c\ (2 \times 2)$ structure up to a coverage of 0.5 ML [4] [1 ML ≡ (number of surface adsorbate atoms) / (number of surface substrate atoms)]. None of these structures was observed to form as a consequence of water exposure on a clean {100}<100> Ni surface at a typical deposition temperature of 600 °C. The clean Ni surface was obtained by depositing a Ni film *in situ* on the textured Ni substrate in ultrahigh vacuum conditions at 600 °C by PLD. RHEED and AES studies revealed that the as-grown films were free of impurities at typical buffer-layer deposition temperatures.

Figure 1a shows a RHEED pattern acquired during the nucleation of CeO_2 on a {100}<100> Ni substrate having the *c* (2 × 2)-S superstructure shown in Fig. 1b. The electron beam is directed along the Ni <100>. The CeO_2 film was grown at 600 °C in the presence of a water partial pressure of 1.5×10^{-5} Torr. The pattern shows diffracted spots that correspond to a (002) biaxially textured CeO_2 film viewed along the <110> crystal direction. This is consistent with the general observation that CeO_2 grows with a 45° in-plane rotation with respect to the {100}<100> Ni substrate. The presence of spots in the RHEED pattern indicates that the nucleation is occurring by formation of islands, which are tall enough to produce a transmission diffraction pattern. The spacing between the RHEED reflections in Fig.1a provides a rough estimate of the in-plane lattice parameter of the film, which is 5.43 ± 0.05 Å.

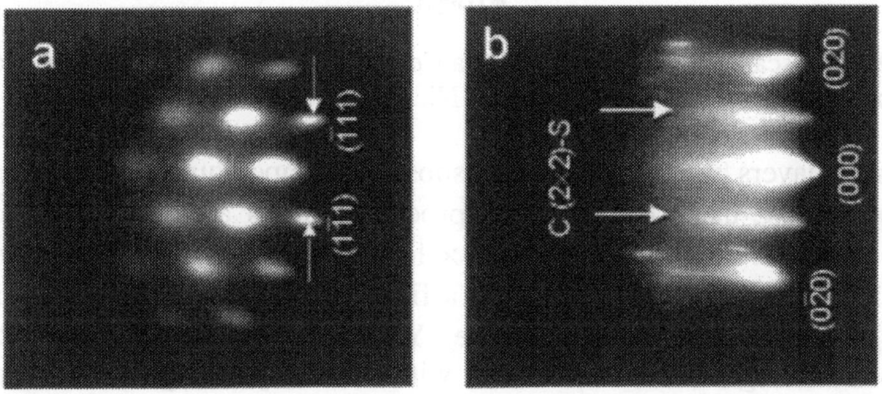

Figure 1. RHEED pattern obtained at 600 °C during the nucleation of a CeO_2 seed layer on a biaxially textured Ni substrate with a c (2 × 2)-S surface structure. The electron beam is directed along the Ni <100> and the spacing indicated by the arrows corresponds to the interatomic distance $a_{CeO_2}\sqrt{2}/2$, (a). Substrate RHEED pattern showing the extra reflections indicating the sulfur c (2 × 2) superstructure, (b).

Figure 2 shows an XRD ϕ–scan of the (111) peak for the completed CeO_2 film, approximately 400 Å thick. We found that the deposition of CeO_2 on the c (2 × 2)-S template consistently produced highly textured (200) films, while deposition on a clean Ni surface always gave rise to (111) oriented films.

Among other oxides that have been successfully used as seed layers, those grown directly on a textured metal by PLD for this study are: $SrTiO_3$ (STO) and

LaMnO$_3$ (LMO), with perovskite and pseudo-perovskite structure, respectively; Y-stabilized Zr$_2$O$_3$ (YSZ), with fluorite structure; and Gd$_2$O$_3$, with RE$_2$O$_3$ structure. For all these oxides, RHEED and X-ray measurements have shown that the S superstructure is needed to obtain (001) epitaxial growth.

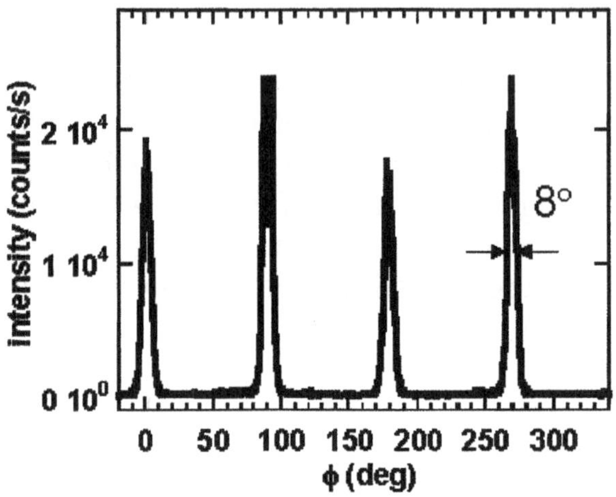

Figure 2. XRD φ-scan of the (111) peak for a CeO$_2$ seed layer deposited on the c (2 × 2) sulfur template chemisorbed on {100}<100> Ni.

YSZ seed layers were deposited at substrate temperatures ranging between 600 °C and 800 °C using the following procedure. After an initial ~100-Å-thick layer was deposited in vacuum ($P_{base} \leq 5 \times 10^{-8}$ Torr), the O$_2$ partial pressure was increased to the value 1×10^{-5} Torr and a final 1200-Å-thick film was grown at the same chosen deposition temperature. YSZ films grown on the *c* (2 × 2) surface showed single (002) orientation, with a (111) pole figure indicating the same degree of grain alignment as the substrate. YSZ films grown on the superstructure-free Ni overlayer showed only the (111) peak in the θ-2θ scan. In this case, the pole figure of the (200) reflection showed 4 different in-plane domains rotated 30° with respect to each other. This epitaxial relation is expected for the nucleation of a threefold symmetric lattice on a square symmetric lattice. The quality of the YSZ seed layers on *c* (2 × 2)/Ni was tested by growing a 0.3-μm-thick YBCO film by the *ex situ* BaF$_2$ method [5] on some of the samples. A 20-nm-thick CeO$_2$ cap layer enabled compatibility of the precursor layer with the YSZ. The resulting YBCO critical current density was 1.15 MA/cm^2 in self-field at 77 K, indicating that a 120-nm-thick YSZ film is a good Ni diffusion barrier.

Although STO is thermodynamically less stable than bulk NiO, it can be easily grown on Ni at oxygen pressures as low as 1×10^{-9} Torr, in conditions at which surface NiO will not form*. The ease of growth of STO on Ni can be attributed in part to the stability of the STO crystal structure in presence of oxygen vacancies [6].

Figure 3a shows the RHEED pattern for a (200) epitaxial STO film grown on the textured Ni substrate with S superstructure. The electron beam is directed along the <100> direction and the epitaxial relationship between STO and Ni is cube-on-cube. The corresponding STO (111) pole figure is shown in Figure 3b. The STO films were grown at a substrate temperature of 700 °C. The oxygen pressure was kept in the range $1 \times 10^{-9} - 1 \times 10^{-8}$ Torr during the deposition of a 20-to-100 Å-thick nucleation layer and then increased to a value of about 1.0×10^{-5} Torr. On a clean (1×1) Ni surface, the STO seed layer grows multi-domain, with the cube axis remaining parallel to the substrate normal, but with two additional in-plane orientations besides the one with STO <100> // Ni <100>.

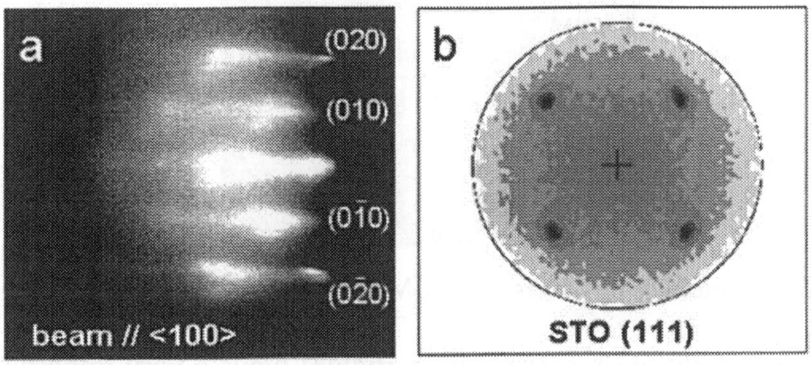

Figure 3. RHEED pattern of an STO film grown epitaxially on a {100}<100> Ni substrate showing the c (2×2) sulfur superstructure, (a). Corresponding (111) logarithmic pole figure of the STO film indicating cube on cube epitaxial relationship with the Ni substrate, (b).

* According to the thermodynamics for the bulk chemical reaction $2Ni + O_2 = 2NiO$, an oxygen pressure in the range of 10^{-9} Torr is sufficient to form NiO at a typical deposition temperature of 600 °C. However, surface contributions to the free energy of formation of NiO appear to be significant. Our RHEED experiments have shown that the surface of {100}<100> Ni is in fact stable in an oxygen partial pressure as high as 5×10^{-7} Torr at a temperature of 600 °C.

Recently $La_{0.7}Sr_{0.3}MnO_3$ (LSMO) and LMO were evaluated as promising single buffer layers for coated conductors. YBCO films deposited by PLD on sputtered LSMO or LMO buffered Ni RABiTS have shown high critical currents [7]. The LMO phase could be nucleated on Ni, by PLD, only at oxygen partial pressures higher than those used for CeO_2. We found, for example, that, for a substrate temperature of 600 to 700 °C, growth in a water partial pressure of 1×10^{-5} Torr gave rise to amorphous films, while a water background of 6 mTorr was sufficient to obtained a crystalline (001) textured film. Figure 4 shows the RHEED patterns of an LMO film, approximately 80 nm thick deposited on *c* (2 × 2)/Ni at a temperature of 600 °C and a H_2O partial pressure of 6 mTorr. The RHEED pattern is consistent with cube on cube texture of the LMO on Ni. The spot-like reflections are indicative of a rough surface morphology.

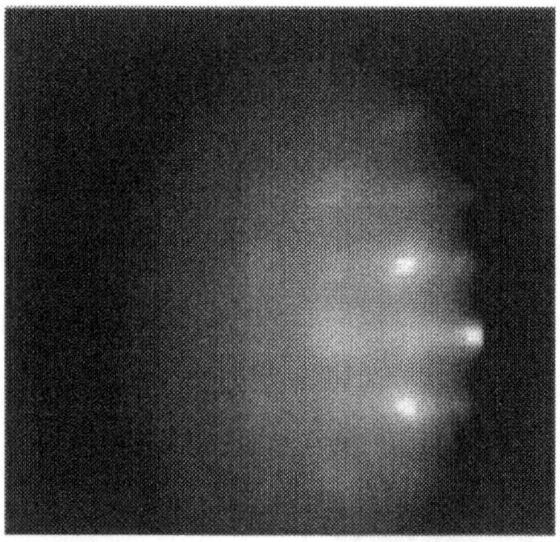

Figure 4. RHEED pattern of an 80 nm-thick LMO film deposited on Ni by PLD.

It is evident that the *c* (2 × 2)-S superstructure plays a very important role in the nucleation of an oxide buffer layer on Ni. We propose that the effect of the S superstructure can be partially explained on the basis of structural and chemical considerations. The S layer behaves like a template that well matches and mimics the arrangement of the oxygen atoms in particular (001) sub-lattice planes of all the seed layer considered in this study, as shown in Figure 5. Sulfur belongs to the VI group and is chemically very similar to oxygen, often exhibiting the same electronic valence. Therefore, it is plausible that during the seed layer deposition the cations easily bond to the S atoms already present on the substrate surface, giving rise to the (001) epitaxial growth of the film, which otherwise would not take place. Ironically, the desired structural and chemical properties of RABiT substrates made in the past few years can be attributed, partially, to the involuntary presence of a sufficient S content in the rolled Ni metal. In cases for which the bulk S content of particular batches of Ni was much lower than 30 wt. p.p.m., the seed layer deposition process produced films with degraded texture or even partially (111) orientation. In those cases, as a recent Auger electron spectroscopy (AES) analysis revealed, S was depleted in the near-surface layer during a high temperature anneal (~1100 °C). This may be the result of the formation and subsequent desorption of SO_2 in the presence of a sufficient partial pressure of oxygen. Sulfur depletion impeded formation of a continuous *c*(2 × 2) layer across the entire Ni surfaces, drastically modifying the oxide film nucleation.

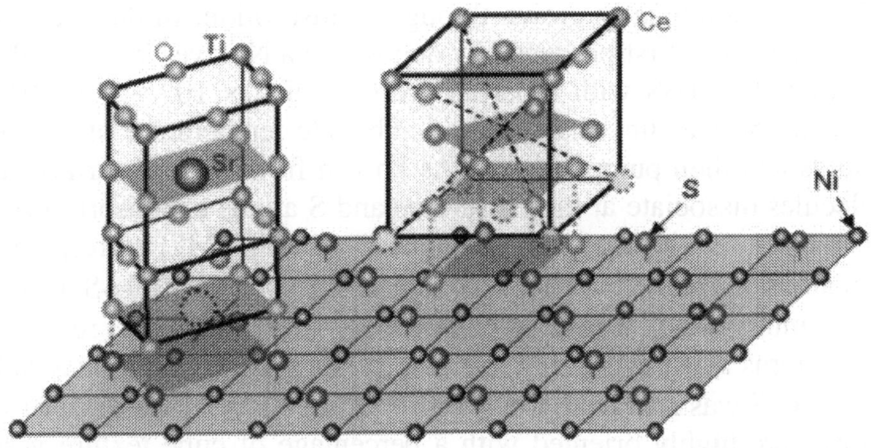

Figure 5. Schematic model for the nucleation of CeO_2 (or YSZ) and STO on a (001) Ni surface with chemisorbed c (2 × 2) sulfur superstructure. The dashed lines indicate correspondence between oxygen sites in (001) planes of the seed layer and sulfur sites on the Ni surface. The seed layer cations impinging on the Ni surface bond easily to the sulfur atoms present on the metal surface, promoting the (001) orientation of the growing film. In the STO case, there is a 1:1 correspondence between oxygen atoms in the SrO plane and sulfur atoms on the Ni surface. Therefore, it is plausible that Ti ions initially bond to the S surface atoms to form the first TiO_2 plane of the STO structure. In the CeO_2 case, two of four oxygen ions per unit cell match the sulfur atoms of the c (2 × 2). During nucleation of CeO_2, oxygen atoms may fill in the empty fourfold Ni hollows and the Ce cations subsequently bond to the template formed by S and O.

QUANTIFICATION AND CONTROL OF THE S-SUPERSTRUCTURE

The above-mentioned considerations illustrate a connection between the quality of the seed layer texture and the degree of coverage of the *c* (2 × 2) superstructure on Ni. To investigate this aspect, we conducted combined AES and RHEED experiments aimed at quantifying the coverage of the $c\ (2 \times 2)$ structure on different Ni samples. During Auger spectroscopy experiments, depending on the material under test and the beam energy, Auger electrons escape and are detected from a layer 5 to 10 Å deep below the atomistic surface. Therefore, the S concentration as deduced from AES can differ from the actual atomic concentration of the top monolayer of the surface 1), because of the uncertainty related to the measurement itself; and 2), because, after segregation, S could exhibit a concentration gradient in the layer measured by the Auger technique. We solved this problem by comparing the AES results obtained on typical samples with those obtained on initially superstructure-free samples on which the $c\ (2 \times 2)$ was intentionally grown through S adsorption. In these experiments, a clean Ni surface was first produced by depositing a Ni layer *in situ*. Subsequently, a small amount of H_2S with a partial pressure of $5 \times 10^{-7} - 1 \times 10^{-6}$ Torr was introduced in the vacuum chamber at a substrate temperature of 700-800 °C for few minutes and then pumped away. As known from several surface studies, the H_2S molecules dissociate at the Ni surface and S atoms chemisorb forming a $c\ (2 \times 2)$ superstructure with a coverage that saturates at 0.5 ML, corresponding to one complete atomic layer of the $c\ (2 \times 2)$-S [8-10]. Exposures to H_2S as low as a few L [1L (langmuir) $\equiv 10^{-6}$ Torr-s] produced very strong $c\ (2 \times 2)$ reflections in the RHEED patterns as shown in Fig. 6a. The sulfur superstructure was stable at 800 °C after the H_2S was removed. CeO_2 and YSZ seed layers deposited *in situ* after S adsorption were highly oriented with a percentage of cube texture very close to 100%, as indicated by the (111) pole figure of Fig. 6b. Longer exposure of the Ni surface to H_2S ($\geq$ 1800 L) did not produce any degradation of the RHEED pattern that could be attributed to adsorption of surface S in excess of 0.5 ML, or formation of a Ni sulfide phase. The as-grown superstructure was found to be stable also after exposure to air at room temperature and consequent re-heating of the sample in vacuum, which reproduced the same initial RHEED pattern. AES experiments performed *in situ* after S adsorption yielded a sulfur signal of about 25% at saturation. The same value was obtained on samples where the $c\ (2 \times 2)$-S had formed consequent to segregation, and that showed RHEED patterns identical to the ones acquired in the S adsorption experiments. From these observations we associated an AES value of ~25% to a full layer of $c\ (2 \times 2)$-S on the Ni surface. Consequently, we were able to quantify the S surface content in cases in which the superstructure did not cover the entire Ni surface and the $c\ (2 \times 2)$ reflections in the RHEED pattern were less intense than in fully covered samples.

We noticed that, in cases for which the S coverage was equal or less than 0.25 ML after the texturing anneal, S appeared to have still formed a $c\ (2 \times 2)$ structure and not a $p\ (2 \times 2)$ as typically observed in S adsorption experiments by others

[11]. In fact, no p (2 × 2) reflections could be identified in RHEED patterns with the electron beam parallel to the Ni <110> direction in these cases.

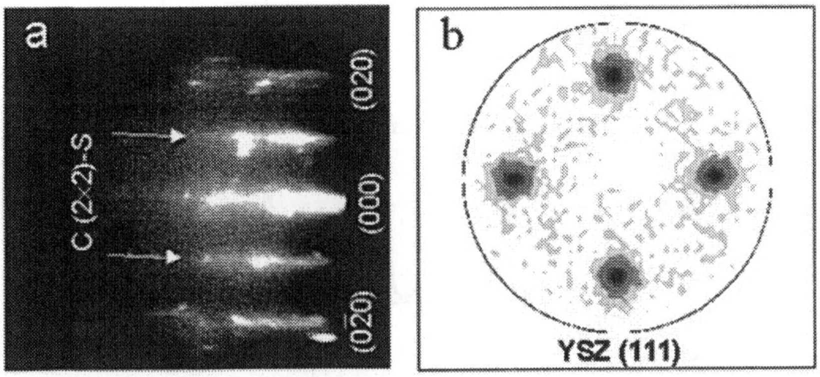

Figure 6. RHEED pattern showing strong c (2 × 2) reflections obtained after exposing a clean {100}<100> Ni surface to few L of H_2S at 800 °C, (a). Logarithmic (111) pole figure of a YSZ seed layer grown *in situ* on the c (2 × 2)-S shown in (a), (b). For this sample the x-ray background signal was eliminated by subtracting a second pole figure obtained after changing the Bragg angle by 1 degree. The calculated percentage of cube texture for this film was 99.96%.

Figure 7 is a comparison of ϕ scans of the (111) peak for 3 different STO films grown on Ni samples showing 1), a complete c (2 × 2) layer; 2), 40% of the full c(2 × 2) coverage; and 3), no superstructure, respectively. We notice that an incomplete c(2 × 2) coverage is not sufficient to obtain full cube texture of the seed layer, and, in the case of 40% coverage, additional peaks indicating crystal domains grown with orientations other than cube on cube, are still present in the ϕ scan.

Figure 8 shows a similar comparison for two CeO_2 films deposited on two different Ni samples cut from the same tape but subjected to two different texturing anneals. In the first case the sample was annealed in high vacuum at 1100 °C for 2 h and exhibited a complete S superstructure. In the second case the anneal was much longer (15 h) and caused partial evaporation of the surface S resulting in a weaker c(2 × 2) superstructure with about 70 % coverage (see insets of Fig. 7). In both cases the Ni substrate showed a FWHM of the (111) peak of 8.2° ± 0.4°. The CeO_2 films were cube textured and showed single (002) orientation on both the Ni substrates. However, it is evident that the degree of grain alignment for the CeO_2 films correlates with the coverage of the c (2 × 2) superstructure on the substrate, and the incomplete S template results in a broadening of the seed layer texture. We found that the texture of the seed layer replicates that of the substrate only when the $c(2 \times 2)$ has a full 0.5 ML coverage.

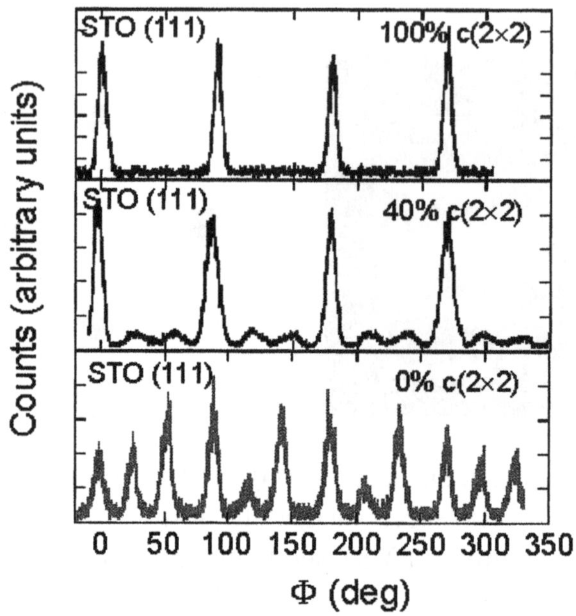

Figure 7. Comparison between ϕ scans of the (111) peak for 3 different STO films grown on {100}<100> Ni substrates having different sulfur coverage: 100% of a c (2×2) layer (or $\theta = 0.5$), top; 40%, middle; 0%, bottom.

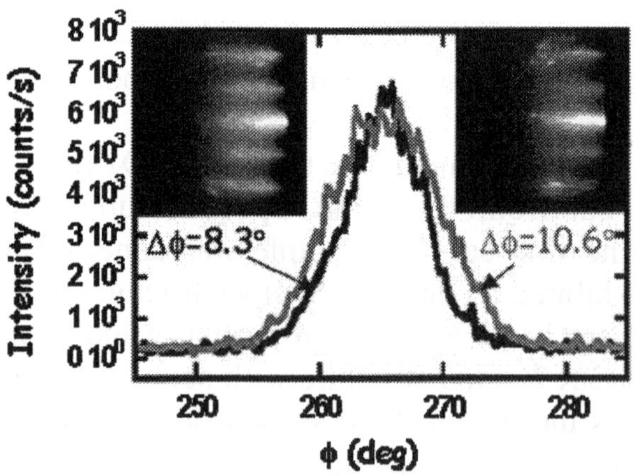

Figure 8. ϕ scans of the (111) reflection of a CeO_2 seed layer grown on a Ni substrate annealed for 2 h (black line), and on a Ni substrate annealed for 15 h (gray line).

Sulfur adsorption experiments provide a method for quantifying the S coverage on the Ni surface after the texturing anneal. But, more importantly, H_2S exposure provides a rapid and efficient way to obtain a full coverage of the *c* (2 × 2)-S on textured Ni, and, consequently, optimize the seed layer texture without being restricted by the amount of S actually present in the bulk and the slow, less efficient segregation process. H_2S exposure has been successfully introduced in a continuous process for texturing and CeO_2 coating of meter-long Ni tapes, yielding highly oriented and reproducible long lengths of RABiTS [12]. YBCO films deposited on short sections of these RABiTS tapes by PLD and *ex situ* BaF_2 method have consistently shown critical currents larger than 1 MA/cm^2.

RESULTS OF SEED LAYER GROWTH ON Ni-3%W AND Ni-13%Cr

Because of the magnetic properties of Ni, coated conductors fabricated using Ni tape show losses in the presence of AC electromagnetic fields, which represents a serious problem for certain applications [13]. Alloying Ni with small quantities of metals like W and Cr results in lower or zero magnetization (the latter being the case of the Ni-13%Cr alloy). In addition, Cr and W add strength to the tape, making handling of lengths easier.

Knowing the importance of the S superstructure for epitaxial seed layer growth on Ni, it is natural to wonder whether the superstructure plays the same role in textured Ni alloys. Figure 9 shows Auger data and a RHEED pattern for a textured Ni-3%W substrate heated from 500 °C to 900 °C in high vacuum. It is evident in this case that, just as for Ni, S surface segregation during texturing has resulted in a *c* (2 × 2) superstructure with complete coverage. The oxygen and carbon signals in the Auger plot derive from air exposure after the texturing anneal, and desorb easily by mild heating. As expected from these observations, high quality buffer layers were deposited on such Ni-W surfaces.

Figure 10 shows the RHEED pattern of a Gd_2O_3 grown by PLD on a textured Ni-3%W substrate at a temperature of 600 °C and a water pressure of 1×10^{-5} Torr. The figure indicates that the seed layer has grown epitaxially with (004) orientation. The weaker streaks marked by the arrows are superlattice reflections originated by the periodicity of the large cubic cell (a = 10.813) of Gd_2O_3. [3]. Fully textured CeO_2, YSZ and Y_2O_3 seed layers have also been deposited on *c* (2 × 2)-S/Ni-3%W, by PLD or other deposition techniques, using the same growth conditions established for deposition on Ni. We would like to point out that, because of the higher reactivity with O_2 of W, as compared to Ni, careful control of background oxygen is important during buffer layer deposition on Ni-W alloys. RHEED observations showed that while the Ni-3%W surface is stable in the presence of a water pressure in the range of 10^{-5} Torr, H_2O pressures in the range of 10^{-4} or higher cause changes in the diffraction patterns that can be interpreted as due to oxide formation. The surface of pure Ni remained stable against oxidation at water pressures as high as tenths of mTorr.

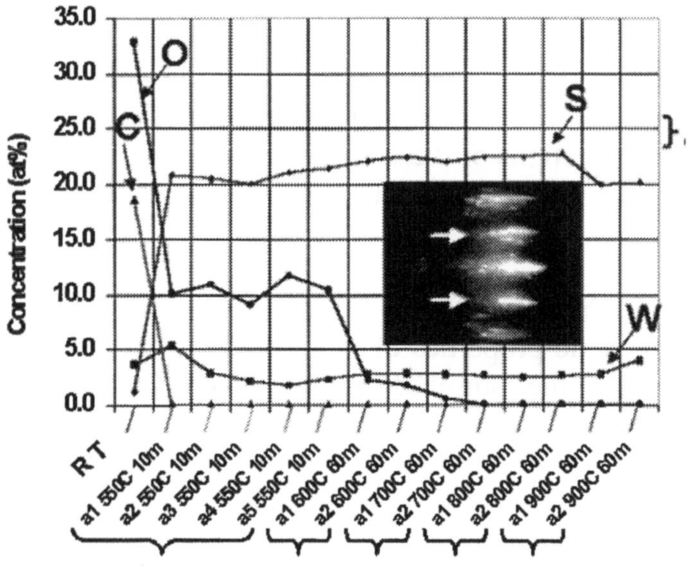

Figure 9. AES data collected on a biaxially textured Ni-3%W alloy substrate after heating at different temperatures and holding for 10 or 60 minutes. The inset shows the RHEED pattern obtained after heating to 550 °C.

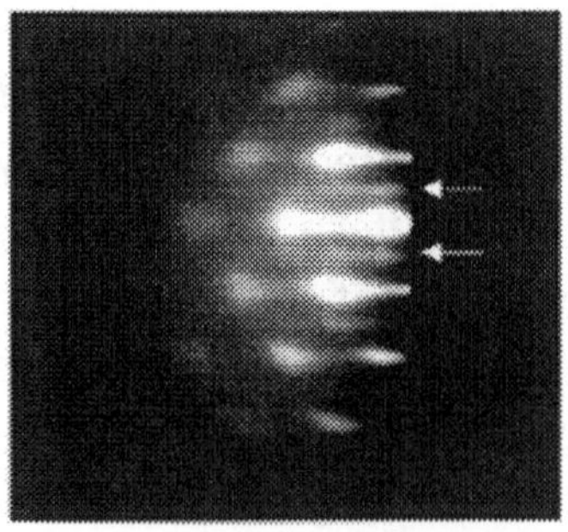

Figure 10. RHEED pattern for a 100 nm-thick G_2O_3 film on Ni-3%W

The reactivity with oxygen is substantially higher for Ni-13%Cr substrates. In fact, attempts to deposit usual seed layers like CeO_2 or Y_2O_3 by several vapor deposition techniques on Ni-13%Cr have failed. A combination of RHEED and Auger measurements on these substrates has shown that C and O species are strongly bound to the Ni-Cr surface after air exposure, and high temperature anneals (700-750 °C) in high vacuum, or 200 mTorr of forming gas, before deposition are necessary to clean the surface and expose the underlying S superstructure. Once cleaned, the Ni-13%Cr surfaces with the top *c* (2 × 2)-S superstructure are stable in background oxygen pressures lower that 10^{-8} Torr. Above this value oxidation of the substrate occurs rapidly, disrupting seed layer growth. As mentioned above, the deposition conditions for most of the seed layers used for RABiTS involve a water pressure of 10^{-5} Torr or higher. Such a water pressure in the deposition chamber gives rise to an equilibrium oxygen pressure in the range of 10^{-7} Torr, as indicated by a quadrupole mass spectrometer installed on our vacuum system, and therefore is sufficient to oxidize the Ni-13%Cr surface. The only successful attempt to oxide seed layer growth on the bare *c* (2 × 2)/Ni-Cr surface by PLD was obtained using YSZ. Unlike other seed layers, YSZ can be deposited from a ceramic target using oxygen partial pressures as low as 10^{-10} Torr. Cube textured YSZ films were grown on Ni-13%Cr at a temperature of 600 °C by ablation of a YSZ target in the base pressure of the vacuum chamber (0.5-1×10^{-8} Torr, corresponding to an O_2 partial pressure of 0.5-1×10^{-10} Torr). In addition to the (002) reflections, these films showed a small (111) component with no preferential in-plane orientation. The (111) component was not observed when the initial part of the YSZ deposition was conducted in a partial pressure of H_2S equal to 2×10^{-6} Torr. Figure 11 shows the logarithmic YSZ (111) pole figure for a 170 nm thick film that was grown using H_2S for the initial layer of 70 nm in thickness. Although the background x-ray signal was subtracted in order to bring out (highlight) possible secondary orientations, the pole figure is clean and the percentage of cube texture is 93%.

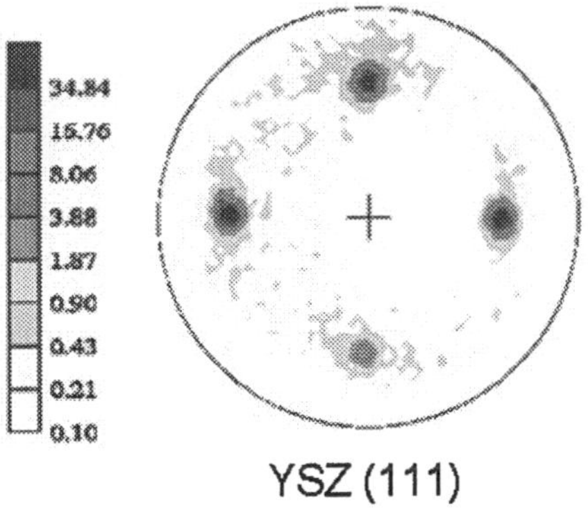

	Ni	YSZ
Δφ	6.0 °	7.4 °
Δω	7.6 °	5.1°

Figure 11. Logarithmic (111) pole figure of a YSZ film deposited on *c* (2 × 2)/ Ni-13%Cr

Despite the low background oxygen pressure, activated oxygen species present in the ablation plume can lead to local formation of Cr-oxide, and consequently partial nucleation of YSZ with (111) orientation. It is useful to recall that Cr-oxide is thermodynamically more stable than YSZ, and therefore its formation is expected for sufficiently high oxygen impingement rates. We think that the presence of H_2S during the first stage of film growth reduces the number of oxygen species in the plume by formation of SO_2 molecules. Consequently, the substrate oxidation rate decreases, allowing a well-textured YSZ film to be deposited. After the YSZ deposition, an 80-nm-thick CeO_2 cap layer was grown in situ at 660 °C in an oxygen partial pressure of 5 10^{-4} Torr. A 200-nm-thick YBCO film grown on this sample by PLD showed a J_c of 0.44 MA/cm^2 at 77 K and in self magnetic field.

In conclusion, the *c* (2 × 2)-S superstructure is not only present on {100}<100> Ni but forms also on the surface of biaxially textured Ni-alloys such as Ni-3%W and Ni-13%Cr. In all cases, the superstructure acts as a template that enables the epitaxial growth of many oxide seed layers on the metal. In addition, a complete coverage of this chemically stable layer is necessary to replicate the substrate texture in the buffer layer. Annealing Ni substrates at 800 °C in the presence of a small amount of H_2S is sufficient to produce a stable c(2 × 2)-S with a coverage of 0.5ML. This simple step allows creating a complete S template independently of the coverage obtained through segregation. Implementing this step in a continuous RABiTS fabrication process allows the initial Ni texturing anneal to be carried out at conditions that yield the best texture (and often involve high temperatures and consequent S evaporation) without being limited by the S segregation process.

Research sponsored by the U.S. Department of Energy under contract DE-AC05-00OR22725 with the Oak Ridge National Laboratory, managed by UT-Battelle, LLC.

REFERENCES

1. C. Cantoni, D. K. Christen, R. Feenstra, D. P. Norton, A. Goyal G. W. Ownby, and D. M. Zehener, *Appl Phys Lett* **79** 3077 (2001)
2. C. Cantoni, D. K. Christen, L. Heatherly, A. Goyal, G. W. Ownby, and D. M. Zehner, *Journal of Materials Research* submitted
3. M. Paranthaman, D. F. Lee, A. Goyal, E. D. Specht, P. M. Martin, X. Cui, J. E. Mathis, R. Feenstra, D. K. Christen, D. M. Kroeger, *Supercond Sci Tech* **12** (5): 319 (1999)
4. P. H. Holloway, *J Vac Sci Technol*, **18**(2) 653 (1981)
5. D. F. Mitchell, P. B. Sewell, and M. Cohen, *Surf. Sci.* **61**, 355 (1976)
6. P. C. McIntyre, *J Appl Phys* **89** 8074 (2001)
7. T. Aytug, M. Paranthaman, B. W. Kang, S. Sathyamurthy, A. Goyal, and D. K. Christen, *Appl. Phys. Lett.* **79** 2205 (2001)
8. M. Perdereau and J, Oudar, *Surf Sci.* **20** 80 (1970)
9. S. Andersson and J. B. Pendry, *J Phys C* **9** 2721 (1976)
10. C. A. Papageorgopoulos, and M. Kamaratos, *Surf Sci* **338** 77 (1995)
11. A. Patrige, G. J. Tatlock, F. M. Leibsle, and C. F. J. Flipse, *Phys Rev B* **48** 8267 (1993)
12. M. M. Kowalewski, F. A. List, L. Heatherly, and D. M. Kroeger, unpublished
13. J. R. Thompson, A. Goyal, D. K. Christen, and D. M. Kroeger, *Physica C* **370** (3) 169 (2002), also cond-mat/0104239.

ION-BEAM-ASSISTED DEPOSITION OF MAGNESIUM OXIDE FILMS FOR COATED CONDUCTORS

T. P. Weber, B. Ma, U. Balachandran
Energy Technology Division
Argonne National Laboratory
Argonne, Il, 60439

M. McNallan
Department of Materials Science and Engineering
University of Illinois at Chicago
Chicago, IL 60612

ABSTRACT

The development of high critical-temperature thin-film superconductors and coated conducting wires is important for electric power applications. To achieve high transport current density, template films are necessary for the successful deposition of biaxially aligned $YBa_2Cu_3O_{7-x}$ (YBCO) on flexible metal substrates. We grew biaxially aligned magnesium oxide (MgO) template films by ion-beam-assisted deposition with electron-beam evaporation. MgO films of ≈100 Å thickness were deposited on Si_3N_4-coated Si substrates at a deposition rate of ≈1.5 Å/sec with an ion flux of ≈110 $\mu A/cm^2$ bombarding the substrate at a 45° angle. To study crystalline structure by X-ray diffraction, we deposited an additional layer of MgO. Good in- and out-of-plane alignment was observed, with (111) ϕ-scan full-width half-maximum (FWHM) of 6.2° and (002) ω-scan FWHM of 2.2°.

INTRODUCTION

High-temperature superconducting wires and tapes have a variety of applications, including transmission wires and microwave devices [1,2]. Fabrication of tapes that can carry the necessary current has been the focus of a great deal of research. Production of high temperature superconducting

$YBa_2Cu_3O_{7-x}$ (YBCO) tapes on nonaligned metallic substrates requires high-quality biaxially textured template layers. Yttria-stabilized zirconia (YSZ) has been used successfully as a template layer when deposited by the ion-beam-assisted deposition (IBAD) process. While superconducting films made with YSZ buffer layers have been successful, with transport current densities $\approx 10^6 A/cm^2$, a limitation to the production process is the time required for YSZ deposition. Due to the competitive growth process of YSZ, films with thickness of ≈8,000 to 10,000 Å are necessary in order to achieve the desired in-plane alignment of ≈13° [3,4]. The excessive time required to deposit such films has driven the search for new buffer-layer materials.

MgO, on the other hand, exhibits a much more rapid growth mechanism. Because texturing of MgO begins at the nucleation stage, its films need only be ≈100 Å thick in order to obtain quality texture [5-7]. Processing times for these films are thus reduced by at least one order of magnitude from those of YSZ films [8]. Using MgO buffer layers is a significant step toward producing high critical-temperature superconductors for practical applications.

We grew biaxially textured MgO films on Si_3Ni_4 substrates with the IBAD process and an e-beam evaporation system. An additional homoepitaxial layer was deposited in order to facilitate film characterization via X-ray analysis; ϕ-scans and ω-scans were used to analyze film texture.

EXPERIMENTAL PROCEDURE

IBAD MgO films were deposited on as-received Si_3N_4 substrates measuring $\approx 1 \times 0.5$cm and mounted on a heatable stage with silver paste. A 3 cm Kaufman-type ion source was used to accelerate argon ions at 750 eV. All substrates were presputtered for 5 minutes before deposition. The angle of incidence, α, for the ion beam was maintained at 45°, the channeling direction for MgO. A Faraday cup was used to measure the beam current density of 110 μA/cm^2. A 10:1 argon-to-oxygen ratio provided a background pressure of 8×10^{-5} torr during IBAD. Vapor flux was provided by e-beam evaporation and monitored with a quartz crystal monitor at 1.5 Å/s for an ion-to-atom ratio of ≈0.9. The IBAD films were deposited to a thickness of 100 Å. The experimental setup is illustrated in Fig. 1 and deposition conditions are given in Table 1.

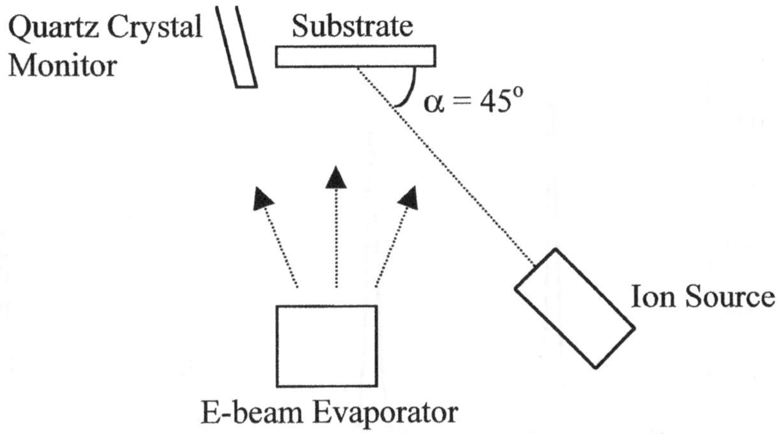

Fig. 1. Experimental setup for IBAD process.

After IBAD MgO deposition, the substrates were heated to ≈300°C before addition of a homoepitaxial layer of MgO deposited without ion fluence. The homoepitaxial layer was necessary in order to characterize the films by x-ray diffraction and is ≈1000 Å thick. Oxygen flow was introduced into the system at a rate of ≈0.35 sccm during deposition. The deposition rate for the homoepitaxial layer was the same as that for IBAD (1.5 Å/s).

Table 1. Deposition conditions for biaxially textured IBAD MgO films.

Operating Pressure	8×10^{-5} torr
Deposition Rate	1.5 Å/s
Ion beam incident angle	45°
Beam energy	750 eV
Ion fluence	110 μA/cm^2
Gas flow (Ar + O_2)	3.35 sccm

RESULTS AND DISCUSSION

Biaxial texture of the IBAD MgO films was characterized by X-ray ϕ-scans and ω-scans. In-plane texture determined by the full-width half-maximum (FWHM) of (111) ϕ-scans is illustrated in Fig. 2. The average FWHM is 6.2°, which compares favorably with the obtainable in-plane texture of YSZ films.

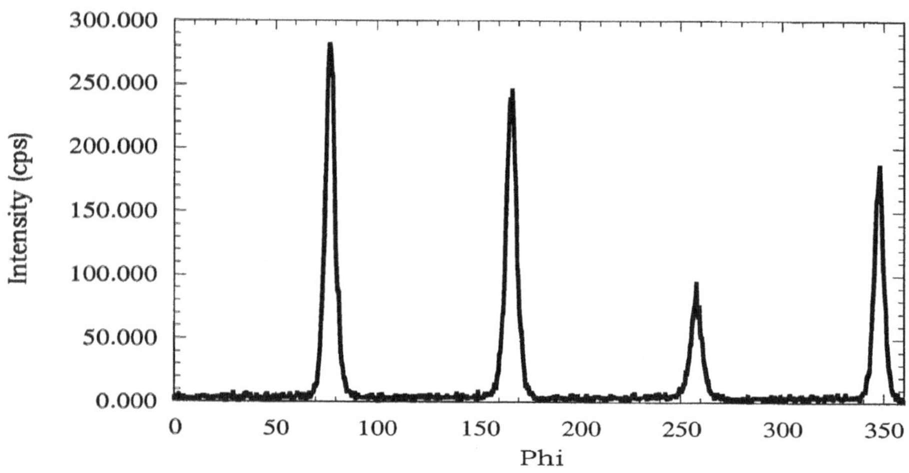

Fig. 2. X-ray diffraction phi-scan of MgO film.

Out-of-plane texture as characterized by the ω-scan of Fig. 3, shows a FWHM of 2.2°. The appearance of high-quality texture in films of only ≈100 Å thickness illustrates the nucleation growth mechanism of IBAD MgO films. Our experience has shown that texture deteriorates rapidly if film thickness deviates more than ≈20 Å from this thickness. The relationship between film texture and thickness of the film was analyzed and described in reference 9; our experiments support these findings. Figure 4 shows diffraction patterns from a series of 2-theta scans performed on films of varying thickness. All films were deposited on Si_3N_4 substrates under the growth conditions mentioned previously. The figure

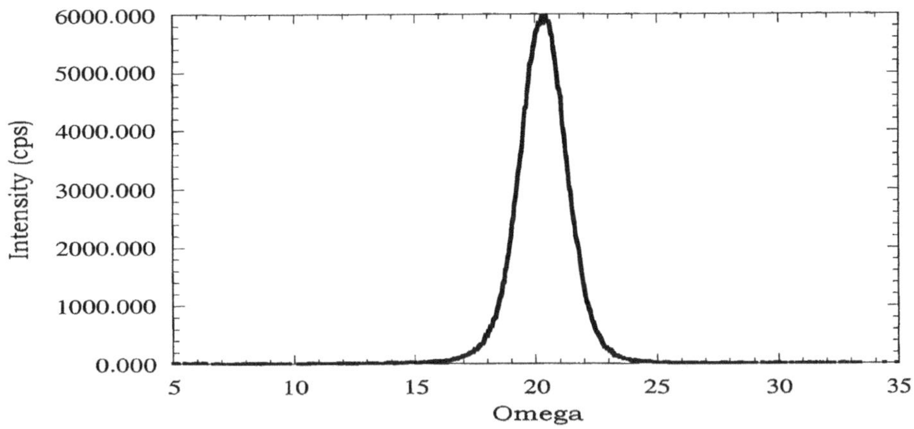

Fig. 3. X-ray diffraction omega-scan of MgO film.

illustrates the presence of the MgO <200> peak at ≈75 Å, followed by a much stronger peak in films of ≈100 Å thickness. In films of ≈125 Å, the <200> peak becomes much less intense and the presence of the <220> peak is evident.

Phi-scans of our films of various thickness also exhibit the broadening of FWHM values to ≈8° for films of ≈80 Å and ≈15° for films of ≈120 Å. The results of our studies of these films agree with the critical thickness for optimal texture of IBAD MgO films at 100 Å thickness.

We also performed studies on deposition temperature for the homoepitaxial layer. Immediately after deposition of the IBAD layer, the films were heated to 300, 400, and 500°C for the subsequent application of the 1000 Å homoepitaxial layer. Our preliminary findings indicate no observable difference in obtainable texture among these deposition temperatures and hence no apparent advantage to depositing at the higher temperatures of 400 and 500°C.

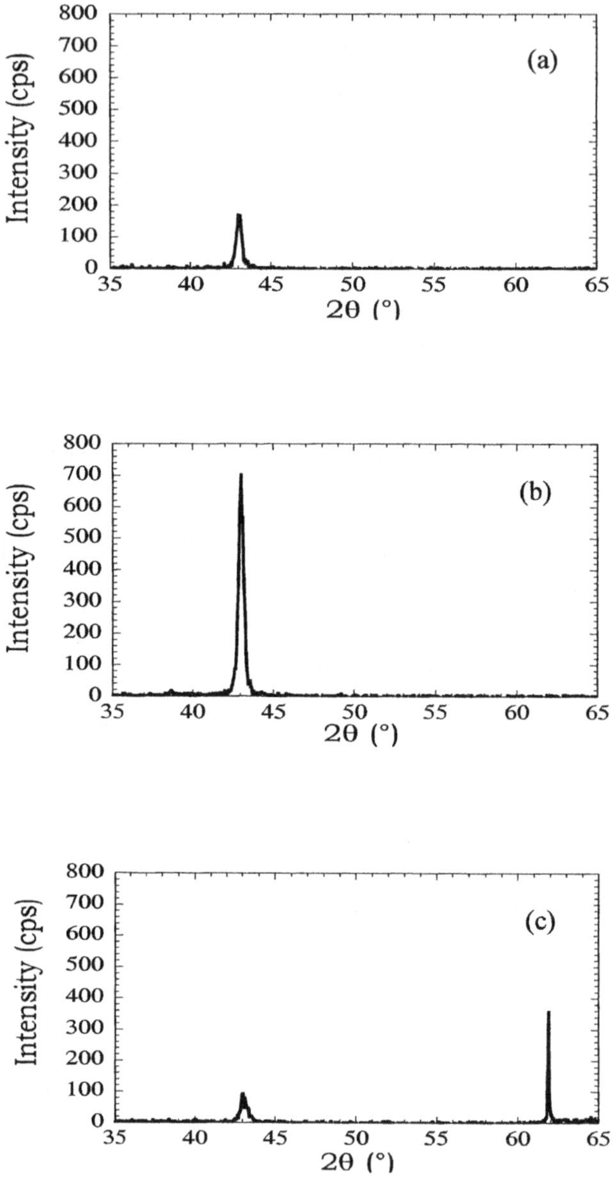

Fig. 4. X-ray 2-theta scans for IBAD MgO films of (a) 75 Å, (b) 100 Å, and (c) 125 Å thickness.

CONCLUSIONS

Biaxially textured MgO films were successfully grown by using the IBAD MgO process. MgO films grown with this approach require much less time than IBAD YSZ films and exhibit favorable texture. We examined texture variation with respect to film thickness and found the in- and out-of-plane texture to be optimal in films of ≈100 Å thickness. Additionally, our research indicates that these films can be fabricated at a homoepitaxial deposition temperature of only 300°C. IBAD MgO films having in-plane texture of 6.2° and out-of-plane texture of 2.2° were observed. This work has demonstrated that high quality textured MgO films can be grown with the IBAD process in very thin films (≈100 Å) in a very time-efficient manner. Such films are excellent candidates for buffer layers in the fabrication of YBCO coated conductors.

ACKNOWLEDGMENT

This work was supported by the U.S. Department of Energy (DOE), Office of Energy Efficiency and Renewable Energy, as part of a DOE program to develop electric power technology, under Contract W-31-109-Eng-38.

REFERENCES

1. J.O. Willis, P.N. Arendt, S.R. Foltyn, Q.X. Jia, J.R. Groves, R.F. DePaula, P.C. Dowden, E.J. Peterson, T.G. Holesinger, J.Y. Coulter, M. Ma, M.P. Maley, and D.E. Peterson, "Advances in YBCO coated conductor technology," *Physica C*, **335**, 73-77 (2000).
2. Y. Iijima, M. Kimura, T. Saitoh, and K. Takeda, "Development of Y-123 coated conductors by IBAD process", *Physica C*, **335**, 15-19, (2000).
3. K.G. Ressler, N. Sonnenberg, and M.J. Cima, "The development of biaxial alignment in YSZ films fabricated by IBAD," *Journal of Electronic Materials*, **25** [1] 35-42 (1996).

4. B. Ma, M. Li, B.L.Fisher, and U. Balachandran, "Ion-beam-assisted deposition of biaxially aligned yttria-stabilized zirconia template films on metallic substrates for YBCO0-coated conductors," *Superconductor Science and Technology*, (March 2002).
5. J.R. Groves, P.N. Arendt, H. Kung, S.R. Foltyn, R.F. DePaula, and L.A. Emmert, "Texture development in IBAD MgO films as a function of deposition thickness & rate," *IEEE Transactions on Applied Superconductivity*, **11** [1] 2822-2825 (2001).
6. J.R. Groves, P. N. Arendt, S. R. Foltyn, R. F. DePaula, E. J. Peterson, T. G. Holesinger, J. Y. Coulter, and Raymond W. Springer, "Ion-beam assisted deposition of biaxially aligned MgO template films for YBCO coated conductors," *IEEE Transactions on Applied Superconductivity*, **9** [2] 1964-1966 (1999).
7. J.R. Groves, P. N. Arendt, S. R. Foltyn, Q. Jia, T. G. Holesionger, H. Kung, R. F. DePaula, P. C. Dowden, E. J. Peterson, L. Stan, and L. A. Emmert, "Development of the IBAD MgO process for HTS coated conductors", *International Workshop on Superconductivity*, 43-46 (2001).
8. C.P. Wang, K.B. Do, M.R. Beasley, T.H. Geballe, and R.H. Hammond, "Deposition of in-plane textured MgO on amorphous Si_3N_4 substrates by IBAD & comparisons with IBAD YSZ," *Applied Physics Letters*, **71** [20] 2955-2957 (1997).
9. J.R. Groves, P. N. Arendt, H. Kung, S. R. Foltyn, R. F. DePaula, L. A. Emmert, and J. G. Storer, "Texture development in IBAD MgO films as a function of deposition thickness and rate," *IEEE Transactions on Applied Superconductivity*, **11** [11] 2822-2825 (2001).

SCALING-UP OF HIGH-Tc TAPES BY MOCVD, SPRAY PYROLYSIS AND MOD PROCESSES

Sandrine Beauquis, Sébastien Donet, François Weiss
Laboratoire des Materiaux et du Genie Physique-ENSPG-INPG,CNRS-UMR5628, Bp46, F-38402 S. Martin d'Hères, France

ABSTRACT

Fabrication of coated conductors is a key aspect of the development of High Temperature Superconductors (HTS). Chemical deposition processes (MOCVD, spray pyrolysis and MOD) become actually the more promising techniques for the production of HTS tapes, due to the good relation between sample quality, mass production and low capital costs of these processes.

We compare the advantages and drawbacks of each process and specify the criteria which may be considered to select the more appropriated one.

Several compounds have been synthesized using the aforementioned techniques, including buffer layers (CeO_2, YSZ, $BaZrO_3$, Y_2O_3...) and the different HTS phases (Y-123, Hg-1223, Tl-1223). Properties of these films and of various architectures will be reported.

INTRODUCTION

The Coated conductor technology requires flexible and mechanically strong textured substrate tapes, with a good chemical compatibility with the deposited superconducting oxide film. The Substrate thickness is generally over 50µm (for a good mechanical behaviour) and the superconducting layer has to be as thick as possible (thickness if possible over 1 µm) in order to increase the overall engineer current flowing in the conductor. The matching between metallic substrates and HTS layers can only be realised by adding adequate buffer layers (CeO_2 and YSZ have actually been identified as good materials for this purpose) [1,2].

We discuss in the present paper the chemical engineering of different Chemical Deposition Processes used for the synthesis of thin or thick superconducting oxide films on RABiTS or IBAD based substrates. The different deposition techniques considered are : i) MOCVD, ii) spray pyrolysis and iii) Metal-Organic Decomposition (MOD). These Chemical routes are very promising for coated conductor production, since they are considered as low cost techniques that can be extrapolated to a large-scale production.

In the present review, the basic discussion, concerning film preparation and film characterisation, is concentrated on YBCO based conductors. The other families of superconducting oxides (Tl and Hg based superconductors) have also been investigated and the main characteristics of these films will also be presented.

For the complex architecture of coated conductors different compounds have been synthesised by the Chemical deposition processes described, from buffer layers (CeO_2, YSZ, Y_2O_3...) to the superconducting phases (Y-123, Hg-1223, Tl-1223).

We present for each of these techniques the basic principle involved, the main results obtained today for the fabrication of HTS thin and thick films and their development for the realisation of coated conductors in reel to reel systems.

CHEMICAL DEPOSITION: BASIC PRINCIPLES

Chemical Deposition processes, used for the formation of HTS films, are based on chemical reactions between inorganic or organic metal containing species, leading to the formation of a thin layer on an appropriate substrate.

These processes can be classified in two main groups :

1.One step processes, (MOCVD) using volatile chemical precursors, where the superconducting oxide phase is formed directly after chemical reactions from the gas phase on the surface of a determined substrate

2.Two step processes, (MOD, sol-gel, spray pyrolysis..) which are based on the deposition of an unreacted precursor film, followed by a recrystallisation step, where solid state reactions induce the formation of the desired superconducting phase.

All these processes are in general simple and less expensive than physical deposition processes. They are commonly used in the production of various coatings and thin films with the main advantage that they allow an easy scaling up to large size and a conformal coverage of complex shapes.

METALORGANIC CHEMICAL VAPOR DEPOSITION (MOCVD) OF HTS

In a MOCVD process thin films are formed on a surface from the thermal decomposition of gaseous precursor molecules. The total pressure used in the gas phase is generally a low pressure, in the range of a few Torrs. This pressure range is commonly attained with classical one stage vacuum pumps and does not need ultra high vacuum equipments, which increase considerably the equipment and maintainance costs in a deposition system.

Precursors for MOCVD should in general meet the following properties:

1) The precursor should have an evaporation rate which is large enough, stable and constant with the time.
2) The precursor molecule should be chemically and thermally stable during the transport through the gas phase to the surface,
3) The precursor should be relatively easily to synthesize
4) The precursor should not be dangerous and should not produce dangerous side products.
5) The precursor must be soluble and stable in a suitable solvent without formation of precipitates (In the case of a droplet derived deposition process)

For MOCVD of HTS materials, chelate compounds are the main candidates, basically β-diketonates compounds are used (they are also named tmhd). They are free of fluorine and have vapor pressures which are high enough for MOCVD experiments.

Since a few years, new liquid delivery systems have been introduced to overcome the problem of thermal stability of Ba complexes (Pulsed Injection (PIMOCVD) [3, 4, 5] Band Evaporation (BEMOCVD) [6]). The main advantages of these liquid delivery systems are :

- A single precursor mixture can be used for multicomponent systems
- Very high growth rates can be obtained

- In PIMOCVD, the solution is injected directly by a micro-valve. The frequency, the opening time and the solution concentration are the main parameters controlling the growth rate.

- In BEMOCVD, a band system is wet with the precursor solution. An evaporator separates the solvent molecules from the precursor molecules and dry precursors are evaporated to the deposition zone. As solvent mostly diglyme or monoglyme are used.

Actually, on the basis of these two new liquid delivery systems, high quality YBCO films have been obtained on static substrate wafers [7, 8, 9].

Fabrication of long Tapes: In order to deposit HTS thin or thick films on moving tapes a special reel-to-reel deposition system has been developed (Fig.1). In this tape casting unit, the tape is wound on two reels, before and after deposition, the reels remaining under vacuum during the whole process. The precursors are dissolved in monoglyme and injected (with a Pulsed Injection system) into an evaporator heated at 250°C. The solution injected is flash volatilized and carried into the deposition chamber by a strong flux of Ar/O_2. The transfer time, between the evaporator and the deposition zone, is very short (1second).

A post annealing furnace is set-up on line in order to control, after deposition, the tape cooling and the oxygen loading of the films.

Figure 1 : Reel-to-reel MOCVD deposition system The substrate wind on a supply reel passes through the deposition chamber to be wound on a winding reel.
A : annealing chamber , B : deposition chamber, C : winding reel, D : supply reel.

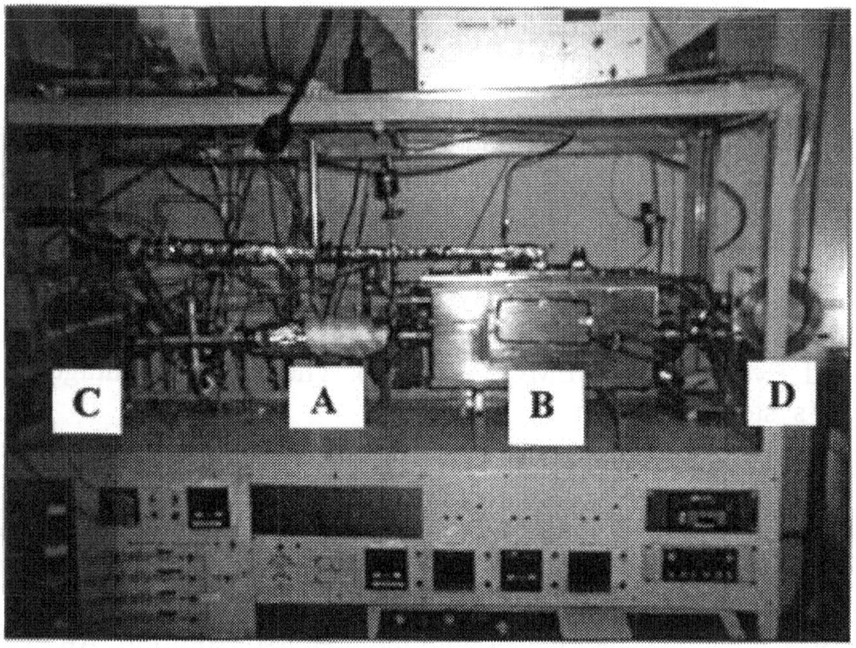

Using this system, buffer layer deposition (YSZ, GdSZ, CeO_2, Y_2O_3, Gd_2O_3...) on 3meters tapes, YBCO (ie: SmBCO) deposition on 2m tapes, as well a combination of complex architectures (CeO_2/ YBCO for example) have been realised.
The experiences were focused on:

- reducing the weak link behaviour of films, high angle grain boundaries
- increasing film thickness and film deposition rates.

High quality YBCO films have been obtained on Ni RABiTS and on SS//YSZ (IBAD) tapes (Fig 3, 4) [10, 11]. Special buffer layer architectures have been developed in order to obtain high-critical current densities (Jc >1MA/cm^2, 77K, OT) as well as a high degree of epitaxial texture.

On RABiTS tapes a successful stacking was realised with Ni//NiO/YSZ/Y_2O_3/YBCO. NiO was prepared by a Substrate Oxidation Epitaxy technique [12], and all the following buffer layers were prepared in-situ by

MOCVD with a final YBCO layer of 750nm. The Jc value reaches 0.9 10^6 A/cm^2 in this stacking.

On Hastelloy substrates, covered with an YSZ IBAD layer, the best results have been obtained for the following architecture : SS//YSZ/CeO_2/YBCO (Fig. 5). Critical current densities , measured on short samples are of Jc= 1.3 10^6 A/cm^2 for a 300nm thick CeO_2 and 750 nm thick YBCO layer.

Figure 3 : Epitaxial growth of YSZ/CeO_2/YBCO on SS/YSZ (IBAD) substrates

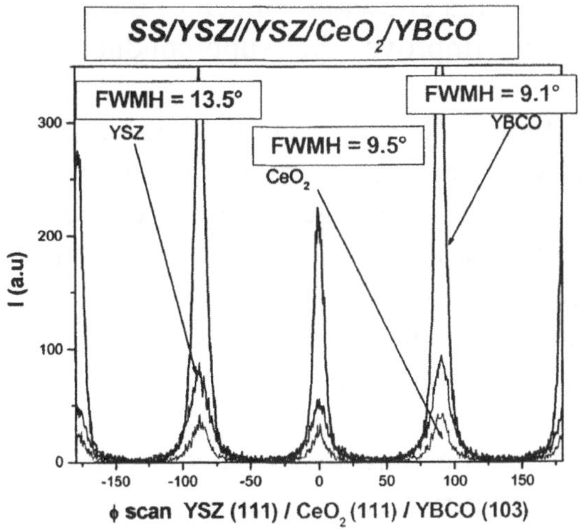

Figure 4 : Tc of YBCO on IBAD and RABITS substrates

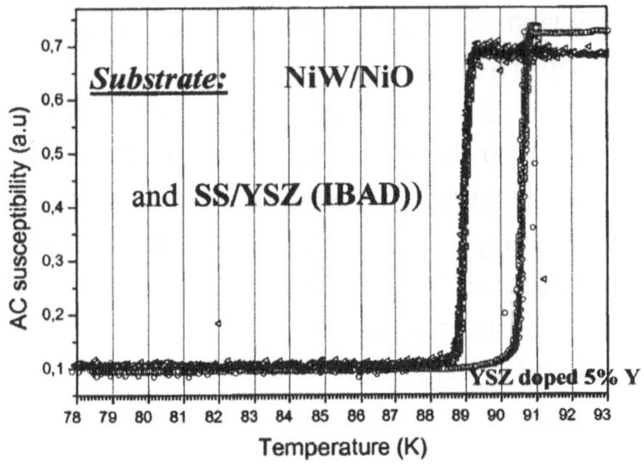

Thicker 2µm HTS layers have also been deposited, based on a multilayer stacking of Y123/Sm123/Y123 (Fig.6) with a Jc value of 1.2 10^6 A/cm^2.

Figure 5 : SEM picture of a SS/YSZ//YSZ/CeO_2/YBCO films

Figure 6 : SEM picture of a 2µm thick Y123/Sm123/Y123 layer

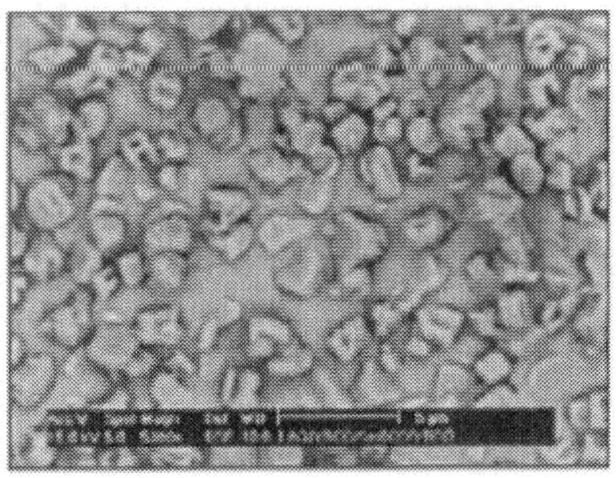

SPRAY PYROLYSIS DEPOSITION

Aerosol routes offer a variety of approaches for film generation. These processes have been used to deposit many materials including ceramic superconductors, simple metal oxides, non-oxide ceramics... at high rates. Experimental procedure involves the pyrolysis on a heated substrate of droplets produced by an aerosol of non-volatile liquid source precursors. This procedure allows in some extend to have a reasonable control over film thickness and over the homogeneity.

In general the standard deposition route for the preparation of HTS materials involves a deposition stage of a non completely reacted film (precursor film), followed by one or several annealing treatments allowing control of the film recrystallisation and a final oxidation step to improve the superconducting properties.

For the spray pyrolysis of HTS materials (YBCO, Tl and Hg based superconductors) aqueous nitrate solutions have been used in most of the cases but sometimes also carboxylates. Nitrate solutions are generally preferred, specially in the case of Ba based superconductors, because they allow to circumvent problems of carbon contamination in the layer.

Aqueous nitrate solutions are prepared either by dissolving $Y(NO_3)_3 \cdot 3H_2O$, $Ba(NO_3)$, and $Cu(NO_3) \cdot 6H_2O$, or $Ca(NO_3)_2 \cdot 4H_2O$ in highly distilled and deionized water to give the suitable cationic ratio for deposition, or from single oxides or carbonates by dissolution in nitric acid. This last procedure gives a better control of stoichiometry because the water content of nitrates is not always exactly known.

YBCO films: YBCO films have been sprayed over single crystalline substrates, MgO, $SrTiO_3$ and on C fibers and wires [13]. After the deposition, films are in an amorphous state and therefore an annealing step at elevated temperature is necessary to build the proper structure. This annealing is performed either with an inert atmosphere or in a mixture of an inert atmosphere with O_2, with a final post-annealing under O_2 atmosphere.

Recent studies have shown that biaxial texturing of YBCO can be reached using nitrate deposition at elevated temperatures on single crystalline substrates and on biaxially textured Ag substrates. Jc in the range 10^4-10^5 A/cm^{-2} were reported [14,15,16].

In the present study, YBCO thick films have been synthesized using the spray pyrolysis method, with the aim to develop an in-situ deposition process. We used an all nitrate precursor approach to grow epitaxial HTS oxide film on

$SrTiO_3(001)$ substrate at a high deposition rate and at high temperature (over 800°C). We performed 1µm thick YBCO films [17, 18] in situ at 850°C in a very short deposition time (4 minutes) corresponding to a very high local deposition rate of 12 µm/hr. The final sintering process is performed at the same temperature for 2 hours under oxygen atmosphere before low temperature oxygen loading (Figure 7). X-ray diffraction (XRD) results (θ-2θ scans and φ scans) show that the films do not have only a strong c-axis texture but also a sharp in-plane biaxial orientation. FWHM values for YBCO (102) and YBCO (005) diffraction lines can be as low as ~1.4° (in-plane epitaxy, Δφ) and ~0.4° (out-of-plane epitaxy, Δω), respectively. Furthermore, the critical temperature (T_c) and the critical current density (**J_c**) of this film are 91K and 1.4 x 10^6A/cm^2 (@ 77K, 0T) (Figure 8).

YBCO thick films have actually been grown on different single crystal substrates with high current performance (**J_c** >1 MA/cm^2) and on Ag single crystals or biaxially textured tapes. The deposition conditions are actually adapted to buffered coated conductor substrates.

The all-nitrate spray pyrolysis technique, using a very low cost equipment and "green" raw materials for the production of YBCO films at high deposition rate (12µm/hr.), encourages very much to attempt the scaling up of this process for producing long length-coated conductors in the future.

A Reel-to-reel equipment is actually under development for this purpose.

Figure 7: SEM photomicrograph of YBCO film on $SrTiO_3$

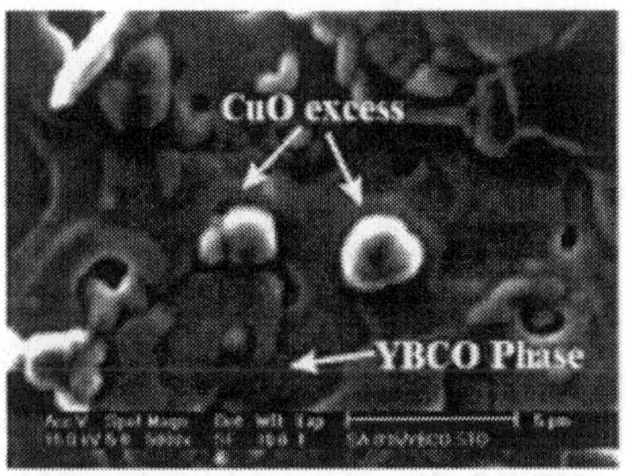

Figure 8 : Tc, Jc and critical current profile of YBCO film on $SrTiO_3$

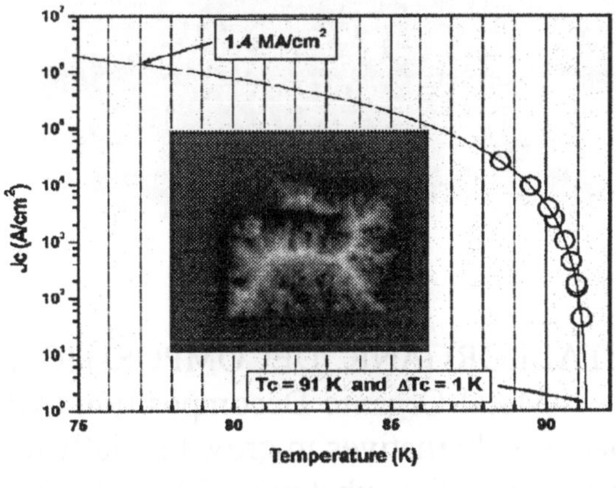

Tl(Hg)-BCCO films: Tl (and Hg) based oxide superconductors have also been synthesized using spray pyrolysis techniques. The synthesis procedure is splitted in two distinct steps :

- deposition of a Ba-Ca-Cu nitrate precursor layers on single crystalline or Ag substrates.
- Tl or Hg introduction into the pre-annealed precursor layer, using a Tl or Hg source from a Tl(Hg)BCCO pellet or from a Tl_2O_3 ($HgCaO_2$) powder.

The second step annealing is carried out at high temperatures (typically 800-860°C) with a careful control of Tl (Hg) partial pressures.
Tl based films have be prepared with very good superconducting properties. The most interesting Tl compound, Tl-1223, can be obtained relatively as a pure phase [19]. The best superconducting properties obtained so far are $T_c = 120$ K, $J_c = 10^5$ Acm^{-2} (77 K, 0 T).

Hg-1223 films present actually the most promising superconducting properties. Films with a thickness of 1-2 μm and T_c~130K have been epitaxially grown by spray pyrolysis on $SrTiO_3$ substrates (Figure 9). X ray diffraction data show a good c-axis orientation, the $CaHgO_2$ phase should disappear with a longer high temperature treatment.. In these films, the critical current density can be as high as: j_c~$4.4 \cdot 10^5$ Acm^{-2} (77 K, 0 T) [20,21].

Figure 9: SEM micrograph of a Hg-1223 thin films.

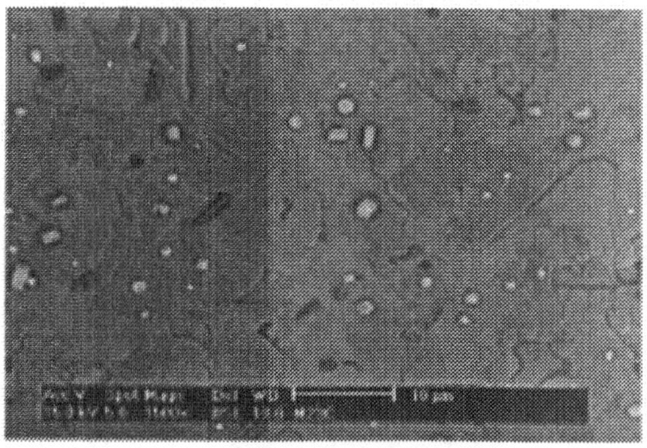

Fig 10: XRD diagram of a Hg-1223 thin films

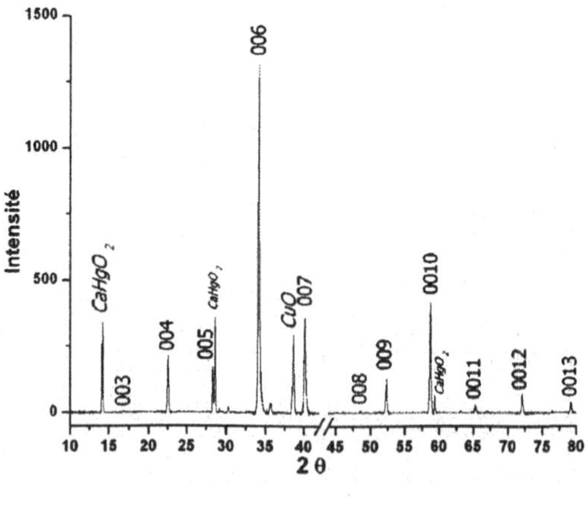

METALLORGANIC DECOMPOSITION (MOD)

Metal-Organic Decomposition (MOD) methods appeared recently as very appealing alternatives to grow biaxially textured YBCO films in combination with textured metallic substrates [22]. It has first been shown that the MOD technique has a very high flexibility in preparing simple oxides buffer layers (CeO_2, YSZ) and binary compounds such as $BaZrO_3$ and $LaNiO_3$. Moreover, the successful deposition of high Jc YBCO at MIT by using Tri-fluoro-acetate (TFA) precursors demonstrates that the chemical processing of new generation of coated conductors can be processed by MOD.

The MOD method consists in combining precursors of the desired metals in the corresponding ratio followed by their thermal decomposition. The decomposition then yields an atomic level mixture of the oxides/carbonates. After thermal treatment the desired mixed metal oxide material is obtained. MOD uses as precursors a metal linked to organic groups, most commonly metal carboxylate and modified metal alkoxides. The MOD ligands are separated from the precursor by thermolysis; therefore, compounds that sublime, evaporate and are not suitable as MOD precursors. Usually, the thermal decomposition leads to the formation of metal carbonates and the corresponding ketone, both of which decompose further following their own thermodynamic characteristics. In this system, relatively stable $BaCO_3$ forms as intermediate compound during the Y-123 synthesis. In order to avoid this problem, trifluoacetate are used as a metal carboxylate. They are forming oxofluorides which are converted to oxides by reacting with the moist atmosphere.
Critical current densities of $Jc=7x10^6 A/cm^2$ and $Jc=1,7x10^6 A/cm^2$ (77k,0T) have been obtained for very thin films (200nm) on single crystalline $LaAlO_3$ substrates and on IBAD substrates with a CeO_2 cap layer respectively [23, 24, 25].

As an alternative MOD process, and following the same idea, which is to prevent the formation of intermediate $BaCO_3$, we have developed an all-iodide deposition route [26]. The solution is spun at 6000rpm for 40sec on *(100)* $LaAlO_3$ single crystals. The thermal treatment applied, consists in a heating ramp at 200°C/hr up to 600°C under pure O_2 (1 bar) after what, the gas is switched to 100ppm O_2 at 800°C for 1hr. The samples are later cooled down at a rate of 200°C/hr in pure O_2 up to 500°C where the sample is maintained for 1h following a furnace cooling to room temperature.

Fig.10 : XRD pattern of Y-123 obtained by the iodine route.

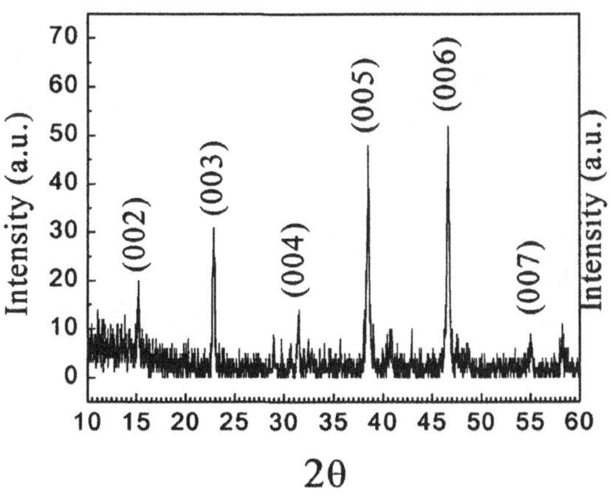

Fig.11 : Rocking curve of (005) peak of Y-123 obtained by the iodine route.

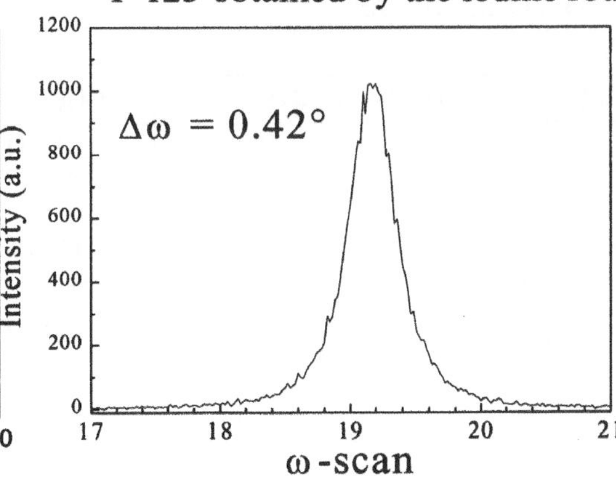

Epitaxial YBCO layers have been obtained (Fig. 10) with a good out of plane orientation of $\Delta\omega = 0.42°$ (Fig.11). These films have a critical temperature of 90K and critical current densities over 10^5 A/cm^2. These first values can certainly be optimised. Nevertheless, the solution handling of the iodide mixture is difficult and raw materials (mainly YI_3) are expensive. Alternative combinations of iodides and carboxylates are actually studied in order to economically scale up this process.

PROCESS RELATED TECHNICAL ISSUES

All the Chemical Deposition Methods presented in this study have shown that they are able to produce high quality HTS films either on single-crystalline (LAO,STO...) substrates or on technical metallic substrates with Jc over 1MA/cm^2 in most of the cases.

The major issues for the economical manufacturing of long length YBCO coated conductors pertain to the capability of uniform, epitaxial deposition at high rates over a large area with a continuous process. Such capability has to meet the other important requirements of proper stoichiometric composition and epitaxial structure. The volumic growth rate Vr (growth speed * covered surface aera A) is a major parameter splitting the different deposition techniques. Vr will fix the whole production time of a given coated conductor as well as its average effective cost.

Table I sums-up the performance of various deposition techniques, extrapolated to the largest deposition zones actually covered, and gives a comparison between Physical deposition techniques and Chemical deposition techniques.

Table I : Deposition technique and deposition rate

	r (µm/hr)	*Vr* (µm.cm^2/hr)	*A* (covered surface area cm^2)
Pulse Laser Deposition	3 - 7	25 - 60	10-20
Magnetron Sputtering	0.3	25 - 100	80 (4")
Thermal Evaporation	1.2	400	320 (8")
PLD - IBAD for buffer		36	
MOCVD	5 -10	250 - 500	50
MOCVD (extrapolated)	40	15000	400
MOD, spray pyrolysis	>40	> 15 000 !!!	> 400 !!

In Chemical Processes, the covered surface area can normally be extended to large sizes by simple chemical engineering techniques and a proper gas phase distribution. The local growth rate can easily be a factor 10 higher than in physical deposition processes. Chemical Processes are therefore much faster than Physical processes and are certainly better adapted for mass production of long length coated conductors. As an example, a $400cm^2$ MOCVD reactor (as shown in Table I) will be able to produce a 1cm wide coated conductor tape transporting 100A (1µm YBCO with $Jc=10^6 A/cm^2$) at a speed of 150m/hr.

For each of the processing option presented, it is finally very important to assess clearly the cost of chemicals, the mass transfer (reactor yield) and reaction kinetic data, the need and cost of diagnostic and control elements and the environmental issues, in order to select the best process for coated conductor production.

Concerning the cost of chemicals, it appears in Table II that the nitrate route is certainly the cheapest one. Equipment (no vacuum), production and maintenance costs of this "green" process strengthen the assess that, among the Chemical deposition techniques, it is actually the only one able to reach the 10EUROS/KA.m criterion fixed for coated conductors.

Table II : Cost comparison of chemicals (raw estimations) for three 1µm thick layers(YSZ, CeO_2 and YBaCuO) on a 1cm wide and 1 meter long tape, with a reactor yield of 50%.

COSTS in EUROS	Nitrate spray pyrolysis	MOCVD (tmhd)	MOD (TFA)
Research cost	0.052	31.3	75
Production cost	0.018	3.1	7.5
Cost / KA.m	0.18	31	75

CONCLUSION

Chemical Deposition Processes have been shown to be, in the present state of the art, good alternatives for the technology development of coated conductors. If good quality YBCO based short samples of conductors can actually be obtained with all of these techniques, strong efforts have always to be made by improving the fabrication pathways, on the quality of substrates and on simplified buffer layer architectures. Easy scaled-up processes and fast deposition rates are the main advantages of Chemical Deposition techniques.These parameters are crucial for the mass production of long lengths coated conductors.

AKNOWLEDGEMENTS.

This work has been supported by the European Community under Brite Euram contracts READY (BRPR-CT-98-0676), SUPERTEXT (BRPR-97-556). Parts of the basic process research were financially supported by the Region Rhône-Alpes.

---

[1]X.D. Wu, S.R. Foltyn, P.N. Arendt *et al.*, "Properties of $YBa_2Cu_3O_{7-\delta}$ thick films on flexible buffered metallic substrates", *Appl.Phys.Lett.*, **67** [16] 2397-9 (1995).

[2] J.O. Willis, P.N. Arendt, S.R. Foltyn, Q.X. Jia, J.R. Groves, R.F. DePaula, P.C. Dowden, E.J. Peterson, T.G. Holesinger, J.Y. Coulter, M. Ma, M.P. Maley, D.E. Peterson, "Advances in YBCO-coated conductor technology". *Physica C* **335** 73|-7 (2000).

[3]J.P. Sénateur, R. Madar, F. Weiss, O. Thomas, A. Abrutis, 93/08838, France, (1993), extended : FR94/0000858, Europe/USA, (1994).

[4]F. Felten, J.P. Senateur, F. Weiss, R. Madar, A. Abrutis, "Deposition of Oxide Layers by computer Controlled "Injection LPCVD ", *J. de Physique IV*, C5 1079 (1995).

[5]J.P. Senateur, F. Felten, S. Pignard, F. Weiss, , A. Abrutis, V. Bigelyte, A. Teiserskis, Z. Saltyte, B. Vengalis, "Synthesis and characterization of YBCO thin films grown by Injection-MOCVD", *Journal of Alloys and Compounds*, **251** 288 (1997).

[6]O. Stadel, J. Schmidt, G. Wahl, F. Weiss, D. Selbmann, J. Eickemeyer, O. Yu. Gorbenko, A. R. Kaul, C. Jimenez, "Scale up of a single source MOCVD system for deposition of YBCO on travelling tapes", *Journal de Physique IV*. **11**, Pr11-233 (2001).

[7]F. Weiss, U. Schmatz, A. Pisch, F. Felten, S. Pignard, J.P. Senateur, A. Abrutis, K. Fröhlich, D. Selbmann, L. Klippe, "HTS thin films by innovative MOCVD processes" *Journal of Alloys and Compounds*, **251** 264 (1997).

[8]J. Lindner, F. Weiss, J.P. Senateur, A. Abrutis, "YBa2Cu3O $_{7-x}$ /$SrTiO_3$//LaAlO heterostructures obtained by injection MOCVD, "Integrated Ferroelectrics", **3(** [1-4] 301-8 (2000).

[9]Jimenez-C; Weiss-F; Senateur-JP; Abrutis-A; Krellmann-M; Selbmann-D Eickemeyer-J; Stadel-O; Wahl-G, "YBaCuO deposition by MOCVD on metallic substrates: a comparative study on buffer layers", *IEEE Trans. App Supercond.*, **1]** [1] Part 3 2905-8 (2001).

[10]O. Stadel, J. Schmidt, G. Wahl, C. Jimenez, F. Weiss, M. Krellmann, D. Selbmann, N.V. Markov, S.V. Samoylenkov, O.Y. Gorbenko, A.R. Kaul, "Continuous YBCO deposition onto moved tapes in liquid single source MOCVD systems", *Physica C*, **341** Part 4 2477-2478 (2000).
[11]S. Donet, F. Weiss, J.P. Senateur, P. Chaudouet, A. Abrutis, A. Teiserskis, Z. Saltyte, D. Selbmann, J. Eickemeyer, O. Stadel, G. Wahl, C. Jimenez and U. Miller "YBCO coated nickel-based tapes with various buffer layers", *Journal de Physique IV,* **11**, Pr11-319-23 (2001).
[12] D. Selbmann, J. Eickemeyer, H. Wendrock, C. Jimenez, S. Donet, F. Weiss, U. Miller, O. Stadel, "NiO layer on Ni RABiTS for epitaxial buffer deposition by LS MOCVD", *J. de Physique IV*, **11** Pr11 239-245 (2001).
[13]M. Jergel, S. Chromik, V. Strbik, V. Smatko, F. Hanic, G. Plesch, S. Buchta, S. Valtyniova, ," Thin YBCO films prepared by low-temperature spray pyrolysis", *Supercond. Sci. Technol.*, **5** 225-230 (1992).
[14]A. Ferreri, J.J. Wells and J.L. MacManus-Driscoll, "Fabrication and post-annealing studies of ultrasonically spray pyrolysed Y-Ba-Cu-O films", *IEEE Trans. App Supercond.,* **11** [1] Part 3 2742-5 (2001).
[15]G.B. Blanchet, C.R. Fincher Jr., " High temperature deposition of HTSC thin films by spray pyrolysis" *Supercond. Sci. Technol.*, **4** [2] 69-72 (1991).
[16]J.L. MacManus-Driscoll, A. Ferreri, J.J. Wells, J.G.A. Nelstrop, "In-plane aligned YBCO thick films grown in situ by high temperature ultrasonic spray pyrolysis" *Supercond. Sci. Technol*, **14** [2] 96-102 (2001).
[17]A. Sin, Z. Supardi, A. Sulpice, P. Odier, F. Weiss, L. Ortega, M. Nunez-Regueiro, "Synthesis by aerosol process of superconductor films and buffer layer materials", *IEEE Trans. Appl. Supercond*, **11** [1] Part 3 2877-2880 (2001).
[18]P. Odier, F. Weiss, Z. Supardi, patent hanging
[19]S. Phok, A. Sin, Z. Supardi, P. Galez, J. L. Jorda and F. Weiss, "Preparation of Tl-1223 and Tl-2223 superconducting films by spray pyrolysis", *Journal de physique IV*, **11** Pr11-157 (2001).
[20]A. Sin, Z. Supardi, P. Odier, F. Weiss, L. Ortega, A. Sulpice, M. Nunez-Regueiro, "Synthesis of $Hg_{0.75}Re_{0.25}Ba_2Ca_2Cu_3O_{8+delta}$ bi-axially textured thin film by the aerosol process", *Thin solid film*, **388** [1-2] 251-5 (2001).
[21]A. Sin, F. Weiss, P. Odier, Z. Supardi, M. Nunez-Regueiro, "Synthesis under in situ pressure control of (Hg,Re)-1223 thick films by aerosol technique", *Physica C*, **341** Part 1 399-402 (2000).

[22] M. W Rupich. et al., "Low cost Y-Ba-Cu-O coated conductors", *IEEE Trans. Appl. Supercond.*, **11**, 2927–30 (2001).
[23]T. Araki, Y. Takahashi, K. Yamagiwa, Y. Iijima, K. Takeda, Y. Yamada, J. Shibata, T. Hirayama, I. Hirabayashi, "Firing conditions for entire reactions of fluorides with water in metalorganic deposition method using triflouroacetate", *Physica C,* **357-360** 991-4 (2001).
[24] Y. Takahashi, T. Araki, K. Yamagiwa, Y. Yamada, S.B. Kim, Y. Iijima, K.Takeda, T. Hirayama, I. Hirabayashi, "Preparation of YBCO films on CeO_2 buffered metallic substrates by the TFA-MOD method", *Physica C,* **357-360** 1003-6 (2001).
[25]Y. Yamada, S.B. Kim, T. Araki, Y. Takahashi, T. Yuasa, H. Kurosaki, I. Hirabayashi, Y. Iijima, K. Takeda, "Critical current density and related microstructures of TFA-MOD coated conductors" , *Physica C,* **357-360** 1007-10 (2001).
[26]I. Matsubara, M. Paranthaman, A. Singhal, C. Vallet, D.F. Lee, P.M. Martin, R.D. Hunt, R. Feenstra, C.Y. Yang, S.E. Babcock, "Preparation of textured YBCO films using all-iodide precursors", *Physica C,* **319**, 127-32 (1999).

TOWARDS AN ALL CHEMICAL SOLUTION COATED CONDUCTOR

O. Castaño, A. Cavallaro, A. Palau, J.C. González, M. Rossell, T. Puig, F. Sandiumenge, N. Mestres, S.Piñol, X.Obradors
Institut de Ciència de Materials de Barcelona, CSIC, 08193 Bellaterra, Spain

ABSTRACT

Chemical solution techniques have been investigated for the growth of oxide buffer layers suitable for coated conductors and for the growth of $YBa_2Cu_3O_7$ thin films. The growth conditions of CeO_2 and $BaZrO_3$ have been investigated and a high quality epitaxy has been demonstrated on single crystalline substrates. Trifluoroacetate precursors have been used for growing $YBa_2Cu_3O_7$ films (thickness of 250nm) on single crystal substrates. The kinetic hindrances for the formation of single phases have been investigated by means of Raman spectroscopy and Fluorine analysis. After optimisation of the deposition and growth conditions very high critical currents have been demonstrated (J_c^{ab}=3.2 10^6 A/cm^2 at 77K and 2,7 10^7A/cm^2 at 5K).

INTRODUCTION

Chemical solution growth of thin films has arisen as a new very exciting opportunity for the development of advanced functional ceramic materials. It has been particularly shown that epitaxial thin films can be grown on single crystalline substrates thus rising new opportunities for many applications where the material anisotropy or the granular character needs to be controlled. This includes ferroelectric-based devices, magnetoelectronic oxides, superconducting materials, etc. [1].

A very outstanding challenge, from the materials preparation point of view, is the preparation of superconducting coated conductors where a biaxial texture of the superconducting oxide needs to be reached on metallic substrates. These metallic substrates can be either textured themselves through metallurgical processes, such as the RABiT technique [2], or they can have an oxide buffer oxide template textured by the IBAD technique, or the so-called Inclined substrate deposition (ISD) method where the substrate is directed at a certain angle of the deposition beam generated by pulsed laser deposition (PLD). Over the past years many experimental techniques have been investigated to obtain $YBa_2Cu_3O_7$ thin films, with a thickness as high possible, on metallic substrates. This includes PLD, sputtering, thermal co-evaporation, CVD, Liquid Phase Epitaxy, etc. Among these techniques those based on a chemical solution growth process appear as very promising in view of the preparation of low cost conductors [4,5]. It´s however a complex issue to reach a full control of all the microstructural factors which may influence the superconducting properties of these conductors. It appears then very appealing to investigate first the influence of the processing parameters on the growth mechanisms and the microstructural development in single crystalline substrates. In this work we present a study of the growth of oxide buffer layers, particularly $BaZrO_3$ and CeO_2, which have a good crystallographic and chemical compatibility with the high T_c oxide superconducting materials, and the growth of $YBa_2Cu_3O_7$ by means of the so called trifluoroacetate (TFA) route [4-7].

EXPERIMENTAL DETAILS

Oxide buffer layers with composition CeO_2 and $BaZrO_3$ have been grown in (100) YSZ single crystalline substrate, in the first case, and in (100) $SrTiO_3$ (STO) and (100) $LaAlO_3$ (LAO) substrates in the second case. The precursor solutions for Ce and Zr were 2,4- pentadionate dissolved in glacial acetic acid and Barium acetate for Ba. The precursor solutions were prepared in a magnetic stirrer and the concentration was modified in order to obtain a spinnable solution with controlled thickness. The spin coater speed was maintained in the 3.000-6.000 r.p.m. range with a constant acceleration. After a drying process at 150-200°C during 30-60 minutes, a high temperature treatment under $Ar/5\%H_2$ atmosphere allowed to grow the layers. The temperature range where single phases were observed was 650°C-900°C.

Growth of $YBa_2Cu_3O_7$ was performed by using trifluoroacetate (TFA) precursor solutions which were prepared by dissolving stoichiometric mixtures of the Y, Ba and Cu acetates in an aqueous solution of 25% of trifluoroacetic acid. The mixture was continuously submitted to a reflux at 75°C for 4 hours. The resulting solution was dried getting a blue glassy residue, which is redissolved in methanol to make a solution of low viscosity. The thin film samples were prepared by coating the substrates with the precursor solution spinned at room temperature at angular speeds up to 6000 r.p.m.. The final film thickness was altered by adjusting the concentration of the mixed-metal precursor solution. In the present case the thickness of the films was $t \approx 250nm$, as determined by transverse SEM observations and interferometric methods.

The wet films were calcined in three stages. First, the organic material was pyrolyzed at 400°C in a moist O_2 atmosphere. The films were then crystallized at high temperatures (750-830°C) in a humid atmosphere (PH_2O=7.3%) and PO_2=20 ppmO_2 .The elapsed time depends on temperature and film thickness. Typically 3 hours allowed to complete the reaction for films of 250 nm. In the last stage the films were oxidized at 450°C in a dry oxygen atmosphere during 1 hour to convert the tetragonal phase in the superconducting orthorhombic phase.

X-ray diffraction (θ-2θ, ω-scan, ϕ-scan and pole figures) was used to determine the phase purity and the crystallographic texture of the films. Micro-Raman spectroscopy allowed to further analyse the kinetics of the formation reaction and to detect the misoriented grains in the films. Fluorine analysis of the exhaust gases in the furnace performed with a selective electrode allowed to determine the optimum duration of the high temperature reaction transforming BaF_2 to $YBa_2Cu_3O_7$ through the formation of HF. SEM and AFM were used to study the microstrutural modifications of the films and the surface roughness. Finally, the superconducting properties were studied through 4-points electrical resistivity measurements and low field dc susceptibility and inductive critical currents measured with a SQUID magnetometer. The critical currents were calculated from the irreversible magnetization using the critical state model which states that $J_c^{ab} = 30\Delta M/2R$, where ΔM is the irreversible magnetization and R the radius of the sample.

RESULTS AND DISCUSSION

Oxide buffer layers

A typical θ-2θ X-ray diffraction pattern of CeO_2 oxide grown on (100)YSZ single crystal is shown in Figure 1, while Figure 2 shows the excellent in-plane epitaxy achieved (FWHM $\Delta\phi$=0.27°). The quality of the out-of plane epitaxy was determined by measuring the ω-scan of the (400) reflection (FWHM $\Delta\omega$=0.23°) (inset of Fig.2) and the uniqueness of the cube on cube texture was verified by measuring pole figures. Similar excellent epitaxy was achieved for films having thickness in the range 20 nm to 100 nm.

The epitaxy achieved in the case of $BaZrO_3$ films grown on STO and LAO single crystals was also of a high quality (Figure 3) (FWHM $\Delta\omega$=0.5° in ω-scan and FWHM $\Delta\phi$=1.17° in ϕ-scan), in agreement with the similar lattice matching in both cases (5.3 % for CeO_2/YSZ vs 6.9 % for BZO/STO). It´s straightforward to note, however, that no residual stresses due this lattice mismatch were detected through X-ray diffraction $\sin^2\chi$ plots performed in both thin film materials, thus we may conclude that these films relax the strong lattice misfit through some kind of microstructural accommodation.

The evolution of the microstructural modifications induced by changes in the growth temperature was followed through AFM observations. Figure 4 display a typical image showing a nanometric grain size ($\Phi\approx$30nm) and a very low porosity and surface roughness (rms≈1 nm). Similar values were achieved in the case of $BaZrO_3$ layers. These microstructural features of the buffer layers could be widely modified through the high temperature thermal treatments thus we may conclude that chemical solution growth techniques allow to prepare oxide layers with excellent performances for the deposition of superconducting layers.

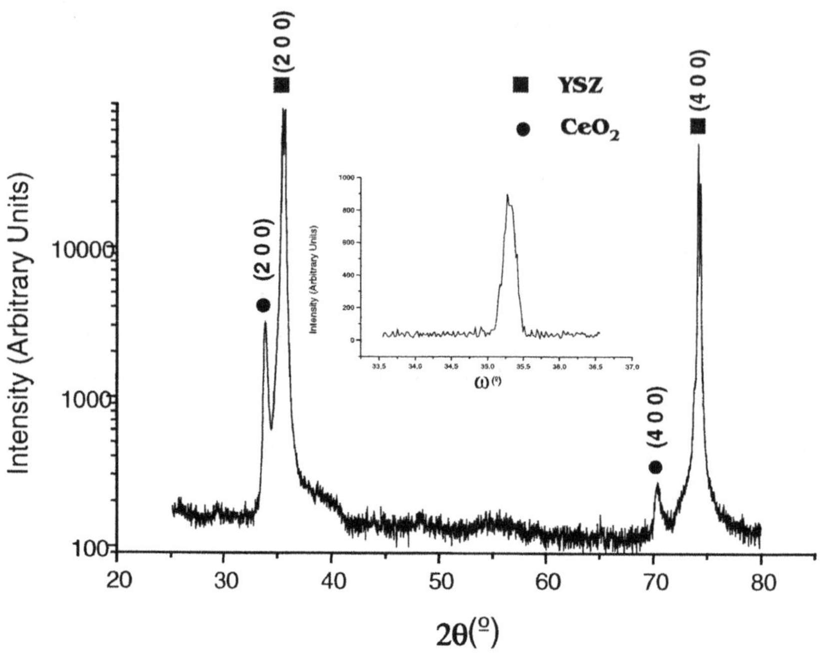

Figure 1.- X-ray diffraction θ-2θ pattern of CeO_2 grown on a YSZ single crystal. Inset: ω-scan of the (400) reflection

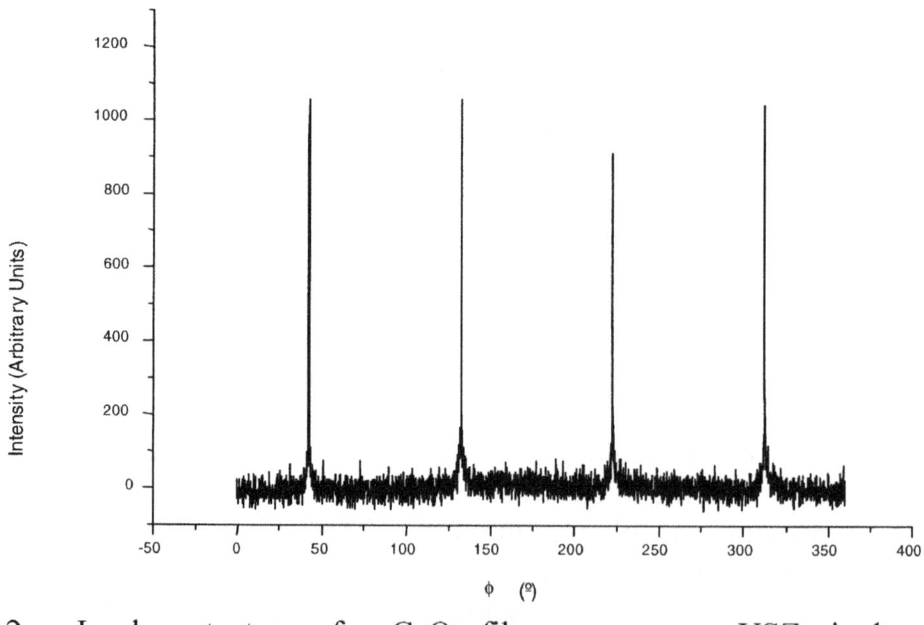

Figure 2 .- In-plane texture of a CeO_2 film grown on a YSZ single crystal measured with the (111) reflection.

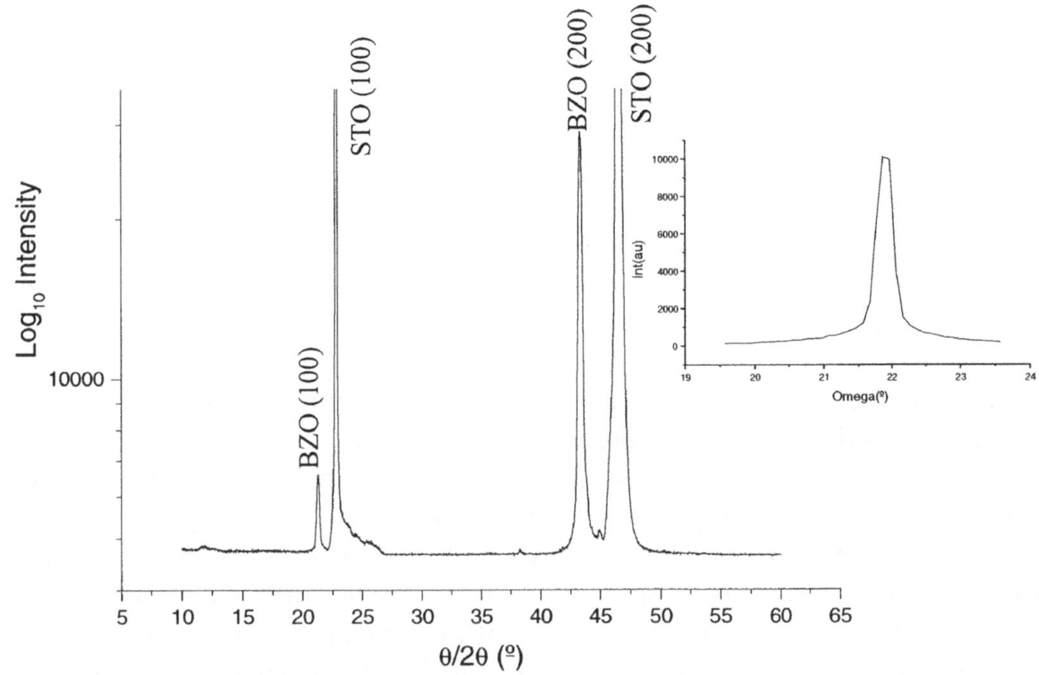

Figure 3.- X-ray θ-2θ diffraction pattern and (inset) out-of-plane texture of $BaZrO_3$ layers deposited on (100) $SrTiO_3$ single crystals.

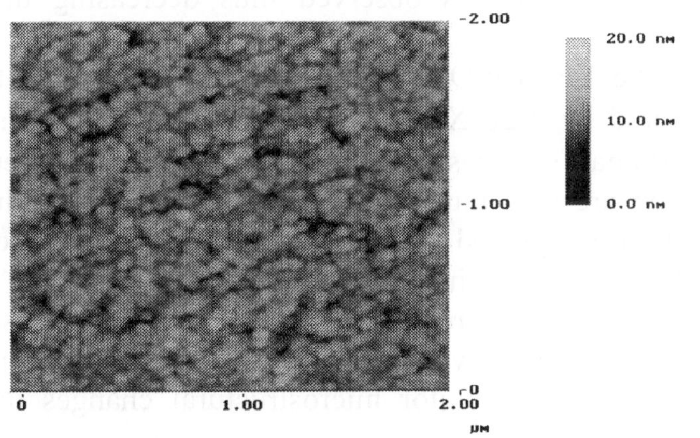

Figure 4.- AFM image of the surface of a CeO_2 thin film grown on a YSZ single crystal.

Superconducting layers

The conversion of the TFA salts into $YBa_2Cu_3O_7$ thin films is a complex process which involves, first, a low temperature decomposition of the metalorganic precursors into nanometric crystals of oxides (CuO, Y_2O_3) and fluoride (BaF_2) [4,5]. This is the pyrolisis process that should be performed in such a way that the metallic ions segregation is minimized and that the film shrinkage does not generate cracks or striations which would degradate the film quality after the growth process. The second reaction step occurs at high temperatures where the decomposition of BaF_2 and its reaction with the oxides leads to the formation of the $YBa_2Cu_3O_7$ phase. The reaction kinetics involves diffusion of the the atomic species diffusion and the elimination of HF, which is controlled through the wet atmosphere within the furnace.

There are multiple processing parameters which require optimisation for the preparation of high quality $YBa_2Cu_3O_7$ thin films and probably not a single combination of them do exist. In our present films we optimised the solution deposition parameters (viscosity, spinning rate) and we performed a slow pyrolisis process (≈20h) to ensure that no degradation of the nanocristalline precursor films occurs. In Figure 5 we show, for instance, a SEM image displaying the formation of striations in the film after the growth process, which are generated when the spinning rate in the coating process is too high. As it can be observed the epitaxy is strongly degraded within these striations where an enhanced porosity, secondary phases and a-axis grains is observed thus decreasing the critical currents of the films.

The formation of the $YBa_2Cu_3O_7$ phase at temperatures within the range 700°C-820°C was monitored by θ-2θ X-ray diffraction, Fluorine analysis in the gas exhaust line and μ-Raman spectroscopy. The processing parameters during this phase conversion step are also multiple. They have a strong influence on reaction kinetics, nucleation energy and grain growth rate and hence they should be widely investigated to control the microstructure. We should mention particularly the temperature, water pressure, gas flow, oxygen pressure and reaction time. In this work we have investigated the relationship among temperature and reaction time to monitor microstructural changes during the growth of films.

Fluorine analysis indicated that the reaction time was doubled when the temperature was reduced from 830°C to 700°C, if the remaining parameters remain fixed. μ-Raman spectroscopy proved very useful to detect small amounts of unreacted phases [8]. Figure 6 show two typical spectra corresponding to an uncomplete reacted sample and to a sample where only the modes of the $YBa_2Cu_3O_7$ phase are observed. In the first case (Fig.6(a)), additionally to the BaF_2 phase,

Figure 5.- SEM image of a $YBa_2Cu_3O_7$ film deposited in a STO single crystal by spin coating at 6.000 r.p.m. The films were reacted at 790ºC (a) Low magnification image where the formation of striations is observed, (b) Magnification of the microstructure observed within the striations where the film has an enhanced porosity, secondary phases and a-axis oriented grains.

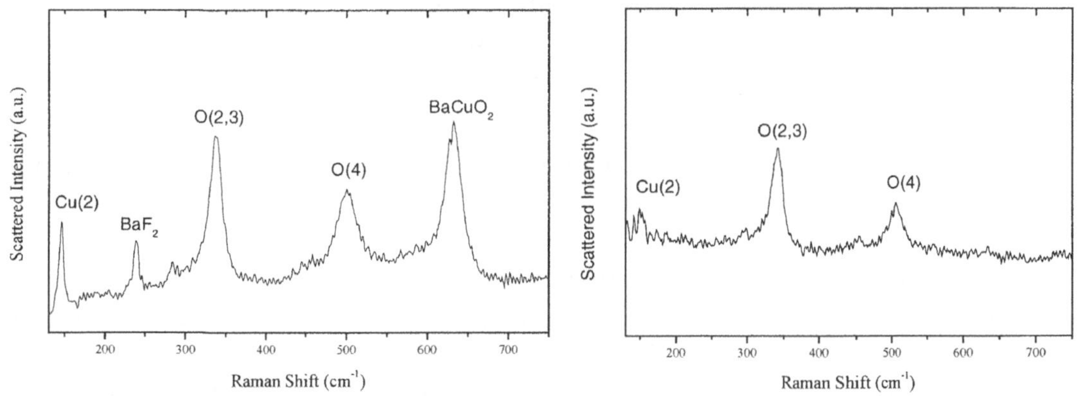

Figure 6.- Raman spectroscopy of $YBa_2Cu_3O_7$ films, (a) sample which has been uncompletely reacted. Note that BaF_2 and $BaCuO_2$ peaks are identified in the spectra, (b) fully reacted sample where only the modes corresponding to $YBa_2Cu_3O_7$ are observed.

we can observe a peak corresponding to $BaCuO_2$. This an impurity that is hardly observed in the X-ray diffraction spectra and which is probably associated to a kinetic hindrance effect promoted by any kind of macrosegregation generated during the deposition or the pyrolisis step. High temperature reaction has been found to promote the disappearance of this impurity.

Figure 7 displays a typical X-ray diffraction spectra corresponding to a pure sample having a high quality texture. ω-scan and ϕ-scan of these samples lead to FWHM values of $\Delta\omega$=0.4° for the (005) reflection and $\Delta\phi$=0.5° for the (115) reflection indicating that good out-of-plane and in-plane biaxial texture have been reached during the growth process.

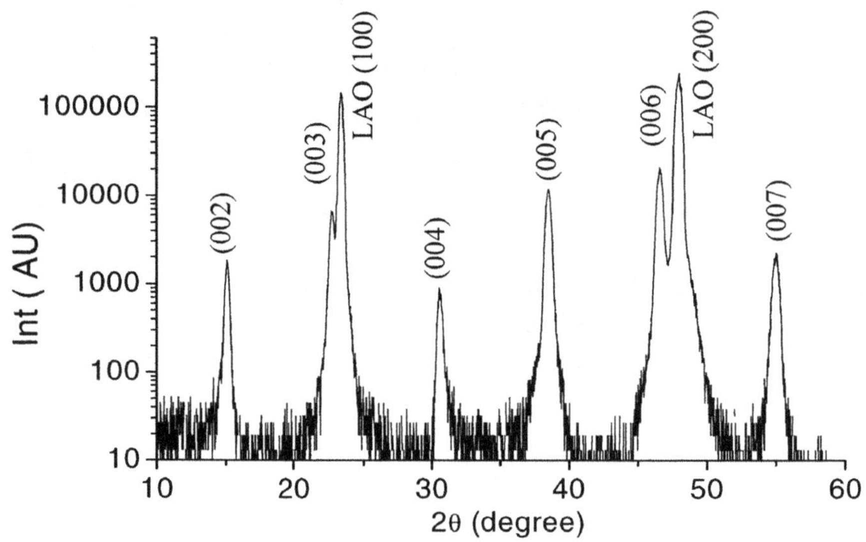

Figure 7.- θ-2θ X-ray diffraction pattern of a $YBa_2Cu_3O_7$ film deposited on a $LaAlO_3$ single crystalline substrate

The increase of reaction temperature during the growth process has proved to have a tremendous influence on the final microstructure and the superconducting properties of the samples. We have found particularly that the porosity of the samples is strongly dependent on this parameter. As a typical example we show in Figure 8 two SEM images of the sample surface of two samples fired at 700°C and at 830°C during the required time for the reaction completion. The differences in the pore concentration in both samples are easily detected. A quantitative indirect determination of this effect was reached through measurements of the electrical resistivity in the normal state. As it may be observed in Figure 9 (a), the slope of the temperature dependence of the resistivity and the residual resistivity are enhanced in samples with a high porosity, thus indicating that the percolative behavior of current is hindered in such disordered films. The resistivity observed at 300K in the samples having a low porosity, $\rho(300K) \approx 300\ \mu\Omega cm$, together with the high resistivity ratio

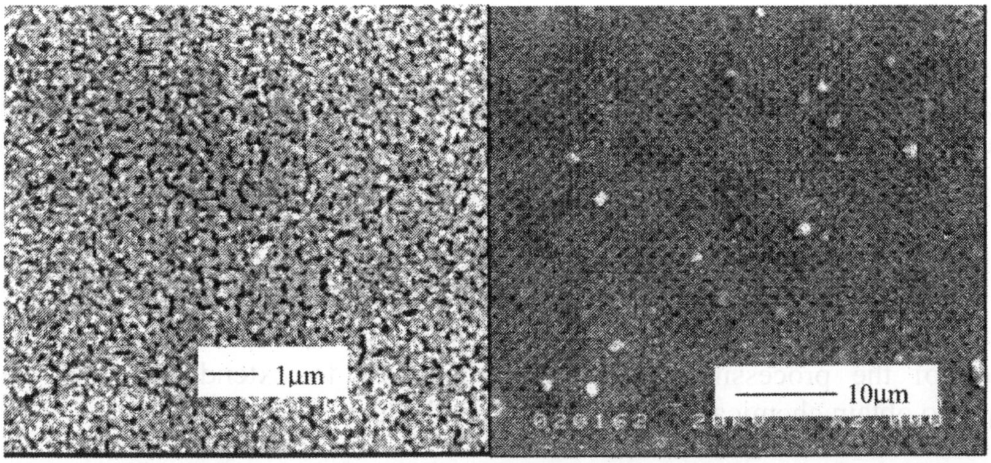

Figure 8.- SEM images of the surface of $YBa_2Cu_3O_7$ films prepared at different temperatures: (a) 700°C, a high porosity is observed, (b) no residual porosity is observed

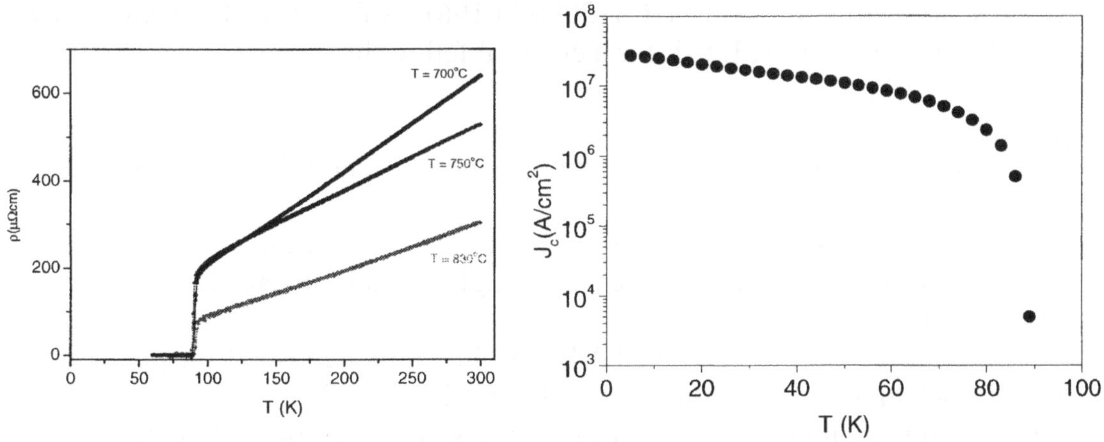

Figure 9.- (a) Temperature dependence of the electrical resistivity in $YBa_2Cu_3O_7$ thin films grown at different temperatures and displaying a very different degree of porosity, (b) Temperature dependence of the critical currents $J_c^{ab}(T)$ in a high quality $YBa_2Cu_3O_7$ thin film

$\rho(300K)/\rho(100K) \approx 3,3$ are typical features of high quality thin films. In this Figure we can also observe that the transition temperature $T_c \approx 90K$ was not modified by the annealing temperature.

Finally, the critical currents of the $YBa_2Cu_3O_7$ thin films were measured inductively by using a SQUID magnetometer. The critical currents were found to be very sensitive to the residual porosity of the films. Very high J_c^{ab} values were measured, instead, in samples showing no residual porosity. As it can be seen in Figure 9(b), at zero field these samples displayed very high critical currents ($J_c^{ab}=3,2\ 10^6\ A/cm^2$ at 77K and $2,7\ 10^7 A/cm^2$ at 5K). To obtain such high performances in thin films usually require the use of vacuum processes and hence our work further support the view that chemical solution growth techniques have the potential to become competitive, from the performance point of view, with these techniques. We note also that in the present work any special refinement process of the metalorganic precursors was applied, as it was previously required [6].

In conclusion, the present work has advanced in the knowledge of the growth mechanisms on single crystalline substrates of oxide buffer layers and superconducting coatings deposited from chemical solutions. The control of the influence of the processing parameters will allow to extend this low cost technique to obtain chemically based coated conductors.

ACKNOWLEDGEMENTS

This work has been supported by the Spanish Science Ministry (MAT 2001-1698 and 2FD97-1628), by the Catalan Pla de Recerca (00206) and the EU-Marie Curie Program (HPMT-CT-2000-00106). A.P. and M.R. acknowledge to C.S.I.C. and J.C.G. to A.E.C.I. for a doctoral fellowship.

REFERENCES

[1] F.F.Lange, "Chemical solution routes to single-crystal thin films", *Science*, **273**, 903 (1996).

[2] D.P.Norton et al, "Epitaxial YBCO on biaxially textured Nickel (001): An approach to superconducting tapes with high critical current density", *Science* , **274**, 755 (1996).

[3] Y.Ijima et al, "In-plane aligned YBCO thin films deposited on polycrystalline metallic substrates", *Appl.Phys.Lett.* , **60**, 769-771 (1992).

[4] P.C. Mclntyre and M.J. Cima, "Heteroepitaxial growth of chemically derived ex situ $YBa_2Cu_3O_{7-x}$ thin films", *J. Mater. Res.*, **9**, 2219-2230 (1994).

[5] J. A. Smith et al.,"High critical current density thick MOD-derived YBCO films", *IEEE Trans. Appl. Supercond.*, **9**, 1531 (1999)

[6] T.Araki et al, "Coating processes for $YBa_2Cu_3O_{7-x}$ superconductor by metaorganic deposition method using trifluoroacetates", *Supercond. Sci. Technol.* **14**, 783-786 (2001)

[7] Y-A. Jee et al, "Texture development and superconducting properties of $YBa_2Cu_3O_x$ thin films prepared by a solution process in low oxygen partial pressure", *Supercond. Sci. Technol.* **14**, 285-291 (2001)

[8] T. Puig et al., "Texture analysis of coated conductors by micro-Raman and synchrotron X-ray diffraction", *Mat. Res. Soc. Symp.* **659**, II5.6.1.-6.6 (2001)

AN ECONOMICAL ROUTE FOR PRODUCTION OF HIGH-QUALITY YSZ BUFFER LAYERS USING THE ECONO™ PROCESS (INVITED)

M. A. Zurbuchen, S. Sambasivan,
I. Kim,* J. Rechner,
J. Wessling, and J. Ji
Applied Thin Films, Inc.
Evanston, IL 60201

S. Barnett
Materials Science and Engineering
Northwestern University
Evanston, IL 60201

B. F. Kang and A. Goyal
Metals and Ceramics Division
Oak Ridge National Laboratory

P. A. Barnes and C. E. Oberly
Power Propulsion Directorate
U.S. Air Force Research Lab., OH

ABSTRACT

A new scalable and economical process has been developed to produce high-quality YSZ buffer layers on metal and alloy RABiTS. Current densities of 1 MA/cm^2 were achieved reproducibly on 0.3 µm-thick $YBa_2Cu_3O_{7-\delta}$ layers deposited on the YSZ buffer using PLD with an intermediate ~20 nm ceria layer. The buffer process essentially involves depositing a ~200 nm-thick epitaxial yttrium zirconium nitride directly on RABiTS, with subsequent conversion to a ~300 nm-thick epitaxial yttria stabilized zirconia film via a simple oxidation step. The new process is termed ECONO™, for Epitaxial Conversion to Oxide via Nitride Oxidation, and is also potentially useful for developing oxide epitaxial templates on non-oxide substrates (e.g., silicon or GaAs) for a broad range of device applications. The key advantage of this approach is that it avoids interference from substrate oxidation during epitaxial growth. For coated conductors, it eliminates both the need for an oxide seed layer and the need for a sulfur superstructure. In addition, the deposition of nitrides by reactive sputtering can be performed at much higher rates relative to oxides and the resulting YSZ films are dense and relatively defect-free. A description of the process and the associated conversion mechanism will be discussed.

* Now at Functional Coating Technology, LLC.

INTRODUCTION

High-temperature superconductor (HTS)-coated conductors represent the next generation of HTS wire technology, with primary application in power generation, power transmission, and compact motors for military and commercial use. HTS tapes, formed by deposition onto biaxially textured metal substrate tapes, are intended to replace copper as a conductor in these applications. The current approach (e.g., RABiTS) being pursued by several research groups worldwide involves deposition of a 10-20 nm-thick a ceria/yttria layer, followed by 200-500 nm-thick YSZ layer, capped with a 10-20 nm-thick ceria layer prior to deposition of the HTS layer.

Initial demonstrations of the proof-of-concept at Oak Ridge National Laboratory (ORNL), now duplicated by other research groups worldwide, utilized a buffer layer configuration of CeO_2 (0.01 μm) / YSZ (0.5 μm) / CeO_2 (0.01 μm) to provide a surface for $YBa_2Cu_3O_{7-\delta}$ (YBCO) epitaxy and to act as a Ni diffusion barrier. The purpose of the first buffer layer is to provide a good epitaxial oxide layer on the reactive, biaxially textured Ni substrate without the formation of undesirable NiO. CeO_2 is special among oxides in its ability to readily form single orientation cube-on-cube epitaxy with Ni. Deposition of CeO_2 was done using a range of deposition techniques under a background of forming gas with small amounts of water vapor (4% H_2O - 96% Ar). Under such conditions, the formation of CeO_2 is thermodynamically preferred over NiO. It is not possible to deposit YSZ under such conditions without undesirable orientations, but in the case of CeO_2, one can readily obtain a single epitaxial-orientation texture. A CeO_2 layer that is thick enough to serve as a chemical diffusion barrier to Ni readily forms microcracks, so a YSZ layer is required. YSZ provides an excellent chemical barrier to Ni diffusion and does not crack when grown thick because there is a significant lattice mismatch between YSZ and YBCO (~5%), leading to a second, 45°-rotated orientation that sometimes nucleates and functions to relieve misfit stresses. A thin CeO_2 layer, which has an excellent lattice match with YBCO (~0.1%), is deposited epitaxially on the YSZ layer to provide an epitaxial template.

The drawbacks of this buffer layer structure are that the deposition of the first CeO_2 layer is quite difficult. Strict control of deposition conditions, in particular the O_2 partial pressure, is required to avoid formation of undesirable NiO. (NiO typically nucleates in mixed orientations and is very weak.) Furthermore, CeO_2 can possess a wide range of oxygen stoichiometry, and is brittle and electrically non-conducting. It will be a challenging engineering task to develop a large-scale continuous process for producing thick (>0.5 μm) crack- and porosity-free oxide films based on a vapor phase process. For example, in a continuous process involving reactive electron beam evaporation of Ce to form CeO_2, the formation of an oxide on the target complicates matters relating to rate of deposition as well as stability of the melt pool. Any change of conditions during deposition is known to have profound affects on the film microstructure. Moreover, oxidation of the biaxially textured metal, even after the successful deposition of CeO_2, can induce undesirable interfacial stresses leading to spallation or further cracking, thus deteriorating the material properties. Microcracks in the oxide buffer layer will adversely affect the epitaxial quality of the growing YBCO film and create weak-links, besides serving as diffusion paths for Ni.

Despite these difficulties, excellent results have been reported by a number of researchers for the current-carrying capacity of HTS films on short samples, the industry is still in search of a low-cost scalable approach that will help produce HTS films in long lengths without compromising on quality and consistency. This has been a significant technical and engineering challenge and has led to substantial R&D investments over the past 5 years. The critical need for second-generation wire technology is a low-cost, scalable, and high-quality biaxially textured yttria stabilized zirconia (YSZ) buffer layer on metal tapes.

A primary concern with the current architecture of the multilayer stack is the precise control needed to deposit the initial ceria/zirconia seed layer. Any oxidation of the underlying substrate during growth of the seed layer results in poor-quality texture. In the case of pure nickel substrates, NiO forms quite readily and interferes with the epitaxial growth of the ceria layer. With non-magnetic substrates (e.g., Ni-Cr or Ni-W), the concern for substrate oxidation is exaggerated by presence of Cr or W and hence requires even more stringent control for seed layer growth. In addition, it is well established that for direct growth of oxide seed layers, a monolayer of 2x2 sulfur superstructure is necessary to promote epitaxy. This necessitates a well controlled annealing procedure in sulfur-containing atmosphere to create the superstructure. For transitioning the current approach to an industrially scalable process, the aforementioned issues present significant technical and engineering challenges, and more importantly, impose a high cost which impedes its deployment as a competitive product.

The performance criterion for superconducting tapes is defined by the texture quality (both in and out of plane) and the microstructural characteristics of the film (YSZ film density and surface morphology). The surface morphology characteristics are especially critical for YBCO films produced using the TFA process,[1] where a surface roughness below 10 nm RMS is desired.

THE ECONO™ PROCESS

The problems encountered with the current buffer layer approach are circumvented by using an innovative and novel process developed at Applied Thin Films, Inc. (ATFI), termed "ECONO" for **E**pitaxial **C**onversion to **O**xide via **N**itride **O**xidation. The ECONO™ process represents a technological breakthrough that has significant implications for the intended application of HTS coated conductors and beyond. This new process provides a distinct cost and performance advantage for fabrication of buffer layers needed for the HTS coated conductor tapes.

Essentially, the process allows the growth of high-quality oxide epitaxial layers on non-oxide substrates. A precursor *epitaxial* nitride layer of composition $Y_{0.2}Zr_{0.8}N$ (YZN) is deposited on the substrate and subsequently converted to an epitaxial yttria stabilized zirconia (YSZ) layer by a simple oxidation step. The process is shown schematically in Fig. 1. The oxidation can be conducted either *in-situ* after nitride deposition or *ex-situ* in a conventional furnace with controlled oxygen atmosphere. As described in more detail below, the conversion from nitride to oxide is very fast and proceeds from the top of the nitride layer which then serves as the template for subsequent oxide conversion downward. The oxidation proceeds in a highly cooperative manner whereby the epitaxy both in- and out- of plane is maintained. The key advantages of the process include: (a) deposition of a high-quality nitride layer which is dense with smooth morphology and relatively free of defects (which are prevalent in oxide deposition by reactive or RF sputtering), (b) does not require a seed layer or a sulfur superstructure, as is the case with the current approach, (c) the nitride can be deposited at relatively high rates compared to oxides, and (d) YSZ, having a larger molar volume than YZN, translates into ~50% thicker films after conversion which reduces the effective deposition time.

It is important to note that even though the above approach can yield a compositionally complex nitride stack, the deposition of all nitride layers can be performed in a single step, in the same chamber, without compromising the process efficiency as compared to YZN single layer deposition.

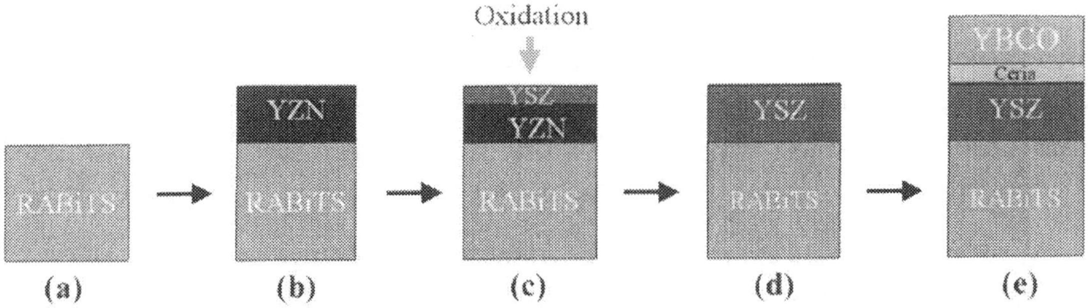

Figure 1. Schematic of YBCO growth on a RABiTS using the ECONO™ process. (a) A bare RABiTS is prepared, (b) $Y_{0.2}Zr_{0.8}N$ is deposited(c) The nitride is then converted to the oxide (yttria stabilized zirconia) by a thermal oxidation step to form an oxide film on the alloy RABiTS, (d). CeO_2 and YBCO are then deposited by the conventional. technique.

To meet the cost target of the final YBCO conductor, costs associated with fabrication of each component must be minimized without compromising film quality. A key parameter affecting the cost of buffer layer production is the deposition rate. Direct deposition of oxide layers by reactive sputtering is relatively slow compared to nitrides. One of the key competitive advantages offered by the ECONO process is the ability to deposit high quality nitrides (dense and uniform with high-quality texture) at high rates.

Broader significance of the ECONO process beyond the use for coated conductors or HTS devices is also apparent and deserves much attention. The ECONO process should be considered as a "generic epitaxial template technology" suitable for a broad range of commercial applications. In addition to RABiTS, the process has been demonstrated for depositing high-quality epitaxial YSZ layers on silicon and sapphire. Reactive sputtering is a well established thin film deposition process routinely used in the semiconductor industry, and can be easily integrated into a manufacturing process. YSZ, with cubic symmetry, is an excellent template layer for subsequent growth of oxide device layers and has been used for deposition of $Pb(Zr,Ti)O_3$, $SrTiO_3$, $(Ba,Sr)TiO_3$, YBCO, RuO_2, and others.[[2],[3]] YBCO films on silicon, using YSZ as the buffer layer, is being pursued by industry for fabricating high efficiency RF filters for telecommunication systems, magnetic resonance imaging systems, bolometers, high speed computing, and others.[4,5]

There are several advantages to using a precursor nitride film. Cubic nitrides can be epitaxially deposited on reactive metals such as Ni in a straightforward manner. The nitride deposition on Ni is done in a nitrogen-containing ambient, and because Ni does not form a nitride, the strict process control required for the current technique is not needed. No separate partial pressure of H_2O is required, so the process is easy to control and scale-up. Nitrides have been used in the tool industry for many years, and their manufacturability in various configurations has already been demonstrated. High-rate deposition (up to 1 μm/min) of nitride coatings is being developed and used in many industrial applications.

Most metal nitrides exhibit a cubic symmetry with a range of lattice parameters suitable for the coated conductor application. Solid solutions of transition metal nitrides and rare earth nitrides can be used to "tailor" lattice parameters, because of extensive compositional range, with solid solubility among most of them. The compositional flexibility and chemical compatibility among these nitrides can be used favorably to tailor several properties required for nitride deposition and resulting oxide buffer layers.

GROWTH OF YBCO / CeO_2 / ECONO™-YSZ / Ni-RABiTS STACKS

Nitride films are grown in a D.C. reactive magnetron sputtering system with an all-metal-sealed chamber and a sample-insertion load-lock system. Base pressures of $\sim 10^{-8}$ Torr are typical. The magnetron sputtering sources (US GUN I) utilize 5 cm-diameter, 3.2 mm-thick sputtering targets. Background pressures of 15-20 mTorr of >99.999 % pure $Ar\text{-}N_2$ were maintained during growth. Sputtering targets consisted of 99.95 % V, 99.95 % Ti, Zr, Y, and 99.95 % $Y_{0.2}Zr_{0.8}$ alloy.

YZN buffer layers were deposited on nickel RABiTS substrates, and were converted to YSZ by an *ex-situ* oxidation step. Standard pulsed laser deposition (PLD) was used to deposit ~20 nm thick ceria layers on selected ECONO™-deposited YSZ / RABiTS substrates to achieve good lattice match with X-ray diffraction (XRD) ϕ-scans of the layers of a representative stack, shown in Fig. 2, reveal the in-plane texture of the layers. Figure 2(a) shows a (111) ϕ-scan of a typical textured Ni-RABiTS, with a full-width at half-maximum (FWHM) of 9.2°. The (111) ϕ-scan of a YSZ layer on this substrate is shown in Fig. 2(b), with a FWHM of 10.0°. Local cube-on-cube epitaxy has resulted in only a slight decrease in preferred orientation of 0.8°. The ϕ-scan of the PLD-deposited CeO_2 layer, Fig. 2(c), shows an *improvement* in texture, with a FWHM of 9.5°. The (001) YBCO layer maintains this texture, as shown in the ϕ-scan in Fig. 2(d), with a 9.5° FWHM. The overall decrease in texture from Ni-RABiTS to YBCO is 0.3°. These data demonstrate the high quality of YBCO that can be grown on an ECONO™-grown buffer layer.

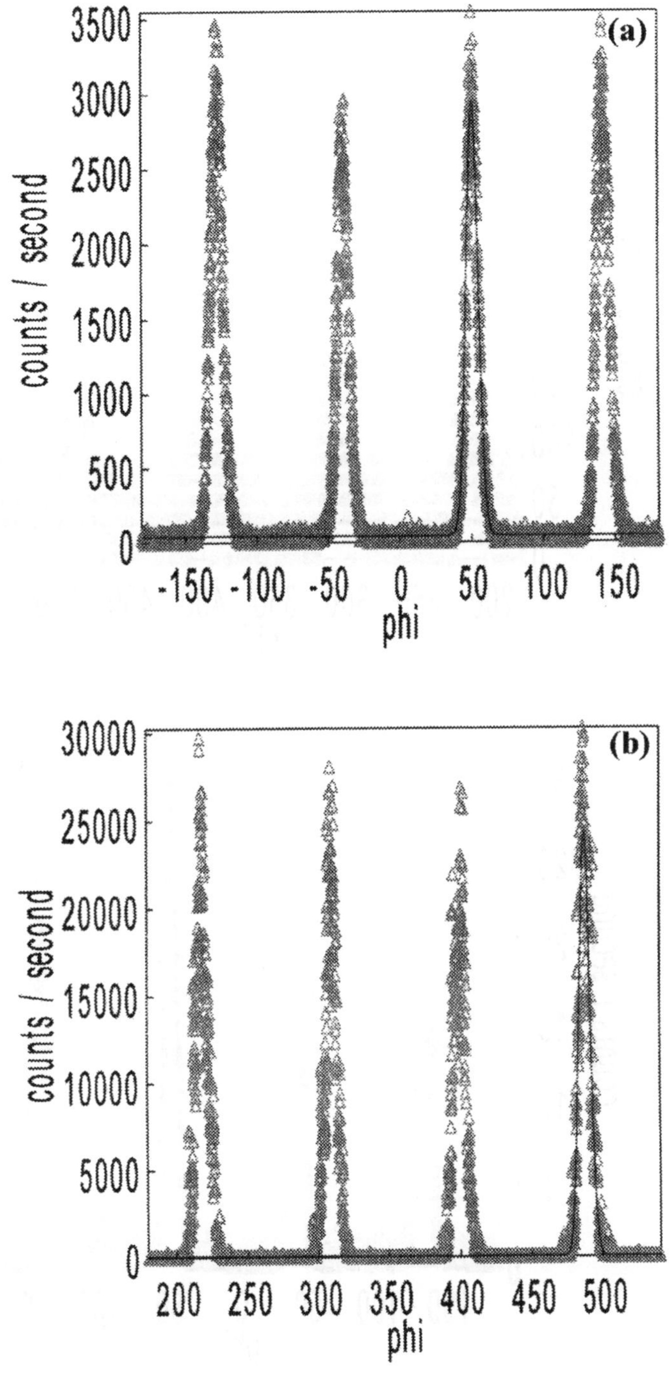

Figure 2. ϕ-scans of (a) Ni RABiTS with FWHM = 9.2° and (b) ECONO™-deposited textured YSZ layer with FWHM = 10.0°, showing the excellent retention of texture of the buffer layer.

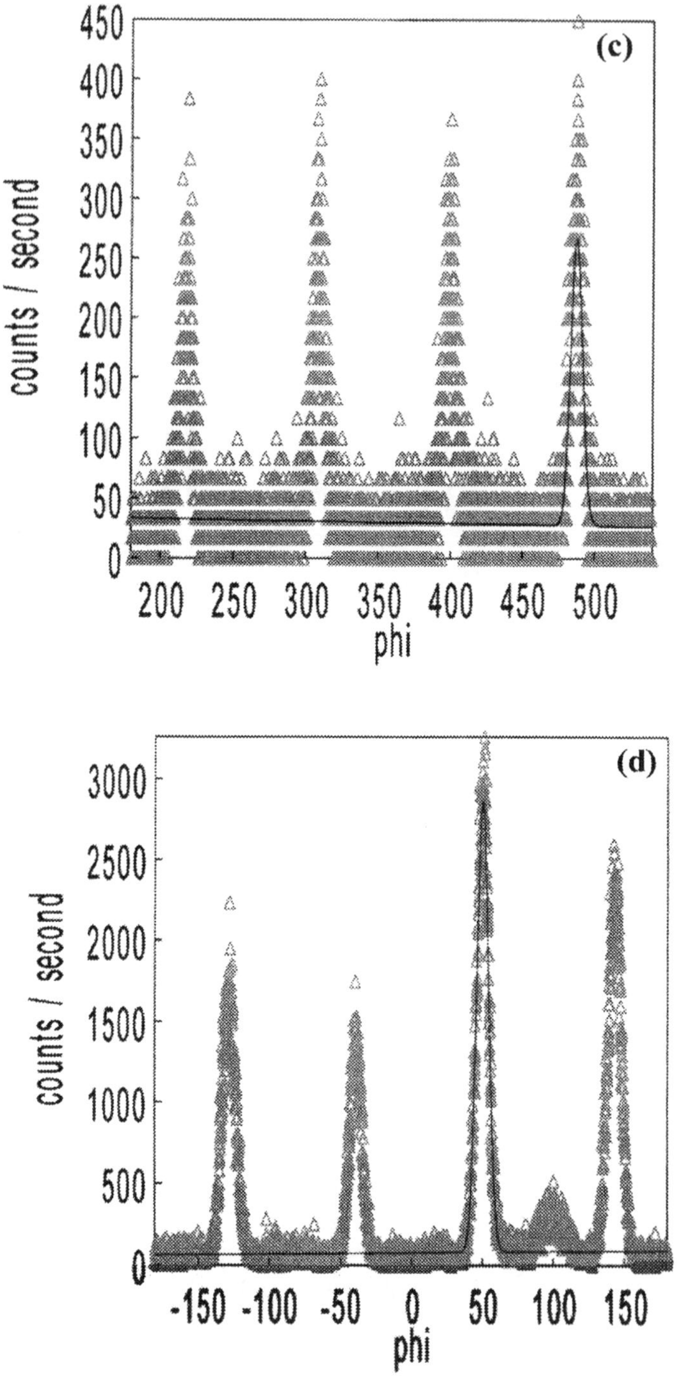

Figure 2 (continued). φ-scans of (c) the CeO2 nickel-diffusion barrier, with FWHM = 9.5°, and *improvement* in texture of 0.5°, and (d) the (001) YBCO film, with FWHM = 9.5°. The decrease in texture through the entire stack is only 0.3°.

The results of electrical transport measurements on a typical film stack of YBCO / CeO_2 / ECONO™-YSZ / Ni-RABiTS, with a 200 nm-thick YBCO layer thickness, are shown in the I-V plot in Fig. 2. Film width was 3 mm. $I_c = 6$ A at 1 μV/cm criterion. J_c of the film is calculated to be 1 MA/cm^2. A second, identically-processed stack showed the same I-V behavior and J_c values.

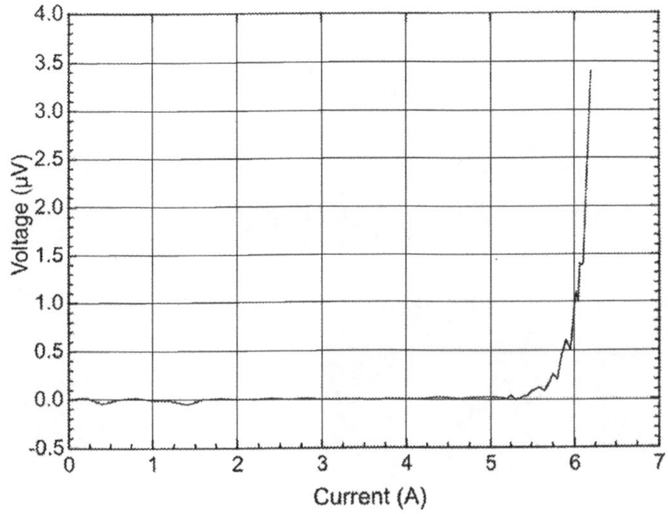

Figure 3. I-V plot measured o a 200 nm-thick YBCO film, grown on a CeO_2 / ECONO™-YSZ / Ni-RABiTS. $J_c = 10^6$ A/cm^2 at 77 K, measured by whole body transport current measurement in self field using a 1 μV/cm criterion.

GROWTH OF YSZ ON SILICON BY ECONO™ PROCESS

Many potential applications of the ECONO™ process to provide an epitaxial oxide template on silicon exist. Thus, the ECONO™ process has also been used to deposit epitaxial YSZ on silicon. A cross-sectional TEM image of a YZN film grown epitaxially on silicon, half-way through the oxygen-nitrogen exchange process, is shown in Fig. 4(a). The conversion front appears quite planar. This is significant considering that YSZ and YZN do not have the same crystal structure, and that a 50 % difference in molar volume exists between the two phases. (The difference in lattice parameter is 13 %.) The two phases are in Fig. 4(b) and (c), taken from the same area, and also in the high-resolution TEM (HRTEM) image in Fig. 4(d). The quality of the YSZ film is strongly dependent upon the initial rate of the oxidation process. The first atomic layer(s) in the YZN film surface to undergo oxide conversion are critical for maintaining epitaxial orientation of the rest of the film.

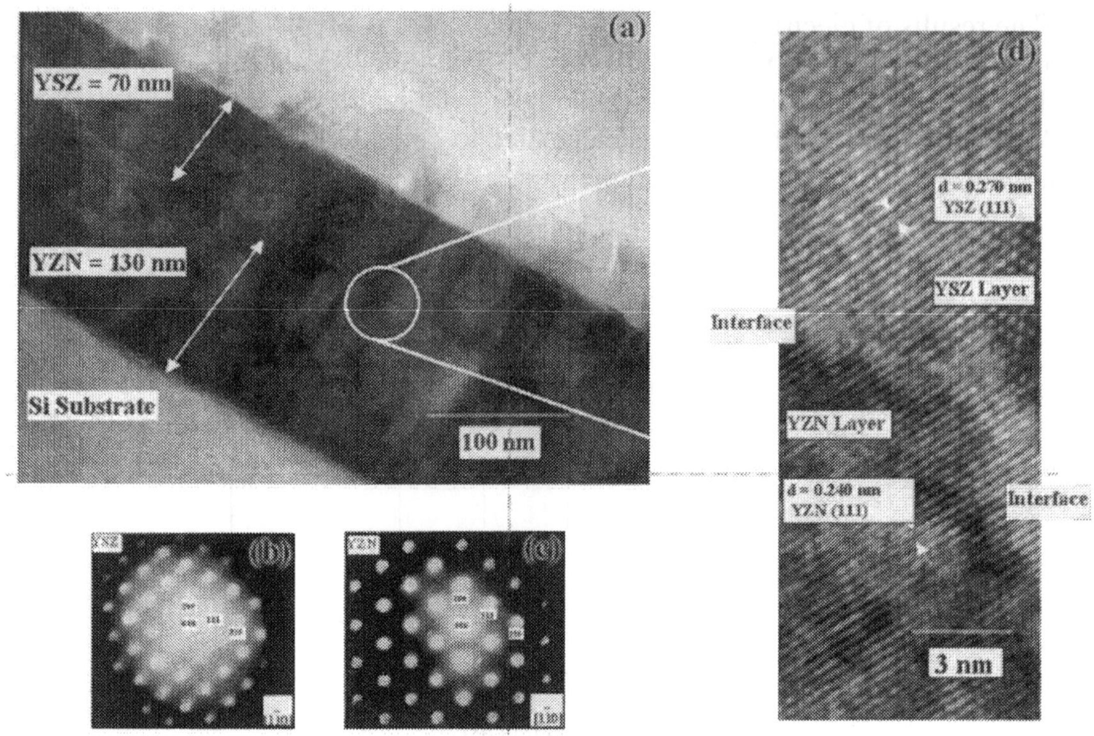

Figure 4. Cross-sectional TEM characterization of a YZN / Si film in the middle of the conversion process to YSZ. (a) The planar growth front is apparent. Epitaxy is maintained despite a 50% difference in molar volume (13 % difference in lattice parameter) between the two phases, as shown by the electron diffraction patterns in (b) and (c), and the HRTEM image of the interface in (d).

The nitrogen distribution within the same film was studied by parallel energy dispersive spectroscopy (PEELS) analysis in the TEM. Figure 5(a) is a cross-sectional TEM image and Fig,. 5(b) is an elemental map of nitrogen over the same area, acquired by setting an electron prism and an aperture to collect only intensity immediately after the nitrogen energy loss edge. A nitrogen-signal line scan is shown at the bottom of Fig. 5(b), clearly revealing a gradient in nitrogen concentration across the YSZ layer. The exact mechanism of nitrogen diffusion through YSZ has not yet been determined.

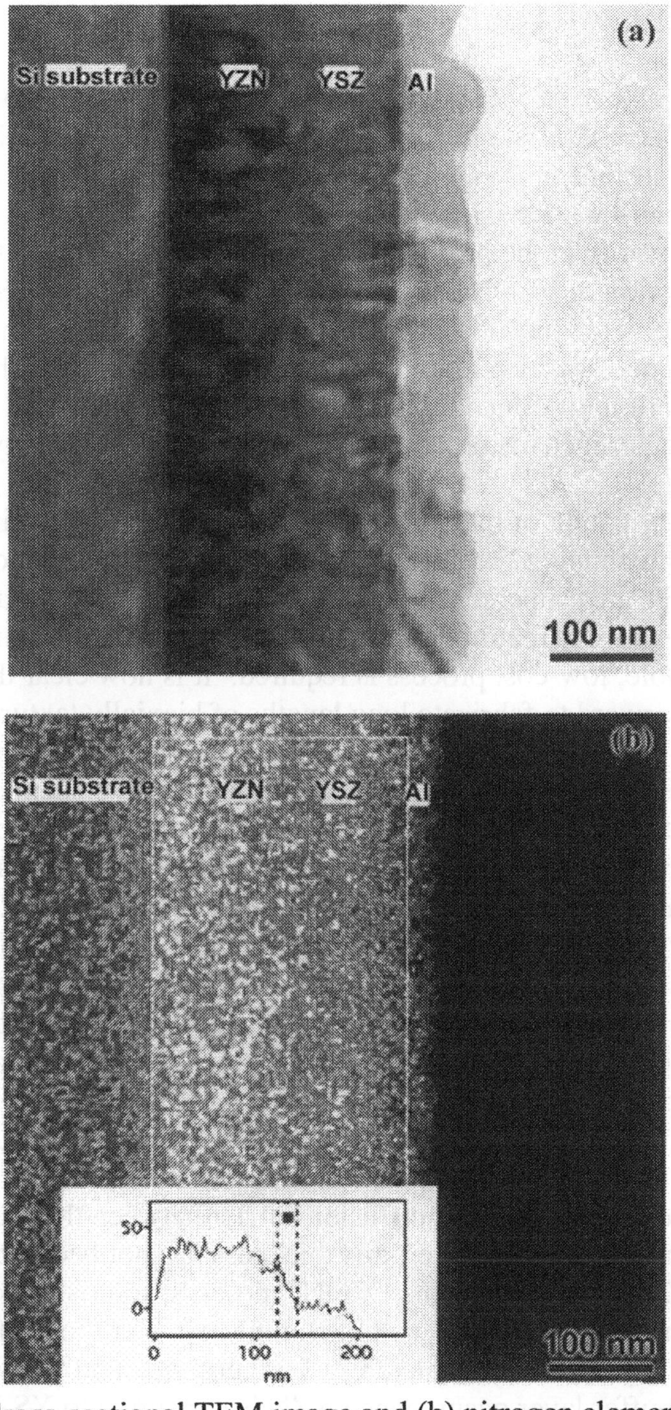

Figure 5. (a) Cross-sectional TEM image and (b) nitrogen elemental map of the same area, collected using PEELS. Brighter regions contain higher concentrations of nitrogen. A nitrogen-concentration line scan is shown at the bottom of (b).

CURRENT RESEARCH FOCUS

The ultimate objective of using the ECONO™ process is to develop a suitable process that produces consistent, reliable, and high texture-quality buffer films with appropriate surface morphology on desirable metal substrate tapes in a continuous manner – with the goal of producing economical superconducting wire. We are currently in the process of upgrading our film deposition system to enable the production of ECONO™ YSZ on lengths of RABiTS. Several buffer layer concepts are being pursued by researchers worldwide with an effort to meet both the cost and performance requirements for tapes in long lengths. The performance criterion is defined by the texture quality (both in and out of plane) and the microstructural characteristics of the film (YSZ film density and surface morphology).

Successful fabrication of biaxially textured superconducting wire based on the coated conductor technology requires optimization of the cost/performance of the HTS conductor. From a superconducting performance standpoint, a long, flexible, single crystal-like wire is required. From a cost and fabrication standpoint, an industrially scalable, low-cost process is required. It is now clear that, although it is fairly straightforward to fabricate long lengths of biaxially textured metals or alloys, it is quite difficult to deposit high-quality oxide buffer layers using low-cost processes. Buffer layers must: (a) provide an effective chemical barrier for diffusion of deleterious elements from the metal to the superconductor, (b) have a high degree of crystallinity via epitaxy with the biaxially textured metal template, and (c) provide a good epitaxial template for superconductor growth.

The above arguments clearly suggest the need for alternate buffer technologies to satisfy the requirements for economic production of superconducting tapes. Our results from Phase I and subsequent efforts show the proposed approach of "epitaxial oxidation of cubic nitrides" is a viable approach for meeting the cost and performance objectives.

The optimum thickness of the nitride layer must be determined. It is desirable to keep the thickness to a minimum for lower production costs, but the YSZ layer must be thick enough to be a good diffusion barrier between the metal and YBCO during YBCO processing. However, this thickness has not been optimized for highly dense YSZ as produced in the ECONO process, so it is reasonable to expect that thickness requirements will be diminished over those of typical RABiTS / buffer layer / YBCO structures. Furthermore, a 50 % increase in thickness of the buffer layer occurs when YZN is converted to YSZ due to the higher molar volume of YSZ compared to YZN. Thus, a 200 nm dense YZN film converts to a 300 nm dense YSZ film.

References

1 A. P. Malozemoff, S. Annavarapu, L. Fritzemeier, Q. Li, V. Prunier, M. Rupich, C. Thieme, W. Zhang, A. Goyal, M. Paranthaman, and D. F. Lee, "Low-Cost YBCO Coated Conductor Technology," presented at the Eucas Conference Sitges, Spain, September 14-17, 1999.

2 S. Jun, S. K. Young, J. Lee, and W. K. Young, Appl. Phys. Lett. **78**, 2542 (2001).

3 Q. X. Jia, C. Kwon, and P. Lu, Integr. Ferroelec. **24**, 57 (1999).

4 Fall 2001 BMDO Update, BMDO-funded Superconductor Research Leads to Revolutionary New Commercial Products (2001).

5 Proceedings of the SPIE, The International Society for Optoelectronic Engineering: High-Temperature Superconducting Detectors, Bolometric and Nonbolometric, Los Angeles, CA (1994).

LATEST DEVELOPMENTS IN USING COMBUSTION CHEMICAL VAPOR DEPOSITION TO FABRICATE COATED CONDUCTORS

Adam C. King, Shara S. Shoup, Marvis K. White, Steve L. Krebs, Dave S. Mattox, Todd A. Polley, Natalie Darnell
MicroCoating Technologies
5315 Peachtree Industrial Blvd
Atlanta, GA 30341

Ken R. Marken, Seung Hong, Bolek Czabaj
Oxford Superconducting Technology
600 Milik St.
PO Box 429
Carteret, NJ 07008-0429

ABSTRACT

The non-vacuum Combustion Chemical Vapor Deposition (CCVD) technique has been utilized to deposit both buffer layers and high temperature superconductors (HTS) on metal substrates that were prepared using the Rolling Assisted Biaxially Textured Substrate (RABiTS^{TM}) process [1]. Previously CeO_2 capped $SrTiO_3$ deposited by CCVD on RABiTS^{TM} substrates have enabled pulsed laser deposition (PLD) yttrium barium copper oxide (YBCO) with current carrying capability of >1MA/cm^2 [2]. MCT has partnered with Oxford Superconducting Technology who supplies the substrates and is in the process of scaling the textured metal to 100 plus meter lengths. Other buffer layers, such as, gadolinium oxide and lanthanum zirconate have also been investigated using CCVD and are reported here. Results indicate high quality epitaxial films of Gd_2O_3 can be grown on nickel and nickel 3% tungsten. Reproducibility, uniformity and scalability of the Gd_2O_3 buffer layer on this nickel alloy have also been proven. Previously it has been reported that CCVD YBCO deposited on single crystal was capable of achieving the benchmark of >1MA/cm^2 [2]. Currently, epitaxial YBCO superconductors were deposited onto a ceria capped $SrTiO_3$ buffer layer on Ni and on a ceria capped Gd_2O_3 buffer layer on Ni-3%W. Experimentation to pass high currents through the HTS films is in progress. Further research to eventually develop long length HTS tapes created solely using the non-vacuum CCVD method will continue at MicroCoating Technologies (MCT).

INTRODUCTION

The use of high temperature superconductors has proliferated over the past few years in several different applications such as transmission cables, motors, etc [3]. Typically, these applications have used bismuth strontium calcium copper oxide (BSCCO) superconductors, but the cost of these superconductors still limits their overall use in the industry. Research on producing lower cost and high performance HTS systems, such as the YBCO superconductor, have continued with some success. Depositing the YBCO material onto a flexible metal tape has proven to be challenging, as has scaling any system to produce high-performance long lengths of this tape. Ideally, once the YBCO system has been commercially scaled, the target cost of the superconducting tape will be \$10-50/kA-m [4]. Many of the costly vacuum-based deposition systems will not be able to achieve this low cost for the production of YBCO, but the non-vacuum CCVD process has this potential.

CCVD can potentially *reduce costs at least tenfold* in comparison to these other systems. As CCVD does not rely on a vacuum to regulate the atmosphere within which the buffer layers and the following superconductor will be deposited, the capital expense is dramatically reduced. Due to the fact that the CCVD process is not a batch process, there exists a significant advantage over most vacuum-based methodologies. The scaling of a typical CCVD system is quite easy and inexpensive. Long lengths of continuous coated tape can be easily achieved for both the buffer and superconducting layers. Adding additional nozzles at a minimal overall expense can also easily increase the deposition rate of said materials. CCVD is a viable processing option for the production of long length HTS tapes.

EXPERIMENTAL DETAILS

In the traditional CCVD process (Fig. 1) [5], precursors are dissolved in a solvent that typically acts as the combustible fuel. This solution is atomized to form submicron droplets by means of the Nanomiser^{TM} technology, proprietary to MCT. These droplets are then convected by an oxygen stream to the flame where they are combusted. A substrate is coated by simply drawing it through the flame plasma. The heat from the flame provides the energy required to evaporate the droplets and for the precursors to react and to vapor deposit on the substrates. Substrate temperature is an independent process parameter that can be varied to actively control the film's microstructure, epitaxy, and oxidation of the nickel-based substrate. In the case of the buffer layer depositions, a localized reducing atmosphere is also used to prevent nickel oxidation, and a solvent system with a

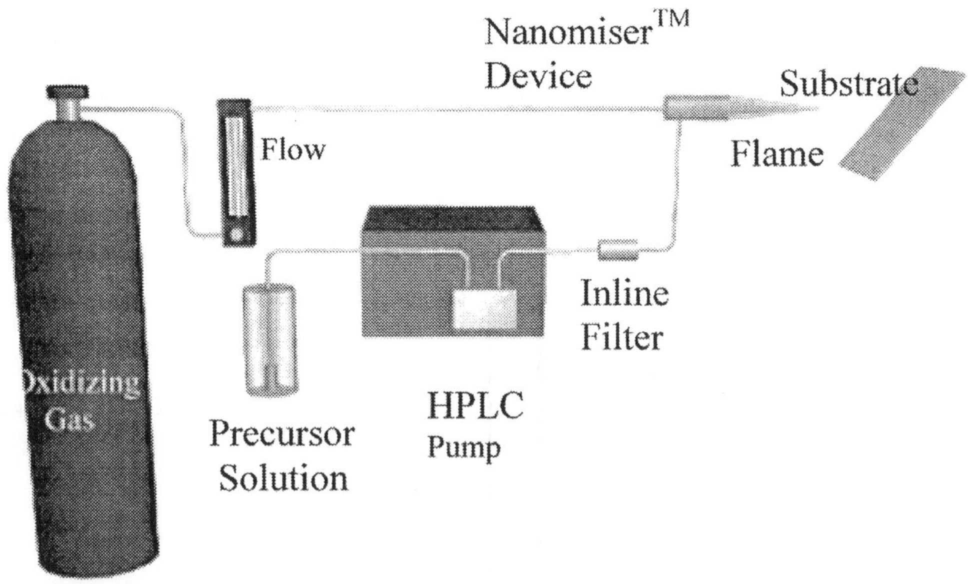

Figure 1. CCVD system showing inexpensive equipment and ease of setup.

lower carbon content is used to minimize carbon deposition in such an atmosphere. Dynamic seals allow for continuous substrate passage between the open air and the localized reducing atmosphere.

One of the largest challenges in tailoring the CCVD system for deposition of YBCO was the removal of all carbon sources from the process. Once formed, barium carbonate is difficult to remove and detrimental to the performance of the superconductor coating. The process was modified to allow for use of aqueous solutions in the deposition of YBCO. A new Nanomiser™ device for the atomization of aqueous solutions and the use of a hydrox flame has resulted in successful YBCO. The locally reducing atmosphere, similar to what is used to deposit buffer layers, is also used to aid in the deposition of YBCO on the metal substrates. The challenge presented now is growing the complex oxide in a reducing atmosphere.

DISCUSSION

Buffer Scaling and Optimization

Experiments have turned from nickel to a new substrate, a nickel alloy that contains 3% tungsten. The switch in substrate emphasis from pure Ni to Ni-3%W was due to its superior strength over Ni and Oxford's success at producing smooth, cube-textured Ni-3%W with lower omega and phi full width half maximum (FWHM) values than for Ni. Several buffer layers have been tested on this substrate including $SrTiO_3$, CeO_2, Gd_2O_3, $LaMnO_3$ and $LaZrO_3$. At the present time, the material that appears to grow with the best orientation is Gd_2O_3. It is also possible to deposit an epitaxial ceria cap on the Gd_2O_3 layer. The X-ray diffraction (XRD) results shown below in Figure 2 and Figure 3 indicate the out-of-plane and in-plane orientation of the sample is quite good. The CeO_2 (111) Phi scan indicates FWHM of 8.36°.

Once it was determined that short samples of the CeO_2/Gd_2O_3 architecture could be deposited onto the nickel alloy, it was then necessary to determine if

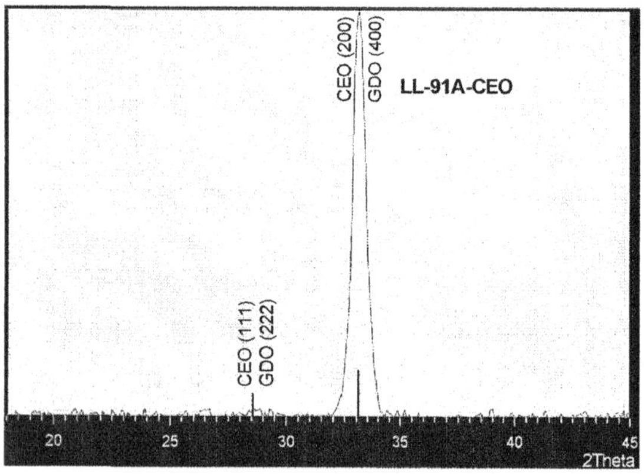

Figure 2. Theta-2theta scan of a CeO_2/Gd_2O_3/Ni-3%W architecture

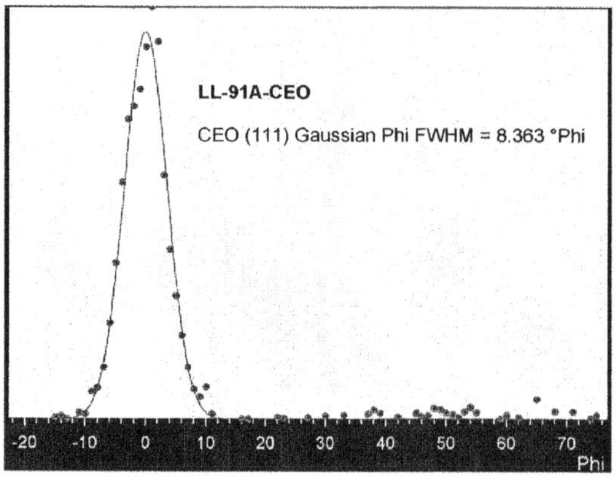

Figure 3. Phi scan of a CeO_2/Gd_2O_3/Ni-3%W architecture

long lengths of this material could be produced with a relatively uniform coating over the length of the sample. For these trials, meter lengths of the CeO_2/Gd_2O_3 architecture were produced on Ni-3%W and then tested at ORNL's user facility with their reel-to-reel x-ray system. The results averaged along the meter are shown in Table I indicating reproducibility was achieved with the CCVD process. Another depiction of this reel-to-reel data is shown in Figure 4.

Table I. Reproducibility of 4 separate meters of CCVD CeO_2/Gd_2O_3/Ni-3%W

Sample	Out-of-Plane % Misorient.	Omega FWHM	In-Plane % Misorient.	Phi FWHM
1	1.42	6.84	0.00	6.84
2	1.33	6.90	0.00	6.82
3	1.53	6.80	0.00	6.62
4	1.54	6.80	0.00	6.58

The microstructure of these buffer layers was also investigated. For this purpose, a scanning electron microscope was utilized to determine the homogeneity of the films and if any major surface defects existed. A typical micrograph of a CeO_2/Gd_2O_3 deposition is shown in Figure 5. The ceria layer forms a dense, homogeneous film over the gadolinium oxide layer. The surface particles are CeO_2/Gd_2O_3 as indicated by energy dispersive spectroscopy (EDS). This surface is a suitable template for the growth of epitaxial as described below. At this time, minimal depositions of YBCO by both PLD and CCVD have been deposited onto these new buffer layers, but preliminary results indicate that

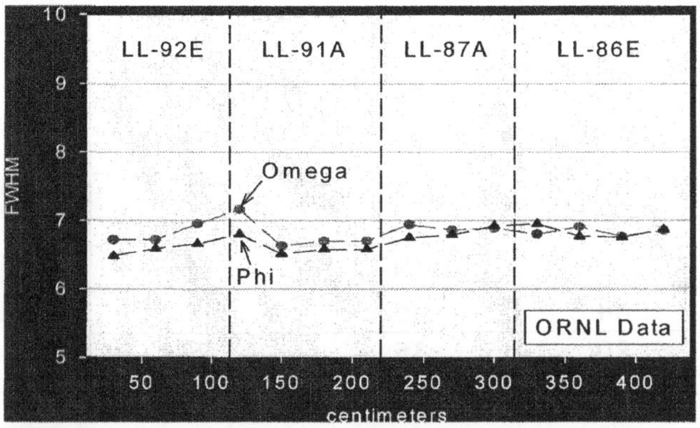

Figure 4. X-ray data over lengths of $CeO_2/Gd_2O_3/$ Ni-3%W architecture buffered tape

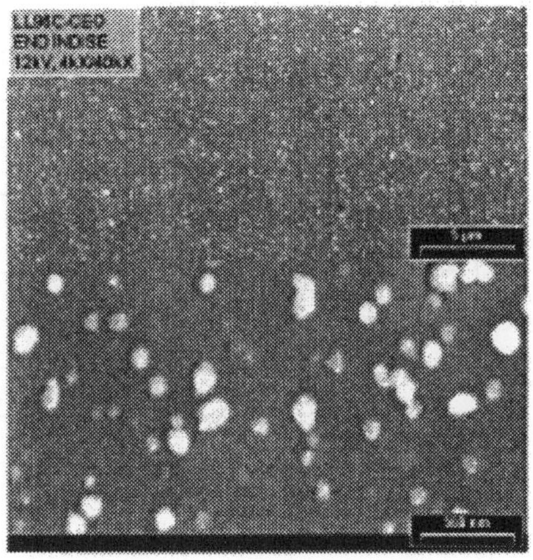

Figure 5. SEM Micrograph of CEO/GDO/Ni-3%W

this architecture is capable of suppressing nickel oxide growth and nickel/tungsten diffusion. The evidentiary support for this is shown in Table II. This table contains analyses from X-ray Photoelectron Spectroscopy (XPS) indicating penetration depth of nickel (presumed to be nickel oxide) and tungsten. The last four samples indicated in Table II are from meter runs mentioned earlier. Although the thinner buffer layers did provide an effective barrier to nickel and tungsten diffusion, the overall thickness of these films will most likely to be increased to provide a more effective buffer layer for later YBCO depositions (LL-77A-YBCO is the only sample that has a YBCO layer deposited on it).

Further development of other buffer layers deposited by CCVD is showing promise for the future. One such layer deposited onto the Ni-3%W substrate is $LaZrO_3$. This buffer layer has been deposited only very recently but has shown epitaxial growth on the Ni-3%W substrate. Initial results show an out-of-plane orientation of ~8.06% misorientation which can be seen in Figure 6. Brief analysis indicates that the composition is slightly off in regards to the ratio of La to Zr, which can easily be adjusted. Based upon past experiments, once the correct composition is achieved, the misorientation should be lowered. Experiments involving $LaZrO_3$, other buffer layers, and various architectures of those buffer layers, will continue. Testing of YBCO films on these new CCVD substrates is currently underway.

Table II. Barrier effectiveness results by XPS of the CeO_2/Gd_2O_3/Ni-3%W architecture

Sample	CeO_2 Thickness (nm)	Gd_2O_3 Thickness (nm)	% Gd_2O_3 Penetration by Ni (nm)	%Gd_2O_3 Penetration by W (nm)
LL-86G	None	243	9	Unmeas.
LL-76G-1	None	180	14	Unmeas.
LL-76G-2	None	315	11	14
LL-77A-YBCO (ORNL-PLD)	32	280	45	8
LL-86E	37	75	39	69
LL-87A	43	84	56	80
LL-92E	38	80	35	73
LL-95C	35	90	28	66

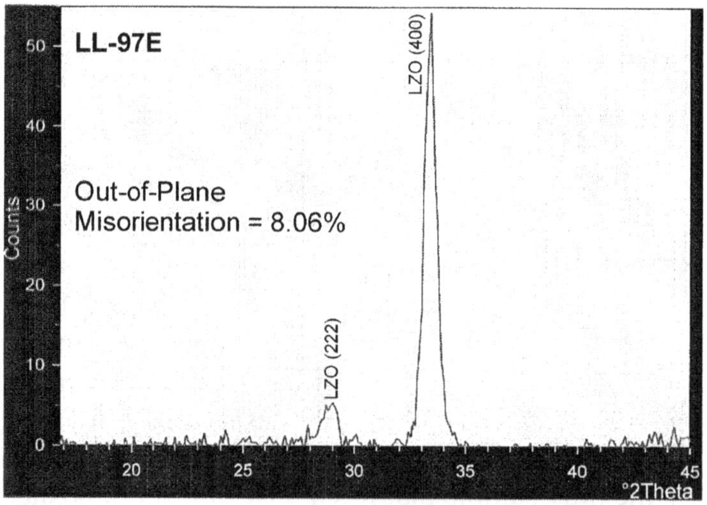

Figure 6. X-ray data of $LaZrO_3$ buffer layer on Ni-3%W

YBCO depositions

Over the period of the past several months, significant progress has been made in depositing CCVD YBCO on various CCVD buffer layers deposited on RABiTS™ textured substrates. It has been reported earlier that YBCO deposited by CCVD had been able to reach the benchmark of >1MA/cm^2 on single crystal substrates. More recently, extensive work has been performed on depositing YBCO on several different CCVD buffered metal tape substrates. CCVD YBCO has been epitaxially grown on CCVD CeO_2 capped $SrTiO_3$ buffer layers on Ni and also on CCVD CeO_2 capped Gd_2O_3 buffer layers on Ni-3%W. Both depositions were performed in a localized reducing atmosphere to limit the growth of nickel oxide into the buffer layer during the YBCO deposition. Lower temperatures also proved to be necessary to decrease the amount of nickel oxide and also to facilitate epitaxial, c-axis oriented growth. The temperature range where the best films were grown in was 700-750°C. Initial experiments show that with elevated temperatures, a-axis growth was present. The results of the CCVD YBCO layers deposited in the 700-750°C temperature range are shown below. YBCO grown on the $SrTiO_3$ buffer layer on Ni shows good orientation with clear evidence of (003) and (005) peaks as seen in Figure 7. The second XRD plot depicted along with the CCVD sample, is a YBCO PLD sample used as a for comparative purposes. The PLD YBCO and CCVD YBCO samples show approximately the same level of NiO growth while the CCVD YBCO layer appears significantly thicker. In Figure 8, the results of the YBCO depositions on CeO_2 capped Gd_2O_3 on Ni-3%W, also show good orientation. A (103) pole figure was performed on the YBCO grown on Gd_2O_3, which is shown in Figure 9. The phi FWHM of this film was 9.68°. A typical micrograph of a CCVD YBCO layer can be seen in Figure 10. The micrograph shows the crystalline grains and the dense microstructure of the CCVD grown YBCO layer.

Though superconducting, these YBCO layers have not yet demonstrated $1MA/cm^2$ capability. Some of the concerns with the efficacy of the YBCO layers arise due to their thickness, their possible oxygen deficiency, possible stresses within the film due to lattice strain and obtaining the optimal composition. The thickness issue can be addressed by simply depositing longer which does not tend to have a deleterious effect on the epitaxy of the films up to relatively thick films. Thickness of the current CCVD YBCO films are ~200nm. The oxygen deficiency of the superconducting layers is still in question but simply because the depositions are done in a locally reducing atmosphere. Initial results show that it may be necessary to use an oxygenation step on these films, but more research is needed. It does appear however, that the largest steps in progress can be made in the region of composition and relieving of stresses within the film.

For some time, EDS data obtained from scanning electron microscopy (SEM) was used to determine whether or not the correct 1:2:3 ratio was being achieved. Some problems with this methodology arose when the YBCO was deposited onto STO. The peak intensities of yttrium overlap those of strontium, and the peak intensities of barium somewhat overlap those of titanium. For these reasons, a derivation of the curves did not yield a relevant compositional analysis. Therefore, in order to obtain more accurate and precise results of compositional changes, an inductively couple plasma mass spectrometer (ICP-MS) was utilized. Initial findings, as compared to PLD deposited films exhibiting critical current densities >1MA/cm^2 and used as standards, did in fact show that CCVD composition was incorrect. Further research will be performed to optimize the composition.

The other most successful route thus far to increasing the current carrying capability of the films was through a high temperature anneal in a low partial pressure of oxygen to alleviate lattice strain. The results from these initial tests prove to be quite promising. Post-anneal electrical testing demonstrated that the YBCO film was able to carry a relatively low current at 77K (critical current density = ~40,000 A/cm^2). In previous experiments, obtaining the initial current was the most important step, as further parameter refinement of the deposition can quickly lead to higher current densities.

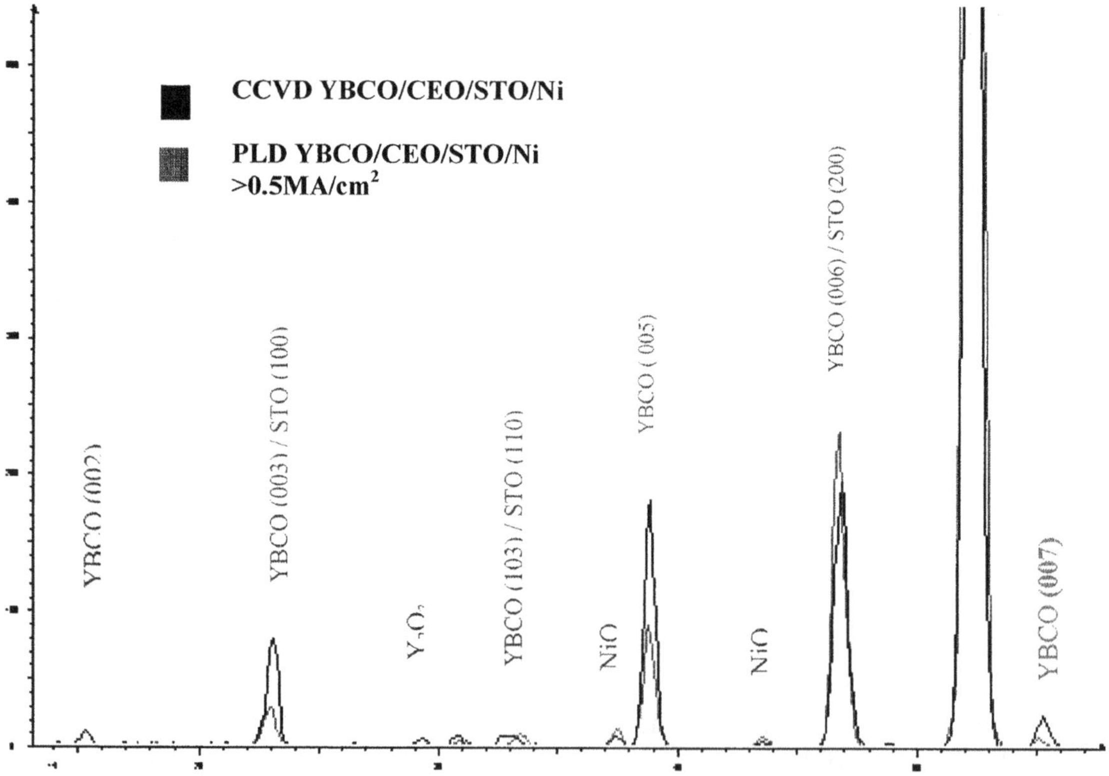

Figure 7. XRD results of CCVD YBCO/CEO/STO/Ni architecture compared to a PLD YBCO standard showing a thicker CCVD YBCO layer with comparable NiO.

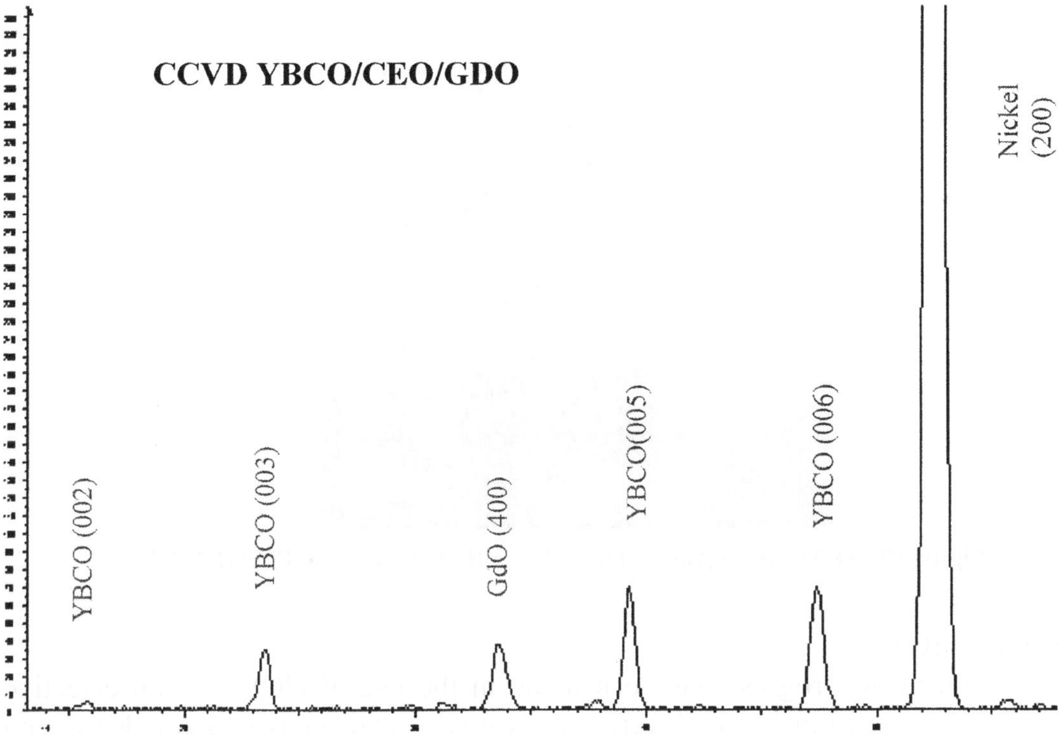

Figure 8. XRD results of a typical CCVD YBCO/CEO/GDO architecture.

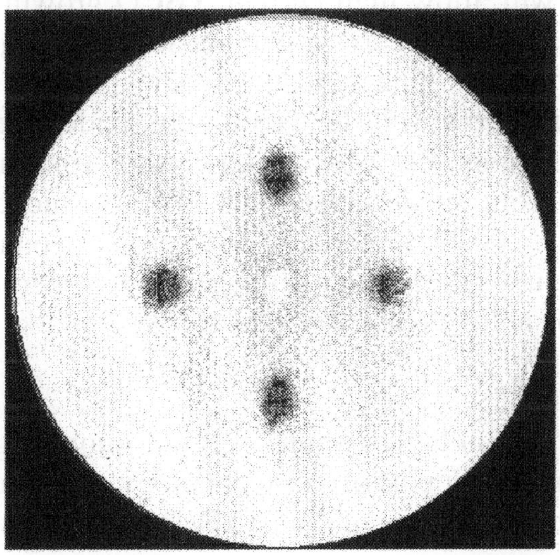

Figure 9. Pole figure of (103) YBCO/CEO/GDO/Ni-W

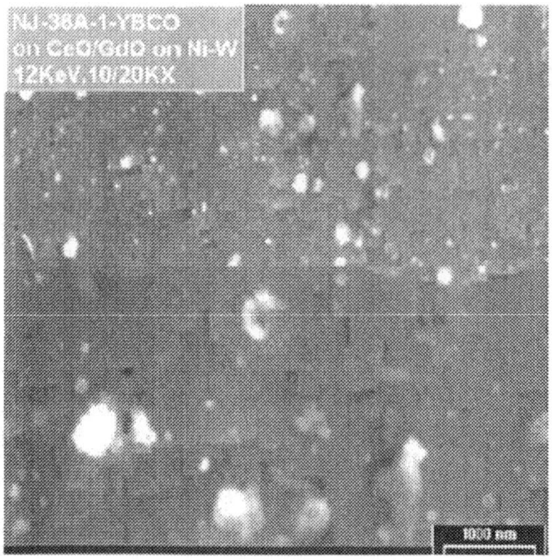

Figure10. SEM Micrograph of a CCVD YBCO layer on CEO/GDO/Ni-W

CONCLUSION

Significant progress has been made in the use of CCVD as an effective process for depositing buffer layers on textured metal tapes and for depositing YBCO on those buffer layers. CCVD buffer layers deposited on textured nickel tape are high quality templates for >1MA/cm^2 PLD films. CCVD YBCO has reached this benchmark on single crystals and will be pursued on CCVD buffered metal substrates. New developments in the textured substrate have led to research on buffer layers that will allow high current YBCO growth on these substrates. Gd_2O_3 depositions have proven to be the most promising, showing excellent growth and uniformity over long lengths of tape. In the near future, YBCO depositions will evaluate how well the gadolinium oxide layer can act as a buffer layer. CCVD depositions of YBCO have also continued to progress as depositions switched from single crystal substrates to nickel and nickel alloys. Initial results of the YBCO material exhibit great epitaxy and with some improvements in density, composition, film stress and perhaps oxygen content, these films will also be able to reach the industry benchmark.

ACKNOWLEDGEMENTS

MicroCoating Technologies would like to acknowledge the Department of Energy on their continued support of our technology and research and ORNL for their help in depositing PLD films on our CCVD and the use of their user XRD facility.

REFERENCES

1. A. Goyal, D.P. Norton, J. Budai, M. Paranthaman, E.D. Specht, D.M. Kroeger, D.K. Christen, Q. He, B. Saffian, F.A. List, D.F. Lee, P.M. Martin, C.E. Klabunde, E. Hatfield, and V.K. Sikka, Appl. Phys. Lett. **69**, 1795 (1996).
2. 2001 Annual Peer Review, F.A. List and D.F. Lee, Oak Ridge National Laboratory report, Washington D.C., August 2001.
3. L.R. Lawrence Jr., C. Cox, D. Broman, High Temperature Superconductivity: The Products and their benefits.
4. Department of Energy Coated Conductor Development Roadmapping Workshop, St. Petersburg, FL, January 2001.
5. A. T. Hunt, W. B. Carter, and J. K. Cochran, Jr., Appl. Phys. Lett. **63** (2), 266 (1993).

GROWTH OF YBCO THICK FILMS ON ND_2CUO_4 BUFFERED SUBSTRATES

Xiaoding Qi, Zainovia Lockman, Masood Soorie, Yura Bugoslavsky, David Caplin and Judith L. MacManus-Driscoll

Department of Materials and Centre for High Temperature Superconductivity, Imperial College, London SW7 2BP, UK

ABSTRACT

A new buffer, Nd_2CuO_4, has been developed for the liquid phase epitaxy (LPE) growth of YBCO thick films. Nd_2CuO_4 is an ideal buffer for the LPE growth of YBCO thick films, since it has a good lattice and thermal-expansion match to YBCO, minimum reaction with the high-temperature CuO:BaO solution, and is not poisoning to superconductivity. Nd_2CuO_4 buffers were also grown from the liquid phase using a number of novel techniques. Nd_2CuO_4 films grown on the surface oxidized Ni foils showed an excellent biaxial texture and a good surface smoothness suitable for subsequent YBCO growth. LPE growth of YBCO on such Nd_2CuO_4/NiO/Ni substrates showed that dense nucleation of YBCO on Nd_2CuO_4 was possible without any REBCO seed layer and there was little reaction between the Nd_2CuO_4 buffer and the high-temperature CuO:BaO solution. T_c's of 90 K and transport J_c's over 10^5 A/cm^2 (77K) have been achieved in such YBCO/Nd_2CuO_4/NiO/Ni coated conductors with the superconducting layer YBCO about 1-2 μm thick Hence, there is a great potential to fabricate YBCO coated conductors via a fast all-LPE route.

INTRODUCTION

$YBa_2Cu_3O_{7-x}$ (YBCO) has a better intrinsic current carrying property in magnetic fields than the Bi-based materials, but this better performance can only be realized when all the crystallites in a film are very well aligned along a same direction, so that the superconducting current can be transferred across the crystallites. In order to achieve a good texture in the YBCO film, a highly textured metallic substrate and its buffers are required. There are currently two main methods to achieve the required texture on the metallic substrates or buffers. The first is called IBAD (ion beam assisted deposition) [1], which gives a highly textured buffer layer on a polycrystalline, randomly oriented Ni-based substrate, and the second is called RABITs (rolling assisted biaxial texturing of substrates) [2], which produces a highly textured Ni-based substrate by mechanical working. At present, the buffers and top YBCO layer are dominantly grown by vapor deposition methods such as pulsed laser deposition (PLD). These vapor deposition methods have a relatively slow growth rate, typically several hundred nanometers per hour, and require a high vacuum system. Although they produced very good samples in short length with Jc's as high as 3 MA/cm^2, none of them has been justified to be viable for making long length coated conductors at reasonable cost.

On the other hand, liquid phase epitaxy (LPE), which has widely been used to grow semiconductor and oxide garnet films during 1970s and 1980s [3], has a much faster growth rate, typically in the order of 1 micron per minute. LPE growth of YBCO on single crystal substrates has demonstrated that Jc's over 1 MA/cm^2 can be maintained in the films of thickness up to about10 microns [4]. This is particularly useful for making thick films for coated conductor applications. However, there is a problem yet to be solved before LPE can be used to grow YBCO on the metallic substrates. That is the relatively high growth temperature, at which most of the metallic substrates and buffers react strongly with the BaO:CuO high temperature solution. A lot of efforts have been made to reduce the growth temperature [5]. However, by reducing the growth temperature, it also reduces the solubility of YBCO in the liquid, which will make the growth more difficult. Therefore, a better method is to search for a special buffer, which is particularly resistant to the attack of the high temperature liquid. Here, we report the development of a new buffer, Nd_2CuO_4, and the initial results of LPE growth of YBCO on Nd_2CuO_4 buffered, surface oxidized Ni substrates.

Nd_2CuO_4 is a tetragonal crystal with $a=b\approx0.394$ nm. It has a lattice mismatch about 2.3% to YBCO, which can be reduced further by doping some smaller rare-earth (RE) ions like Yb^{3+} or Ce^{4+}. By doping Ce^{4+}, this compound becomes conductive. A conductive buffer layer is desired for the coated conductor application because in the event of any breakdown in the superconducting layer, a conductive buffer can reduce the impulse significantly. Nd_2CuO_4 forms in the pseudo binary system of Nd_2O_3-CuO and the phase diagram is shown in Fig. 1, which was determined by differential thermal analysis (DTA), high temperature crystallization and X-ray diffraction (XRD). Nd_2CuO_4 has a much larger solubility in the self flux CuO compared to YBCO in the 2BaO:5CuO flux. Therefore, it is relatively easy to grow from the liquid phase. Since its composition falls under the RE-Ba-Cu-O system, Nd_2CuO_4 neither poisons superconductivity nor reacts strongly with the barium cuprate solution at high temperature. It is a suitable buffer for LPE of YBCO.

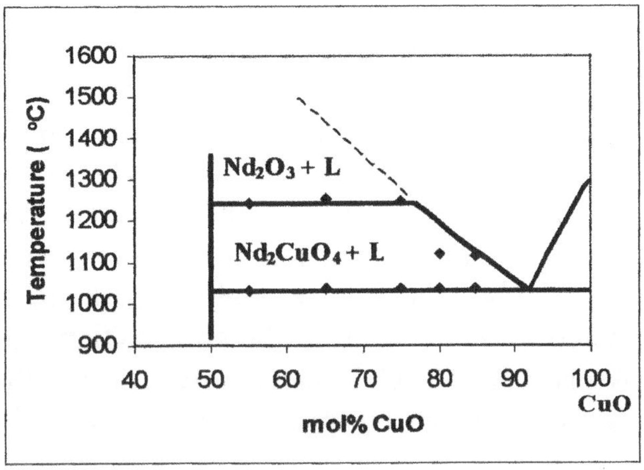

Fig.1 Pseudo-binary phase diagram of Nd_2CuO_4-CuO

EXPERIMENTS

The substrates used in this study were RABITs Ni with a biaxial texture about 10°. The first buffer was NiO, which was formed in flowing O_2 at 1200 °C for 12 hours. NiO was found to grow pseudomorphically on the (002) Ni foils with a biaxial texture around 10° and a thickness around 40 μm.[6]. On top of NiO was Nd_2CuO_4, which was prepared by a number of liquid phase processing methods [7]. The first method was the conventional top dipping LPE [3]. A high temperature solution of 20%Nd_2O_3:80%CuO was prepared at 1200 °C and then quickly cooled down to about 1150 °C, at which the substrates were dipped into the liquid and rotated at 60 rpm. After growing in liquid for about 5 minutes, the substrates were pulled clear and brought to room temperature in about half an hour.

In the second method, a precursor with composition around 30%Nd_2O_3:70%CuO was screen-printed on the substrates using an organic binder and then heated up to 1200 °C to reach a single liquid state. This temperature was kept for about 20 minutes and then cooled down to room temperature at 300 °C/hr. Because of the large surface area to volume ratio and preferential evaporation of CuO, the liquid became supersaturated rapidly, leading to the nucleation and growth of Nd_2CuO_4 on the substrates..

In a third method, a growth system similar to the zone-refining furnace was used. A hydraulically pressed and sintered source rod of 40%Nd_2O_3:60%CuO was fed into the focus of an infrared beam to produce a molten zone at one end. When the increasing size of the molten zone exceeded the limit of the surface tension, the liquid dropped down onto the substrate below, which was maintained at a suitable temperature (~1100 °C) for the epitaxial growth induced from the highly textured substrate. Because of the extremely good wetting property of the high-temperature cuprate solution, the liquid drops quickly spread out on the substrates. Actually, the wetting was so good that not only the top side of the substrate was completely covered, but also the back side of the substrate.

LPE growth of YBCO on the Nd_2CuO_4/NiO/Ni substrates was carried out in 1% pO_2 using the top dipping method. The experimental details about LPE of YBCO can be found elsewhere [8]. The high-temperature solution used in this experiment was prepared at 1010 °C from the composition Y:Ba:Cu= 1:12:30 and the LPE growth took place at 950 °C. Typical growth time in the liquid was around 60 seconds. After the growth, the films were annealed in flowing O_2 for 120 hours.

RESULTS AND DISCUSSIONS

Although surface oxidized NiO had a good biaxial texture, its surface was quite rough. The typical roughness was around 200nm. Considering such a roughness, liquid phase processing is probably the best way to grow Nd_2CuO_4 on top of NiO. Especially, the high-temperature cuprate solution had an extremely good wetting property, which could easily creep into all the groves between the NiO grains. Shown in Fig.2 is the scanning electron microscope (SEM) image of the cross section of the Nd_2CuO_4/NiO/Ni structure. The interface between the Nd_2CuO_4 and NiO was quite smooth and Nd_2CuO_4 could grow into all the gaps.

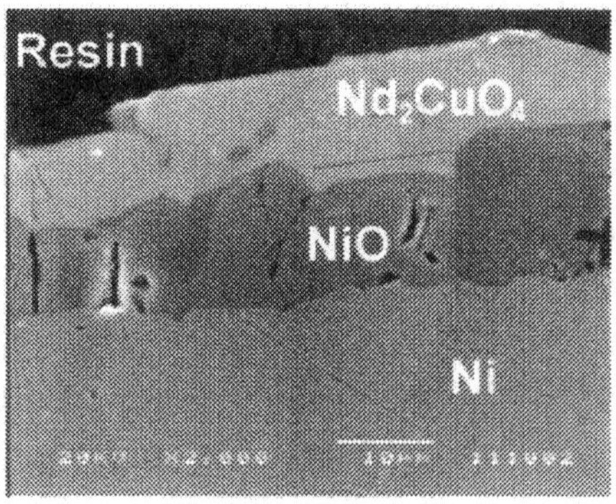

Fig.2 SEM image of cross-section

X-ray diffraction (XRD) showed that the Nd_2CuO_4 films grown on NiO/Ni were c-oriented with high crystallinity and a very good biaxial texture [7]. XRD pole figure showed that the full width at half maximum (FWHM) was only about 5°, which was better than the original texture in NiO. Nd_2CuO_4 can grow relatively easily on NiO because there is a good lattice match between these two compounds, as illustrated in Fig.3. The

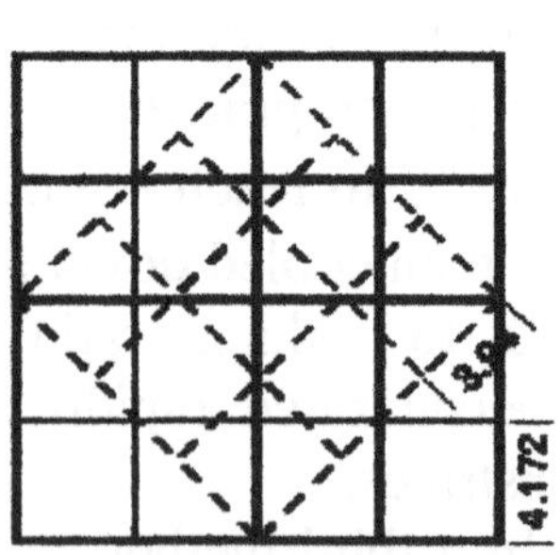

Fig.3 Epitaxial relation between Nd_2CuO_4 and NiO

a/b-axis is about 0.394nm for Nd_2CuO_4 and 0.417 nm for NiO, therefore, 3 unit cells of Nd_2CuO_4 can grow along two units of the [110] axis of NiO with lattice mismatch as close as 0.3%. Such an epitaxial relation has been confirmed by the XRD pole figure as well as the observation of surface morphology.

Although three liquid phase processing methods have been tried to grow Nd_2CuO_4, the second method was so far studied in most detail. Most of the results presented above were measured from the samples made by this method. The first method had the simplicity and the ease of control, but it was found that the available supersaturation was not high enough to achieve a dense nucleation on NiO, therefore it was difficult to achieve 100% coverage, although the texture obtained was excellent. The great advantage of the third method was that before the growth, the high temperature solution was prepared at a different place from the substrate, therefore the contact time between the substrate and liquid was much shorter and hence, it could minimize any potential reaction at the interface. In addition, in this method a higher temperature could be used to prepare liquid and at the same time, the temperature of the substrate could be kept below 1200 °C, which was the requirement not to destroy the texture in NiO. By using a higher temperature to make liquid, a smaller amount of flux was required, so the growth could be easier and quicker. Recently, we found that it might not need to oxidize Ni before the growth, a textured NiO was possible to form during the process of making Nd_2CuO_4 by this method. Therefore, there is a potential to make Nd_2CuO_4/NiO/Ni in one process. More detailed study about this method is currently being carried out.

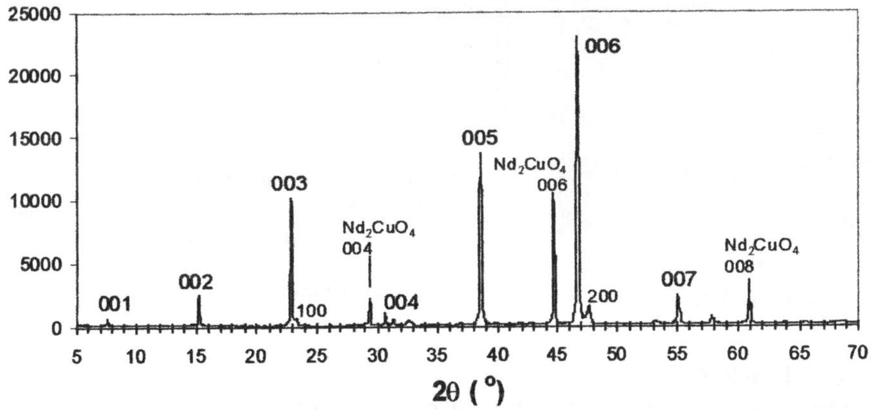

Fig.4 XRD θ-2θ scan of the sample YBCO/Nd_2CuO_4/NiO/Ni

Considering the much lower Y solubility in the liquid, the top YBCO layer on Nd_2CuO_4/NiO/Ni was grown by the conventional top dipping LPE. The growth of

YBCO on the Nd_2CuO_4 was found to be relatively straightforward. It was found that dense nucleation of YBCO on Nd_2CuO_4 was possible without any REBCO seed layer and there was little reaction between the Nd_2CuO_4 buffer and the high-temperature CuO:BaO solution. A XRD θ-2θ scan of the sample YBCO/Nd_2CuO_4/NiO/Ni is shown in Fig.4, which indicates that the films was dominantly c-oriented except small amount of a/b-oriented grains, which was also observed on the surface by optical microscopy. High X-ray counts in the Fig.4 (scanning rate=0.5 sec/step) also show a good crystallinity of the YBCO film, which is one of the benefits of films grown from the liquid phase. The optical micrograph of the surface is shown in Fig.5. The magnified inset at bottom-left corner shows a good alignment of the grains. XRD pole figure conformed the excellent biaxial texture in the YBCO thick films with FWHM=2.5°. The 45° tilted grains were consistent with the epitaxial relation discussed above. T_c's of 90 K and transport J_c's over 10^5 A/cm^2 (77K) have been achieved in such YBCO/Nd_2CuO_4/NiO/Ni coated conductors with the superconducting layer YBCO about 1.5-2 μm thick. A R-T plot is presented in Fig.6, which shows a sharp superconducting transition with "zero" resistance at 86 K, suggesting that there was little contamination of Ni.

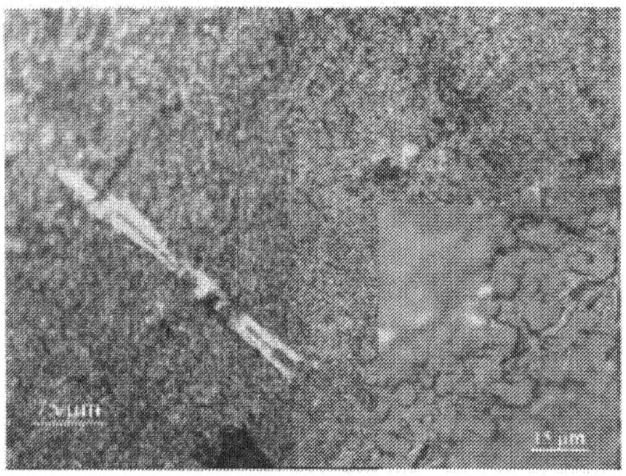

Fig.5 Optical micrograph of the surface of YBCO/Nd_2CuO_4/NiO/Ni

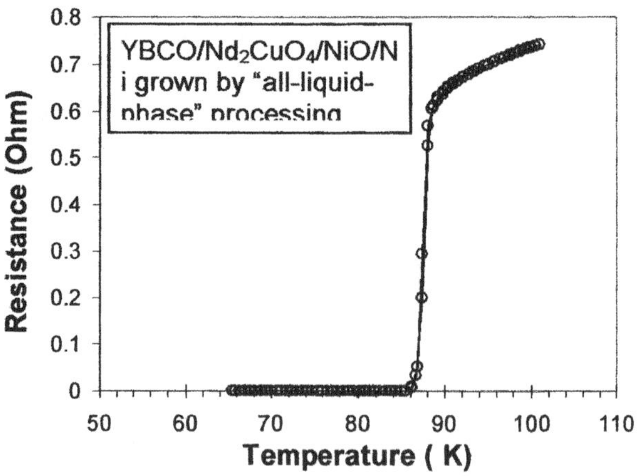

Fig.6 R(T) curve of YBCO/Nd_2CuO_4/NiO/Ni

CONCLUDING REMARKS

A new buffer, Nd_2CuO_4, has been developed for the LPE growth of YBCO coated conductor. There are a few potential liquid phase processing methods that can be used for fast production of Nd_2CuO_4 on surface oxidized RABiTs Ni. LPE growth of YBCO showed that dense nucleation of YBCO on Nd_2CuO_4 was possible without any REBCO seed layer and there was little reaction between the Nd_2CuO_4 buffer and the high-temperature CuO:BaO solution. T_c's of 90 K and transport J_c's over 10^5 A/cm^2 (77K) have been achieved in such YBCO/Nd_2CuO_4/NiO/Ni coated conductors with the superconducting layer YBCO about 1-2 μm thick Therefore, there is a great potential to fabricate YBCO coated conductors via a fast all-LPE route.

ACKNOWLEDGEMENTS

We are indebted to EPSRC of the UK for funding this work. Dr. W. Goldacker and Dr. R. Nast of Karlsruhe Forschungszentrum are acknowledged for provision of the (200) Ni foils. Prof. Don McK Paul and Dr. Geetha Balakrishana at Warwick University are thanked for providing crystal growth facilities for the development of a novel horizontal LPE process.

REFERENCES

1. Y. Iijima, N. Tanabe, O. Kohno, Advances in Superconductivity IV, Proc.4th Int. Symp. Supercond. (ISS), pp.679-82, 1992, Tokyo, Springer Verlag
2. J.D. Budai, R.T. Young, B.S. Chao, Appl. Phys. Lett. **62**(15),1836-38(1993)
3. R.L. Moon; Liquid Phase Epitaxy, In: B.R. Pamplin, editor, Crystal Growth, 2nded., Oxford, Pergamon Press, 1980, pp.421-61.
4. S. Miura, K. Hashimoto, F. Wang, Y. Enomoto and T. Morishita; Physica C **278**, 201(1997).
5. X. Qi and J.L. MacManus-Driscoll, *Current Opinion in Solid State & Materials Science*, **5**, 291-230(2001)
6. Z. Lockman, X. Qi, A. Berenov, R. Nast, W. Goldacker and J.MacManus-Driscoll, Physica C: Superconductivity, **351**, 34-37(2001)
7. X. Qi, M. Soorie, Z. Lockman and J L MacManus-Driscoll, Journal of Materials Research, **17**(1), 1-4(2002)
8. X. Qi and J.L. MacManus-Driscoll; J. Crystal Growth **213**, 312(2000).

MICROSTRUCTURAL CHARACTERISATION OF HIGH J_C, YBCO THICK FILMS GROWN AT VERY HIGH RATES AND HIGH TEMPERATURES BY PLD

A. Berenov, N. Malde, Y. Bugoslavsky, L.F. Cohen and J. L. MacManus-Driscoll
Center for High Temperature Superconductivity, Imperial College of Science Technology and Medicine, Prince Consort Road, London SW7 2BP, U.K.

S. J. Foltyn
Superconductivity Technology Center, Los Alamos National Laboratory, Los Alamos, New Mexico 87545, U.S.A.

ABSTRACT

In order to achieve scaleability in processing of IBAD conductors for commercial applications, rapid YBCO film growth rates are required. In this work scanning Raman spectroscopy, x-ray diffraction, scanning electron microscopy and vibrating sample magnetometry have been used to study high rate grown YBCO films. The films were grown by PLD on (100) $SrTiO_3$ using a high power industrial laser at growth temperatures from 750°C to 870°C and growth rates up to 4 micron/minute. The films showed high degree of in-plane and out-of-plane alignment. Two YBCO layers with extended "c" lattice parameters observed in the films were caused by a cation disorder. High values of T_C and J_C (90 K and 10^6 A/cm^2, respectively) were observed despite presence of structural disorder in the films.

INTRODUCTION

Over the last year significant efforts have been made to demonstrate the feasibility of long length coated HTS conductors. Different preparation techniques (e.g. IBAD, LPE, CVD) are being evaluated for possible scalability with the aim to achieve fast growth rate and attractive values of critical current densities (>1 MA/cm^2). However fast growth rates can lead to the crystalline disorder (especially Y/Ba cation disorder [1]), poor crystallinity (both in-plane and out-of-plane) and possible formation of metastable phases during deposition, with reduction in both T_c and J_c.

Recently, the Los Alamos group has reported preparation of YBCO thick films using an industrial laser [2]. Independently, Jo *et al.* [3] and Suh *et al.* [4] reported preparation of YBCO thin films at deposition rates around 0.7 μm/min by electron beam co-evaporation and pulsed laser deposition (PLD), respectively. Jo *et al.* [3] observed that YBCO phase was composed from two distinctive layers. Whereas Suh *at al.* [4] reported that fast deposition rate at relatively low substrate temperature of 700°C led to cation disorder in Y/Ba sub lattices and the formation of poorly conductive cubic YBCO phase. Ohmatsu *et al.* [5] reported fabrication of long length YBCO/YSZ/Ni tapes by fast growth PLD in a reel-to-reel system. In order to form epitaxial YSZ bluffer layer on non-textured Ni substrate, the inclined substrate deposition was used. A linear dependence between the YBCO deposition rate and the power of the laser was observed with highest deposition rate of 3.8 μm/min achieved at 200W. In addition an improvement of in-plane texture of YSZ was observed at high laser power. Satisfactory values of J_c $2x10^5$ A/cm^2 within 11m tape length were reported.

In this work YBCO thick films were grown at fast deposition rates. In order to decrease the amount of cation disorder in the films, high growth temperatures were employed. The microstructure and superconducting properties of the films were studied.

EXPERIMENTAL

1-μm thick YBCO films were grown by PLD on (100) single crystal $SrTiO_3$ substrates with the high growth rate of around 4 micron/min [6]. The growth temperatures were measured by optical pyrometer and are presented in Table 1. Several pieces of the films were etched in 1% bromine solution in methanol.

The films were characterised by XRD, SEM and Raman spectrometry. Magnetisation measurements were performed using a vibrating sample magnetometer and critical current densities were calculated using Bean model [7].

Table 1. Preparation conditions and properties of studied YBCO thick films.

Sample	Growth Temperature °C	'*c*' parameter, Å		$\frac{(005)}{(007)}$ ratio	T_C (onset), K	J_C at 77K (0 T), A/cm^2
PD574B	751	11.695	11.679	6.95	92.1	$1.1x10^6$
PD571	802	11.708	11.691	6.40	91.5	$1.8x10^6$
PD572	861	11.739	11.714	4.02	91.5	$8.7x10^5$
PD576	872	11.743	11.721	3.94	90.2	$1.8x10^6$

RESULTS AND DISCUSSION

XRD analysis of the films showed intensive (*00l*) and small (*103*) YBCO reflections. In addition, epitaxialy grown $BaCuO_2$ was present in the films grown at the two highest temperatures, namely 861°C and 872°C. The YBCO peaks were asymmetric and presumably consisted of two overlapping peaks as shown in Figure 1. The degree of asymmetry and thus the relative intensity of the overlapping peaks seem to be temperature dependent. Bromine etching decreased the intensity of the peak located at higher angle (see inset in Figure 1). Consequently two YBCO layers were present in the films: a top layer with shorter "c" axis and a bottom layer with larger "c" axis. Similarly Jo *et al.* [8] observed a faulted surface YBCO layer (c=11.68Å) and a well crystalline YBCO layer close to the film/substrate interphace (c=11.74 Å) in the films grown by electron beam co-evaporation. Those distinctive layers caused significant asymmetry of YBCO x-ray reflections.

Two sets of (00l) YBCO reflections were obtained by deconvolution of XRD data and two values of 'c' lattice parameters were calculated using Nelson-Riley extrapolation function (Table 1). "c" lattice parameter for both YBCO phases increased with the growth temperature (Figure 2). This increase can be due to changes in oxygen stoichiometry and/or cation disorder in the films. The method

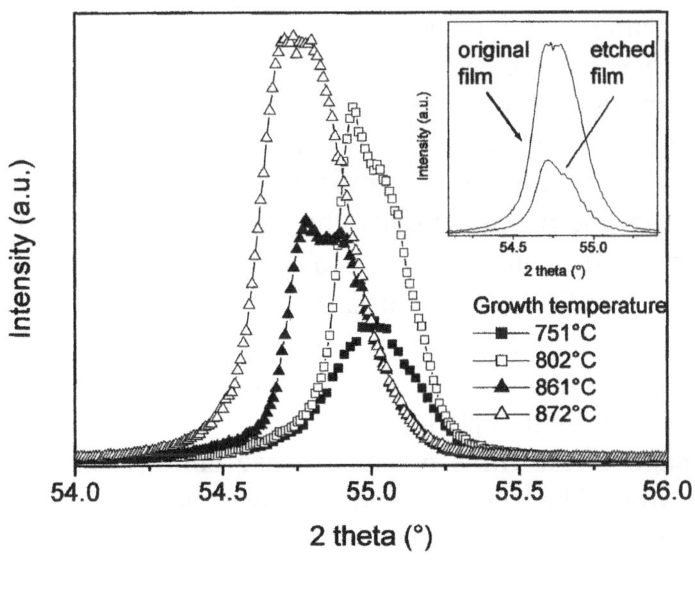

Figure 1. XRD plots of (*007*) YBCO peak of the films. Inset shows (007) YBCO peak of the film grown at 872 °C before and after bromine etching.

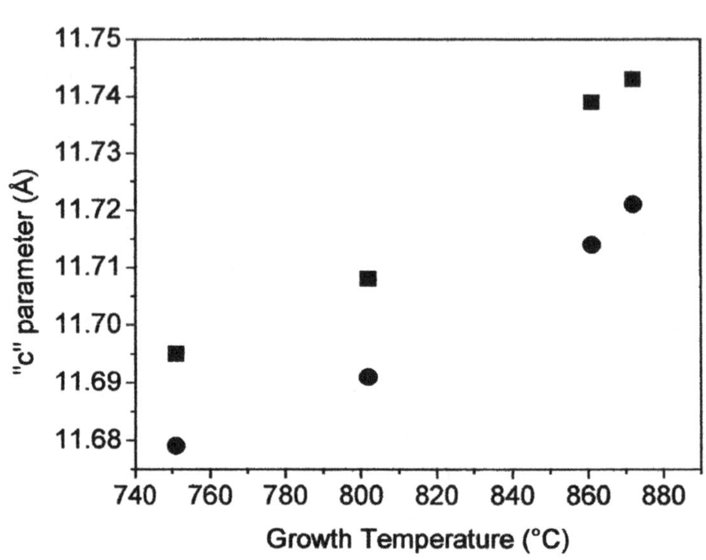

Figure 2. "c" lattice parameters of YBCO thick films as a function of growth temperature.

to assess the amount of disorder in YBCO by measuring integrated intensity of (*005*) and (*006*) peaks [9] could not be used as the (*006*) peak overlaps with (*002*) $SrTiO_3$ peak. However, the ratio of integrated intensity of (*005*) and (*007*) peaks is also valid for estimating cation disorder. The (*005*)/(*007*) ratio decreased with the increase of '*c*'-lattice parameter suggesting that a significant amount of cation disorder was present in the films (Table 1). Hence, the enhanced '*c*' value is not due to deoxygenation of the films. The Raman measurements and magnetisation measurements did not reveal any significant changes in oxygen contents as discussed later. From the rocking curves, the FWHM of the (*007*) peak was 0.6° for all the films. The Φ scans of (*012*) peaks showed sharp a-b alignment with FWHM of 0.6-1.0°. Additionally splitting of the peaks during the Φ scans was observed in the films grown at the temperatures higher than 800°C. This splitting of the peaks was attributed to the YBCO twinning caused by orthorhombic-tetragonal phase transformation [10].

The SEM image of the films revealed dense uniform surface covered with spherical outgrowths. The number and average diameter of the outgrowths increased with the growth temperature. Some a-b aligned grains of YBCO were also present.

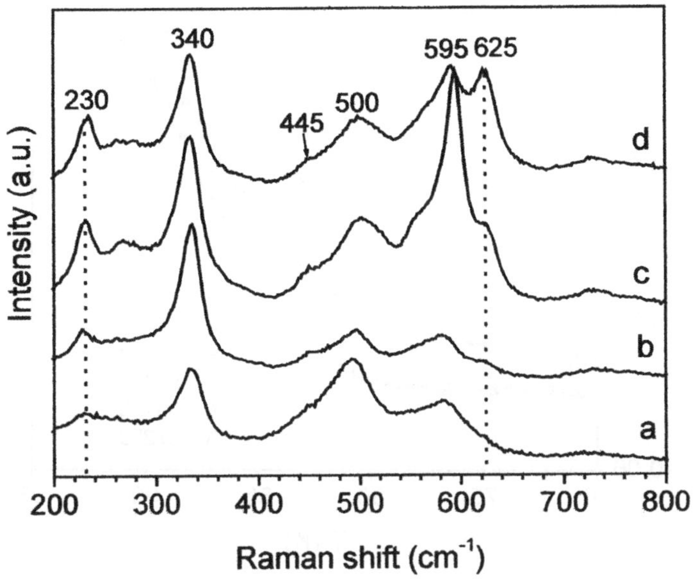

Figure 3. Room temperature Raman spectra of the YBCO films. Growth temperatures were 751°C (a), 802°C (b), 861°C (c), 871°C (d).

The Raman spectra of the films are shown in Figure 3. The modes observed at 147 (not shown), 340, 445 and 500 cm^{-1} correspond to the vibrations along the c-axis of Cu, planar oxygen in CuO_2 (out of phase), planar oxygen in CuO_2 (in-phase) and apical oxygen in orthorhombic YBCO, respectively. The origin of 595 cm^{-1} mode is caused by a loss of transitional symmetry upon defect formation in YBCO [1]. The intensity of the two peaks at 230 and 625cm^{-1} increased with growth temperature (Figure 3). These peaks were assigned to a $BaCuO_2$ [1,11]. A significant variation of the peak intensities across the sample surface was observed in the films grown at 861 and 872°C. This lack of reproducibility was attributed to the presence of partial melting leading to inhomogeneity. The position of the apical oxygen vibration was in the range 494 – 496 cm^{-1} and did not significantly change across the film surface.

The values of T_C and J_C at 77K and 0 T are given in Table 1. The onset of T_C only slightly decreased when the growth temperature was increased (Figure 4). The magnetic transition became broader presumably due to the increase of inhomogeneity in the films caused by cation disorder. At the same time no clear trend of J_C with the growth temperature was observed, although J_C values remained ~ 1 MA/cm^2. Field dependences of critical current densities of the films are shown in Figure 5. Slight enhancement of J_C in magnetic field of up to 4 T was observed in the film grown at 872°C.

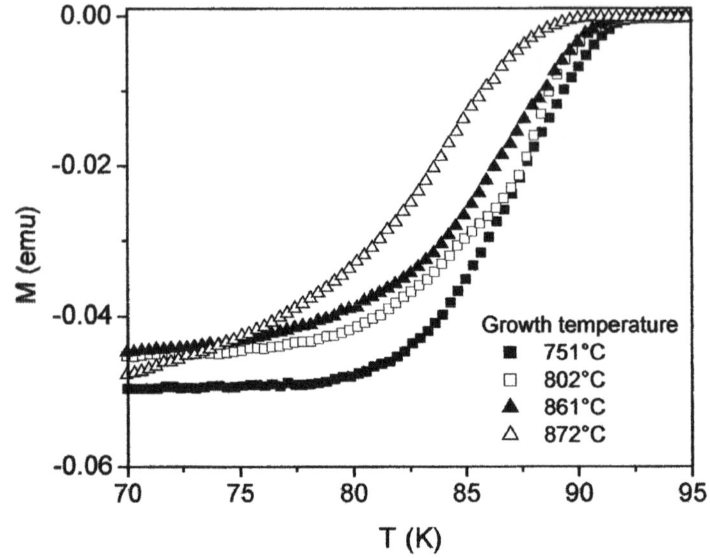

Figure 4. Temperature dependence of the magnetisation moment. Films were cooled in zero field and warmed in 1 mT.

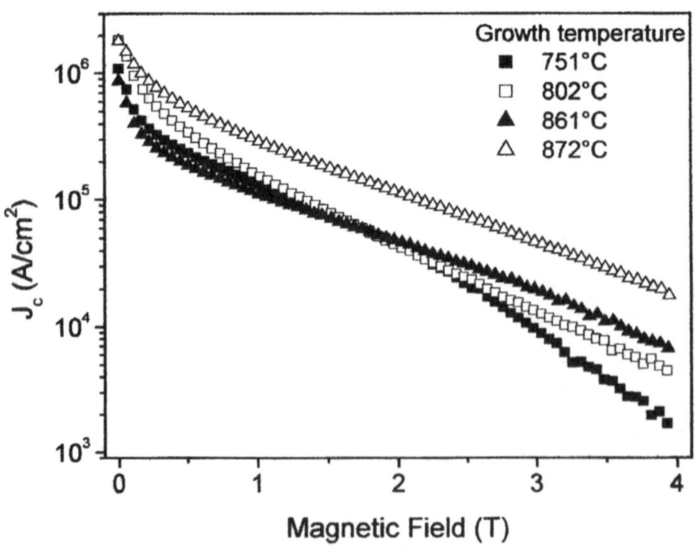

Figure 5. Magnetic field dependences of critical current density of YBCO films at 77K. Field was applied perpendicular to the film surface.

CONCLUSIONS

YBCO films were prepared by high temperature (751 - 872°C) PLD at growth rate of upto 4 micron/min. The films showed high degrees of in-plane and out-of-plane texture. The $BaCuO_2$ (quenched liquid) phase was observed in amounts increasing with temperature. The relationship between the ratio of integrated intensity of (005) and (007) x-ray peaks and 'c'-lattice parameters indicated the presence of increasing cation disorder with temperature. Observed high values of T_C (around 90 K) and J_C (~ 1 MA/cm^2) were independent of growth temperature and/or the amount of cation disorder.

ACKNOWLEDGMENTS

Financial support from EPSRC is gratefully acknowledged.

REFERENCES

[1] G. Gibson "A Probe of Structural Defects in YBCO Thin Films using XRD and Raman Microscopy", PhD. Thesis, University of London, 1999.

[2] S.R. Foltyn, E.J. Peterson, J.Y. Coulter, P.N. Arendt, Q.X. Jia, P.C. Dowden, M.P. Maley, X.D. Wu and D.E. Peterson, "Influence of deposition rate on the properties of thick $YBaCu_3O_{7-\delta}$ films", *Journal of Materials Research* **12** [11] 2941-2946 (1997)

[3] W. Jo, L.S-J. Peng, W. Wang, T. Ohnishi, A.F. Marshall, R.H. Hammond, M.R. Beasley and E.J. Peterson, "Thermodynamic stability and kinetics of Y-Ba-Cu-O film growth at high rates in atomic and molecular oxygen", *Journal of Crystal Growth* **225** 183-189 (2001).

[4] J.D. Suh, G.Y. Sung, K.Y. Kang, H.S. Kim, J.Y. Lee, D.K. Kim, C.H. Kim, "Cubic Y-Ba-Cu-O thin films by high speed pulsed laser deposition", *Physica C* **308** 251-256 (1998).

[5] K. Ohmatsu, K. Muranaka, S. Hahakura, T. Taneda, K. Fujino, H. Takei, Y. Sato, K. Matsuo, Y. Takahashi, "Development of in-plane aligned YBCO tapes fabricated by inclined substrate deposition", *Physica C* **351-360** 946 –951 (2001).

[6] S. Foltyn, Q. Jia, T. Holesinger, P. Arendt, H. Kung, Y. Coulter, J. F. Smith, in DOE Annual Peer Review Proceedings, Washington (2000).

[7] Bean C.P., *Physical Review Letters* **8** 250 (1962).

[8] W. Jo, T. Ohnishi, J. Huh, A.F. Marshall, R.H. Hammond, M.R. Beasley, "Growth of high critical current $YBa_2Cu_3O_{7-\delta}$ films with liquid-fluxed microstructures by the in-situ electron beam method"; pp. 77-79 in Extended Abstracts of 2001 International Workshop on Superconductivity co-sponsored by ISTEC and MRS, Hawaii, 2001.

[9] J.L. MacManus-Driscoll, J.A. Alonso, P.C. Wang, T.H. Geballe, J.C. Bravman, *Physica C* **232** 288 (1994).
[10] D. Schweitzer, T. Bollmeier, B. Stritzker and B. Rauschenbach, "Twinning of $YBa_2Cu_3O_7$ thin films on different substrates", *Thin Solid Films* **280** 147-151 (1996).
[11] P. Gomez, J. Jimenez, P. Martin, J. Pigueras and F. Dominguez-Adame, "Cathodoluminescence and microRaman analysis of oxygen loss in electron irradiated $YBa_2Cu_3O_{7-x}$" *Journal of Applied Physics*, **74** [10] 6289-6292 (1993).

$YBa_2Cu_3O_{7-\delta}$ FILMS THROUGH A FLUORINE FREE TMAP MOD APPROACH

Yongli Xu and D. Shi
Department of Materials Science, University of Cincinnati, Cincinnati, OH 45221, USA

A. Goyal, M. Paranthaman, N.A. Rutter, P.M. Martin, and D. M. Kroeger
Oak Ridge National Laboratory, Oak Ridge, TN 37831, USA

ABSTRACT

$YBa_2Cu_3O_{7-\delta}$ (YBCO) films were fabricated via a fluorine free metal trimethylacetate (TMA) proponic (P) acid and Amine based sol-gel route (TMAP) and spin-coat deposited on single crystal (001) $LaAlO_3$ (LAO) and (001) $SrTiO_3$ (STO) with a focus on optimizing the processing parameters in the non-fluorine chemical solution deposition (CSD). Trimethylacetate salts of copper and yttrium and barium-hydroxide were used as the precursors, which were dissolved in proponic acid and Amine based solvents. A 180 ppm oxygen partial pressure and water vapor atmosphere were employed for the pyrolysis at 745°C for one hour. A critical transition temperature (T_{c0}) of 90K and a critical transport current density (J_c) of 0.5 MA/cm^2 (77 K and self-field) were demonstrated for the YBCO film on (001) oriented LAO substrates with a thickness of 300 nm. A FWHM of 0.6° for the (005) omega scan shows good out-of-plane texture. In (113) phi scans, the peaks are 90° apart with a FWHM of 1.4° showing good in-plane textures. A good in-plane texture is also shown in (113) and (115) pole figures. The TMAP approach is promising for high current density and high film quality.

INTRODUCTION

Chemical solution deposition (CSD) is one of the deposition techniques used to fabricate YBCO films. CSD has some advantages over vapor deposition of YBCO such as precise control of composition, high speed, and low-cost. Metal Organic Deposition (MOD), one of the popular CSD methods, involves the coating of an organic precursor solution on a substrate followed by thermal decomposition to form the final desired compound. Epitaxial nucleation and growth can occur when the process is carried out on lattice matched single crystal substrates. Trifluoroacetate (TFA) MOD is well established as a promising method for the fabrication of high J_c (over 1 MA/cm^2) YBCO films[1,2] and has been applied on biaxially textured metal substrates like RABiTS[3,4]. The interest in fluorine-containing precursors for YBCO arises because it is believed that non-fluorine precursors might result in the formation of stable $BaCO_3$ at the grain boundaries[5]. The use of TFA salts appears to avoid the formation of $BaCO_3$ because the stability of barium fluoride is greater than that of barium carbonate and fluorine can be removed during the high temperature anneal (>700°C) in a humid, low oxygen partial pressure environment[2]. Nonetheless several factors maintain interest in a fluorine-free precursor MOD approach. The most important being that removal of fluorine at high temperatures is a non-trivial process. There appear to be many issues related to fluid-flow and complicated reactor designs may be required for scale-up. In this work, we report on the fabrication of high quality YBCO films using a fluorine-free precursor. The transport critical current density of these films is over 0.5 MA/cm^2 at 77 K and self-field.

EXPERIMENTAL

Films were prepared by the following procedure[6,7]: yttrium trimethylacetate, barium hydroxide, and copper trimethylacetate were dissolved into proponic acid and amine solvents in a controlled stoichiometry (Y:Ba:Cu=1:2:3) to form a dark green solution with a total ionic concentration of 0.5-0.8 M. After filtration, this solution was applied onto the substrates in a spin-coater at speed of 2000-5000RPM. A 200-250°C baking step was used between multi-coatings on a hot plate to make thicker films. Burn out was carried out in the quartz tube furnace at 400°C for 10 hours in a humid pure oxygen atmosphere. These intermediate films were then subject to a high temperature anneal in which they were heated to 700 - 860°C at 10-40°C/min and held in a low oxygen partial pressure for 1-2hrs. Following the hold, the furnace atmosphere was switched to dry, and later to dynamic dry oxygen. All films were slow-cooled in oxygen to 500°C where they were held for 0.5-3hrs and furnace-cooled to room temperature in flowing oxygen. Humid gas was introduced into the furnace in the burned-out stage and prior turning on pure oxygen in the annealing stage. J_c was measured in the usual four-point transport method at 77 K and zero field. In addition, for selected samples, texture was analyzed by x-ray diffraction, surface morphology was characterized by scanning electron microscopy (SEM) and TGA was used for the thermal decomposition analysis.

RESULT AND DISCUSSION:

Fig. 1. shows Thermal Gravimetric Analysis (TGA) results for the copper trimethylacetate and yttrium trimethylacetate precursors. For the copper trimethylacetate precursor, the weight reduces quickly when the temperature increases beyond 200°C, and then become constant after 250°C, which demonstrates the full decomposition of copper trimethylacetate at that temperature. For yttrium trimethylacetate precursor, the decomposition occurs over a wider range and is finished at about 520°C. Weight losses are structure-related thermal processes and reflect the binding energies. Different atmospheres, for example, air and pure oxygen, have minor effects on the shapes and temperatures of TGA curves both for copper trimethylacetate and yttrium trimethylacetate precursor.

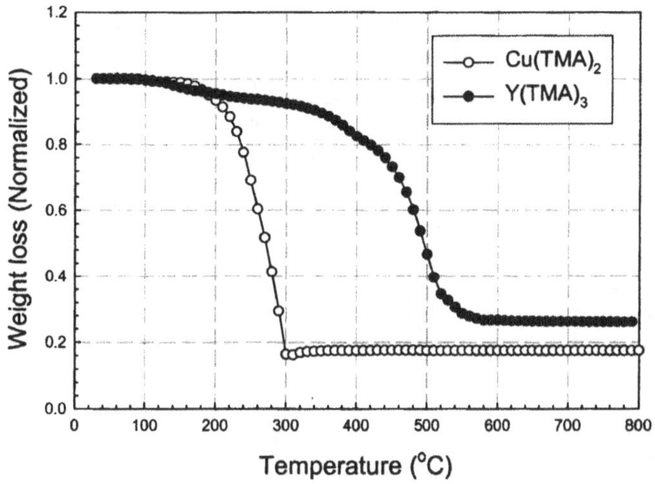

Fig. 1. Weight loss in TGA analysis for copper trimethylacetate and yttrium trimethylacetate precursors, atmosphere: N_2 with Po_2=100-200ppm, ramp: 10°C/min.

CuO is easily evaporated at elevated temperature in dry atmosphere. The decomposition temperature for copper trimethylacetate (250°C) is lower than those of yttrium trimethylacetate (520°C) and barium hydroxide[8] (490°C). In considering CuO formation during the decomposition of copper trimethylacetate, keeping the baking temperature below 250°C is reasonable to prevent the loss of CuO between multi-coatings. But when the baking temperature is less than 150°C, later coatings dissolve away former ones easily. So normally 200-250°C for 3-5 minutes was used for the baking process.

In the burn out stage, several temperatures (310-600°C) were selected to control reactions of the components in precursor decomposition. 400°C was used as an optimized temperature for the burn out stage because rough surface morphologies of YBCO film were observed routinely on samples burned out below 400°C (for example, 310°C) and above 400°C (for example, 500°C and 600°C). For this research, four temperatures 725°C, 745°C, 780°C, and 820°C were used for high temperature annealing. Fig. 2 shows the θ-2θ scans of YBCO films annealed at 745°C and 820°C respectively.

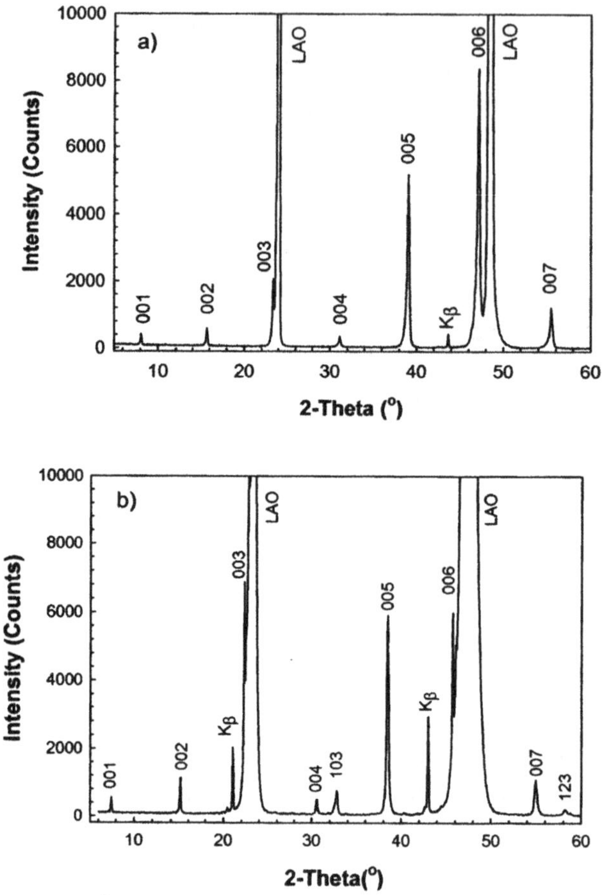

Fig. 2. θ-2θ scans of YBCO thin film on (001) LAO single crystal substrates burned out at 400°C in wet O_2 and high temperatures annealed in low oxygen partial pressure (Po_2=180ppm) for 50min wet and 10min dry at different temperatures: (a) 745°C, (b) 820°C

Only substrate peaks can be detected for the films burned out at temperatures below 400°C, indicating that the film is amorphous. YBCO (00*l*) peaks can be identified for samples annealed at 700°C and are stronger after annealing at 725°C for one hour. The (103) peak is seen when the temperature is beyond 780°C, especially for the sample annealed at 820°C as shown in Fig. 2 b). All (00*l*) YBCO peaks are shown in Fig. 2 a). The absence of other YBCO peaks confirms that the YBCO has good out-of-plane texture. The values of FWHM (Full Width at Half Maximum) of the (005) rocking curves are small and nearly constant (0.6°) for the samples annealed at 725°C, 745°C, and 780°C, demonstrating high quality *c*-oriented texture of these YBCO films. In the (113) phi-scan, Fig. 3, peaks are 90° apart with a FWHM of 1.4° showing good in-plane texture for a sample annealed at 745°C for one hour. This sample also had the highest YBCO intensities both for ω-scans and ϕ-scans.

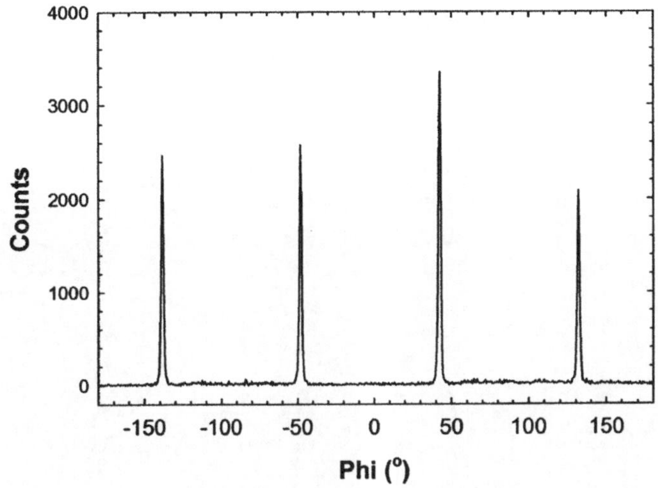

Fig. 3 The 113 ϕ-scan of YBCO films on LAO (001) substrate burned out at 400°C and annealed at 745°C for one hour in humid Ar/O_2 atmosphere with Po_2=180ppm

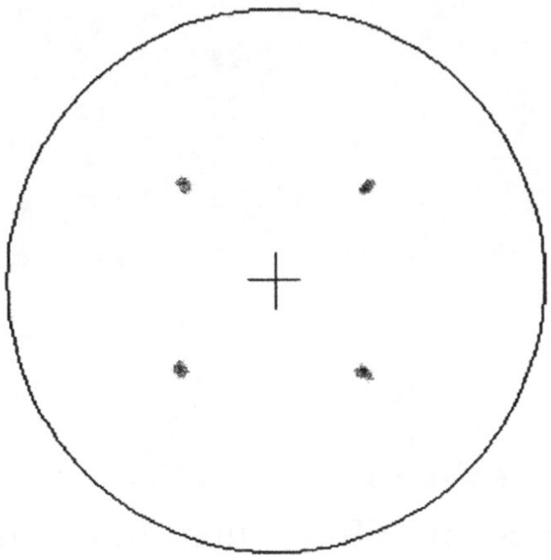

Fig. 4 The YBCO (115) pole figure of 400°C burned out film and annealed at 745°C/1hr in humid Ar/O_2 atmosphere with Po_2=180ppm

Pole figures were obtained to get a more complete examination of the degree of epitaxy. For 745°C annealed sample, only four small poles located on the correct positions on the pole figure plot, Fig. 4, confirms that the YBCO phase grows epitaxially on the LAO substrate. The T_c measurement shows a sharp transition with zero resistance at 90 K for the sample annealed at 745°C for one hour, but broader transitions were detected for samples annealed at 725°C and 780°C. A semiconductor transition behavior was shown for the 820°C annealed sample. The transport critical current density (J_c) of 0.5 MA/cm^2 was obtained at 77 K in self-field for the 745°C annealed sample, whilst very limited currents can go through the YBCO films annealed at 725°C and 780°C. No current can go through the film annealed at 820°C for one hour.

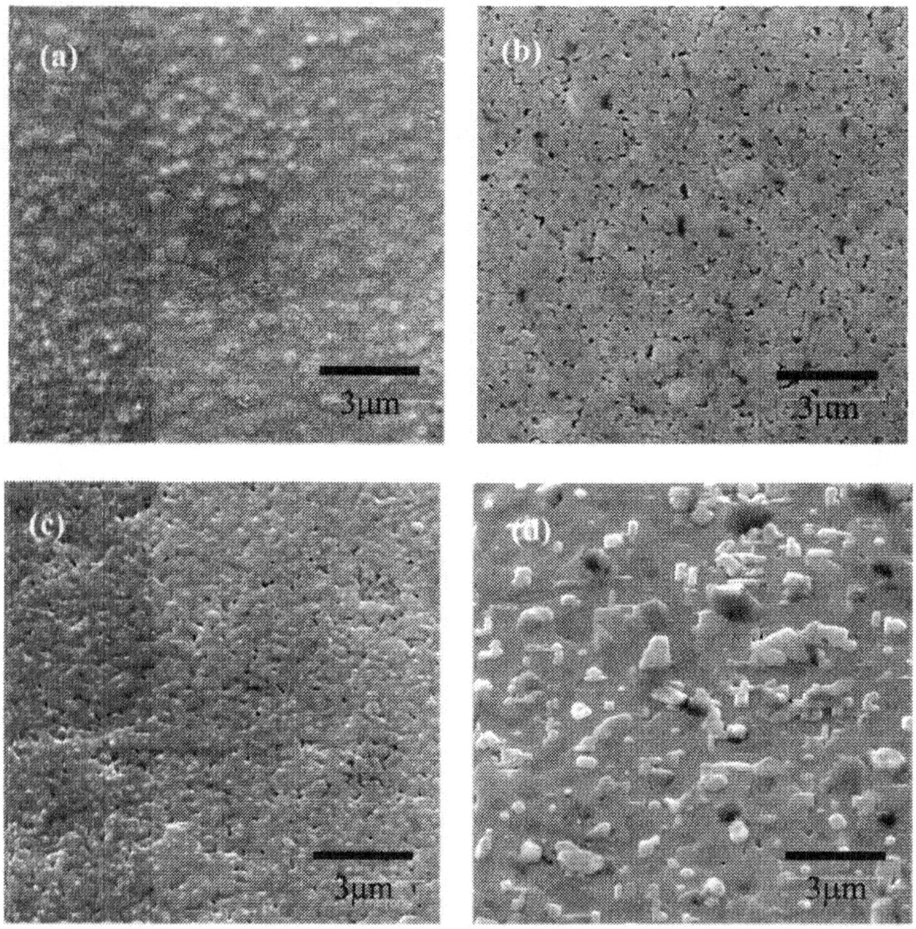

Fig. 5. SEM morphologies of YBCO thin films heat-treated at different temperatures. (a) 600°C burn out in wet oxygen; (b), (c), and (d): 400°C burn out in wet oxygen and annealed at 725°C/1hr, 745°C/1hr, and 820°C/1hr respectively in Po_2=180ppm wet atmosphere.

Though no crystallization is detected on the sample heat treated below 600°C, the SEM top view morphology, Fig. 5 (a), shows the nuclei like dots started at 600°C or below. No composition difference can be identified by EDS (Energy Dissipation Spectra) on dots and other areas. Associated with the decomposition of the copper trimethylacetate and yttrium trimethylacetate precursors in the TGA analysis (Fig. 1), it is believed that these dots are the decomposition products formed initially in the green films. As the decomposition of the copper trimethylacetate occurs at relative low temperature, and the resistivity (calculated from resistance of quenched sample during 600°C burn out) can go as low as 0.1-0.01ohm-cm, so correlating these dots with CuO is reasonable. The distribution density of these dots is comparable with that of grains; thus they might be the places for the further reactions and/or nucleations of YBCO phase. When increasing treatment temperature, these dots increase in size and finally connect to each other forming the grain like areas shown in Fig. 5 (b).

The surface morphology of the 725°C and 745°C annealed samples is relatively smooth, although some pores decorate the surface. From the SEM micrograph Fig. 5 (b), it is clear to identify that the microstructure of the 725°C annealed sample is not fully developed. This is also confirmed both by XRD data and the results of electrical property measurements. The underdeveloped YBCO phase is the reason for the low J_c and relatively broad transition temperature in the T_c measurement. Irregularly shaped surface segregates are observed commonly on samples annealed higher than 820°C as shown in Fig. 5 (d). The rough surface is obviously related to the over reaction among the components and with the environment such as humid gas, or to the decomposition of YBCO phase as reported in the literature[9]. With the rough surface coupled with texture changes shown in XRD data, no good electrical properties can be expected on samples annealed at temperature beyond 820°C.

The morphology of the 745°C annealed sample is similar to that reported by R. Feenstra *et al*[9]. in e-beam derived YBCO films with post annealing. Enhanced growth of plate-like grains, smooth surface, absence of large pores and second phase particles, and the absence of *a*-axis oriented grains indicate the potential for high critical current density.

SUMMARY

A fluorine-free Metal Organic Deposition (MOD) method was employed for the fabrication of $YBa_2Cu_3O_{7-\delta}$ films. A T_c of 90 K and a transport critical current density (J_c) of 0.5 MA/cm^2 (77 K and self-field) were demonstrated for the YBCO film on (001) oriented LAO substrate annealed at 745°C for one hour in humid furnace gas with oxygen partial pressure of 180ppm. There is no evidence that $BaCO_3$ is formed as an intermediate compound during decomposition of the precursors in well-prepared samples. SEM morphology reveals that feature development on top of the film occurs much earlier than crystallization takes place (as revealed by XRD data). Annealing temperature is critical for the YBCO phase development in low oxygen partial pressure and humid furnace gas. Undeveloped phases and poor texture at relatively low annealing temperature are the reasons for the low current densities. The YBCO phase seems to decompose at relatively high annealing temperatures (820°C) resulting in a rough surface and a semiconductor behavior.

ACKNOWLEDGEMENT

This work was sponsored by the U.S. Department of Energy, Office of Energy Efficiency and Renewable Energy - Superconductivity Program for Electric Systems. The research was performed at the Oak Ridge National Laboratory, managed by U.T.-Battelle, LLC for the USDOE under contract No. DE-AC05-00OR22725.

REFERENCE

[1]A. Gupta, R. Jagannathan, E.I. Cooper, E.A. Geiss, J.I. Landman, and B.W. Hussey, Appl. Phys. Lett. **52**(24), 2077 (1988)

[2]Paul C. McIntyre, Michael J. Cima, John A. Smith, Jr., and Robert B. Hallock, J. Appl. Phys. **71**(4), 1868-77, (1992)

[3]A. Goyal, D.P. Norton, J.D. Budai, M. Paranthaman, etc., Appl. Phys. Lett. **69**(12) (1996) 1795. A. Goyal et al., US Patents: 5, 739, 086; 5, 741, 377; 5, 846, 912, 5, 898, 020.

[4]A.P. Malozemoff, S. Annavarapu, L. Fritzemeier, Q. Li, V. Prunier, M. Rupich, C. Thieme, W. Zhang, A. Goyal, M. Paranthaman, and D.F. Lee, Superconductor Sci. & Technol. **13**, 473-476 (2000).

[5]F.Parmigiani, G. Chiarello, and N.Ripamonti, Physical Review B, **36**(13) (1987) 7148.

[6]Yongli Xu, Donglu Shi, Shaun McClellen, Relva Buchanan, Shixin Wang and L.M. Wang, IEEE Trans. Appl. Supercond. **11**(1), 2865-2868, (2001)

[7]Donglu Shi, Yongli Xu, Shaun McClellen, and Relva Buchanan, Physica C **354** (2001) 71-76

[8]Peir-Yung Chu and Relva C. Buchanan, J. Mater. Res., **9**(4), 844-51, (1994)

[9]R. Feenstra, T.B. Lindemer, J.D, Budai, and M.D. Galloway, J. Appl. Phys. **69** (9), 1 May (1991) 6569-85

EFFECT OF TRANSVERSE COMPRESSIVE STRESS ON TRANSPORT CRITICAL CURRENT DENSITY OF Y-Ba-Cu-O COATED Ni and Ni-W RABiTS TAPES*

Najib Cheggour, Jack W. Ekin, and Cameron C. Clickner
National Institute of Standards and Technology, Boulder CO 80305, USA

Roeland Feenstra, Amit Goyal, Mariappan Paranthaman, and Noel Rutter
Oak Ridge National Laboratory, Oak Ridge TN 37831, USA

ABSTRACT

Transport properties of yttrium-barium-copper-oxide (YBCO) coatings on both pure-nickel and nickel-3at% tungsten rolling-assisted-biaxially-textured substrates (RABiTS) were tested under transverse compressive stress up to 180 MPa. Transport critical-current densities (J_c) were measured on samples having an initial J_c in the range of 1 MA/cm^2 at 76 K and self magnetic field. Transverse compressive stress can cause significant degradation of J_c in YBCO deposited on *pure*-nickel RABiTS unless appropriate work-hardening treatment or a sufficient frictional support are provided to the sample. Preliminary results obtained for YBCO on nickel-3at% tungsten *alloy* RABiTS suggest that the electromechanical properties of these conductors may be improved if such material is used for the substrate. Stress-strain characteristics were measured on nickel and nickel-3at% tungsten RABiTS substrate materials at 76 K. The tensile yield strength, Young's modulus, and proportional limit of elasticity of both materials were determined. Scanning electron microscopy was used to also investigate the microstructure of the coated-conductor samples after electromechanical testing. Mechanical cracks with multi-patterned orientations were observed in the ceramic layers. The data indicate that the formation of cracks in the ceramic layers initiated by the *yielding of the substrate material*, is the primary cause for the degradation of J_c in the samples investigated.

INTRODUCTION

Highly textured yttrium-barium-copper-oxide (YBCO) films deposited on buffered flexible metallic substrates exhibit critical-current densities (J_c) that are in excess of 1 MA/cm^2 at liquid-nitrogen temperature and self magnetic field[1-3]. These coated conductors may therefore potentially be used in the construction of electrical devices such as underground power-transmission lines, transformers, motors, generators, and magnetic separators. Strong grain-alignment is essential for achieving such high J_c values[4-9]. This requirement can be met by using, for example, the rolling-assisted-biaxially-textured substrates (RABiTS) or the ion-beam-assisted-deposition (IBAD) technologies. Since the ceramic coatings are inherently brittle, the tolerance of the coated conductors to mechanical stresses can affect the ultimate performance of devices made of these tapes. Hence the study of their electromechanical properties is crucial to both guide conductor development and provide critical data for the design of specific applications[1,10-15]. When a power-transmission cable is bent, each tape-element is subjected to bending strain and also to transverse compressive stress as the tapes in the different layers of the cable are pushed against each other and against the wall of the conduit. In this work we present experimental results on the effect of transverse compressive stress on J_c, obtained in thin (0.3 μm)YBCO films deposited on pure-nickel and nickel-3at% tungsten RABiTS. The microstructure of the samples after electromechanical testing is shown and reveals cracking of the ceramic films due to stress application. We also report stress-strain characteristics of pure-Ni and Ni-3at%W substrate materials and compare their tensile yield strength, Young's modulus and proportional limit of elasticity. These results will provide insight into the role played by the yield strength of the substrate in determining the electromechanical properties of these conductors, as well as provide a preliminary comparison between YBCO coatings on pure-Ni and Ni-3at%W substrate materials.

EXPERIMENTAL PROCEDURE

Nickel and Ni-3at% tungsten rods were thermomechanically processed to obtain biaxially textured substrates[16]. The buffer layers were grown on the substrates by reactive evaporation and RF magnetron sputtering. Thereafter the YBCO layer was deposited on the buffered substrates using the *ex situ* BaF_2 method[17]. Samples with pure-Ni (sample 1 to 3) had a structure, from outside in, as YBCO | CeO_2 | yttria-stabilized zirconia (YSZ) | CeO_2 | Ni. Sample 4 incorporated a Ni-3at%W substrate and had an architecture as YBCO | CeO_2 | YSZ | Y_2O_3 | Ni-W. The tapes were then coated with a layer of silver (1 μm thick) and annealed in oxygen. An additional silver layer (10 μm thick) was deposited by thermal evaporation to insure low contact resistivity between the sample the current leads. For samples 1 to 3, the thickness of the pure-Ni subtrate was 50 μm, the buffer layers less than 1 μm, and the superconductor layer about 0.3 μm. Sample 4 had a Ni-3at%W substrate 50 μm thick, while the thickness of the buffer and YBCO layers were kept unchanged. The samples had a width of about 3 mm to 3.5 mm and a length of about 2.5 cm.

The apparatus for conducting J_c vs. transverse stress (σ_T) measurements has been described elsewhere[18]. Two copper-leads and a pair of voltage-taps were soldered to the sample. The sample was mounted onto a flat stainless steel block (bottom anvil), located at the bottom end of the apparatus. The current-leads were designed and soldered to the apparatus in such a way that one copper-lead was stationary and the other was free to flex[19]. The flexible copper-lead insured that the sample was stress-free during the cooling of the apparatus from room temperature to the operating temperature. The top anvil, also flat and made of stainless steel, was attached to the probe via a biaxially gimbaled pressure-foot so that this anvil conformed to the sample and bottom anvil surfaces. Measurements were carried out in liquid nitrogen at 76 K. Voltage-current curves were taken as a function of σ_T up to 180 MPa in self-field, and values of J_c determined at an electrical-field criterion of 1 μV/cm. The values of J_c were measured to within ± 1 % and transverse compressive stress to within ± 2 %. The samples investigated had initial values of J_c in the range of 1 MA/cm^2 at 76 K and self-field.

The apparatus for measuring stress-strain characteristics was constructed to investigate very soft substrates[20]. The procedure for mounting samples on the probe was adapted in such a way to avoid work-hardening of the substrates prior to stress-strain measurements. Samples were 50 μm thick, 1 cm wide and 30 cm long. The apparatus was connected to a servo-hydraulic actuator, equipped with a 1300 N load-cell and a linear variable differential transformer (LVDT) to measure stress and strain respectively. The apparatus was inserted, under load-control, into a liquid-nitrogen cryostat to insure stress-free cooling of the sample. To check the accuracy of measurements, the Young's modulus was determined for stainless-steel tapes about 80 μm thick and 3 mm wide and was found to be 176 GPa at 293 K, 200 GPa at 76 K and 210 GPa at 4 K. These values are consistent with those for type 304 stainless-steel sheets. The errors in the determination of the Young's modulus and yield strength are mainly due to the estimation of the sample dimensions, particularly its thickness. This uncertainty can be estimated to about 10%, depending on the thickness variation along the substrate's length.

TRANSVERSE COMPRESSIVE STRESS RESULTS

Two modes of application of transverse compressive stress to the sample were used. In the *monotonic-loading* mode, stress was applied to the sample and gradually increased without releasing the load between measurement steps. In contrast the *load-unload* mode consisted of applying stress of a certain value to the sample, then releasing it (with both anvils keeping physical contact with the sample) before reapplying it at a higher value. Sample 1 and 2 were measured in the *monotonic-loading* mode and *load-unload* mode respectively, and presented very different behavior (Fig. 1). Sample 1 showed no J_c degradation up to about 120 MPa, while Sample 2 presented a significant J_c degradation amounting to about 28 % at 100 MPa and 39 % at 180 MPa. In the *monotonic-loading* mode, the sample has a strong frictional support from the pressing anvils. This support prevents the sample from expanding laterally. In contrast, the frictional support is significantly reduced when operating in the *load-unload* mode. In-plane expansion becomes possible, which may lead to yielding of the substrate material and cracking of the buffer and YBCO layers.

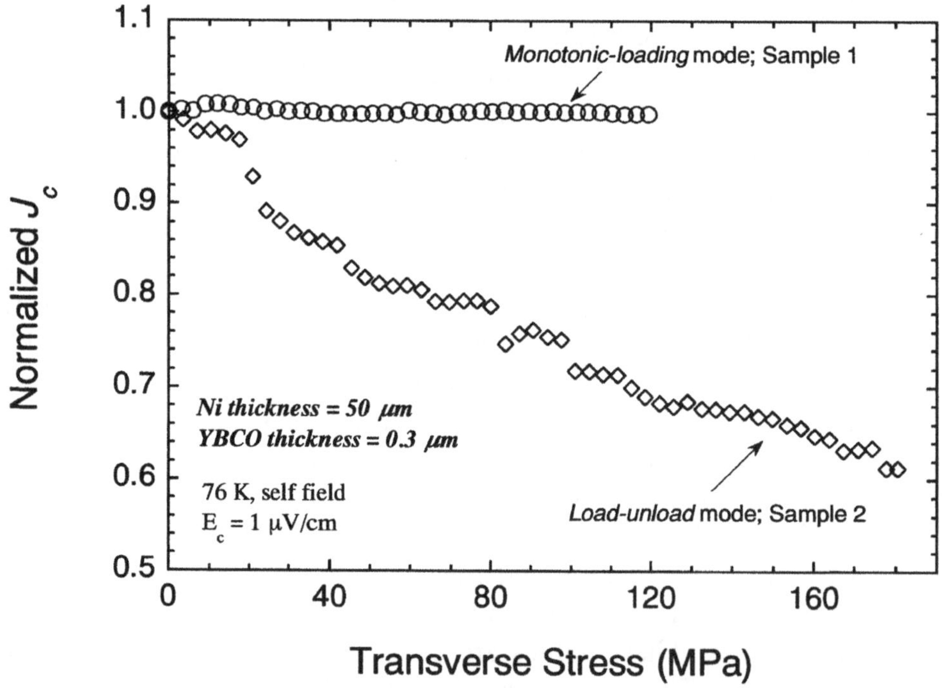

FIGURE 1. Effect of transverse stress on J_c in YBCO films on *pure*-nickel RABiTS. The results obtained in two modes of measurements illustrate the role played by friction between the sample and pressing anvils.

In Fig. 2 we show the benefit of work-hardening the sample substrate and present a preliminary comparison between the response to transverse stress of the YBCO coatings on pure-Ni and Ni-3at%W substrate materials. Sample 3 was pressed first monotonically to 160 MPa, then unloaded and measured in the *load-unload* mode up to 180 MPa. In the monotonic pressing, J_c did not degrade (result not shown in Fig. 2). During this operation the pure-Ni substrate work-hardened, which improved substantially the tolerance of this sample to transverse stress as compared to Sample 2 which did not receive the work-hardening treatment. Sample 3 only showed 6 % degradation of J_c at 100 MPa, and 12 % at 180 MPa. Sample 4, which had a Ni-3at%W substrate, exhibited the best electromechanical performance in the *load-unload* mode, as the degradation of J_c amounted only to about 2% at 100 MPa, and 7.5 % at 180 MPa without a work-hardening treatment.

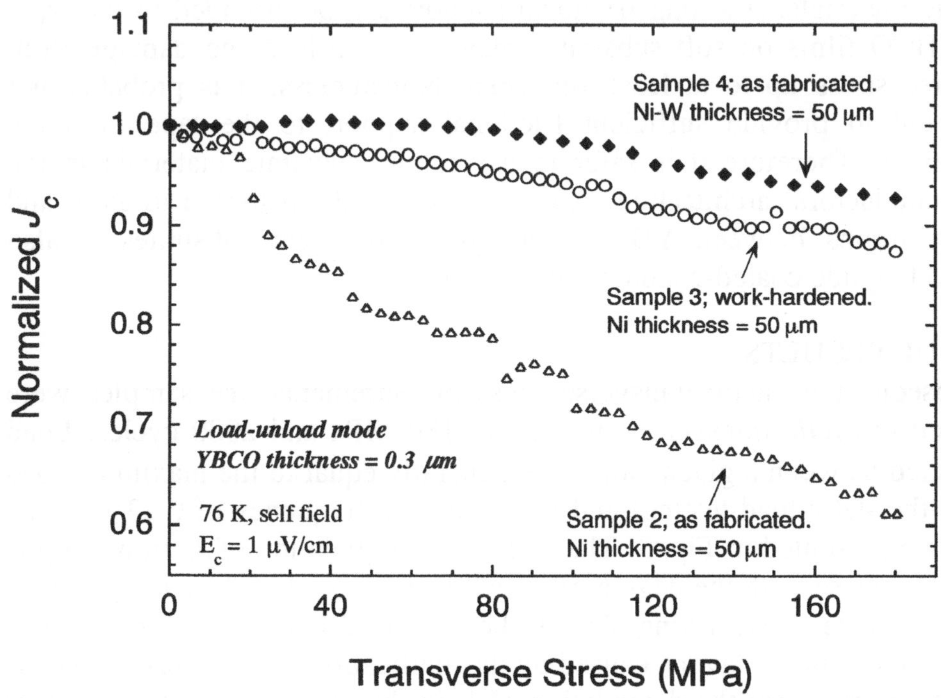

FIGURE 2. Effect of transverse stress on J_c in YBCO films on pure-Ni and Ni-3at%W *alloy* RABiTS. The results illustrate the benefits of work-hardening pure-Ni substrates and provide a preliminary comparison between YBCO coatings on pure-Ni and Ni-3at%W substrate materials.

These results provide evidence that the mechanical properties of the substrate material play a predominant role in determining the response of these samples to transverse compressive stress. This appears to be a valid conclusion at least for the samples measured, which had in common the initial values of J_c that were in the range of 1 MA/cm^2 at 76 K and self-field, and therefore probably a similar grain-alignment structure. More comprehensive data are still required to draw definitive conclusions.

As will be presented below, the tensile yield strength, Young's modulus and proportional limit of elasticity of Ni-3at%W substrates are substantially higher than those of pure-Ni substrates. The mechanical properties of work-hardened pure-Ni substrates are also expected to be superior, as compared to pure-Ni substrates' properties. For the samples measured, the higher these properties are, the more the transverse load needs to be to initiate cracking of the ceramic layers. These results do not rule out the usefulness of soft substrate materials. If strong frictional support can be provided to the tape, even YBCO films on soft substrates may suffer little or no damage from transverse stress up to at least 160 MPa. Nonetheless, it is probably not convenient to provide sufficient frictional support to the tape in many applications. Therefore it is safer to use robust substrate materials in the coated-conductors' architecture. A comparison of the response to axial and bending strains between YBCO coatings on different substrates is also required to guide coated-conductor development.

FATIGUE RESULTS

Subsequent to static transverse-stress measurements, the samples were subjected to *cyclic* transverse-stress up to 180 MPa and 2000 cycles. Load was cycled between a given value (σ_T, generally equal to the maximum load used in the static-load testing) and near zero at a frequency of 0.33 Hz. The results are presented in Fig. 3. The additional degradation of J_c from fatigue tends to increase with the degree of initial damage produced by the previous static transverse-stress testing. As will be presented below, the formation of cracks in the ceramic layers, caused by the application of transverse stress, is the primary cause for the degradation of J_c in the samples investigated. It is therefore possible that the more the sample is pre-cracked, the more it becomes sensitive to fatigue cycles. Most likely, fatigue tends to propagate the existing cracks which results in a further degradation of J_c. When a sample is free from cracks (Sample 1), fatigue does not seem to induce a noticeable damage to the YBCO layer.

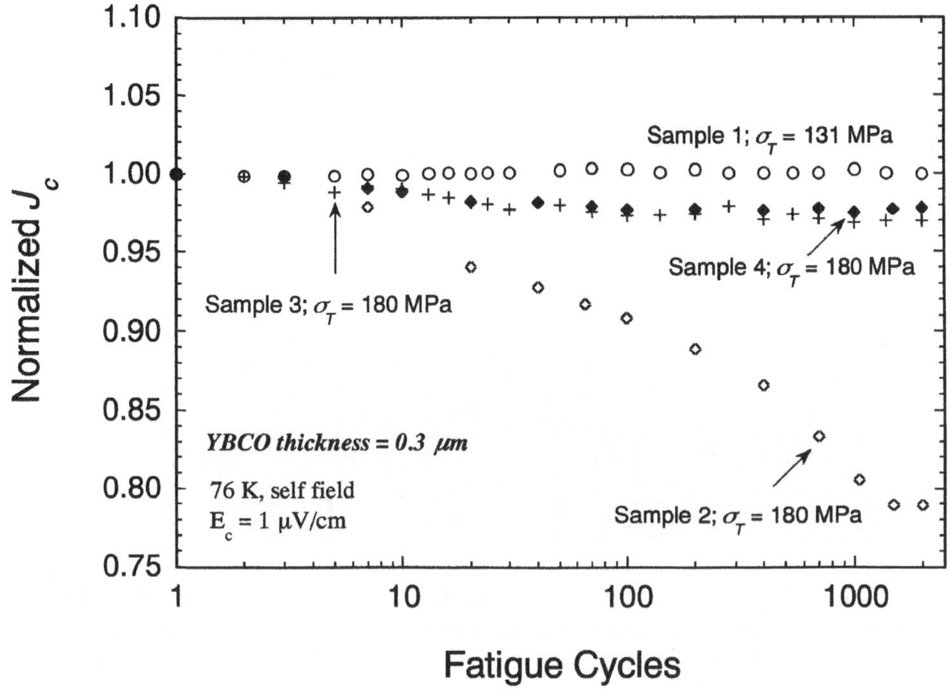

FIGURE 3. Effect of fatigue on J_c in YBCO films on pure-Ni and Ni-3at%W RABiTS. Fatigue tests were carried out on the samples after the static transverse-stress measurements. Fatigue has more effect on J_c in samples containing cracks previous to testing.

MICROSTUCTURE STUDY

After electromechanical testing, silver was etched away to expose the YBCO layer. A solution of 25%H_2O_2 + 25%NH_4OH + 50%H_2O was used for silver etching[14]. The microstructure of this layer was examined using scanning electron microscopy (SEM), and regions with cracks were found in it (Figs. 4-6). For samples with pure-Ni substrates, the cracked regions were distributed in patches of a few micrometers to about 600 μm wide (Fig. 4). In the case of Sample 4 that has a Ni-3at%W substrate, cracks were mostly located near the sample edges, with a group of diagonal cracks extending about 400 μm towards the middle of the sample (Fig. 6). Cracks had multi-patterned orientations with respect to the direction of the electrical-current flow. The percentage of the affected areas correlates well with the total J_c degradation in each sample.

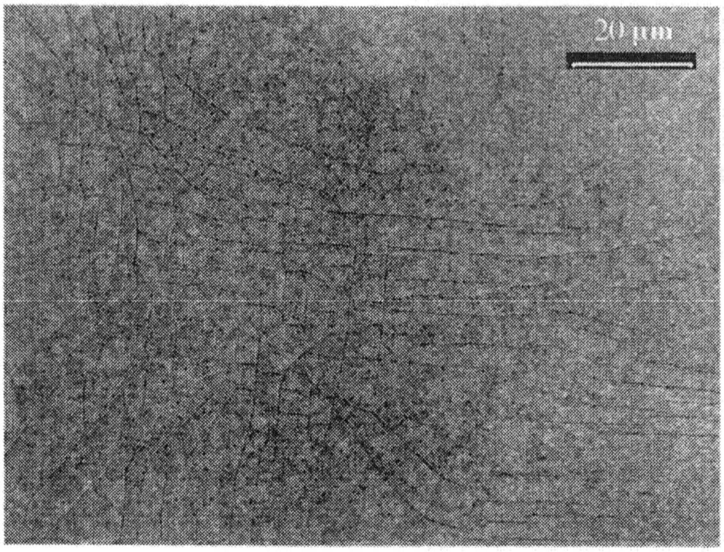

FIGURE 4. Scanning electron micrograph of YBCO film on buffered pure-Ni RABiTS substrate after electromechanical testing, showing a mixed pattern of mechanical cracks throughout the sample, both longitudinal and transverse to the electrical-current flow direction. The vertical axis of the image coincides with the direction of the electrical current applied to the sample.

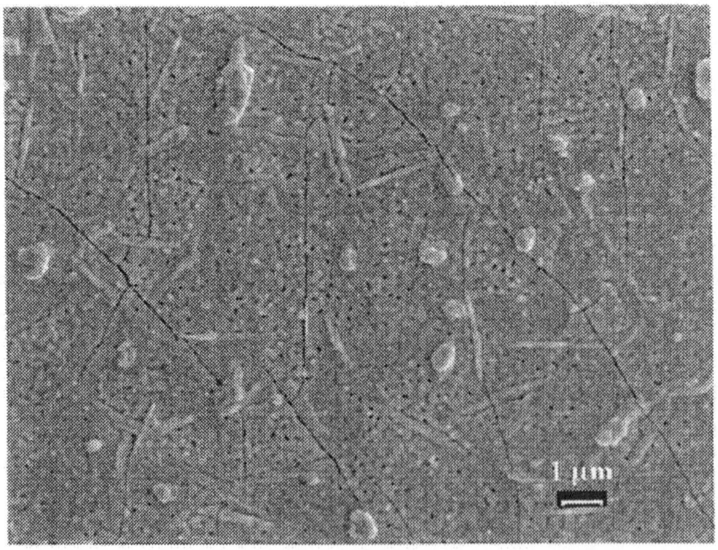

FIGURE 5. Scanning electron micrograph of YBCO film on buffered Ni-3at%W RABiTS substrate after electromechanical testing, showing very fine cracks located mostly near the sample edges. The vertical axis of the image coincides with the direction of the electrical current applied to the sample.

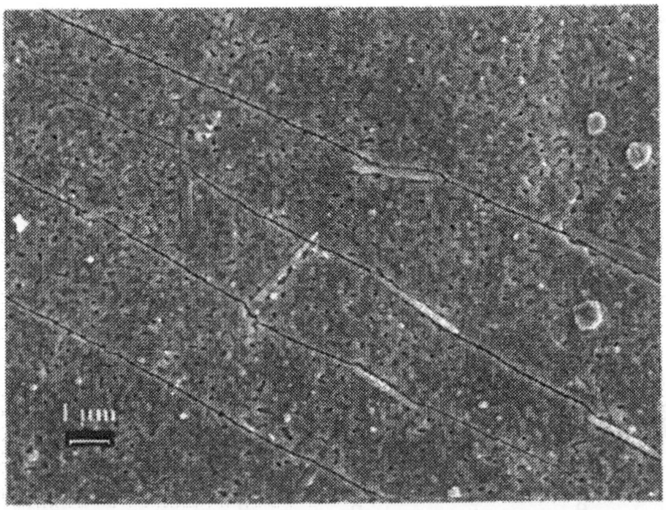

FIGURE 6. Scanning electron micrograph of YBCO film on buffered Ni-3at%W RABiTS substrate after electromechanical testing, showing diagonal cracks with respect to the electrical-current flow direction. The vertical axis of the image coincides with the direction of the electrical current applied to the sample.

MECHANICAL PROPERTIES OF PURE-Ni AND Ni-3at%W SUBSTRATES AT 76 K

Stress-strain characteristics of pure-Ni and Ni-3at%W substrates at 76 K are presented in Figs. 7a and 7b. The samples investigated received the same heat-treatment as the ones used for buffer and YBCO depositions. Three samples of each material were measured. Average values over the three samples of the tensile yield strength, Young's modulus and proportional limit of elasticity are summarized in Table I. The tensile yield strength of Ni-3at%W is more than a factor of three higher than that of pure-Ni, while the Young's modulus and proportional limit of elasticity are nearly a factor of two higher. When the sample was plastically deformed, the tensile load was released and reapplied to evaluate the degree of sample's work-hardening (Figs 7a, 7b). This operation was repeated several times up to a strain of about 1 %. For the samples presented in Figs. 7a and 7b, the Young's modulus of the second to the sixth slope was estimated and an average value determined (Table I). The change of the Young's modulus before and after plastic deformation for pure-Ni substrates is substantial (~64 %) as compared to that for Ni-3at%W substrates (~3 %). Also note that in the plastic regime, the stress increases with strain at a higher rate for pure-Ni than for Ni-3at%W. Therefore the work-hardening of pure-Ni is greater than that of Ni-3at%W. The difference in the mechanical properties between the two materials at 76 K is also depicted in Fig.8 for direct comparison.

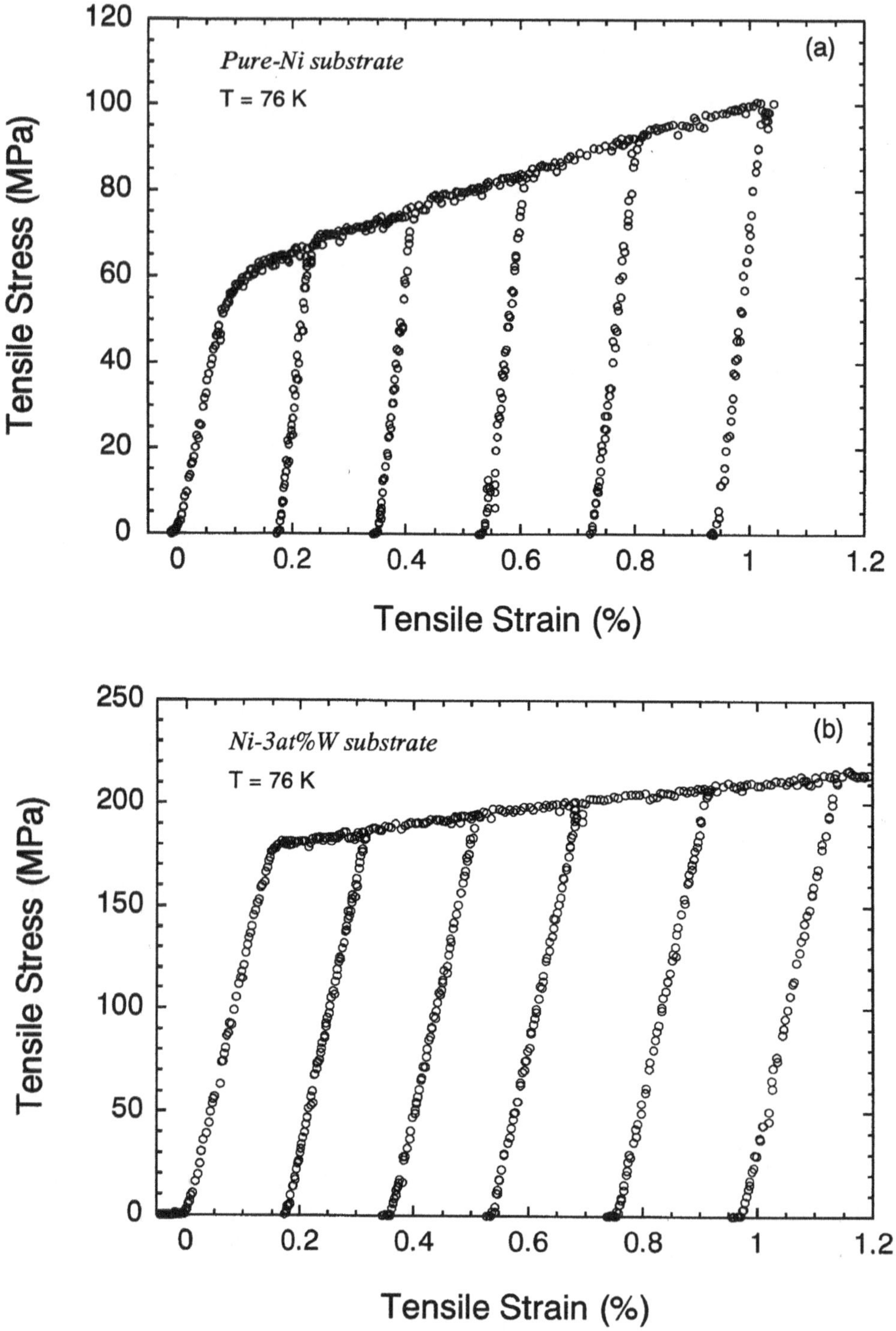

FIGURE 7. Stress-strain characteristics at 76 K of annealed pure-Ni (a) and Ni-3at%W (b) substrates.

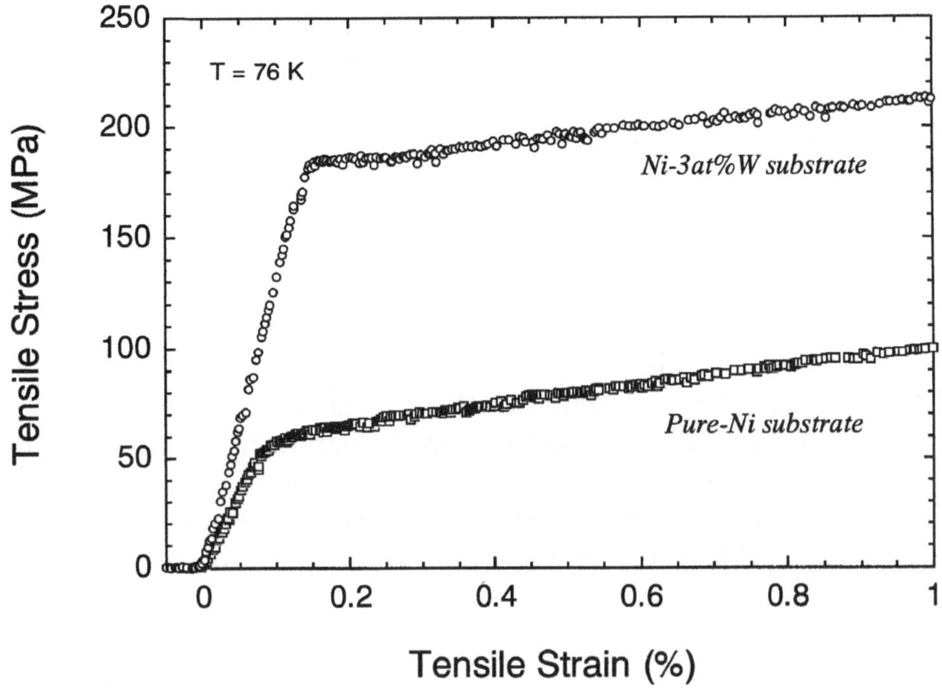

FIGURE 8. Comparison of stress-strain curves at 76 K of annealed pure-Ni and Ni-3at%W substrates.

Table I. Mechanical properties at 76 K of annealed pure-Ni and Ni-3at%W substrates.

	Pure-Ni	Ni-3at%W
Proportional limit of elasticity, ε_p (%)	0.08	0.15
Tensile Yield Strength (MPa)	55	183
Young's Modulus, E (GPa); 1st stress-strain slope	72	128
Young's Modulus, E (GPa); Average of second to sixth stress-strain slopes (Fig. 7)	118	132

CONCLUSION

The critical-current degradation from transverse compressive stress in YBCO coated conductors on buffered pure-nickel and nickel-3at% tungsten RABiTS has been investigated on samples that had initial J_c values in the range of 1 MA/cm^2 at 76 K and self-field. Transverse stress can cause significant degradation of J_c in YBCO deposited on pure-nickel RABiTS unless a sufficient frictional support is provided to the sample. Appropriate work-hardening of pure-Ni substrates or the use of Ni-3at%W substrates improve the electromechanical performance of these conductors.

The microstructure of the samples after electromechanical testing has been studied and has shown cracking of the YBCO layer. For samples on pure-Ni substrates, these defects were randomly distributed. In contrast for the sample on Ni-3at%W, the cracks are more confined near the sample edges. The mechanical properties of pure-Ni and Ni-3at%W substrates have also been presented. These results are given to guide the development of YBCO coated conductors.

ACKNOWLEDGMENTS

This research was supported in part by the U.S. Department of Energy, Office of Energy Efficiency and Renewable Energy, Office of Power Technologies—Superconductivity Program, and the U.S. Department of Energy, High Energy Physics Program.

REFERENCES

[1]X.D. Wu, S.R. Foltyn, P.N. Arendt, W.R. Blumenthal, I.H. Campbell, J.D. Cotton, J.Y. Coulter, W.L. Hults, M.P. Maley, H.F. Safar, and J.L. Smith, "Properties of $YBa_2Cu_3O_{7-\delta}$ thick films on flexible buffered metallic substrates," *Appl. Phys. Lett* **67**, pp. 2397-2399 (1995).

[2]A. Goyal, D.P. Norton, J.D. Budai, M. Paranthaman, E.D. Specht, D.M. Kroeger, D.K. Christen, Q. He, B. Saffian, F.A. List, D.F. Lee, P.M. Martin, C.E. Klabunde, E. Hartfield, and V.K. Sikka, "High critical current density superconducting tapes by epitaxial deposition of $YBa_2Cu_3O_x$ thick films on biaxially textured metals" *Appl. Phys. Lett* **69**, pp. 1795-1797 (1996).

[3]D.P. Norton, A. Goyal, J.D. Budai, D.K. Christen, D.M. Kroeger, E.D. Specht, Q. He, B. Saffian, M. Paranthaman, C.E. Klabunde, D.F. Lee, B.C. Sales, and F.A. List, "Epitaxial $YBa_2Cu_3O_7$ on biaxially textured nickel (001): An approach to superconducting tapes with high critical current density" *Science* **274**, pp. 755-757 (1996).

[4]D. Dimos, P. Chaudhari, J. Mannhart, and F.K. LeGoues "Orientation dependence of grain-boundary critical currents in $YBa_2Cu_3O_{7-\delta}$ bicrystals," *Phys. Rev. Lett.* **61**, pp. 219-222 (1988).
[5]D. Dimos, P. Chaudhari, and J. Mannhart, "Superconducting transport properties of grain boundaries in $YBa_2Cu_3O_7$ bicrystals," *Phys. Rev. B* **41**, pp. 4038-4049 (1990).
[6]Y. Iijima, N. Tanabe, O. Kohno, and Y. Ikeno, "In-plane aligned $YBa_2Cu_3O_{7-x}$ thin films deposited on polycrystalline metallic substrates," *Appl. Phys. Lett* **60**, pp. 769-771 (1992).
[7]A. Goyal, D.P. Norton, D.M. Kroeger, D.K. Christen, M. Paranthaman, E.D. Specht, J.D. Budai, Q. He, B. Saffian, F.A. List, D.F. Lee, E. Hatfield, P.M. Martin, C.E. Klabunde, J. Mathis, and C. Park, "Conductors with controlled grain boundaries: An approach to the next generation, high temperature wire," *J. Mater. Res.* **12**, pp. 2924-2940 (1997).
[8]D.T. Verebelyi, D.K. Christen, R. Feenstra, C. Cantoni, A. Goyal, D.F. Lee, M. Paranthaman, P.N. Arendt, R.F. DePaula, J.R. Groves, and C. Prouteau, "Low angle grain boundary transport in $YBa_2Cu_3O_{7-\delta}$ coated conductors," *Appl. Phys. Lett* **76**, pp. 1755-1757 (2000).
[9]D.T. Verebelyi, C. Cantoni, J.D. Budai, D.K. Christen, H.J. Kim, and J.R. Thompson, "Critical current density of $YBa_2Cu_3O_{7-\delta}$ low-angle grain boundaries in self-field," *Appl. Phys. Lett* **78**, pp. 2031-2033 (2001).
[10]H.C. Freyhardt, J. Hoffmann, J. Wiesmann, J. Dzick, K. Heinemann, A. Isaev, F. Garcia-Moreno, S. Sievers, and A. Usoskin, "YBaCuO thick films on planar and curved technical substrates," *IEEE Trans. Appl. Supercond.* **7**, pp. 1426-1431 (1997).
[11]C. Park, D.P. Norton, J.D. Budai, D.K. Christen, D. Verebelyi, R. Feenstra, D.F. Lee, A. Goyal, D.M. Kroeger, and M. Paranthaman, "Bend strain tolerance of critical currents for $YBa_2Cu_3O_7$ films deposited on rolled-textured (001) Ni," *Appl. Phys. Lett* **13**, pp. 1904-1906 (1998).
[12]C.L.H. Thieme, S. Fleshler, D.M. Buczek, M. Jowett, L.G. Fritzemeier, P.N. Arendt, S.R. Foltyn, J.Y. Coulter, and J.O. Willis, "Axial strain dependence at 77 K of the critical current of thick YBaCuO films on Ni-alloy substrates with IBAD buffer layers," *IEEE Trans. Appl. Supercond.* **9**, pp. 1494-1497 (1999).
[13]J.Yoo, and D. Youm, "Tensile stress effects on the critical current densities of coated conductors," *Supercond. Sci. Technol.* **14**, pp. 109-112 (2001).
[14]J.W. Ekin, S.L. Bray, N. Cheggour, C.C. Clickner, S.R. Foltyn, P.N. Arendt, A.A. Polyanskii, D.C. Larbalestier, and C.N. McCowan, "Transverse

stress and fatigue effects in Y-Ba-Cu-O coated IBAD tapes," *IEEE Trans. Appl. Supercond.* **11**, pp. 3389-3392 (2001).

[15]N. Cheggour, J.W. Ekin, C.C. Clickner, R. Feenstra, A. Goyal, M. Paranthaman, D.F. Lee, D.M. Kroeger, and D.K. Christen, "Transverse compressive stress, fatigue, and magnetic substrate effects on the critical current density of Y-Ba-Cu-O coated RABiTS tapes," *Adv. Cryog. Eng.*, in press (2002).

[16]M. Paranthaman, C. Park, X. Cui, A. Goyal, D.F. Lee, P.M. Martin, T.G. Chirayil, D.T. Verebelyi, D.P. Norton, D.K. Christen, and D.M. Kroeger, "$YBa_2Cu_3O_{7-y}$—coated conductors with high engineering current density," *J. Mater. Res.* **15**, pp. 2647-2652 (2000).

[17]R. Feenstra, T.B. Lindemer, J.D. Budai, and M.D. Galloway, "Effect of oxygen pressure on the synthesis of $YBa_2Cu_3O_{7-x}$ thin films by post-deposition annealing," *J. Appl. Phys.* **69**, pp. 6569-6585 (1991).

[18]J.W. Ekin, "Effect of transverse compressive stress on the critical current and upper critical field of Nb_3Sn," *J. Appl. Phys.* **62,** pp. 4829-4834 (1987).

[19]P.E.Kirkpatrick, J.W. Ekin, and S.L. Bray, "A flexible high-current lead for use in high-magnetic-field cryogenic environments," *Rev. Sci. Instrum.* **70**, pp. 3338-3340 (1999).

[20] C.C. Clickner, J.W. Ekin, and N. Cheggour, to be published.

PHASE AND MICROSTRUCTURE CHANGE OF HIGH CRITICAL CURRENT DENSITY TFA-MOD YBCO COATED CONDUCTOR

Yutaka Yamada, Takeshi Araki, Haruhiko Kurosaki, Toyotaka Yuasa, Yuh Shiohara and Izumi Hirabayashi
ISTEC-SRL
Mutsuno 2-chome 4-1, Atsuta-ku, Nagoya 456-8587, Japan

Yasuhiro Iijima and Takashi Saito
Fujikura Ltd.
1-5-1, Kiba, Koto-ku, Tokyo 135-8512, Japan

Junko Shibata and Yuichi Ikuhara
Tokyo University
2-11-16 Yayoi, Bunkyo-ku, Tokyo 113-8656, Japan

Takeharu Katoh and Tsukasa Hirayama
Japan Fine Ceramic Center
Mutsuno 2-chome 4-1, Atsuta-ku, Nagoya 456-8587, Japan

ABSTRACTS

For TFA-MOD process in which we obtained high J_c values up to 12.6MA/cm^2 on $LaAlO_3$ substrates and 2.9MA/cm^2 on metallic IBAD substrates, phase formation and microstructure change during heat-treatment were investigated using resistivity measurements, AFM and TEM. Precursor consisted of fine CuO and amorphous Y-Ba-Cu-O-F matrix was further oxidized up to 600°C and then transformed into Y123 above 670°C where the intermediate phases of $(Y,Ba)(O,F)_2$ and (Ba,Cu) poor phase such as $Y_2Cu_2O_5$ phase were formed at the same time. A fraction of these phases remained un-reacted on the surface of the film and also resulted in rough surface morphology. Although high J_c have been attained, we have much room for further enhancement in J_c. The magnetic field dependence of J_c was also measured and exhibited superior property up to 9.5T at 77K to that of the conventional NbTi conductor at 4.2K.

1. INTRODUCTION

Among many processes for an YBCO coated conductor, TFA-MOD (metal organic deposition using trifluoroacetate) is one of the most promising processes because of its cost efficiency [1] and high critical current density, J_c. TFA-MOD method is principally consisted of only coating and heat-treatment process without vacuum chamber. It is performed in a non-vacuum atmosphere while other processes such as PLD (Pulsed Laser Deposition) method require a complicated vacuum system. Furthermore TFA-MOD process leads to a high critical current density, J_c, over 1 MA/cm^2 at 77K [1-3]. Many groups obtained promising results using textured Ni substrates [1] and IBAD substrates [3-5]. The J_c reached 1 to 2 MA/cm^2 at 77K and 0T. Recently the effort for a long YBCO conductor is also being carried out by improving the heat-treatment furnace for the continuous production system.

However, understanding of the reaction process in TFA-MOD YBCO still remains ambiguous especially in terms of the Y123 formation and microstructural change. For further development, we need such basic understanding.

In this paper, we report the resistivity measurement results, which give some insight on the Y123 formation mechanism. Together with XRD and TEM data, we describe phase and microstructure change during heat-treatment. Especially we concluded that super-current did not flow in the whole cross section of the YBCO film because much un-reacted phases still remained near the surface. Actually some improvements on microstructure by optimizing heat-treatment process resulted in high J_c values above 12MA/cm^2. The J_c-B characteristic at 77 K was also measured and exhibited a superior property at high magnetic fields. These results imply that TFA-MOD YBCO coated conductor is a strong candidate for a practical coated conductor.

2. EXPERIMENTAL

The coating solutions were made from Y, Ba and Cu acetates [3]. During synthesis, special attention was paid to the reduction of the impurities including H_2O and acetic acid. $LaAlO_3$ (LAO) single crystals of 10 x 10 mm^2 and CeO_2/YSZ buffered Hastelloy tapes (IBAD) [6] 5-10 x 10 mm^2 were spin-coated or dip-coated with the coating solutions. The coated films were heat-treated at 200°C to 400°C in wet oxygen gas (calcination process) and fired at 800°C in a wet mixture gas of argon and oxygen (firing process). The oxygen and water concentrations were typically 1000 ppm and 4.2%, respectively. The electric resistivity was measured during the firing process at 800°C. The calcined film on the LAO was patterned and Au wires for voltage taps were attached. Critical currents were measured by the four-probe transport method at 77.3 K and in the magnetic fields up to 9.5 T. The criterion of I_c was 5. V/cm.

3. RESULTS AND DISCUSSION

3.1 Resistivity Measurement and Reaction in TFA Process

In the TFA-MOD process, first we obtained a calcined film heat-treated at 200°C to 400°C in wet oxygen gas. Figure 1 shows the microstructure of the calcined film. This was consisted of YBaCuOF matrix and fine CuO grains about 10nm in diameter. The coating film before calcination was M-CF_3COOH (M=Y, Ba, Cu), metal trifluoroacetate. This was transformed to YBaCuOF and CuO. Only Cu was oxidized to CuO at the low temperature.

This calcined film was fired at 800°C for 1hour, which was the best condition to obtain a high J_c in our laboratory. The resistivity of the film during firing was measured as shown in Fig. 2. The heat-treatment was done first in an atmosphere of wet Ar-1000 ppm O_2 gas with humidity of 4.2% and then the gas was changed to dry Ar/O_2 gas in the terminate stage at 800°C. Finally the sample was oxygenated in a pure O_2 gas at 450°C.

The resistivity was first decreased (region A in Fig. 2) due to the evaporation of chemicals including ethylene glycol or sintering of Ag particles of the Ag paste in the voltage taps for the resistivity measurement. Then, the resistivity was increased when the temperature was increased to 600°C (B). This was due to the further oxidation of the matrix of the calcined film. Above 600°C the resistivity first decreased slowly (C) and then decreased rapidly above 670C° (D). The decrease of the resistivity means the formation of the aligned Y123 phase [7] with low resistivity among the phases observed in TFA-MOD process.

From the X-ray measurements [8] and the TEM study [9], the film before complete reaction, heat-treated at 775°C, included CuO, (Y, Ba)(O, F)$_2$ and $Y_2Cu_2O_5$ phases. As reported in detail by McIntyre et al.[2,8] and Solovyov et al.[10], Y123 was formed through the reaction of H_2O and these phases. The microstructure of the resultant film was highly aligned especially in the vicinity of the interface between the film and the substrate. This highly orientation resulted in high J_c as reported by the authors [3-5,7]. Normally the high J_c was obtained at 800°C for 1hour. However, the

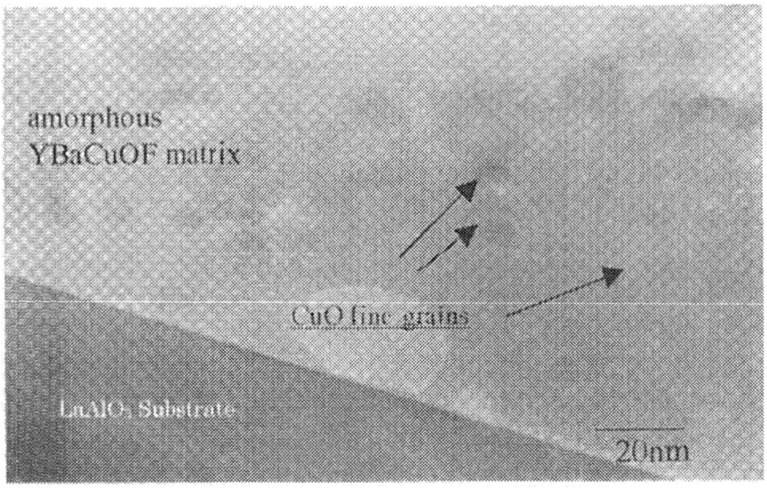

Figure 1. TEM of TFA-MOD YBCO on $LaAlO_3$ substrate. Matrix is amorphous Y-Ba-Cu-O-F and grains are CuO 10nm in diameter.

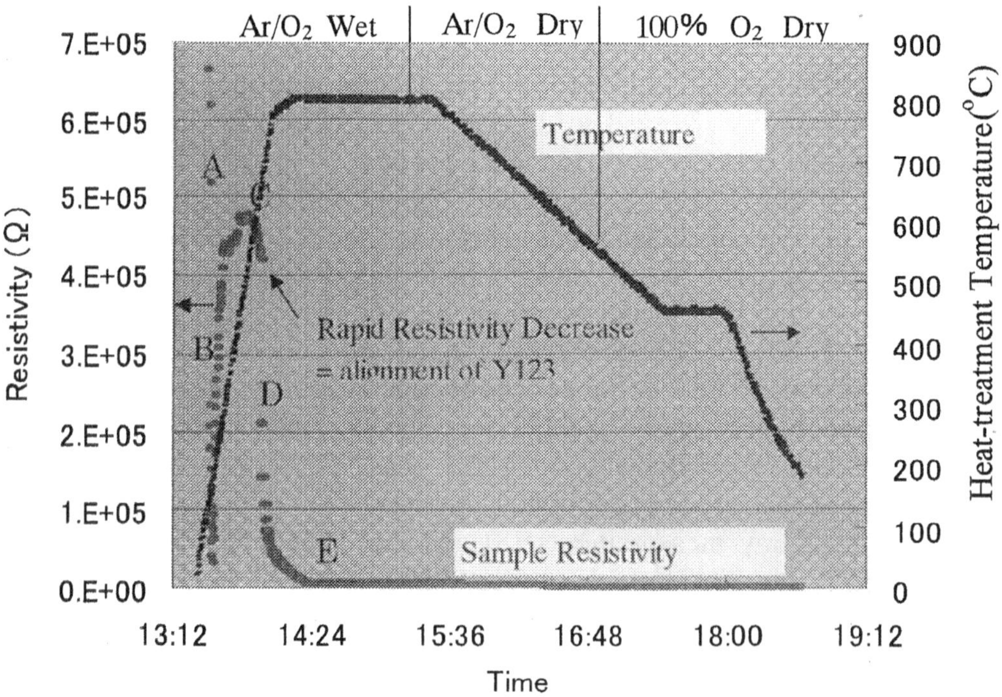

Figure 2. Resistivity change during TFA process.

formation of Y123 and the alignment of Y123 occurred at much lower temperatures around 670°C from the rapid decrease of resistivity in Fig. 2.

Figure 3 is the XRD data at 700°C for different heat-treatment times. A calcined film showed only CuO or CuO_x peaks in YBaCuOF amorphous matrix as confirmed by TEM in Fig.1. After 10min the matrix YBaCuOF formed $(Y,Ba)(O,F)_2$. In 30min, this peak and the Y123 peak began to develop with the formation of (Ba,Cu) poor phase (compared to Y123) such as $Y_2Cu_2O_5$. This continued up to 60min. In 120min, the $(Y, Ba)(O,F)_2$ peak vanished while the CuO and $Y_2Cu_2O_5$ phases remained. This means the following reaction occurred: YBaCuOF +CuO → $(Y,Ba)(O,F)_2$ +CuO + $Y_2Cu_2O_5$ with H_2O and O_2 → Y123+ CuO + $Y_2Cu_2O_5$.

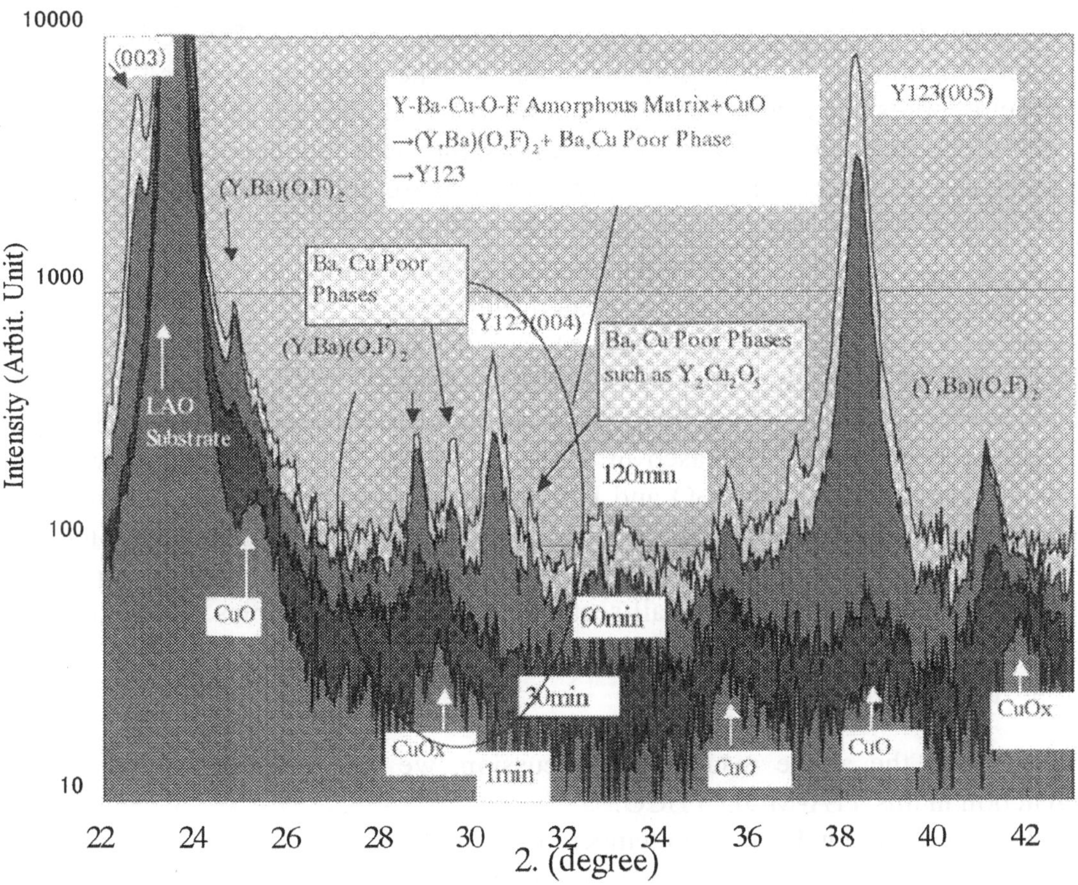

Figure 3. XRD patterns for various heat-treatment times at 700°C.

The un-reacted CuO and (Ba,Cu) poor phase is due that the Y123 formation reaction does not precede because the each grain such as CuO and $Y_2Cu_2O_5$ become too large. Actually large un-reacted $Y_2Cu_2O_5$ and CuO grains were observed by TEM of TFA sample [9] and ex-situ sample [11]. Very recently BaO phase other than these phases was also observed in the vicinity of the surface by TEM [12]. These facts mean that we have still possibility to increase J_c by optimizing the microstructure and thus the heat-treatment process. As described later, we actually obtained improved J_c of 12MA/cm^2 [13] from the former 7MA/cm^2 [14]. This was achieved by shortening the calcination time. Appropriate calcination resulted in small CuO grains and subsequent homogeneous Y123 formation reaction in firing process.

At the temperature of 800 °C, the formation of the aligned Y123 grains was completed in a few tens minutes as shown in Fig. 2. The resistivity reached minimum in 10 to 20min (region E). Therefore, the oriented Y123 structure finished forming at this minimum. After this annealing, the gas was changed to dry Ar/O_2 and the temperature was decreased. Then, the resistivity decreased with the temperature decrease and the oxygen content of the gas. The resistance decreased finally to 30 Ω.

Surface morphology was observed by AFM as shown in Figs. 4 (a) and (b) for a calcined film and a heat-treated film at 800°C for 15min. While the surface was very smooth in a calcined film, the surface became rough only in 15min. This means that the reaction to form Y123 was significantly proceeded in 15min, which corresponds to the resistivity data in Fig. 2. The resistivity was much decreased in 15min at 800°C. The rough surface was due to the reaction with H_2O and HF gas phases. For the transport current density, the effective cross sectional area is considered to be small from the above observations of AFM and un-reacted phases in Fig.3. In other words, super-current flows only a small section of TFA-MOD samples. Therefore, we can increase J_c by increasing the fraction of effective cross section for transport current.

From the above results and discussion, we can describe the whole reaction in this TFA-MOD YBCO:

In the calcination from room temperature to 400 °C,

$$CF_3COO\text{-}M(Y, Ba, Cu) + H_2O + O_2 \rightarrow (Y, Ba,Cu)\text{-}(O,F) + CuO \quad \text{(precursor)}.$$

In subsequent firing process from 400°C to 800°C,

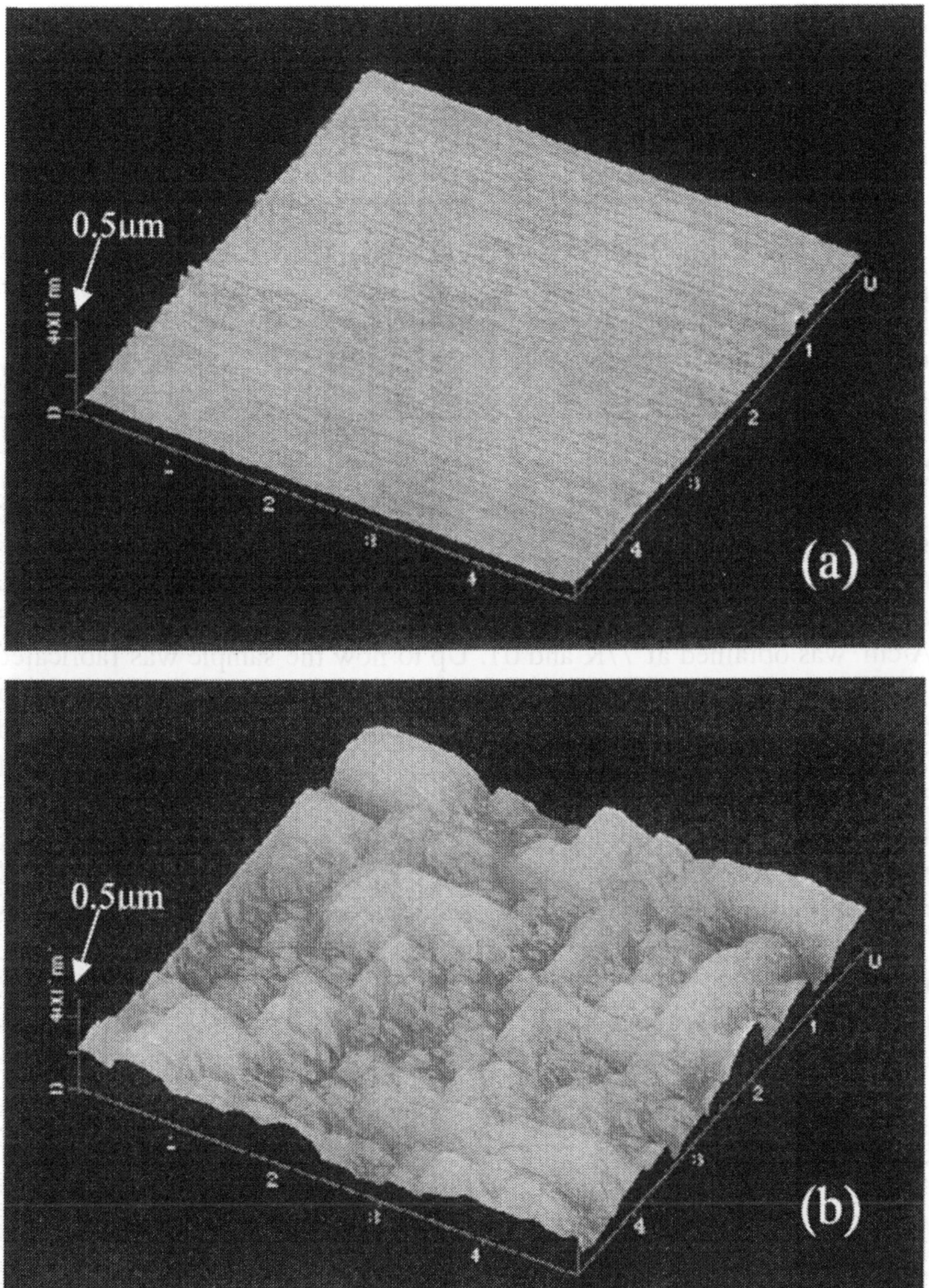

Figure 4. AFM surface morphology of (a) calcined film and (b) heat-treated film at 800°C for 15min.

(Y, Ba, Cu)-(O, F) + CuO (precursor) + H_2O + O_2
→further oxidization to CuO < 600°C
→formation of $(Y,Ba)(O,F)_2$ phase and $Y_2Cu_2O_5$ phase
→amorphous YBCO+ HF.
→$YBa_2Cu_3O_{6.1}$ formation and alignment + HF. > 670°C

According to the studies by Suenaga et al. [10,11] and Hirayama et al. [9], amorphous YBCO are observed just on the aligned Y123 phase and below the $Y_2Cu_2O_5$, CuO and $(Y,Ba)(O,F)_2$ phases. Thus, amorphous YBCO is considered to be an intermediate phase to form Y123 phase, which may also decrease the resistivity between 600 to 670 °C.

3.2 TFA-MOD YBCO on IBAD

IBAD was well known as one of the effective method to obtain an aligned buffer layers such as YSZ on the metallic substrate such as Hastelloy [6]. Using dip coating method [15], TFA-MOD YBCO/IBAD coated conductor was fabricated as shown in Fig. 5. The transport J_c of 1.3 MA/cm^2 was obtained at 77K and 0T. Up to now the sample was fabricated in 10cm long [13], which was withdrawn at a speed of 18m/hr during dip coating. The obtained J_c value of the conductor was 1.3 MA/cm^2 at 77K and 0T. On the other hand, spin coating sample exhibited higher value of 2.5MA/cm^2. This ascribes that more homogeneous cross-section may be obtained by spin coating than by dip coating. These high J_c values imply that TFA-MOD YBCO/IBAD is a strong candidate for the practical coated conductor. Figure 6 showed the cross sectional area of TFA-MOD samples on IBAD tape. The thickness of sputtered CeO_2 layer was about 1μm and that of YBCO was 2000Å. The texturing was investigated in the order of the depositing process as shown in Fig. 7. The original IBAD tape exhibited a good texturing, FWHM of $\Delta\phi$ of 9.1°. After depositing CeO_2 layer, the in-plane alignment was improved to 5.2° of $\Delta\phi$. Finally $\Delta\phi$ was 6.0° after deposition of YBCO. This superior texturing resulted in high Jc of TFA-MOD samples.

Figure 5. Dip Coated TFA-MOD YBCO on metallic IBAD substrate with J_c of 1.3 MA/cm^2 at 77K and 0T.

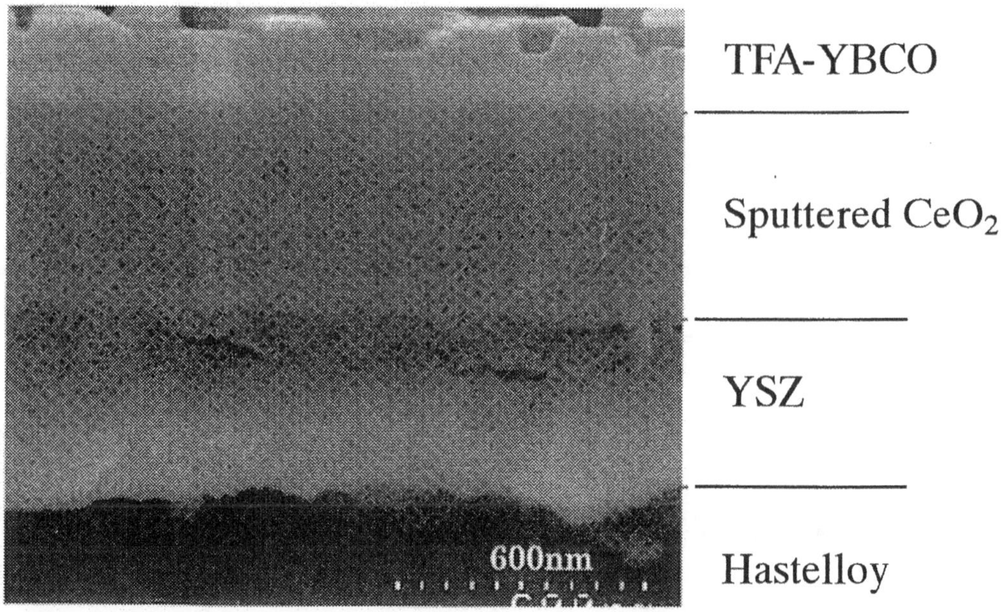

Figure 6. SEM micrograph of the cross sectional area of TFA -MOD YBCO on CeO_2 buffered IBAD substrate.

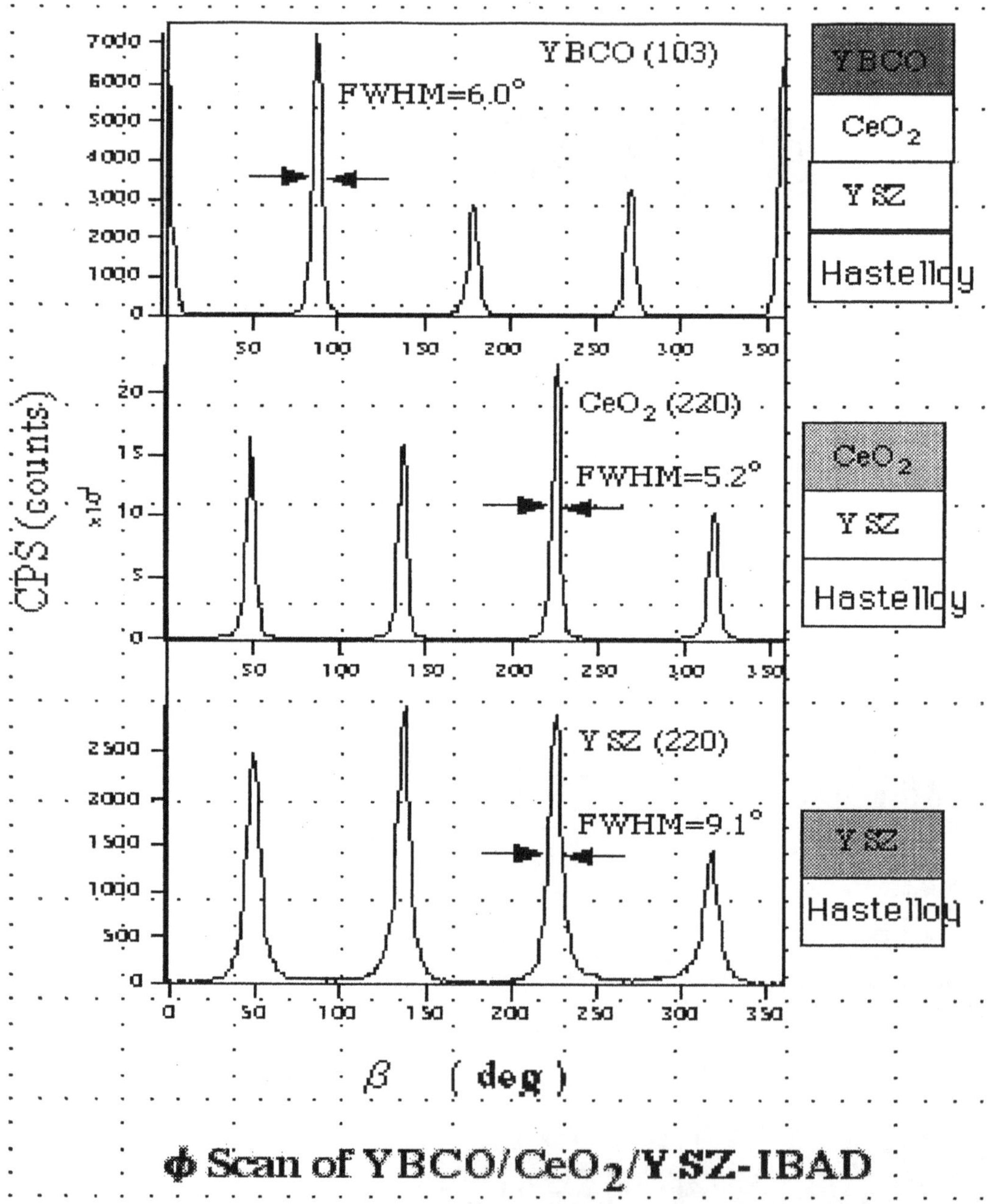

Figure 7. Variation of the degree of texturing for each layer of TFA -MOD YBCO on CeO_2 buffered IBAD substrate.

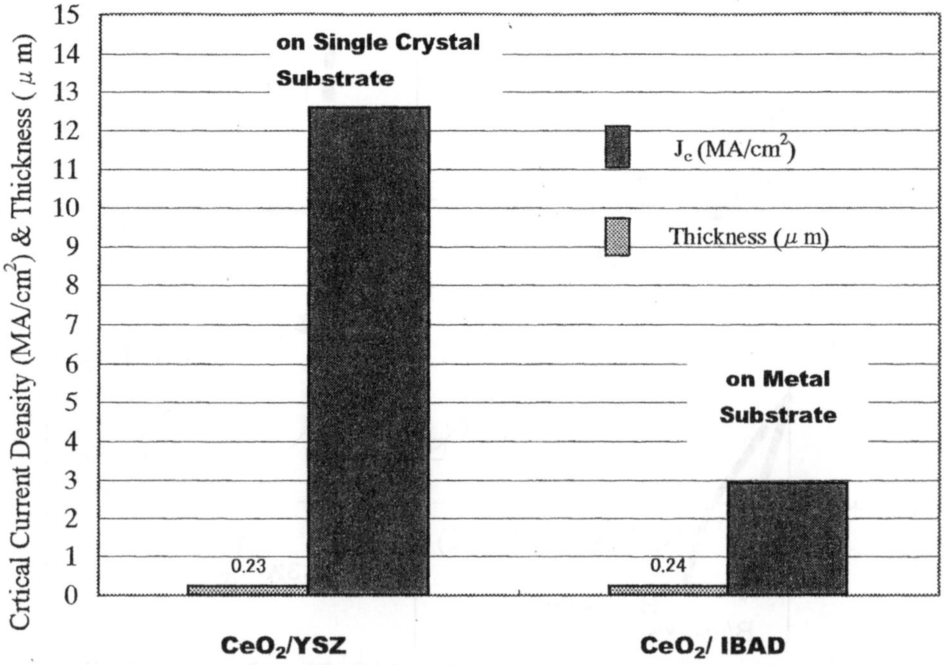

Figure 8. Critical current densities at 77K and 0T achieved in our laboratory for the metallic IBAD and single crystal substrates.

3.3 Critical Current Density

We have reported high J_c values of TFA-MOD YBCO samples up to now. Recently we improved J_c further by optimizing the heat-treatment process, especially calcination stage [13]. Although normal calcination time used by us [3] and McIntyre et al. [2] was more than 20hrs at 200°C to 400°C, the time was shortened to about 9 hours. This resulted in fine CuO grains in a calcined film. This was considered to bring about the homogeneous reaction to form Y123 in the firing stage. Furthermore, through efforts to reduce the amount of impurities such as H_2O and acetic acid in the coating solution, high values up to 12.6 MA/cm^2 were obtained in the case of LAO single crystals substrate as shown in Fig.8. On the other hand, the J_c value on metallic IBAD substrate, 2.9MA/cm^2, was below one fourth of that on the single crystal.

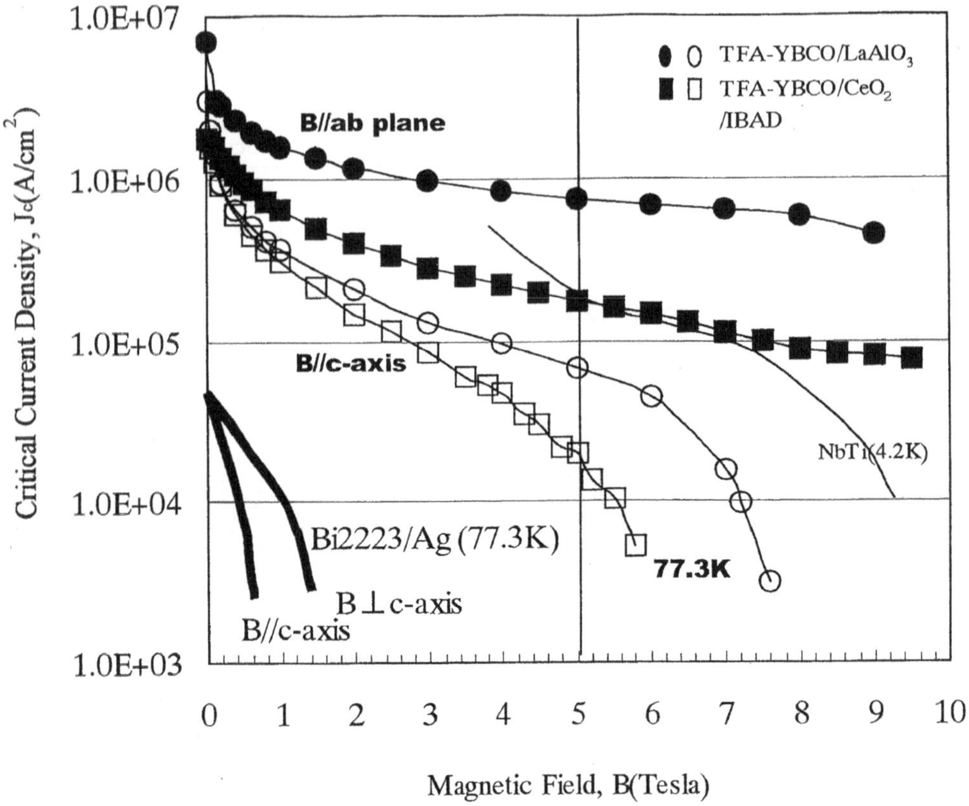

Fig. 9. J_c –B characteristic at 77.3 K for TFA-MOD YBCO on CeO_2 buffered IBAD-YSZ and on single crystal of $LaAlO_3$. The data is compared with the typical values of NbTi wires at 4.2 K and Ag sheathed Bi2223 wires.

Two typical samples on a LAO single crystal and a CeO_2 buffered IBAD-YSZ tape were measured for the magnetic field dependence of J_c at 77 K. The J_c values at 0T were 7 MA/cm^2 and 1.7 MA/cm^2 for the samples on the LAO and the IBAD, respectively. The magnetic field dependence of J_c is shown in Fig. 9, compared with other practical superconducting wires of NbTi and Ag sheathed Bi-2223. The TFA-YBCO/IBAD conductor exhibited high J_c even in a high magnetic field; 20 kA/cm^2 at 5 T parallel to the c-axis and 170 kA/cm^2 at 5 T parallel to the ab-plane. The latter is comparable to J_c of the most popular conductor of NbTi at 4.2 K. The J_c values for the film on LAO was much higher than these values especially for the fields parallel to the ab plane, which was almost $1MA/cm^2$ at 77K and 5T.

4. CONCLUSIONS

In the TFA-MOD process, resistivity measurements, TEM and XRD data suggested that the Y123 phase alignment occurred below 700 °C and was completed at an early stage of the firing at 800 °C from the resistivity minimum. CuO grains well developed first, then, high Jc was obtained after improving the calcination process resulted in a small CuO grains before firing. Then, we obtained high J_c on LAO substrate more than $12MA/cm^2$ at 77K and 0T from the former values of 7 to $8MA/cm^2$. The J_c value on metallic substrate of IBAD exhibited $2.9MA/cm^2$. Using the dip coating favorable for a practical process, 10cm long sample was also obtained. Together with the superior magnetic field dependence of J_c, the TFA-MOD YBCO conductor is considered to be promising as the next generation superconductor.

ACKNOWLEDGEMENTS

This work is supported by the New Energy and Industrial Technology Development Organization (NEDO) as Collaborative Research and Development of Fundamental Technologies for Superconductivity Applications.

REFERECNCES

[1] A.P. Malozemoff, S. Annavarapu, L. Fritzemeier, Q. Li, V. Prunier, M. Rupich, , C. Thieme, W. Zhang, A.Goyal, M. Paranthaman, D.F. Lee, Supercond. Sci. Technol. **13** 473 (2000).

[2] P.C. McIntyre, M.J. Cima and Ng Man Fai, J. Appl. Phys. **68** 4183 (1990).

[3] T. Araki, K. Yamagiwa, S.B. Kim, K. Matsumoto and I. Hirabayashi, Advances in Superconductivity, **XII** 610 (1999).

[4] T. Araki, Y. Takahashi, K. Yamagiwa, T. Yuasa, Y. Iijima, K. Takeda, S.B. Kim, Yutaka Yamada and I. Hirabayashi, IEEE Transactions on Applied Superconductivity, **11** 2869 (2000).

[5] Yutaka Yamada, S.B. Kim, T. Araki, T. Yuasa, Y. Takahashi, H. Kurosaki, Y. Shiohara, I. Hirabayashi, Y. Iijima, and K.Takeda, Physica C, **357-360** (2001) 1007.

[6] Y. Iijima, M. Hosaka, N. Tanabe, N. Sadakata, T. Saitoh, O. Kohno and K. Takeda, Applied Superconductivity **4** 475 (1996).

[7] Yutaka Yamada, T. Araki, H. Kurosaki, S.B. Kim, T. Yuasa, Y. Shiohara, I. Hirabayashi, Y. Iijima, T. Saitoh, J. Shibata, Y. Ikuhara, T. Katoh and T. Hirayama, in Proceedings of CEC/ICMC Conference, Madison, USA, July 16-20 (2001).

[8] P.C. McIntyre, and M.J. Cima, J. Mater. Res. **9** 2219 (1994).

[9] T. Hirayama, J. Shibata, T. Araki, Yutaka Yamada, K. Yamagiwa, I. Hirabayashi, T. Izumi, Y. Shiohara and Y. Ikuhara, in Extended Abstracts of 2001 International Workshop on Superconductivity, 2001, p.251.

[10] V.F. Solovyov, H.J. Wiesmann, Li-jun Wu, M. Suenaga and R. Feenstra, IEEE Trans. on Applied Supercond. **9** 1467 (1999).

[11] Li-jun Wu, Y. Zhu, V.F. Solovyov, H.J. Wiesmann, A.R. Moodenbaught, R.L. Sabatini and M. Suenaga, submitted to J. Mater. Res (2000).

[12] T. Hirayama, private communication.

[13] T. Arakai, to be presented in Applied Superconductivity Conference in Houston, USA, (2002).

[14] Yutaka Yamada, S.B. Kim, T.Araki, T. Yuasa, Y. Takahashi, H. Kurosaki, Y. Shiohara, , I. Hirabayashi, Y. Iijima and K. Takeda, Mat. Res. Soc. Symp. Proc., **659** II4.7.1 (2001).

[15] H. Kurosaki, T. Araki, T. Yuasa, S.B. Kim, Yutaka Yamada, I. Hirabayashi, Y. Iijima and T. Saitoh, in Extended Abstracts of 2001 International Workshop on Superconductivity, p.131 (2001).

GROWTH KINETICS AND TEXTURE OF SOE NiO/Ni AND Ni-BASED ALLOYS RABiTS

Z. Lockman[A], X. Qi[A], W. Goldacker[B], R. Nast[B], B. deBoer[C], A.Kursumovic[D], R. Tomov[D], R. Hühne[D], J.E. Evetts[D] B.A. Glowacki[D], J. L. MacManus-Driscoll[A]

[A] Department of Materials, Imperial College of Science Technology and Medicine, Prince Consort Road, London, SW7 2BP, UK
[B] Forschungszentrum Karlsruhe, Technik und Umwelt, Hermann- von-Helmholtz- Platz 1 76344, Eggenstein-Leopoldshafen, Karlsruhe, Germany
[C] Institut für Festkörper-und Werkstoffforschung, Helmholtzstraße 20, 01069 Dresden, Germany
[D] Department of Materials Science and Metallurgy, University of Cambridge, Pembroke Street, Cambridge CB2 3QZ, UK

Abstract

We report on basic studies of a new coated conductor architectures: Ni/NiO/$Nd_{2-x}Ce_xCuO_4$/YBCO. Various Ni compositions were explored including pure Ni, Ni-Cr, and Ni-Fe, and a new liquid phase processing (LPP) technique to make $Nd_{2-x}Ce_xCuO_4$. Fundamental studies of surface oxidation epitaxy of Ni and Ni-alloys were undertaken. It was shown that a very thin NiO layer can be formed on pure Ni with excellent texture and low surface roughness. Similar, laterally grown surface oxides form for larger thicknesses. For Ni-Cr, a ~5micron thick NiO surface layer with excellent texture and surface smoothness can be produced. For Ni-Fe, a smooth, thin Ni_2FeO_4 was formed underneath a surface (111) oxide which can be removed. The $Nd_{2-x}Ce_xCuO_4$ layer was fabricated by LPP to have a surface roughness of a few 10's of nm. The layer was highly out-of-plane textured and could be completely in-plane textured by adjusting the Ce concentration.

Introduction

Since the discovery of high T_c superconductors in 1986, there has been an enormous effort worldwide to establish a scaleable method for the fabrication of coated conductors on metallic tapes. Although short length samples with J_c over 1 MA/cm^3 have been prepared by a number of vapour deposition methods, e.g. pulsed laser deposition (PLD), they are so far failed to produce long length conductors at a reasonable cost. An alternative route to vapour deposition methods is LPE, which has been widely used for the growth of semiconductor and garnet thin films during 1970s and 1980s. The advantages of LPE compared to vapour methods are the very fast growth rate, typically 1-10 μm/min, and the capability of growing thick films without degrading the structural perfection[1]. However, LPE growth of REBCO thick films on metallic substrate is much more difficult. The high growth temperature largely increases the risk of reaction between the substrate and the film. The solubility of RE's in the BaO:CuO solution is very low, therefore nucleation and growth is more difficult than the traditional LPE of semiconductors and garnets. Compared to PLD and other vapor deposition methods, LPE has a much lower available supersaturation and hence a much smaller driving force for growth. As a consequence, a close lattice matched substrate is required. At present, most of the buffer layers were developed for the use in the vapor deposition methods and therefore are not suitable for LPE growth. Nd_2CuO_4, a tetragonal crystal with $a=b\approx0.394$ nm, is a good potential buffer for LPE. As a part of the RE-Ba-Cu-O system[2] Nd_2CuO_4 neither poisons superconductivity nor reacts strongly with the barium cuprate solution at high temperature. It has a fairly close lattice match and thermal expansion match to REBCO and NiO. In this paper, we report on the growth of $Nd_{2-x}Ce_xCuO_4$ thick films on surface oxidised Ni-alloy substrates by scaleable methods. We also report on the basics of the surface oxidation of different Ni-alloys.

Experimental

100μm thick pure nickel, Ni-10%Cr, Ni-13%Cr and Ni-Fe were prepared by RABiTS method[3]. A summary of all Ni and Ni-alloys used for the SOE is shown in Table I. The tape surface was finished by a standard electro-polishing or chemical etching method (5%HF). The foils were then ultrasonically cleaned in acetone and methanol.

Table I Summary of Ni and Ni-foils used for SOE

Foils	Used by	Thickness (μm)	Supplier
Pure Nickel	Imperial College	80-100	Forschungszentrum, Karlsruhe [4]

Pure Nickel	Cambridge	100	CONTEXT [5]
Ni-10%Cr	Imperial College	100	Forschungszentrum, Karlsruhe [4]
Ni-13%Cr	Imperial College	100	Institut für Festkörper - und Werkstoffforschung [6]
Ni-Fe	Cambridge	50	CONTEXT [5]

Oxidation for pure Ni was done at high temperature (1000-1300°C) in two different oxidation atmosphere a) In air (Cambridge), b) In oxygen (Imperial). Oxidation was carried out in an evacuable high temperature tubular furnace, enabling oxidation in a controlled atmosphere. The progress of oxygenation of the NiO scale was monitored by mass and thickness increase. Mass was measured by a precise microbalance after each partial oxidation. Thickness was measured both by a micrometer and using secondary electron microscope (SEM) or optical microscope. The texture analysis was carried out by x-Ray diffraction techniques: θ-2θ scans, rocking curves and pole figures. Optical microscopy was used to monitor the development of microstructure during film growth.

Oxidation of Ni-Cr alloys was done isothermally in air at 1050°C. Details of the oxidation procedures could be found in our previous study[7].

In order to systematically study high temperature oxidation process of Ni-Fe, a number of Ni-Fe tapes were oxidised using O_2 and Ar/O_2 mixtures in a UHV chamber, tapes being heated resistively using a direct current. There are two methods of carrying out the oxidation which turn out to show important differences. One way is to preheat the tape in vacuum, then introduce the Ar/O_2 gas mixture, or alternatively to introduce the required pressure of gas into the chamber, then ramp up the current. Oxidation experiments were carried out at temperatures between 500°C and 1100°C, in each case preheating the tape before introducing the oxidising atmosphere. Note that the temperature quoted is the initial temperature of the bare Ni-Fe tape, and will be reduced when the surface is oxidised. Below around 950°C a continuous oxide is grown, which is dark grey in appearance at sufficient thicknesses. At higher temperatures however, macroscopic oxide grains become visible on the surface due to secondary recrystallisation and abnormal grain growth.

Results and Discussion

1. SOE of RABiTS Pure Nickel: At low temperatures, NiO film growth is affected by the grain boundaries in the "parent" Ni-phase which has an average grain size of about 30 μm. The growth is faster around grain boundaries creating humps on the NiO surface. The grain size of the oxide film is in the order of 1 μm. At higher temperatures, the grain size of the NiO film increases being >20 μm at 1250°C on textured Ni. The NiO growth has a specific flat surface that is a consequence of predominantly volume diffusion of growing species. At very high temperatures, >1300°C, highly concave growth around grain boundaries makes the film very rough[8].

1a) Air Oxidation Studies in Cambridge: A working window for SOE NiO/Ni was established from the ratio of XRD intensities I(200)/I(111)NiO. For samples oxidised for approximately two hours in air at different temperatures the results are shown in Fig. 1-a. As can be seen, there is a pronounced peak around 1260°C, at somewhat higher temperature than found in oxygen oxidation (refer next section on oxygen oxidation) presumably due to using air rather than 100% oxygen[9]. As expected sharp in-plane alignment is achieved at the "peak" of the working window (Fig. 1-b). In Fig. 1-c the pole figure shows that an acceptable in-plane orientation can be achieved in the range of <20°C. It should be noted, however, that at temperatures >1270°C, increased (220) peak intensities are present that add-up to overall deterioration of the NiO/Ni epitaxy.

The sample mass per unit area, Δm/A, for NiO/Ni films grown for two hours, both in pure oxygen and in air, is shown in Fig. 2. The sample mass increases due to formation of NiO film in the same manner as oxide thickness. These data are corresponding to data on epitaxial development shown in Fig. 1. The oxide film thickness was calculated here, but corresponds closely to experimentally measured values that were performed on some of the samples. As can be seen, an almost perfect epitaxy that was achieved at around 1260°C corresponds to ~25 μm thick NiO layer. The kinetics of NiO film growth in air at different temperatures versus time was investigated in more detail in Ref[10]. This kinetics was found shown to obey expected parabolic (√t) law. The activation energy calculated from the kinetic data was 220 kJ/mol for temperatures above 1100°C, corresponding to that for volume diffusion of Ni in NiO in excellent agreement with literature data. On the other hand a grain growth led to the epitaxial development. This grain growth was attributed to the grain boundary diffusion[10]. There are three distinct temperature regimes affecting the growth. At lower temperatures (<1000°C), close to grain boundary dominated diffusion in NiO, a predominant (111) orientation was found in this regime. At higher temperatures (1150-1270°C) a flat oxide surface was obtained on both textured and as-rolled Ni tapes, which indicate uniform volume diffusion during the growth. However, both (111) and (100) oriented NiO grains are nucleated initially and continue to grow competing with each other. This lateral growth occurs by competitive growth of the (001) and (111) NiO grains. Matsumoto et al.[11] obtained cube on cube (NiO on Ni) epitaxial relationship with FWHM for (111) Ni and (111) NiO typically 6-8 and 12-14 degrees respectively.

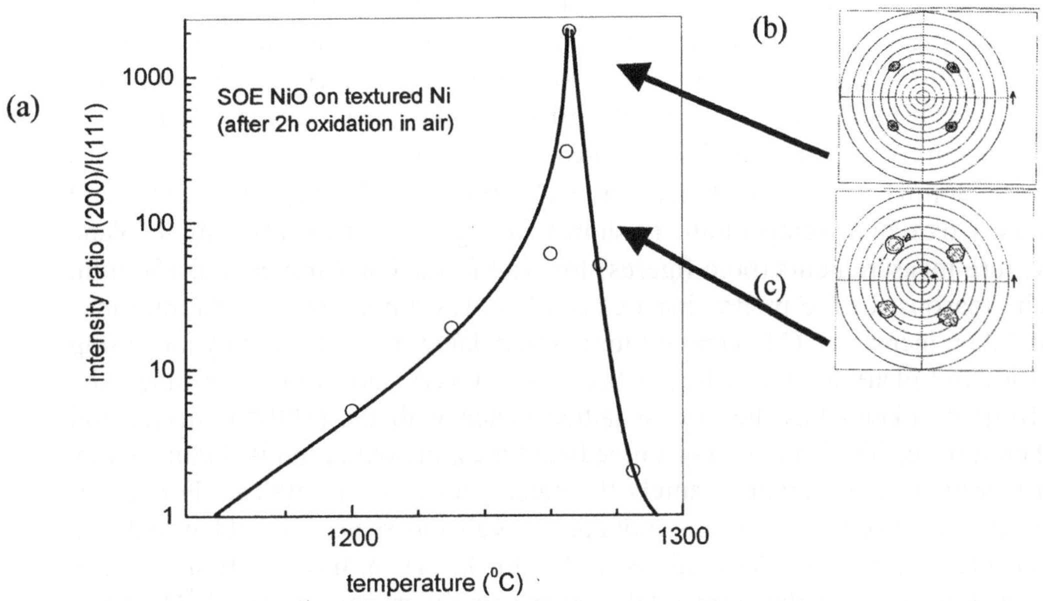

FIG. 1. (a) Effective working window for SOE NiO/Ni in air; (b) pole figure for a narrow temperature range of excellent in-plane texture, I(200)/I(111)>1000, and (c) and satisfactory in-plane texture , with I(200)/I(111)>50.

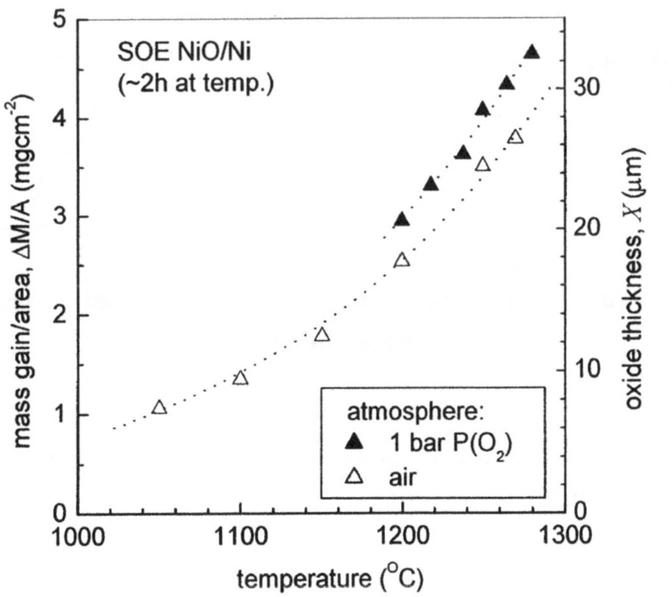

FIG. 2. Mass gain/area for NiO growth on textured Ni after ~2h oxidation in air and in pure oxygen (1 bar) atmosphere.

1b) Oxygen Oxidation Studies at Imperial College: In accordance to our previous studies[12] oxidation of RABiTS pure Ni was done for a constant time of 2 or 3 hours at different temperatures (1000-1300°C) in flowing oxygen. Single phase (100)NiO was found to be formed in a narrow temperature range near 1250°C. Having achieved this 'optimum temperature', isothermal oxidation at various oxidation times were performed. Formation of single phase (100)NiO was seen at a very short oxidation time of 0.2-10 min (oxide thickness < 6μm). We lose the cube oriented NiO when the oxide thickness is 10-30 μm. Cube oriented NiO formed again after the thickness of the scales are > 30μm i.e. after oxidation for > 150 min. A plot of I(100)/I(111)NiO versus oxidation temperature is shown in Fig. 3 with an inset of in plane texture measurements (pole figures) for NiO formed at 2 min and at 150 min. The reasons for the predominance (100)NiO texture at short oxidation times and then increasing (111) texture for intermediate times followed by increasing (100)NiO plane for much longer times is: for very short oxidation times, the (100)NiO planes has the closest lattice match with the (100)Ni surface and therefore readily forms[13]. For immediate times, the oxidation is dominated by the growth of the grains, namely the close packed (111) grains. For longer times, the flux of Ni cations moving through the scales of NiO is reduced considerably because the oxide is far too thick. Then, the growth and lateral spread of grains at the surface takes over namely growth of (100)NiO grains. These grow laterally along fast <111> directions, consuming (111)NiO grains.

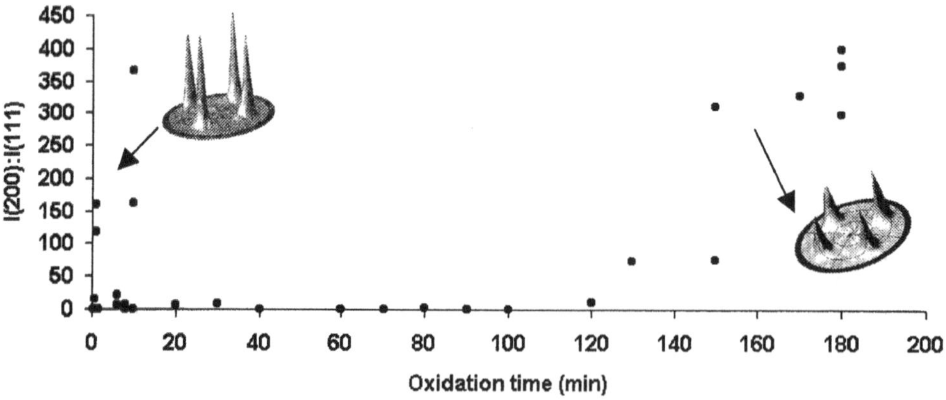

FIG. 3. I(200)/I(111)NiO versus oxidation time (min). Inset is (111) NiO pole figure showing sharp in-plane texture.

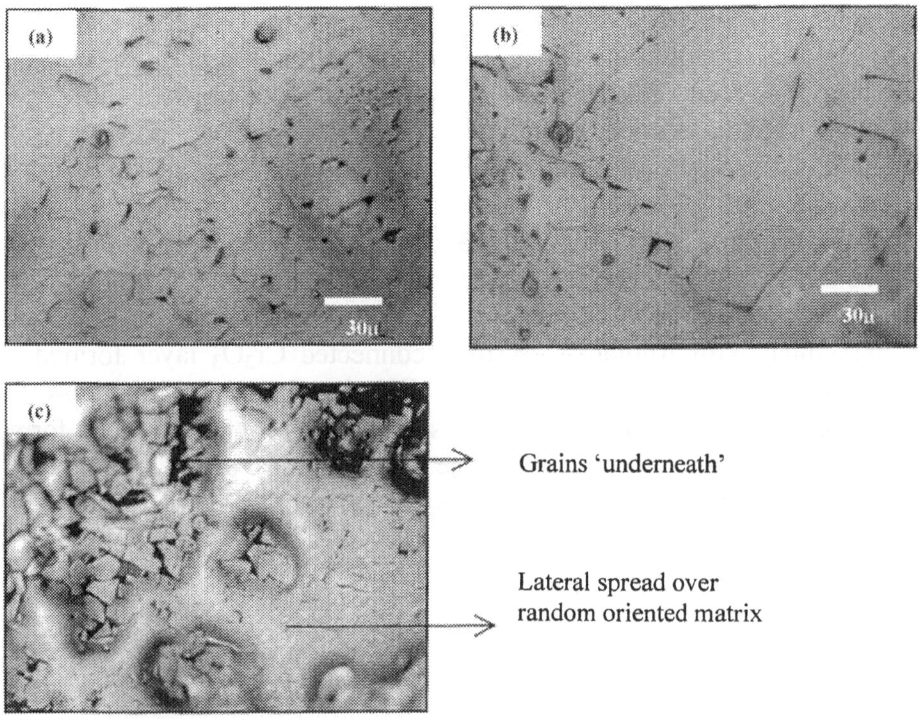

FIG. 4. SEM micrographs of (a) 2 min (b) 180 min (c) 180 min oxidised Ni showing abnormal grains

Surface microstructures of the textured oxide are shown in Fig. 4. Fig. 4-c shows the lateral oxide growth confirming our conclusion on the mechanism of texture evolution of the NiO. For a well textured NiO normally, large flat grains growing in a matrix of randomly oriented grains (for short oxidation time oxide) or square grains which coalesce together to form flat surface for oxide formed after 150 min of oxidation. The root mean square (rms), roughness of the oxide is roughly 150nm for the flat surface oxide. The formation of the large highly oriented grains at a very narrow temperature range and for specific time is concluded to result from the formation of abnormal grains growth of surface grains.

2. SOE of RABiTS Ni-alloys: NiO is always the first oxide to be formed on Ni-Cr alloys[14] thermal oxidation of Ni-Cr at high pO_2 (> 10^{-10} atm at 1000°C). On Ni-Fe, mixed oxides are formed at high temperature oxidation mainly $NiFe_2O_4$.

2a) SOE of Ni-10% and 13% Cr: For Ni-10%Cr and 13%Cr, oxidation was done isothermally at 1050°C as per earlier Ni-superalloy oxidation studies[15]

for different times. The out of plane texture was similar to that shown in Fig. 3 for pure Ni. For both alloys, the optimum out-of-plane texture of cube oriented NiO was obtained for shortest oxidation time of few minutes, although a strong texture still remained after longer time. For an oxidation times of 1 min < t < 40 min, the NiO layer showed both 0° and 45° in-plane orientation with respect to the underlying oxide. For an optimum oxidation time, the NiO thickness was ~ 5μm and was sharply textured both in and out-of-plane, as shown in Fig. 5. Formation of Cr_2O_3 plays an important role in the scale behaviour on Ni-Cr alloys. NiO always forms the outer oxide layer in agreement with Ahmad[14]. A fully connected Cr_2O_3 layer formed at the interface of NiO and Ni acted as a diffusion barrier to suppress further oxidation. The textured NiO is rather smooth (rms ~ 200nm) when formed at short oxidation time. A typical surface microstructure of the high quality NiO is shown in Fig. 6-a and 6-b is the cross section of oxides on Ni-Cr.

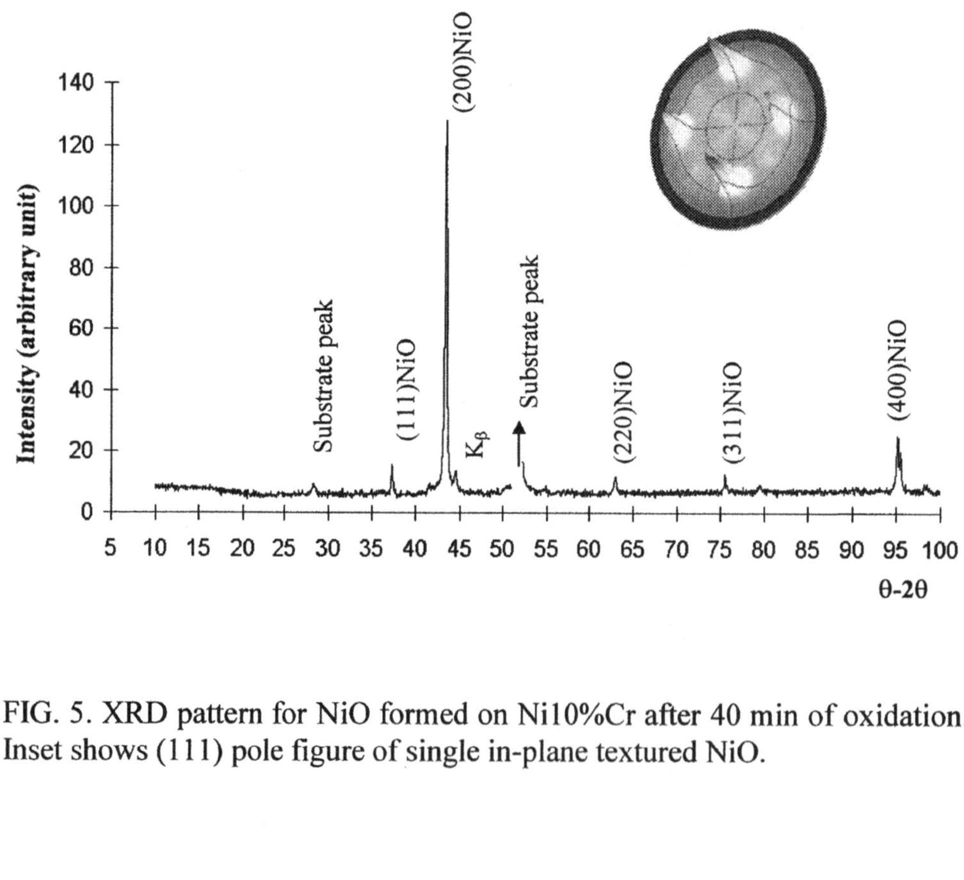

FIG. 5. XRD pattern for NiO formed on Ni10%Cr after 40 min of oxidation. Inset shows (111) pole figure of single in-plane textured NiO.

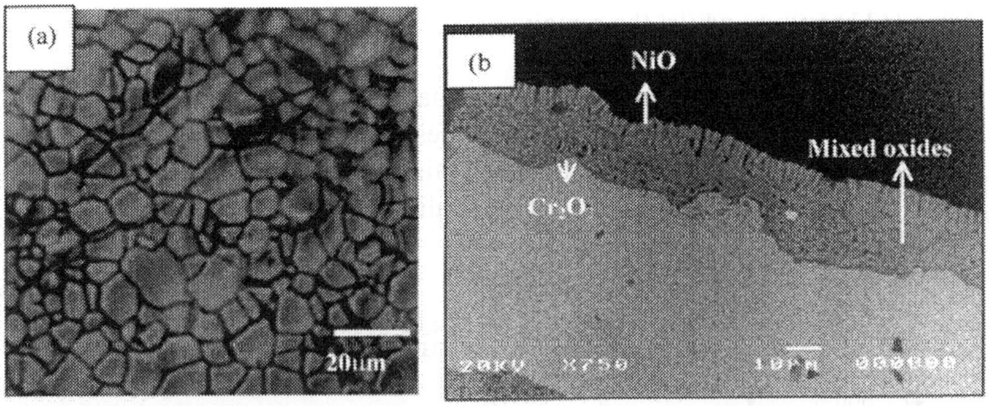

FIG. 6. (a) Optical surface micrograph of NiO formed on Ni-10%Cr. (b) Cross section of NiO on Ni-Cr.

2b) SOE of NiFe 50/50 Tape: Fig. 7 shows x-ray traces for the bare Ni-Fe substrate and a typical sample oxidised at around 850°C for 60 minutes. The oxide is cubic and the lattice parameter is calculated as a=(8.352±0.010) Å. The important thing to note about the texture of the oxide is that it is a combination of (100) and (111), rather than being randomly oriented. The (311) and (440) peaks which have high intensities in the powder pattern do not appear in the θ-2θ scan. It is likely that the oxide is neither entirely $NiFe_2O_4$ nor Fe_3O_4, but has some intermediate composition, best described as $(Ni_xFe_{1-x})^{2+}Fe^{3+}{}_2O_4$. This oxide has a spinel structure. A further feature of Fig. 7 is the fact that after oxidation, a large shoulder is produced on the high angle side of the (200) NiFe peak. This indicates that there is an excess of Ni over Fe in the remaining alloy, consistent with the formation of the iron-rich oxide detailed above. Although a mixed (111) and (100) layer is evident, the degree of (100) nature varies with temperature, with no (100) oriented material at low temperatures, a maximum around 800°C, and then less at higher temperatures.

In order to study a layered structure of the oxide, a sample was intentionally made to spall after the oxidation. This is done by rapidly quenching the oxidising tape by turning off the heating current suddenly. Complete spalling happens reliably only if the oxide layer is sufficiently thick, which is probably associated with the elastic strain produced in the film due to a large volume change. As the oxide spalls in large pieces, it is possible to x-Ray both the oxide flakes and the underlying tape. These θ-2θ scans are shown in figure 1a and b and clearly show that the oxide which was removed has a (111) texture whilst the remaining tape is covered with an oxide which is (100) oriented.

This fact that a thin (100) layer remains on the surface may be utilised to consistently produce cube textured oxides on the surface of the Ni-Fe tape. If the tapes are further oxidised they maintain the cube texture, and the rocking curve of Fig. 8 demonstrates the excellent alignment.

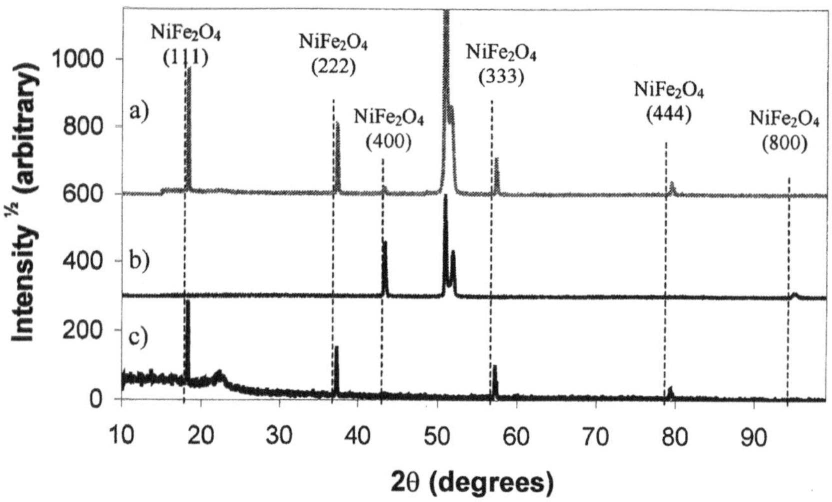

FIG. 7. The (111) oxide (c) may be completely removed from the oxidised tape (a) to leave behind a NiFe tape which has a cube textured oxide layer (b).

The atomic force microscope images of Fig. 9. indicate that the oxide layer is rough, which could be a problem for subsequent superconductor deposition. The (111) layer which is produced initially is very rough, but when removed, it leaves a much smoother cube textured oxide layer. Unfortunately, the re-oxidation process results in a rough surface again.

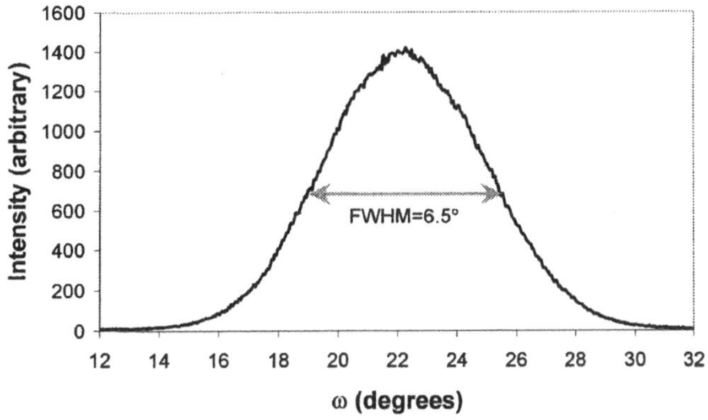

FIG. 8. Figure 2 {400} rocking curve showing excellent alignment after re-oxidation of Ni-Fe tape

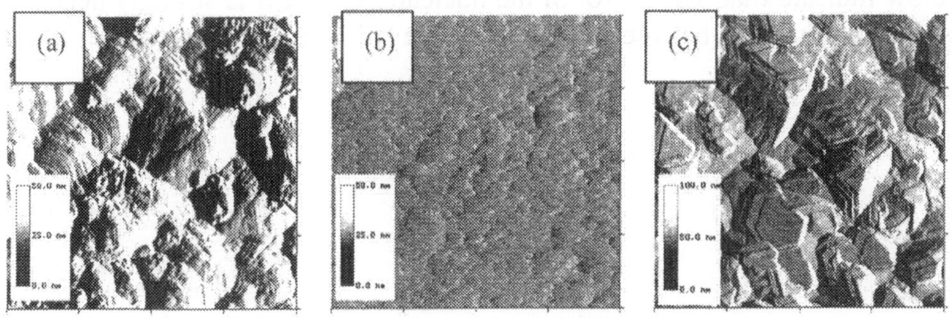

FIG. 9. Atomic force micrographs of a 10 x 10 μm^2 region of the surfaces of the oxide layers. a) is the (111) oxide, b) the surface (1000) oxide and (c) the thicker (100) oxide

3a. Liquid phase processing of $Nd_{2-x}Ce_xCuO_4$ on Ni / NiO: For the film growth of $Nd_{2-x}Ce_xCuO_4$ a precursor with a composition of 50 wt% CuO and 50 wt% (Nd_2O_3 + CeO_2) was used. The material was printed on the oxidized Ni tape and heated up to a temperature of 1300°C in order to melt this composition completely according to the phase diagram[16] .The samples were kept on this temperature for about 10 min and than fast cooled down to a temperature of about 1240°C to initiate the nucleation. After keeping it for several minutes at these temperature the sample was further cooled down slowly with a rate between 5 and 10°/min leading to further growth of $Nd_{2-x}Ce_xCuO_4$ on the substrate. Optical microscopy showed large grains with a smooth surface (Fig. 10). X-ray diffraction measurement revealed a almost completely c-axis growth of the $Nd_{2-x}Ce_xCuO_4$ and a decrease of the out-of-plane alignment to a value of about 3° FWHM in comparison to 5° of the NiO layer. However texture measurements showed a variety of in-plane orientations resulting from large grains with single crystalline quality rotated around the substrate normal. These results combined with those in section 3b below indicate that the control of the nucleation process is strongly dependent on lattice mismatch between the NiO and the $Nd_{2-x}Ce_xCuO_4$ layers.

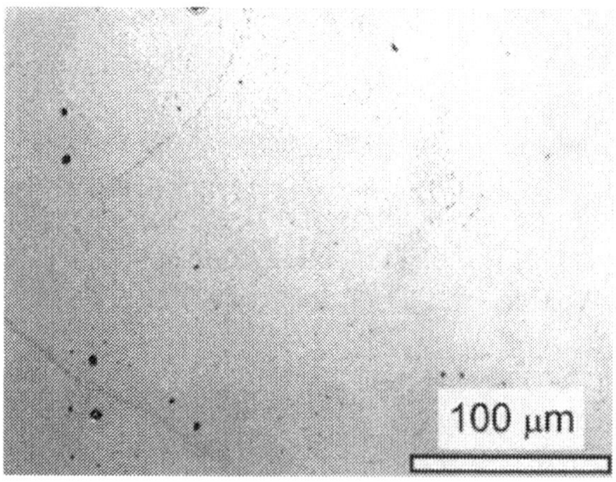

FIG. 10. Optical image of a $Nd_{2-x}Ce_xCuO_4$ film grown by liquid phase epitaxy on surface oxidized biaxial textured Ni tapes showing a flat surface

3b. Liquid phase processing of $Nd_{2-x}Ce_xCuO_4$ on Ni / NiO: The $Nd_{2-x}Ce_xCuO_4$ (x=0-0.15) buffer layer on top of NiO was grown from liquid phase using a number of liquid-phase processing (LPP) methods. The first method used was the traditional top dipping LPE growth. The composition of the high temperature solution for the LPE growth was 20%Nd_2O_3:80%CuO. The liquid was prepared at 1200°C and the LPE growth was carried out at about 1150 °C. In a second method, a precursor with composition around 30%Nd_2O_3:70%CuO was screen-printed on the substrates using an organic binder and then heated up to 1200 °C to reach a single liquid state. This temperature was kept for about 20 minutes and then cooled down to room temperature at 300 °C/hr. Because of the large surface area to volume ratio and the faster evaporation of CuO than Nd_2O_3, the liquid became supersaturated rapidly, leading to the nucleation and growth of Nd_2CuO_4 on the substrates.

In the third method, a growth system similar to the traditional zone-refining furnace was used. A hydraulically pressed and sintered source rod of 40%Nd_2O_3:60%CuO was fed into the focus of an infrared beam to produce a molten zone at one end. When the increasing size of the molten zone exceeded the limit of the surface tension, the liquid dropped down onto the substrate below, which was maintained at a suitable temperature (*ca* 1100-1200 °C) for the epitaxial growth induced from the highly textured substrate. Because of the extremely good wetting property of the high-temperature cuprate solution, the liquid drops quickly spread out on the substrates.

A XRD pole figure of $Nd_{2-x}Ce_xCuO_4$ (x=0 for this sample) grown on NiO/Ni is shown in Fig.11, which shows an every good biaxial texture with FWHM=5°. There was a 45° angle between the *a/b*-axes of the Nd_2CuO_4 and NiO, indicating that the epitaxial relation between this two compounds was [100]//[110]. The lattice parameter of the *a/b*-axis was about 0.394nm for Nd_2CuO_4 and 0.417 nm for NiO, therefore, 3 unit cells of Nd_2CuO_4 could grow along two units of the [110] axis of NiO with lattice mismatch as small as 0.3%. A cross-sectional SEM of the Nd_2CuO_4/NiO/Ni structure is shown in Fig.12. The interface between the Nd_2CuO_4 and NiO was quite smooth and Nd_2CuO_4 could grow into all the groves between NiO grains. Therefore, LPP is a good method to grow $Nd_{2-x}Ce_xCuO_4$ on the relatively rough surface of NiO.

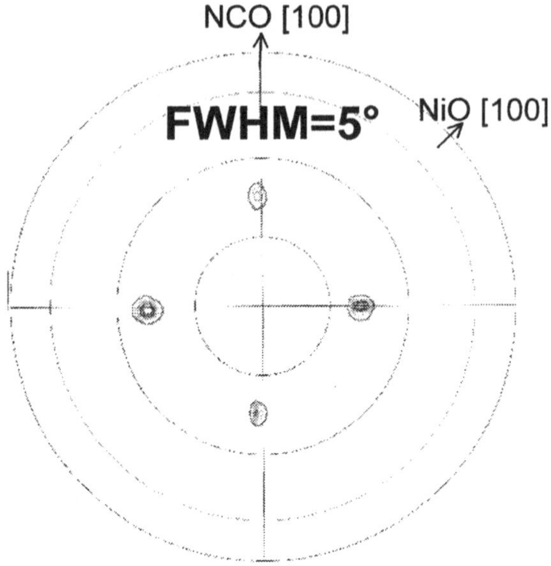

FIG. 11 Fig.11 (103) pole figure of Nd2CuO4 on NiO/Ni

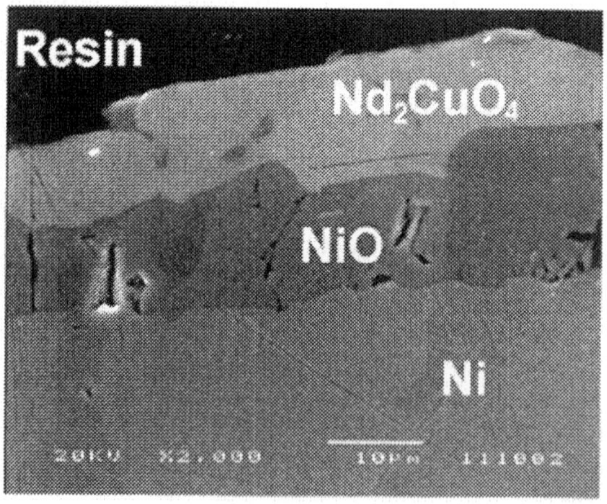

FIG. 12. SEM image of cross-section

Conclusions

We report on basic studies of a new coated conductor architectures: Ni/NiO/$Nd_{2-x}Ce_xCuO_4$/YBCO. Thin, highly textured and smooth NiO layers could be formed on pure Ni, Ni-Cr and Ni-Fe alloys. A smooth dense $Nd_{2-x}Ce_xCuO_4$ was formed on the surface of the NiO.

Acknowledgements

This research was funded by the EPSRC.

References

[1] X. Qi and J.L. MacManus-Driscoll; Journal of Crystal Growth, 213,(2000), 312

[2] X. Qi, M. Soorie, Z. Lockman, J. L. MacManus-Driscoll, Journal of Materials Research 17 (1) (2002) 1-4.

[3]E D Specht, A Goyal, D F Lee, F A List, D M Kroeger, M Paranthaman, R K Williams and D K Christen, Supercond. Sci. Technol. **11** No 10, (1998) 945-949

[4] W. Goldacker and R.Nast, private communication, (2001)

[5] A. Tuissi and E. Villa, CNR-TEMPE, Corso Promessi Sposi 29, 23900 Lecco, Italy

[6] J. Knauf, R. Semerad, W. Prusseit, B. DeBoer, J. Eickemeyer, IEEE Trans. on App. Supercon. 11, (2001), 2885 -2888

[7] Z. Lockman, X. Qi, A. Berenov, W. Goldacker, R. Nast, B. deBoer, B. Holpfapzel, & J. L. MacManus-Driscoll, Surface Oxidation of Cube-textured Ni-Cr for the formation of a NiO Buffer Layer for Superconducting Coated Conductor, (2002) to be published in Physica C

[8] P. Kofstad, in Nonstoichiometry, Diffusion, and Electrical Conductivity in Binary metal Oxides, Wiley-Interscience, N. York 1972.

[9] Z. Lockman, W. Goldacker, R. Nast, B. de Boer and J.L. MacManus-Driscoll, IEEE Trans. Appl. Supercond. 11, (2001), 3325.

[10] A. Kursumovic, R. Tomov, R. Hühne, B.A. Glowacki, J.E. Evetts, A. Tuissi, E. Villa and B. Holzapfel' submitted to Physica C

[11] K. Matsumoto, S-B. Kim J-G. Wen and I. Hyrabayashi, IEEE Trans Appl. Supercon., 9, (1999), 1539.

[12] Z. Lockman, W. Goldacker, R. Nast, B. de Boer and J.L. MacManus-Driscoll, Physica C, 351, 1, (2001), 34-37

[13] N.N. Khoi, W.W. Smeltzer and J.D. Embury, " Growth and Structure of NiO on Ni Crystal Faces", Journal of Electrochemical Society, 122 (11) (1975) 1495-1502.

[14] B. Ahmad and P. Fox, Oxidation of Metals 52 (1/2) (1999) 113 – 138 and G. Calvarin, R. Molins, and A.M. Huntz, "Oxidation Mechanism of Ni-20Cr Foils and Its Relation to the Oxide- Scale Microstructure", Oxidation of Metals 53 (1/2) (2000) 25-48.

[15] G. C. Wood and T. Hodgkiess, "Characteristic Scales on Pure Nickel-Chromium Alloys at 800-1200°C, Journal of the Electrochemical Society, 113 no. 4 (1966) 319-327

[16] Tanaka, I., Komai, N. and Kojima, H., Physica C, 190 (1990), 112-113

YBCO/YSZ/HASTELLOY SUPERCONDUCTING TAPES BY IBAD MAGNETRON DEPOSITION

S. Gnanarajan and N. Savvides
CSIRO Telecommunications and Industrial Physics
Bradield Road, WestLindfield NSW 2070
Australia

ABSRACT

Superconducting YBCO/YSZ/hastelloy tapes were fabricated by depositing epitaxial YBCO films on biaxially aligned YSZ layers on polished hastelloy substrates. YSZ buffer layers were deposited by ion beam assisted magnetron deposition. The degree of biaxial alignment in YSZ and YBCO films was determined for x-ray pole figures and phiscans. YSZ layers of different thicknesses reveal biaxial alignment develops at thickness as small as 100 nm and enhanced biaxial alignment was observed in homoepitaxial YSZ layers. The best YBCO films had biaxial alignment phiscan FWHM of 9°–10° and critical current density, J_c(77 K, 0 T) = (0.9 – 1.2) MA cm^{-2}.

INTRODUCTION

Ion beam assisted deposition (IBAD) is well known to influence the growth and properties of thin films. Recently IBAD techniques have been used to deposit biaxially aligned oxide films on technical substrates to be used as lattice matched substrates or templates for subsequent epitaxial growth [1,2]. A major driving force for this work is to develop commercially viable and scalable process to fabricate long lengths of high-current superconductor tapes that use heat-resistant metal substrates[3-6]. One such process makes use of biaxially aligned oxide buffer layers on hastelloy metal tape to fabricate $YBa_2Cu_3O_7$ (YBCO) tapes. Suitable oxide buffer layers include yttria stabilized zirconia (YSZ) [1-3, 7], cerium oxide (CeO_2) [5, 8], magnesium oxide (MgO) [6] and praseodymium oxide Pr_6O_{11} [9]. The buffer layer serves both as a template and as a barrier to diffusion metal species during the high-temperature deposition.

The critical current density, J_c, of YBCO tapes is strongly influenced by the degree of the biaxial alignment of the buffer layer. In general the crystalline quality of the buffer layer must be similar to a single crystal so that the buffer layer has pure [001] axis alignment (normal to the substrate plane) and excellent in-plane alignment ([010] and [100] axes in the substrate plane). The in-plane alignment is usually assessed by determining the full width at half maximum (FWHM) $\Delta\phi$ of x-ray ϕ-scans [1-4]. Previous studies have shown short lengths of YBCO tapes having a high J_c (~10^6 A/cm^2) suitable for commercial applications can be produced provided the biaxially aligned oxide buffer layer is highly aligned with $\Delta\phi < 14^o$ [2].

The basic IBAD technique involves a physical vapor source which provide the depositing species and a Kaufman type ion beam source which is directed towards the growing film at a specific angle of 55° to the normal of the film plane [1,2, 7-12]. The ion bombardment causes [001] axis growth normal to the film plane and also induces in plane alignment of the [010] and [100] axes.

Iijima.et.al [11] and Freyhardt et.al[12] have investigated the development of the biaxial alignment of the IBAD YSZ with increasing film thickness and showed that it is difficult to achieve $\Delta\phi < 10^o$ in relatively thin buffer layers by simple IBAD techniques. For YBCO/YSZ/hastelloy tapes the YBCO film is grown epitaxially over the YSZ buffer layer. Thus the degree of biaxial alignment imparted to the YBCO film is determined primarily by the biaxial alignment at the surface of the YSZ buffer layer. We have investigated the development of the biaxial alignment of IBAD YSZ buffer layer as a function of thickness and developed a novel scheme whereby an additional YSZ layer is grown epitaxially over the IBAD-YSZ layer to form a bi-layer buffer architecture (i.e. epi-YSZ/IBAD-YSZ) on hastelly substrates. Thin films of YBCO were grown onto the buffered substrates to form YBCO tapes. Here we report measurements of the degree of biaxial alignment as function of the thickness of the IBAD and epi YSZ buffer layers and show that the epi-YSZ/IBAD-YSZ/hastelloy architecture leads to YBCO tapes that have enhanced biaxial alignment with $\Delta\phi < 14^o$.

EXPERIMENTAL

The IBAD system consisted of an UHV chamber fitted with an unbalanced magnetron sputter gun to provide the physical vapour source and a 30 mm diameter Kaufman-type ion beam source to bombard the growing YSZ film with energetic Ar^+ ion [7, 14]. The ion beam source was operated using pure Ar gas (2 sccm) and its collimated beam was directed at an angle of 55° to the normal of the substrate plane to bombard the film with Ar^+. The ion energy and arrival rate ratio of ions to atoms were varied in the ranges 300-400 eVand 0.6–0.8 to achieve optimum film properties.

The planar magnetron was fitted with crystalline YSZ target and operated at 200 W rf power (13.56 MHz). The discharge gas species was Ar/O2 mixture (mass flow ratio 50:1) at a pressure of 0.1 Pa. The hastelloy substrates were held onto a high-temperature metal block substrate heater, which located opposite the target at a distance of 50 mm.

Both IBAD and epi YSZ layers were deposited under the same magnetron discharge conditions. The epi-YSZ layers were grown at a rate of 12 nm/min. During IBAD a significant amount of condensable material is re-sputtered by the ion beam. Consequently the deposition rate for the IBAD-YSZ layers was reduced to approximately 6 nm/min. Detailed investigations to assess the quality of YSZ layers and their suitability for YBCO tape production established that YSZ films free of micro cracks and pin holes that retained their integrity during subsequent processing into YBCO tapes could be produced at temperatures in the range of 300 – 500°C. In this study we deposited the IBAD-YSZ at 300°C and epi-YSZ over layer at 500°C. Note that during the epi-YSZ growth the ion beam is switched off. To enhance the adatom mobility of the condensing species we applied a small negative dc bias (20-50 V) to the hastelloy substrate.

The superconducting YBCO films (300 nm thick) were deposited onto the buffered hastelloy substrates at 730°C in a separate chamber. We used a non ion assisting unbalanced magnetron to sputter stoichiometric YBCO target in an Ar/O2 gas mixture at a pressure of 50 Pa [14]. This method generally yields YBCO films on single crystal MgO(100) and SrTiO3(100) substrates which have $Jc = (2\text{-}4) \times 10^6$ A/cm^2 [14].

The crystalline structure and orientation of all the films were determined by x-ray θ-2θ diffraction and their in-plane biaxial alignment was determined by X-ray phi scan measurements using Philips X'pert diffraction system.

RESULTS AND DISCUSSION

IBAD YSZ films

The biaxial alignment of the IBAD YSZ films was investigated with increasing film thickness. The thickness of the films varied from 100 to 900 nm. The in-plane biaxial alignment of the films was studied using YSZ(111) phi -scans (Fig.1). The phi-scan of the film with thickness of 100 nm indicates it has biaxial alignment with FWHM of about 37.8°. Fig.1 shows the biaxial alignment increases rapidly with increasing thickness up to about 300 nm to 17° FWHM and above 300 nm, biaxial alignment of the films increases slowly to 15.4° FWHM at 900 nm. The variation of FWHM with thickness found in our experiments were similar to the results reported by Iijima.et.al [11] and Freyhadt.et.al [12], however in our films the biaxial

Figure I. YSZ(111) ϕ - scans with increasing IBAD YSZ film thickness

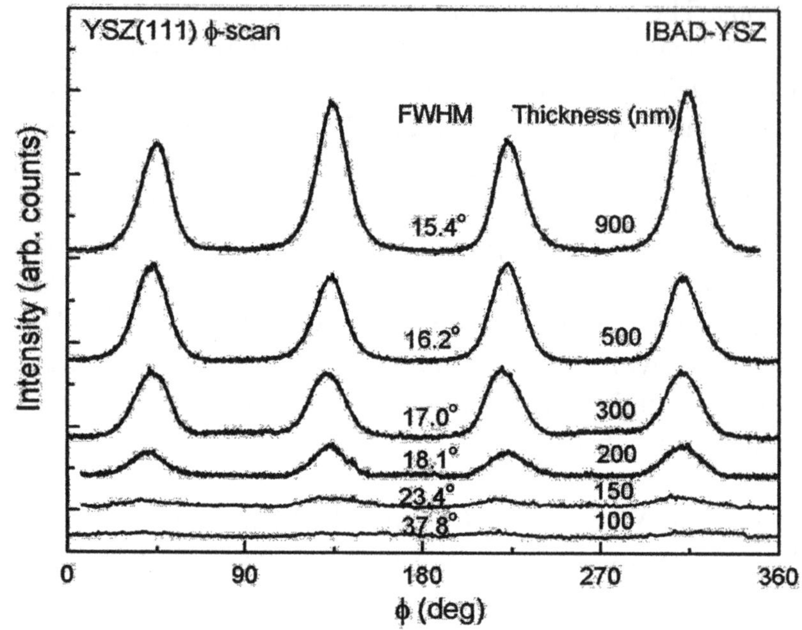

Figure II. YBCO(103) ϕ - scans with increasing IBAD YSZ film thickness

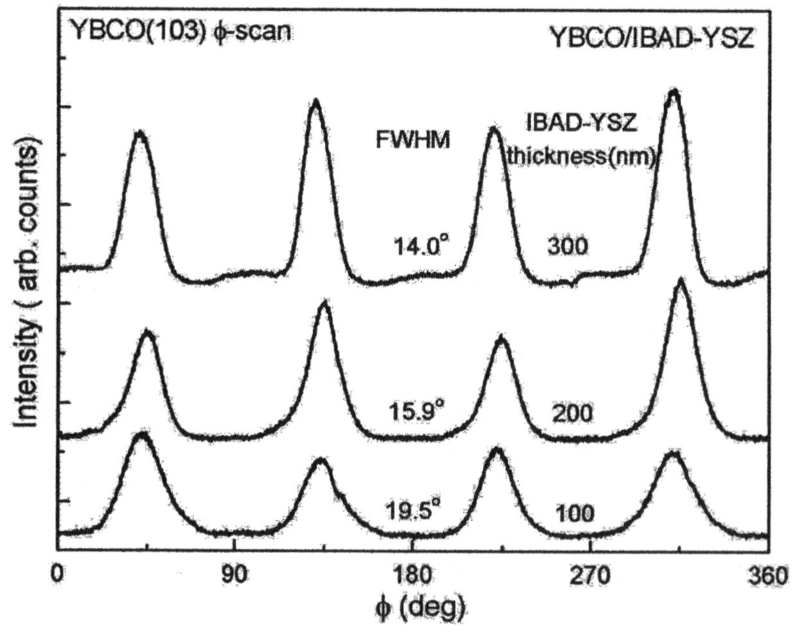

alignment increases more rapidly with increasing thickness below 300 nm. The phi-scan measurements mentioned above can only be considered as the mean biaxial alignment over the whole YSZ film rather than the biaxial alignment on the surface of the film.

To determine the biaxial alignment on the surface the IBAD YSZ films, we deposited YBCO films on these substrates. The biaxial texture of the YBCO film is considered as the biaxial texture of the YSZ film on the surface. The biaxial alignment of the YBCO films were determined using YBCO(103) phi-scans. The YBCO(103) phi-scans of the YBCO films on the 100nm, 200nm and 300nm IBAD YSZ films were shown in Fig.2. The YBCO film on 100 nm thick YSZ film had YBCO(103) phi-scan FWHM of 19.5°. The biaxial alignment of the YBCO films on the IBAD YSZ films increases with increasing thickness of the IBAD YSZ films up to about 300 nm to FWHM of 14° and saturates for further increase in thickness. It is clear that there is little improvement in texture above 300 nm. Hence for films above 300nm thickness IBAD YSZ is essentially growing epitaxially.

Homoepitaxial YSZ/IBAD YSZ films

The epi YSZ layers were grown on 300 nm IBAD YSZ films. The biaxial alignment of the epi YSZ/IBAD YSZ films was investigated as a function of epi layer thickness of 50 – 400 nm. For the 50nm epi YSZ film on IBAD YSZ film, YSZ(111) phi-scan FWHM was about 16.6° (Fig.3). The biaxial alignment improved as the thickness of the epi layer increased up to 200 nm and no further improvement was observed above 200 nm. The biaxial alignment of the 200 nm epi YSZ/300nm IBAD YSZ layers have FWHM of 13.4° which is slightly better than the FWHM of the YBCO film (14°) of the thickest IBAD YSZ layer.

The YBCO films were deposited on the epi YSZ layers to determine the biaxial alignment on the surface of the epi YSZ films. Again the biaxial alignment of the YBCO films were determined by the YBCO(103) phiscans (Fig.4). As expected the biaxial alignment of the YBCO film on 50 nm epi YSZ layer is as good as the biaxial alignment of the YBCO film on the thickest IBAD YSZ film. The YBCO layer on the 200 nm epi YSZ film has FWHM of less than 9.1°. The YBCO films with biaxial alignment phiscan FWHM of 9°–10° had critical current density, J_c(77 K, 0 T) = (0.9 – 1.2) MA cm^{-2}.

Figure III. YSZ(111) ϕ - scans with inceasing epi YSZ film thickness

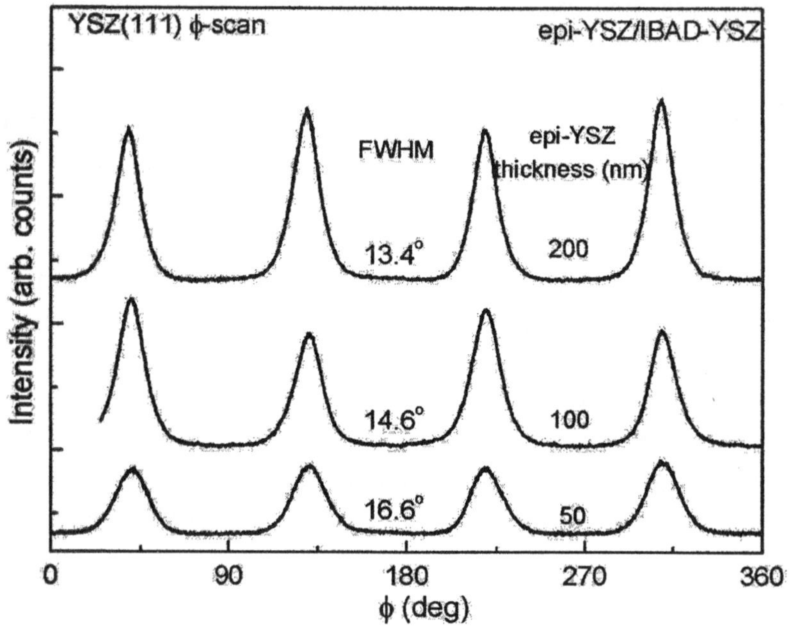

Figure IV. YBCO(103) ϕ - scans for increasing epi-YSZ film thickness

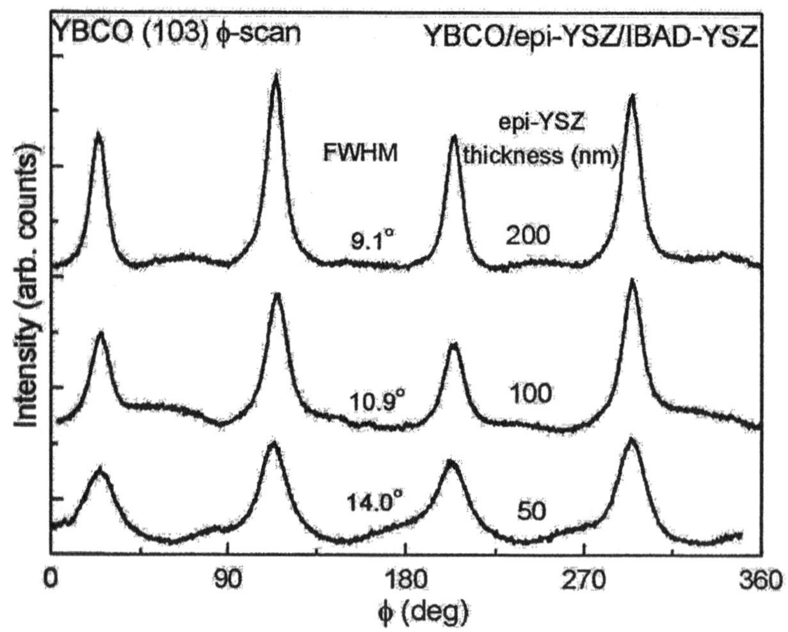

Fig.5 shows FWHMs of the all the films plotted against the YSZ film thickness. It is clear there is significant increase in the biaxial alignment on the surface of the typical epi YSZ layer (FWHM = 9.1°) in comparison to the typical surface of the IBAD YSZ layers (FWHM = 14°. This process not only increases the critical current density in the YBCO film by a factor of two but also decreases the time needed to deposit the buffer layer.

Various mechanisms have been proposed to explain IBAD induced biaxial alignment in thin films. Ressler.et.al [15] mechanism is based on ion induced damage to the top layers of the film due to Ar^+ ion penetration into the top few unit cells. According to Iijima.et.al [11] after the early stage of growth the orientations in the directions of [010] and [100] develop by collaboration of selective growth of top layers and homoepitaxy on grains in the film which were already biaxially aligned. The enhancement in biaxial alignment observed in our results due to homoepitaxial growth appears to be consistent with the above models. The homoepitaxy seems to occur on the good biaxially aligned grains and it also heals the damage done to the top layer of the film by the Ar^+ ions, which leads to the enhanced biaxial alignment for the YBCO films grown on the homoepitaxial YSZ layer.

Figure V. YSZ thickness vs ϕ - scan FWHM for IBAD YSZ, epi YSZ and YBCO films

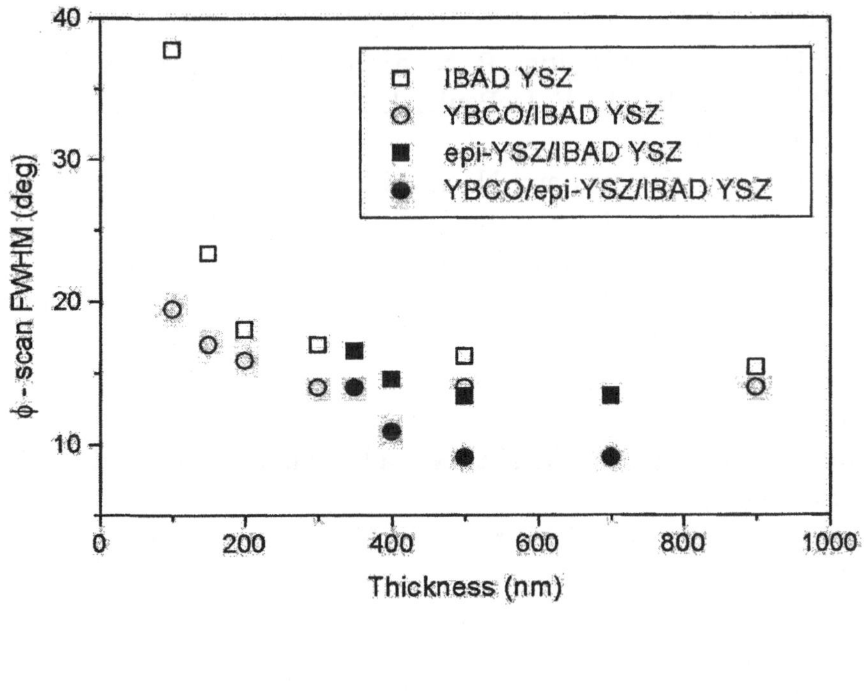

CONCLUSIONS

Biaxial alignment of ion beam assisted magnetron deposited YSZ films were investigated as a function thickness. The biaxial alignment of the YBCO film on the 100 nm thick IBAD YSZ film had FWHM of 19.5°. The improvement in the biaxial alignment of the YBCO film on IBAD YSZ film saturates to a value of about 14° at about 300 nm IBAD YSZ thickness

YBCO films grown on epi-YSZ/IBAD YSZ films develop better biaxial alignment than the YBCO films grown on IBAD YSZ films. The enhancement increases with increasing homoepitaxial layer thickness up to about 200 nm with FWHM of about 9°. The YBCO films with biaxial alignment phiscan FWHM of 9°–10° had critical current density, J_c(77 K, 0 T) = (0.9 – 1.2) MA cm^{-2}.

REFERENCES

[1]Y. Iijima, N. Tanabe,O. Kohno and Y. Ikeno, Appl. Phys. Lett. **60** (1992) 769.

[2]X. D. Wu, S. R. Foltyn, P. Arendt, J. Townsend, C. Adams, I. H. Campell, P. Tiwari, Y. Coulter and D. E. Peterson, Appl. Phys. Lett. **65** (1994) 1961.

[3]N. Savvides, S. Gnanarajan, J. Herrmann, A. Thorley, A. Katsaros and A. Molodyk, Mat. Res. Soc. Symp. Proc. **616** (2000) 199.

[4]Y. Iijima and K.Matsumoto, Supercond. Sci. Technol. **13** (2000) 68.

[5]J. E. Mathis, A. Goyal, D. F. Lee, F. A. List, M. Paranthaman, D. K. Christen, E. D. Spect, D. M. Kroger and P. M. Martin, Jpn. J. Appl. Phys **37** (1998) 1379.9

[6]T. G. Holesinger, S. R. Foltyn, P. N. Arendt, H. Kung, Q. X. Jia, R. M. Dickerson, P. C. Dowden, R. F. DePaula, J. R. Groves and J. Y. Coulter, J. Mater. Res. **15** (2000) 1110

[7]S. Gnanarjan, A. Katsaros and N. Savvides, Appl. Phys. Lett. **70** (1997) 2816.

[8]S. Gnanarajan and N. Savvides, Thin Solid Films **350** (1999) 124.

[9]C. P Wang, K. B. Do, M. R. Beasley, T. H. Geeballe and R. H Hammond, Appl. Phys. Lett. **71** (1997) 2995.

[10]V. Betz, B. Holzapfel, D. Raouser and L. Shultz, Appl. Phys. Lett. **71** (1997) 2952.

[11]Y. Iijima, M. Hosaka, N. Tanabe, N. Sadakata, T. Saitoh and O. Kohno, J. Mater. Res. **13** (1998) 3106.

[12]J. wiesmann, J. Dzick, J. Hoffmann, K. Heinemann and H. C. Freyhardt, J. Mater. Res. **13** (1998) 3149.

[13]N. Savvides and A. Katsaros, Appl. Phys. Lett. **62** (1993) 528.

[14]N. Savvides and A. Katsaros, Thin Solid Films **228** (1993) 182.

[15]K. G. Ressler, N. Sonnenberg, M. J. Cima, J. Am Ceram. Soc. **80** (1997) 2637.

GROWTH KINETICS AND TEXTURE OF $(Nd,Ce)_2CuO_4$ / NiO BUFFERS ON Ni-BASED RABITS

Z. Lockman, X. Qi and
J.L. MacManus-Driscoll
Department of Materials
Imperial College of Science Technology and Medicine
Prince Consort Road
London, SW7 2BP, UK

W. Goldacker and R. Nast
Forschungszentrum Karlsruhe
Technik und Umwelt
Hermann-v-Helmholtz-Platz 1
76344 Eggenstein-Leopoldshafen
Karlsruhe, Germany

B. deBoer
Institut für Festkörper-und
Werkstoffforschung, Helmholtzstraße 20
Dresden, Germany

A. Kursumovic, R. Tomov*,
R. Hühne, J.E. Evetts and
B.A. Glowacki
Department of Materials Science and Metallurgy
University of Cambridge
Pembroke Street
Cambridge CB2 3QZ, UK

R. Major
Carpenter Technology (UK) Ltd,
Napier Way,
Crawley, RH10 2RB, UK

ABSTRACT

We report on basic studies of a new coated conductor architecture: $Ni/NiO/Nd_{2-x}Ce_xCuO_4$/YBCO. Various Ni compositions were explored including pure Ni, Ni-Cr, and Ni-Fe, and a new liquid phase processing (LPP) technique to make $Nd_{2-x}Ce_xCuO_4$ (NCCO) buffer layers was utilized. Fundamental studies of surface oxidation epitaxy of Ni and Ni-alloys were undertaken. It was shown that a very thin NiO layer can be formed on pure Ni with excellent texture and low

* corresponding author: rit21@cam.ac.uk

surface roughness (~80nm). For Ni-Cr, ~5micron thick NiO surface layer with excellent texture and surface finish (roughness ~200nm) can be produced. For Ni-Fe, a smooth, thin Ni_2FeO_4 was formed underneath a surface (111) oxide, which was then successfully removed. The (NCCO) layer was fabricated by LPP to have a surface roughness of a few 10's of nm.

INTRODUCTION

Since the discovery of high T_c superconductors in 1986, there has been an enormous effort worldwide to establish a scaleable method for the fabrication of coated conductors on metallic tapes. Although short length samples with J_c over 1 MA/cm^2 have been prepared by a number of vapor deposition methods, e.g. pulsed laser deposition (PLD), they are so far failed to produce long length conductors at a reasonable cost. An alternative route to vapor deposition methods is LPE, which has been widely used for the growth of semiconductor and garnet thin films during 1970s and 1980s. The advantages of LPE compared are the very fast growth rate, typically 1-10 μm/min, and the capability to grow thick films without degrading the structural perfection[i]. However, LPE growth of REBCO thick films on metallic substrate is much more difficult. The high growth temperature largely increases the risk of reaction between the substrate and the film. The solubility of RE's in the BaO:CuO solution is very low, therefore nucleation and growth is more difficult than the traditional LPE of semiconductors and garnets. Compared to PLD and other vapor deposition methods, LPE has a much lower available supersaturation and hence a much smaller driving force for nucleation. As a consequence, a close lattice matched substrate is required. At present, most of the buffer layers were developed for the use in the vapor deposition methods and therefore are not suitable for LPE growth. Nd_2CuO_4, a tetragonal crystal with $a=b\approx0.394$ nm, is a good potential buffer for LPE. As a part of the RE-Ba-Cu-O system[2] Nd_2CuO_4 neither poisons superconductivity nor reacts strongly with the barium cuprate solution at high temperature. It has a fairly close lattice match and thermal expansion match to REBCO and NiO. In this paper, we report on the growth of $Nd_{2-x}Ce_xCuO_4$ thick films on surface oxidised Ni-alloy substrates by scaleable methods. We also report on the basics of the surface oxidation of different Ni-alloys.

EXPERIMENTAL

100μm thick pure nickel, Ni-10%Cr, Ni-13%Cr and Ni-Fe were prepared by RABiTS method[3]. A summary of all Ni and Ni-alloys used for the SOE is shown in Table I. The tape surface was finished by a standard electro-polishing or chemical etching method (5%HF). The foils were then ultrasonically cleaned in acetone and methanol.

Table I Summary of Ni and Ni-based tapes and SOE conditions

Foils	thickness (μm)	anneal.temp. (°C)	anneal.time (min)	atmosphere	heating	source
Ni	80-100	1000-1300	0.2-180	O_2/air	radiant	4, 5
Ni-10%Cr	100	1050	0.2-180	air	radiant	4, 8
Ni-13%Cr	100	1050	0.2-180	air	radiant	6
Ni-50%Fe	13-25	500-1100	0.2-60	Ar/O_2	ohmic	7, 9

The oxidation conditions of different Ni and Ni-based alloys are given in Table I. The progress of oxygenation was monitored by mass and thickness increase. Mass was measured by a precise microbalance after each partial oxidation. Thickness was measured both by a micrometer and using secondary electron microscope (SEM) or optical microscope. The texture analysis was carried out by X-ray diffraction techniques: θ-2θ scans, rocking curves and pole figures. Optical microscopy was used to monitor the development of microstructure during film growth.

In order to systematically study high temperature oxidation process of Ni-Fe, a number of Ni-Fe tapes were oxidized using O_2 and Ar/O_2 mixtures in a UHV chamber, tapes being heated resistively using a direct current. There are two methods of carrying out the oxidation, which turn out to show important differences. One way is to preheat the tape in vacuum, then introduce the Ar/O_2 gas mixture, or alternatively to introduce the required pressure of gas into the chamber, then ramp up the current. Below around 950°C a continuous oxide is grown, which is dark gray in appearance at sufficient thickness. At higher temperatures however, macroscopic oxide grains become visible on the surface due to secondary re-crystallization and abnormal grain growth.[9]

RESULTS AND DISCUSSIONS

Pure Nickel: At low temperatures <1000°C, NiO film growth is influenced by the grain boundaries in the "parent" Ni-phase which has an average grain size of about 30 μm. The growth is faster around grain boundaries creating humps on the NiO surface. The grain size of the oxide film is in the order of 1 μm. At higher temperatures, the grain size of the NiO film increases being >20 μm at 1250°C after 2h on textured Ni. The NiO growth has a specific flat surface that is a consequence of predominantly volume diffusion of growing species. At very high temperatures, >1300°C, highly concave growth around grain boundaries makes the film very rough[10].

Evolution of texture with temperature: A working window for SOE NiO/Ni was established from the ratio of XRD intensities I(200)/I(111)NiO. For samples oxidized for approximately two hours in air at different temperatures the results are shown in Fig. 1-a. As can be seen, there is a pronounced peak around 1260°C, at somewhat higher temperature than found in oxygen oxidation presumably due to using air rather than 100% oxygen[11]. As expected, a sharp in-plane alignment is achieved at the "peak" of the working window, Fig.1-b. In Fig.1-c the pole figure shows that an acceptable in-plane orientation can be achieved in the range of ΔT<20°. It should be noted, that at temperatures >1270°C, increased (220) peak intensities are present that add-up to overall deterioration of the NiO/Ni epitaxy.

The sample mass per unit area, Δm/A, for NiO/Ni films grown for two hours, both in pure oxygen and in air, is shown in Fig. 2. The sample mass increases due to formation of NiO film in the same manner as oxide thickness. These data correspond to data on epitaxial development shown in Fig. 1. The oxide film thickness was calculated here, but corresponds closely to experimentally measured values that were performed on some of the samples. As can be seen, an almost perfect epitaxy that was achieved at around 1260°C corresponds to ~25 μm thick NiO . The kinetics of NiO film growth in air at different temperatures versus time was investigated in more detail in Ref[12]. This kinetics was found to obey expected parabolic (√t) law. The activation energy calculated from the kinetic data was 220 kJ/mol for temperatures above 1100°C, corresponding to that for volume diffusion of Ni in NiO in excellent agreement with literature data. There are three distinct temperature regimes affecting the growth. At lower temperatures (<1000°C), due to boundary dominated diffusion a predominant (111)NiO orientation was found. At higher temperatures (1150-1270°C) a flat oxide surface was obtained on both textured and as-rolled Ni tapes, indicating uniform volume diffusion during the growth. However, both (111) and (100) oriented NiO grains are nucleated initially and continue to grow competing with each other. Thus lateral growth occurs by competitive growth of the (001) and (111) NiO grains.

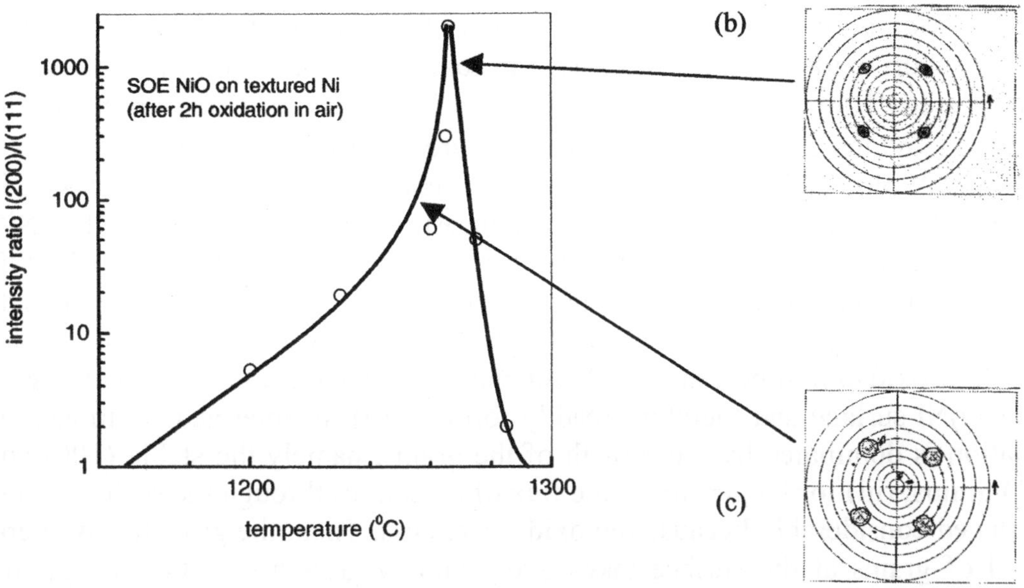

FIG. 1. (a) Effective working window for SOE NiO/Ni in air; (b) pole figure for a narrow temperature range of excellent in-plane texture, I(200)/I(111)>1000, and (c) satisfactory in-plane texture , with I(200)/I(111)>50.

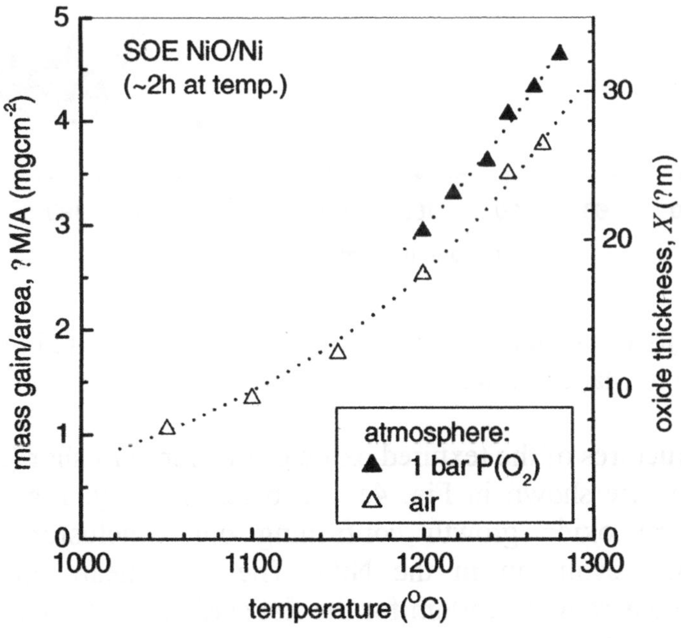

FIG. 2. Mass gain/area for NiO growth on textured Ni after ~2h oxidation in air and in pure oxygen (1 bar) atmosphere.

Evolution of texture with time: Formation of single phase (100)NiO was seen at a very short oxidation time of 0.2-10 min (oxide thickness < 6μm). We lose the cube oriented NiO when the oxide thickness is 10-30 μm. Cube oriented NiO formed again after the thickness of the scales are > 30μm i.e. after oxidation for > 150 min. A plot of I(100)/I(111)NiO versus oxidation temperature is shown in Fig. 3 with an inset of in plane texture measurements (pole figures) for NiO formed at 2 min and at 150 min. The reasons for the predominance (100)NiO texture at short oxidation times and then increasing (111) texture for intermediate times followed by increasing (100)NiO plane for much longer times is: (i) for very short oxidation times, the (100)NiO planes has the closest lattice match with the (100)Ni surface and therefore readily forms[13]; (ii) for intermediate times, the oxidation is dominated by the growth of the grains, namely the stable (100) and (111) grains; (iii) for longer times, the flux of Ni cations through the scales of NiO is reduced considerably because the oxide thickness. Then, the growth and lateral spread of grains at the surface takes over, namely growth of (100)NiO grains. These grow laterally along fast <111> directions, consuming (111)NiO grains.

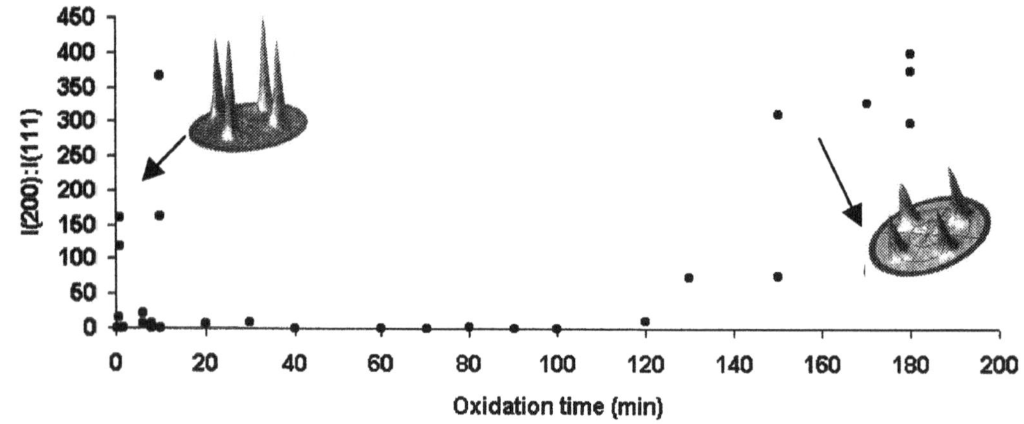

FIG. 3. I(200)/I(111)NiO versus oxidation time (min). Inset is (111) NiO pole figure showing sharp in-plane texture.

Surface microstructures of the textured oxide grown for very shor (2 mni) and long times (180 min) are shown in Fig. 4a and b respectively. Fig. 4-c shows regions of the lateral oxide growth confirming our conclusion about the mechanism of texture evolution in the NiO. The root mean square (rms), roughness of the oxide is roughly 150nm for the flat surface oxide. The formation of the large highly oriented grains at a very narrow temperature range and for specific time is concluded to result from the competitive growth of the surface grains.

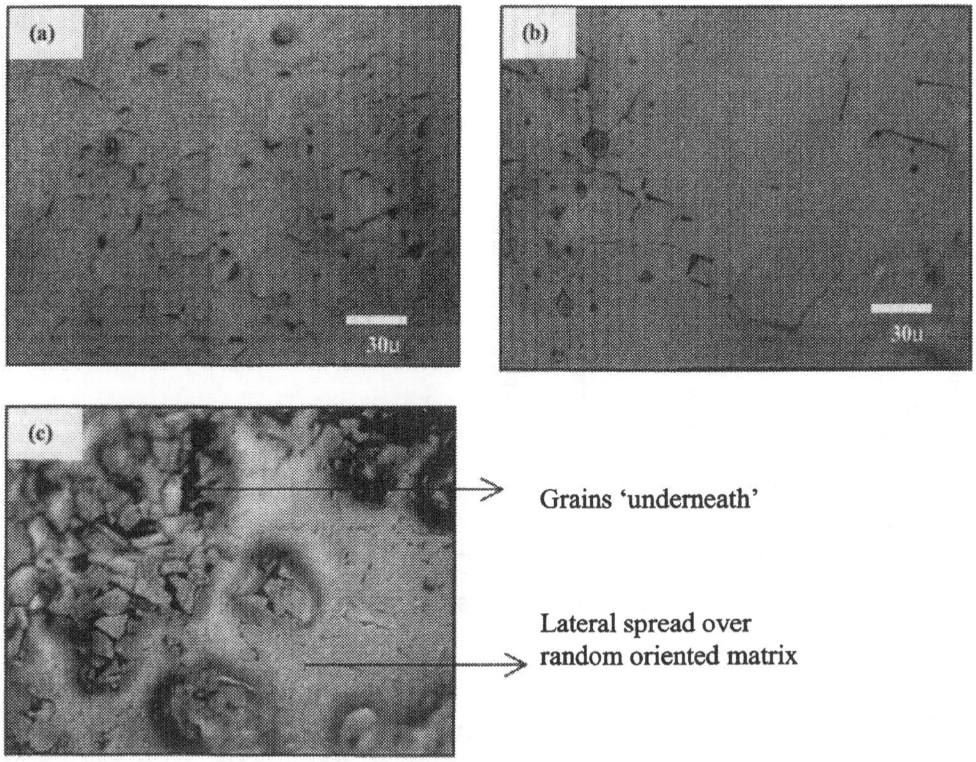

FIG. 4 SEM.micrographs of oxidized Ni (a) 2 min, (b) 180 min showing disrupted growth..

Ni-10%Cr and Ni-13%Cr: For Ni-10%Cr and 13%Cr, oxidation was done isothermally at 1050°C as per earlier Ni-superalloy oxidation studies[14]. The out-of-plane texture was similar to that shown in Fig. 3 for pure Ni. For both alloys, the optimum out-of-plane texture of cube oriented NiO was obtained for the shortest oxidation time of a few minutes, although a strong texture still remained after longer times. For an oxidation times of 1 min< t <40 min, the NiO layer showed both 0° and 45° in-plane orientations with respect to the substrate. For an optimum oxidation time, ~ 5µm thick NiO sharply textured both in and out-of-plane was produced, as shown in Fig. 5. Cr_2O_3 plays an important role in the scale behaviour on Ni-Cr alloys. NiO always forms the outer oxide layer in agreement with Ahmad[15]. A fully connected Cr_2O_3 layer formed at the interface of NiO and Ni acted as a diffusion barrier to suppress further oxidation. The textured NiO is rather smooth (rms ~ 200nm) when formed for short oxidation times. A typical surface microstructure of the high quality NiO is shown in Fig. 6-a, and 6-b is the cross section of oxides on Ni-Cr.

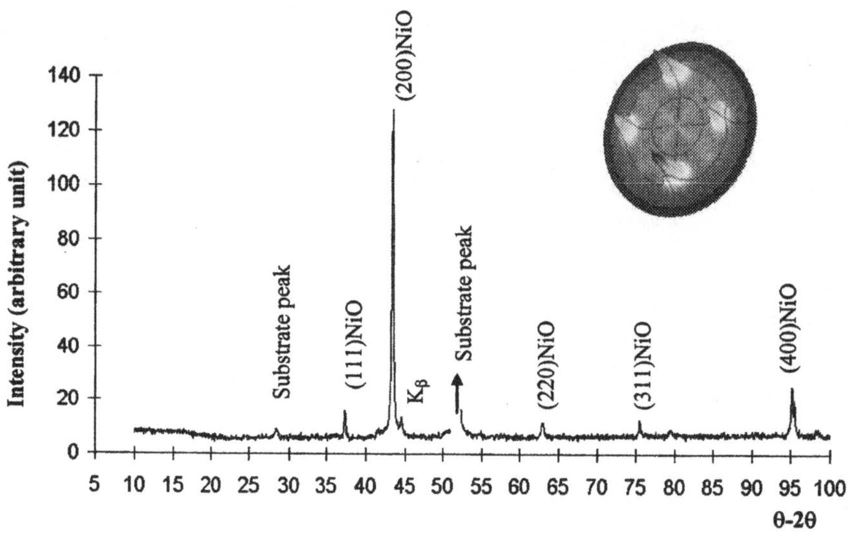

FIG. 5. XRD pattern for NiO formed on Ni10%Cr after 40 min of oxidation

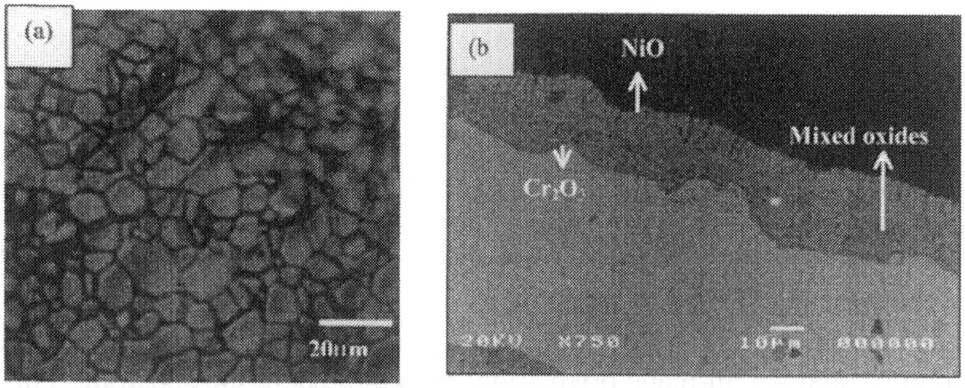

FIG. 6 (a) Optical surface micrograph of NiO formed on Ni-10%Cr. (b) Cross-section of NiO on Ni-Cr.

Ni-50%Fe tape: Fig. 7 shows x-ray traces for the bare Ni-Fe substrate and a typical sample oxidised at around 850°C for 60 minutes. The oxide is cubic and the lattice parameter is calculated as a= 8.352 Å. The important thing to note about the texture of the oxide is that it is a combination of (100) and (111), rather than being randomly oriented. The (311) and (440) peaks, which have high intensities in the powder pattern, do not appear in the θ-2θ scan. It is likely that the oxide is neither entirely $NiFe_2O_4$ nor Fe_3O_4, but has some intermediate composition, best described as $(Ni_xFe_{1-x})^{2+}Fe^{3+}{}_2O_4$. This oxide has a spinel structure. A further feature of Fig. 7 is the fact that after oxidation, a large shoulder is produced on the high angle side of the (200) NiFe peak. This indicates that there is an excess of Ni over Fe in the remaining alloy, consistent with the formation of the iron -ich oxide detailed above. Although a mixed (111) and (100) layer is evident, the degree of (100) nature varies with temperature, with no (100) oriented material at low temperatures, a maximum around 800°C, and then less at higher temperatures.

In order to study a layered structure of the oxide, a sample was intentionally made to spall after the oxidation. This is done by rapidly quenching the oxidizing tape by turning off the heating current suddenly. Complete spalling happens reliably only if the oxide layer is sufficiently thick, which is probably associated with the elastic strain produced in the film due to a large volume change. As the oxide spalled in large pieces, it was possible to take x-ray spectra from both the oxide flakes and the underlying tape. These θ-2θ scans are shown in figure 1b and c and clearly show that the oxide which was removed has a (111) texture whilst the remaining tape is covered with an oxide which is (100) oriented.

FIG. 7 X-ray diffraction of surface oxidized Ni-50%Fe tapes. The (111) oxide (c)

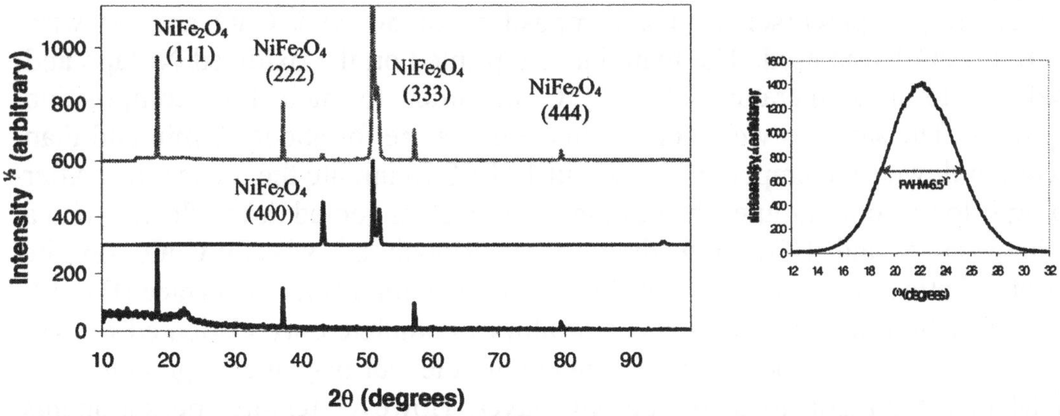

may be completely removed from the oxidized tape (a) to leave behind a NiFe tape, which has a cube textured oxide layer (b). (The inset on the right shows {400}-rocking curve with excellent grain alignment after re-oxidation of Ni-Fe tape).

This fact that a thin (100) oriented layer remains on the surface may be utilized to consistently produce cube-textured oxides on the surface of the Ni-Fe tape. If the tapes are further oxidized they maintain the cube texture, and the rocking curve of Fig. 7 demonstrates the excellent alignment. The atomic force microscope images of Fig. 8 indicate that the oxide layer is rough. The (111) layer, which is produced initially, is very rough, but when removed, it leaves a much smoother cube textured oxide layer. Unfortunately, the re-oxidation process results in a rough surface again.

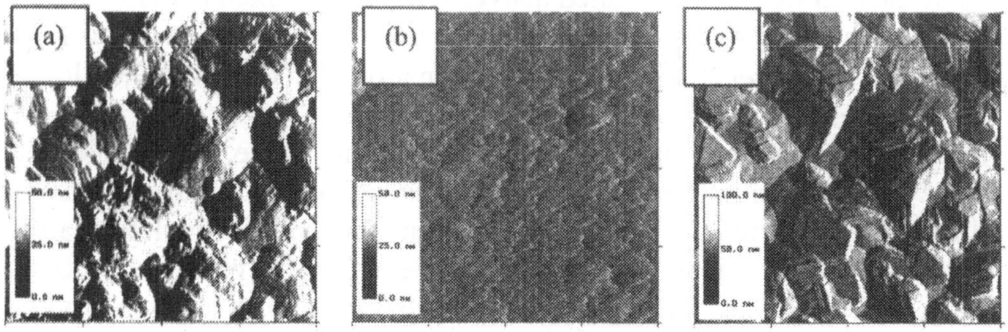

FIG. 8. Atomic force micrographs of a 10 x 10 μm^2 region of the surfaces of the oxide layers: a) the (111) oxide, b) the surface (100) oxide and (c) the thicker (100) oxide obtained after re-oxidation.

Liquid phase processing of $Nd_{2-x}Ce_xCuO_4$ on Ni / NiO: For the film growth of $Nd_{2-x}Ce_xCuO_4$, a precursor with a composition of 50 wt% CuO and 50 wt% (Nd_2O_3 + CeO_2) was used. The material was printed on the oxidized Ni tape and heated up to a temperature of 1300°C in order to melt this composition completely. The samples were kept at this temperature for about 10 min and than fast cooled down to a temperature of about 1240°C to initiate the nucleation. After keeping it for several minutes the samples were further cooled down slowly with a rate between 5 and 10°/min leading to the growth of $Nd_{2-x}Ce_xCuO_4$ on the substrate. Optical microscopy showed large grains with a smooth surface (Fig. 9). X-ray diffraction measurements revealed almost complete c-axis oriented growth of the $Nd_{2-x}Ce_xCuO_4$ and an improvement in the out-of-plane alignment - 3° FWHM in comparison to 5° of the NiO layer. However texture measurements showed a variety of in-plane orientations resulting from the large grains with single crystalline quality rotated around the substrate normal. These results combined with those in next section indicate that the control of the nucleation process is strongly dependent on lattice mismatch between the (100)NiO and the $Nd_{2-x}Ce_xCuO_4$ layers.

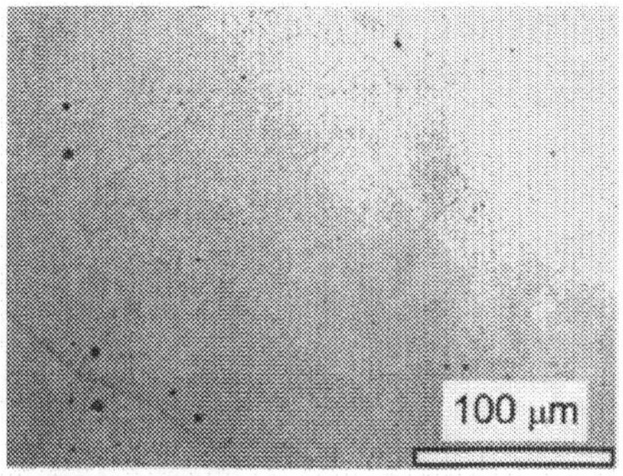

FIG. 9. Optical image of a $Nd_{2-x}Ce_xCuO_4$ film grown by liquid phase epitaxy on surface oxidized biaxial textured Ni tapes showing a flat surface

Liquid phase processing of Nd_2CuO_4 on Ni/NiO: The Nd_2CuO_4 buffer layer on top of NiO was grown from liquid phase using a number of liquid-phase processing (LPP) methods. The *first method* used was the traditional top dipping LPE growth. The composition of the high temperature solution for the LPE growth was 20%Nd_2O_3: 80%CuO. The liquid was prepared at 1200°C and the LPE growth was carried out at about 1150 °C. In a *second method*, a precursor with composition around 30%Nd_2O_3: 70%CuO was screen-printed on the substrates using an organic binder and then heated up to 1200 °C to reach a single liquid state. This temperature was kept for about 20 minutes and then cooled down to room temperature at 300 °C/hr. Because of the large surface area to volume ratio and the faster evaporation of CuO than Nd_2O_3, the liquid became supersaturated rapidly, leading to the nucleation and growth of Nd_2CuO_4 on the substrates.In the *third method*, a growth system similar to the traditional zone-refining furnace was used. A sintered source rod of 40%Nd_2O_3: 60%CuO was fed into the focus of an infrared beam to produce a molten zone at one end. When the increasing size of the molten zone exceeded the limit of the surface tension, the liquid dropped down onto the substrate below, which was maintained at a suitable temperature (1100-1200°C) for the epitaxial growth induced from the highly textured substrate. Because of the extremely good wetting property of the high-temperature cuprate solution, the liquid drops quickly spread out on the substrates.

An XRD pole figure of Nd_2CuO_4 (x=0 for this sample) grown on NiO/Ni is shown in Fig.10. In the absence of Ce the in-plane texture of the Nd_2CuO_4 layer appeared to be significantly improved. There is a 45°-angle rotation between the *a/b*-axes of the Nd_2CuO_4 and NiO, indicating that the epitaxial relation between these two compounds was [100]//[110]. The measured lattice parameter a=0.394nm for Nd_2CuO_4 and 0.417 nm for NiO. Therefore, 3 unit cells of Nd_2CuO_4 align with two units of the [110] axis of NiO. The lattice mismatch is only 0.3%. A cross-sectional SEM of the Nd_2CuO_4/NiO/Ni structure is shown in

Fig.11. The interface between the Nd_2CuO_4 and NiO was quite smooth and Nd_2CuO_4 grew into all the groves between NiO grains.

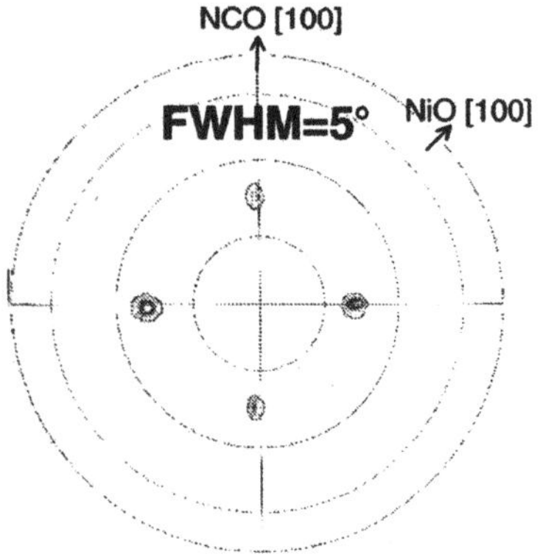

FIG. 10 (103) pole figure of Nd_2CuO_4 on NiO/Ni

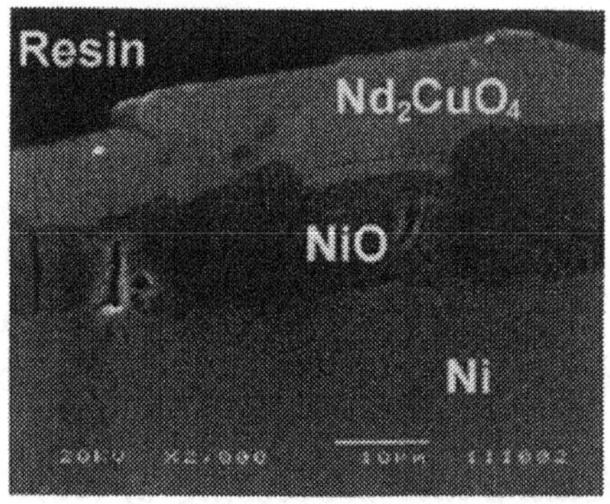

FIG. 11 SEM cross-section image of Ni/NiO/$Nd_{2-x}Ce_xCuO_4$ architecture.

CONCLUSIONS

We report on basic studies of a new coated conductor architectures: Ni/NiO/$Nd_{2-x}Ce_xCuO_4$/YBCO. Thin, highly textured and relatively smooth NiO layers could be formed on pure Ni, Ni-Cr and Ni-Fe alloys. A smooth dense well-textured $Nd_{2-x}Ce_xCuO_4$ (x=0-0.15) was formed on the surface of the NiO by LPP methods.

ACKNOWLEDGEMENTS

Presented research was conducted jointly between Department of Materials Science and Metallurgy at University of Cambridge and Department of Materials in Imperial College in London and was funded by the EPSRC. The support from ABB-Research, Zurich is also gratefully acknowledged.

REFERENCES:

[1] X. Qi and J.L. MacManus-Driscoll, "Liquid-phase epitaxial growth of REBa2Cu3O7-d (RE=Y,Yb,Er) thick films at reduced temperatures", *Journal of Crystal Growth*, **213** 312-318 (2000).

[2] X. Qi, M. Soorie, Z. Lockman and J.L. MacManus-Driscoll, "Rapid growth of $Nd_{2-x}Ce_xCuO_4$ thick films as a buffer for the growth of rare-earth barium cuprate-coated conductors", *Journal of Materials Research*, **17** [1] 1-4 (2002).

[3] E.D Specht, A. Goyal, D.F Lee, F.A List, D.M Kroeger, M. Paranthaman, R. K Williams and D.K Christen, "Cube-textured nickel substrates for high-temperature superconductors ", *Supercond. Sci. Technol.* **11** (10) 945-949 (1998).

[4] R. Nast, W. Goldacker B. Obst and G.Linker, "Recrystallisation kinetics of biaxially textured Ni substrates for YBCO coated conductors," *Inst. Phys. Conf. Ser.,* **167** 411-414 (2000).

[5] A. Tuissi and E. Villa, CNR-TEMPE, Corso Promessi Sposi 29 23900 Lecco, Italy

[6] J. Knauf, R. Semerad, W. Prusseit, B. DeBoer and J. Eickemeyer, "YBaCuO – deposition on metal tape substrates", *IEEE Trans. on App. Supercon.*, **11** 2885 - 2888 (2001).

[7] B.A. Glowacki, M. Vickers, N. Rutter, E. Maher, F. Pasotti, A. Baldini and R. Major "Texture development in long lengths of NiFe tapes for superconducting coated conductor', *Journal of Materials Science*, **37** 157- (2002).

[8] Z. Lockman, X. Qi, A. Berenov, W. Goldacker, R. Nast, B. deBoer, B. Holpfapzel, and J. L. MacManus-Driscoll, "Surface oxidation of cube-textured Ni-Cr for the formation of a NiO Buffer Layer for Superconducting Coated Conductor", to be published in *Physica C.*

[9] N.A. Rutter, B.A. Glowacki, J.H. Durrell, J.E. Evetts, H. te Lintelo, R. De Gryse and J. Denul, "Formation of native cube textured oxide on a flexible NiFe tape substrate for coated conductor application," *Inst. Phys. Conf. Ser.*, **167** 407-410 (2000).

[10] P. Kofstad, in Nonstoichiometry, "Diffusion, and electrical conductivity in binary metal oxides", Wiley-Interscience, N. York 1972.

[11] Z. Lockman, W. Goldacker, R. Nast, B. de Boer and J.L. MacManus-Driscoll, "Study of thermal oxidation of NiO on pure Ni, Ni-10%Cr and Ni-V9% tapes", *IEEE Trans. Appl. Supercond.* **11** 3325 (2001).

[12] A.Kursumovic, R. Tomov, R. Hühne, B.A. Glowacki, J.E. Evetts, A. Tuissi, E. Villa and B. Holzapfel , submitted to *Physica C*

[13] N.N. Khoi, W.W. Smeltzer and J.D. Embury, " Growth and structure of NiO on Ni crystal faces", *Journal of Electrochemical Society*, **122** (11) 1495-1502 (1975).

[14] G.C.Wood and T.Hodgkiess, "Characteristic scales on pure Nickel-Chromium Alloys at 800-1200°C", *Journal of the Electrochemical Society*, **113**(4) 319-327 (1966).

[15] B.Ahmad and P.Fox, "STEM analysis of the transient oxidation of a Ni-20Cr alloy at high temperatures", *Oxidation of metals* **52**(1/2) 113-138 (1999).

CA DOPING OF YBCO THIN FILMS

A. Berenov, J.L. MacManus-Driscoll, D. MacPhail
[a]Department of Materials, Imperial College, Prince Consort Road, London, SW7 2BP, UK

S. Foltyn
Superconductivity Technology Center, Los Alamos National Laboratory, Los Alamos, New Mexico 87545, USA

ABSTRACT

Ca-gels were applied to PLD YBCO thick films followed by high temperature post-annealing. Parameters of Ca diffusion in YBCO (diffusion coefficients, activation energy, etc) as a function of temperature were determined by depth profiling SIMS. Fast Ca diffusion along the grain boundaries was observed. The effect of Ca-doping on T_c and J_c was studied by VSM. Ca distribution in the samples, induced by preferential diffusion along the grain boundaries, did not appear to change superconducting properties within the grains but showed increased J_C across the grain boundaries.

INTRODUCTION

It is well known that the properties of grain boundaries in YBCO differ drastically from those of bulk material. It has been shown both experimentally and theoretically that the grain boundaries in YBCO superconductors are depleted of carriers as compared to the bulk [1]. This depletion leads to weak intergrain links and limit critical currents in superconductor. Recently, the promise for improved grain boundaries has been shown in YBCO thin films which have been doped throughout with Ca [2, 3]. In this study the effect of Ca-doping of grain boundaries in YBCO has been investigated in an attempt to increase hole concentration on the grain boundaries and attainable J_C.

EXPERIMENTAL

Starting materials were $YBa_2Cu_3O_{7-\delta}$ thin films grown by PLD on (100) $SrTiO_3$ [4]. A dilute (0.05M) solution of $Ca(NO_3)_2$ (99.99+%) was dropped on the surface of the specimens. Samples were annealed in air at different preparation conditions (Table I). All the films annealed at 870°C were later oxygenated in flowing air at 500°C for 2 days. Several samples were not doped and used as reference.

The Ca concentration profiles were measured by SIMS (Atomika 6500). The depth of the crater was measured by an optical surface profilometer. The raw data was converted into the Ca diffusion profiles, by calculating the isotopic fraction from the raw intensities and converting the sputter time to depth assuming a constant sputtering rate. Diffusional profiles were fitted using the following solution of Fick's second law:

$$C(x,t)/C_0 = \exp\left(-\frac{x^2}{4D_{bulk}t}\right) \quad (1)$$

where $C(x,t)$ is the concentration of Ca as a function of distance from the surface, x, for an annealing time, t. D_{bulk} is the bulk diffusion coefficient, and C_o is the concentration at the surface. In several diffusional profiles, short-circuiting tails were observed which were attributed to the fast Ca diffusion along the grain boundaries. The product of grain boundary diffusion coefficient and grain boundary width was calculated by the method proposed by Chung and Wuensch [5].

Samples were characterized by XRD, SEM. Magnetization was measured by VSM. Critical current densities were calculated using Bean model [6]. Tc was determined from magnetic permeability measurements.

RESULTS AND DISCUSSIONS

XRD scans of starting PLD films showed only highly intensive (00l) peaks with c-axis oriented perpendicular to the film surface. Pole figures on (111)

Table I. Preparation conditions and results of measurements

Sample	Ca-doped	Doping conditions	"c" parameter, Å	D_{bulk}, cm^2s^{-1}	$D_{g.b.}\delta$, cm^3s^{-1}	T_C, K
YE102	yes	870°C, 30 min		NA	NA	89.5
YE103	yes	870°C, 20 hrs	11.698	NA	NA	92.0
YE104	no	870°C, 20 hrs	11.693	NA	NA	92.8
YE105	yes	870°C, 6 hrs	11.681	NA	NA	91.6
YE202	yes	870°C, 2 hrs	11.778	6.2×10^{-14}	4.2×10^{-18}	92.6
YE203	no	870°C, 2 hrs	11.752	NA	NA	NA
YE204	yes	798°C, 6 hrs	11.753	2.7×10^{-14}	3.1×10^{-18}	NA
YE205	yes	700°C, 24 hrs	11.767	7.9×10^{-15}	NA	NA
YE501	yes	700°C, 45 min		1.3×10^{-14}	NA	NA

YBCO peak showed in-plane alignment of the film. No secondary phases were observed in starting thin films. After doping peaks belong to CaO and BaO were observed. The intensity and presumably the amount of BaO increased with the duration of anneal. Interestingly, no Y-rich phases were observed in Ca doped samples and no BaO was present in undoped samples. Therefore it is possible to conclude that some Ca is substitute onto the Ba site in $YBa_2Cu_3O_{7-\delta}$. This observation is supported by the structural study of $Y_{1-x}Ca_xBa_2Cu_3O_{7-\delta}$ which showed that Ca substituted onto Y site at low Ca concentrations (<11%), whereas at higher concentrations, Ca also substituted onto Ba site [7]. Values of the "c" lattice parameters are given in Table 1. No clear correlation of the duration of doping with the "c" parameter was observed.

The values of the Ca chemical diffusion coefficient as well as the grain boundary diffusion coefficient are given in Table I. The Ca-diffusion coefficient in PLD films is about 20 times lower than in the bulk YBCO [3]. This discrepancy is attributed to the anisotropy of Ca diffusion along the a-b planes and c- direction in $YBa_2Cu_3O_{7-\delta}$. Analogous anisotropy in the ion diffusion was observed for oxygen [8] and barium diffusion [9] in $YBa_2Cu_3O_{7-\delta}$. The activation energy for Ca chemical diffusion along c-direction is ~ 113 kJ/mole. This value is smaller than the activation energies of tracer diffusion of Ba and Y in $YBa_2Cu_3O_{7-\delta}$, but

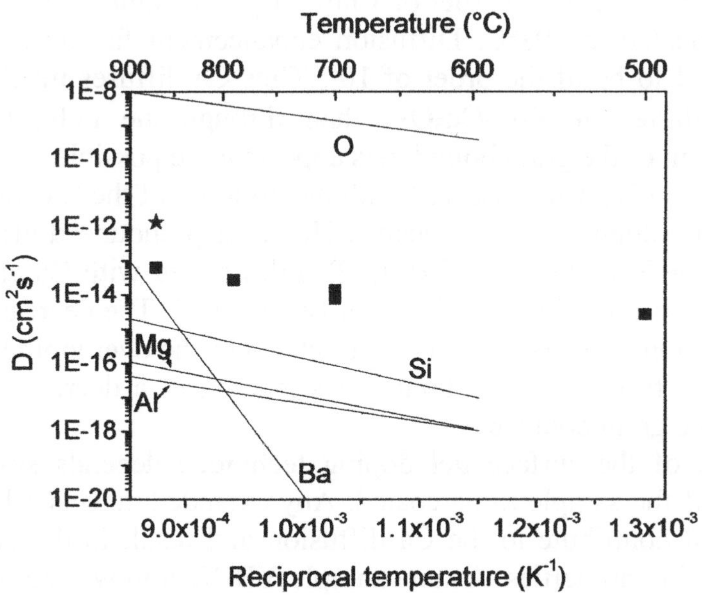

Fig. 1. Arrhenius plot of Ca diffusion coefficient in $YBa_2Cu_3O_{7-x}$ thin films (■). Solid lines are literature data on ion diffusion in YBCO [8,9].

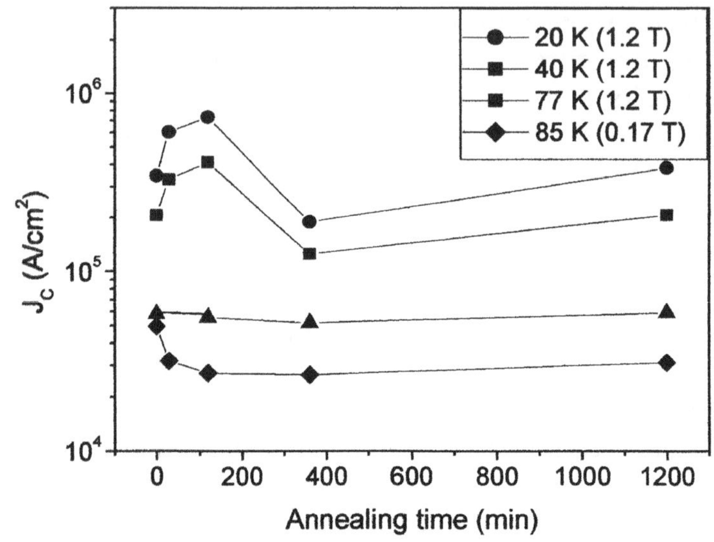

Figure 2. Effect of annealing time on the critical current densities of Ca doped films in annealed air at 870°C.

compares reasonably well with the activation energy for the chemical diffusion of Ag and Co in $YBa_2Cu_3O_{7-\delta}$ [9].

The grain boundary diffusion coefficient was calculated assuming the width of the grain boundary to be of the order of 1 nm [10]. The value was 3.1×10^{-11} cm^2s^{-1} for the film annealed at 798°C. Diffusion enhancement factors ($D_{bulk}/D_{g.b.}$) are roughly estimated to be of the order of 10^3. Thus Ca diffuses much faster along the grain boundaries in $YBa_2Cu_3O_{7-\delta}$ than through the bulk. Consequently, preferential doping of the grain boundaries appears to be possible.

No correlation of T_C was observed with the duration of the Ca anneal implying that no significant doping of grain occurs. However pronounced effect of the Ca annealing time on Jc was observed (Fig. 2). Film doped with Ca at 870°C for 2 hrs, showed an increase of J_C by a factor of two at 40 K. The decrease of the J_C at prolonged annealing times is thought to be related to the overdoping of the grains with Ca supplied from the grain boundaries leading to a decrease of T_C in the grains close to the grain boundary.

The success of the surface gel doping technique depends strongly on the microstructure of the sample to be coated. Any extended defects other than grain boundaries could contribute to the Ca diffusion and result in the degradation of the superconducting properties. As an example YBCO film was cut in three parts. Two resulting films were doped with Ca (YE607 and YE606) and one left as a reference (YE605). All films were annealed at 500°C in oxygen. As expected J_C

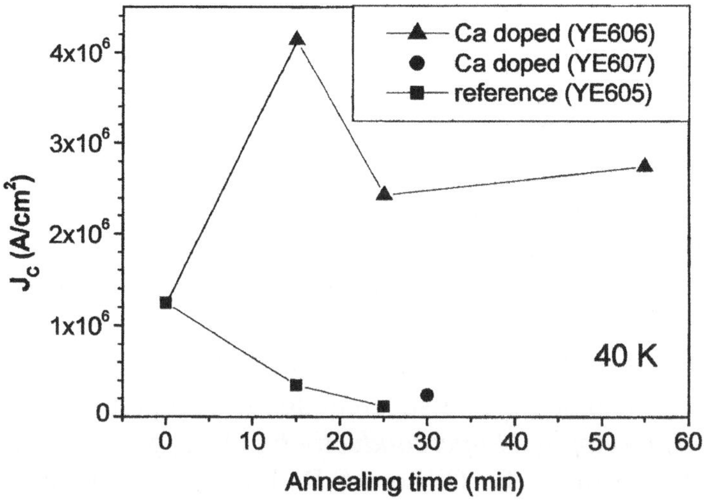

Figure 3. Effect of annealing time on the critical current densities of films annealed in oxygen at 500°C.

for YE607 film increased with annealing and later decreased with the optimal annealing time of 15 min (Fig. 3). At the same time J_C for another film (YE606) decreased with annealing and followed the data of the reference film (YE605). Short annealing time required to optimally dope the film at 500°C is surprising, as it required 2 hrs to achieve analogous effect at 870°C where much faster diffusion rates are expected. It is thought that the pores observed by SEM on YE607 caused higher Ca doping rate due to gel penetration in the pores and/or faster diffusion along the pores as compared to the grain boundaries. The decrease of J_C in YE605 and YE606 films after annealing in oxygen is currently unclear.

CONCLUSIONS

Ca chemical diffusion was studied in $YBa_2Cu_3O_{7-\delta}$. Fast diffusion along the grain boundaries was observed in c-oriented thin films. The optimum duration of Ca annealing increased J_C by a factor of 2 in epitaxial $YBa_2Cu_3O_{7-\delta}$ thin films at 40 K. The doping /diffusion technique developed in this work allows preferential doping of grain boundaries in $YBa_2Cu_3O_{7-\delta}$.

ACKNOWLEDGEMENTS

The authors are grateful EPSRC for financial support.

REFERENCES

[1] H. Hilgenkamp, C.W. Scheider, R.R. Schulz, B. Goetz, A. Schmehl, H. Bielefeldt and J. Mannhart, "Modifying electronic properties of interfaces in high-T_c superconductors by doping", *Physica C*, **326-327** 7-11 (1999).

[2] A. Schmehl, B. Goetz, R.R. Schulz, C.W. Schneider, H. Bielefeldt, H. Hilgenkamp and J. Mannhart, "Doping-induced enhancement of the critical currents of grain boundaries in $YBa_2Cu_3O_{7-\delta}$", *Europhysics Letters*, **47** [1] 110-115 (1999).

[3] A.V. Berenov, R. Marriott, S.R. Foltyn, J.L. MacManus-Driscoll, "Effect of Ca-doping on grain boundaries and superconducting properties of $YBa_2Cu_3O_{7-\delta}$", *IEEE Transactions on Applied Superconductivity*, **11** [1] pt 3 3780-3783 (2001).

[4] N. Malde, L.F. Cohen, G. Gibson, S.R. Foltyn, J. Nelstrop, J.J.Wells, A. Berenov, J.L. MacManus-Driscoll, "Raman and X-ray characterisation of structural disorder in YBCO thick films grown at high rates for IBAD conductors", *Institute of Physics Conference Series*, **167** 415-418 (1999).

[5] Yong-Chae Chung and B.J. Wuensch, "An improved method, based on Whipple`s exact solution, for obtaining accurate grain-boundary diffusion coefficients from shallow solute concentration gradients", *Journal of Applied Physics*, **79** [11] 8323-8229 (1996).

[6] C.P. Bean, *Physical Review Letters*, **8** 250 (1962).

[7] G.Böttger, H. Schwer, E. Kaldis, K. Bente, "Ca doping of $YBa_2Cu_3O_{7-\delta}$ single crystals: structural aspects", *Physica C*, **275** 198–204 (1997).

[8] J.L. Routbort, N. Chen, K.C. Goretta, S.J. Rothman, p. 569 in *High-Temperature Superconductors: Materials Aspects. Procedings of the ICMC'90 Topical Conference on Materials of High-Temperature Superconductors*. Edited by H.C. Freyhardt, R. Flükiger, M. Peuckert. Informationsgesellschaft, Verlag, 1991.

[9] N. Chen, S.J. Rothman, J.L. Routbort, K.C. Goretta, "Tracer diffusion of Ba and Y in $YBa_2Cu_3O_x$", *Journal of Material Research*, **7** [9] 2308-2316 (1992).

[10] J.R. Farver, R.A. Yund, D.C. Rubie, "Magnesium grain boundary diffusion in forsterite aggregates at 1000 ° -1300 °C and 0.1 MPa to 10 GPa", *Journal of Geophysical Research*, **99** [B10] 19809-19819 (1994).

DEMONSTRATION OF HIGH CURRENT DENSITY YBCO FILMS ON ALL SOLUTION BUFFERS

M. Parans Paranthaman, S. Sathyamurthy, H.Y. Zhai, H.M. Christen, S. Kang, and A. Goyal
Oak Ridge National Laboratory, Oak Ridge, TN 37831, USA

ABSTRACT

Chemical solution deposition has been emerged as a viable, low-cost, non-vacuum process for fabricating long lengths of $YBa_2Cu_3O_{7-\delta}$ (YBCO) coated conductors. Single, epitaxial buffer layers of $La_2Zr_2O_7$ (LZO) have been grown directly on biaxially textured and strengthened Ni-W (3 at.%) substrates by a solution process. Smooth, epitaxial, and crack-free buffer layers with desired thickness were produced using multiple coating. YBCO films grown on 60 nm thick LZO-buffered Ni-W substrates using pulsed laser deposition yielded very high critical current densities of 1.9 MA/cm^2 at 77 K and self-field. These results are comparable to those typically obtained on identical substrates using our standard architecture of $CeO_2/YSZ/Y_2O_3/Ni$. The dependence of critical current density, J_c on thick YBCO films grown on 60 nm thick LZO layers is investigated.

INTRODUCTION

Since the discovery of high temperature superconductivity in cuprate based superconductors, much progress has already been made, from the near-term commercialization of the first-generation bismuth strontium calcium copper oxide (BSCCO) superconductor tapes to the continuing advancement in second-generation conductors coated with yttrium barium copper oxide (YBCO). The development of second-generation coated conductors continues to show the steady improvement toward the long-length processing capabilities. The U.S. Department of Energy's vision is, *"Low-cost, high-performance YBCO Coated Conductors will be available in 2005 in kilometer lengths. For applications in liquid nitrogen, the wire cost will be less than \$ 50/kA-m, while for applications requiring cooling to temperatures of 20-60 K the cost will be less than \$ 30/kA-*

m." One of the important critical needs to achieve this goal is to develop a simple low-cost buffer layer technology on strengthened Ni-alloy substrates.

Metal-organic decomposition (MOD) based solution precursor approaches have been examined as a potential low-cost process for manufacturing YBCO coated conductors. This is mainly due to the *speed/cost* advantage of this process. Recently, ORNL and American Superconductor Corporation have reported critical currents of 118 A with a standard deviation of less than 3% over a 1-m length of centimeter-wide RABiTS tape with the architecture of YBCO/CeO_2/YSZ/Y_2O_3/Ni/Ni-W using solution-based metal-organic decomposition (MOD).[1] Although there has been significant progress in the development of low-cost manufacturing technologies for select components of the YBCO coated conductors, key deposition technologies must still be improved. Even though the oxide buffer layers are much thinner than the YBCO layer, the buffer deposition accounts for over 20% of the total conductor cost – the single highest projected cost element in the entire conductor. Hence, the coated conductor technology dictates that a considerable and concentrated effort must be devoted to the development of higher rate, high quality buffer layer deposition processes as alternatives to the already established methods.

The main focus of this work is to develop a chemical solution deposition process to grow high quality lanthanum zirconium oxide, $La_2Zr_2O_7$ (LZO) buffer layers on strengthened Ni-W substrates. LZO has a cubic pyrochlore structure with a close lattice match with YBCO (less than 1% lattice mismatch). Recently, we have demonstrated the growth of high current density YBCO films on both solution LZO seed layers[2,3] and all solution LZO layers[4] grown on biaxially textured Ni substrates. Here we report the results of using a single LZO buffer layer deposited on strengthened Ni-W (3 at. %) substrates with reduced magnetism for YBCO coated conductors.

EXPERIMENTAL PROCEDURE

The LZO precursor solution with 0.25 M cation concentration was prepared by refluxing a stoichiometric mixture of lanthanum isopropoxide and zirconium n-propoxide in 2-methoxyethanol in a Schlenk-type apparatus. As-rolled Ni-W (3 at. %) tapes were cleaned by ultrasonification in iso-propanol. The tapes were then annealed in either reducing atmosphere (Ar-H_2 4%) or in high vacuum at 1300 °C to get the desired cube texture. The metal substrates were coated with LZO film by spin coating the precursor solution at 2000 rpm for 30 s, followed by annealing at 1100 °C for 1 h in a flow of 1 atm Ar-H_2 4% gas mixture. The process was repeated three times to get 60 nm thick LZO layers. On the LZO-buffered Ni-W substrates, pulsed laser deposition was used for YBCO deposition, conducted at 780 °C and pO_2 of 120 mTorr. After deposition, the samples were first cooled to 500 °C at a rate of 5 °C/min; then the O_2 pressure was increased to

550 Torr, and the samples were cooled to room temperature. YBCO thickness was varied from 200 nm to 1500 nm.

The films were analyzed by X-ray diffraction techniques. SEM micrographs were taken using a Hitachi S-4100 field emission microscope. The thickness of both buffer layers and YBCO were determined by Rutherford backscattering spectroscopy. The resistivity and transport critical current density, J_c, were measured using a standard four-probe technique. The voltage contact spacing was 0.4 cm. Values of J_c were calculated using a 1 μV/cm criterion.

RESULTS AND DISCUSSION

Typical θ-2θ scans for 200 nm thick YBCO films grown on 60 nm thick solution LZO buffered Ni-W substrates are shown in Figure 1. These scans indicate the presence of c-axis aligned films. It is interesting to note that there is

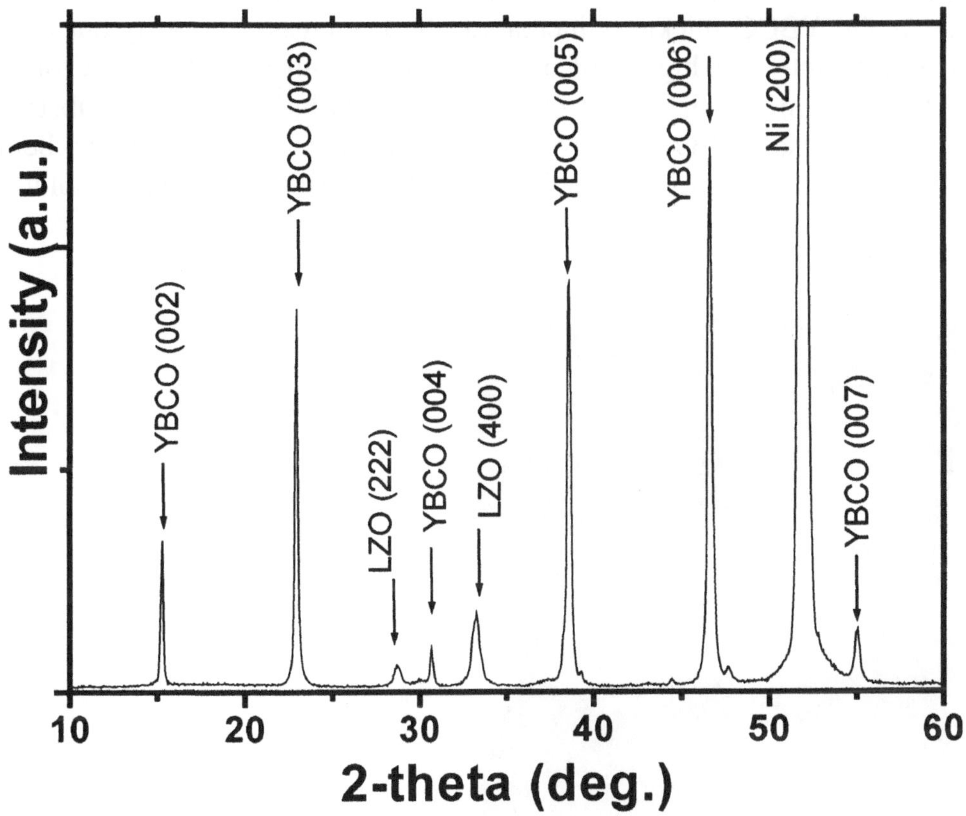

FIGURE 1. A typical θ-2θ scan for a 200 nm thick YBCO on solution LZO buffered Ni-W substrates. The YBCO film has a preferred c-axis orientation.

no NiO present in the film. However, small amounts of (222) LZO peaks were present. In certain cases, (222) LZO peaks were absent. Detailed RHEED studies are being conducted to determine the texture of the surface layers and the results will be published later. Detailed XRD results from ω and ϕ scans on YBCO (200 nm)/LZO (60 nm)/Ni-W revealed good epitaxial texturing. The full width at half maximum (FWHM) values for Ni (002) and YBCO (006) are 5.7° and 6.8°, and those of Ni (111) and YBCO (103) are 7.7° and 9.3°, respectively. The YBCO (103) pole figure revealed the presence of a single four-fold cube texture. SEM micrograph for a 200 nm thick PLD YBCO film on 60 nm thick LZO-buffered Ni-W substrate is shown in Figure 2. The microstructure of the YBCO film looks like a typical dense PLD film. These films had zero resistance temperature, T_c close to 90 K. The field dependence of J_c for the same film is shown in Figure 3. The zero-field J_c was close to 1.9 MA/cm^2 at 77 K. J_c at 0.5 T was about 25 % of the zero-field J_c. This indicates the presence of a strongly linked YBCO film. The dependence of critical current, I_c and J_c on YBCO thickness is shown in Figure 4. As the YBCO thickness increased, the I_c increased steadily but the J_c decreased significantly. A high I_c of over 70 A/cm-width was measured for 1.1 µm thick YBCO film grown on just 60 nm thick LZO-buffered Ni-W substrate. Optimization of thick YBCO film growth is in progress. Efforts are also being made to demonstrate the growth of solution MOD YBCO layers on all solution buffers.

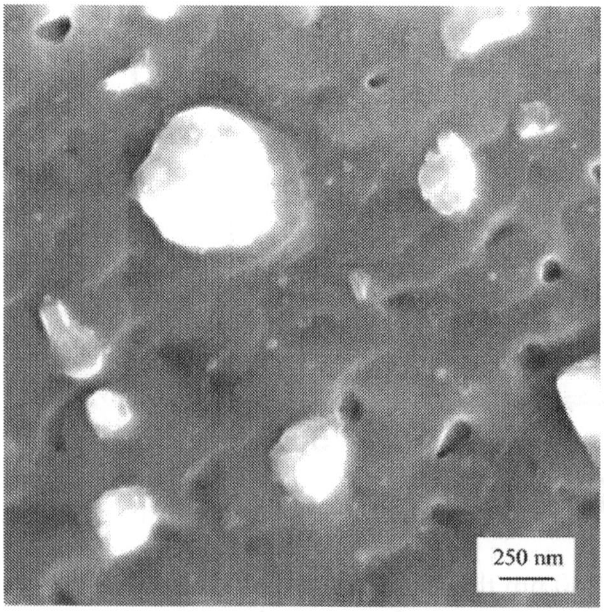

FIGURE 2. SEM micrograph for a 200 nm thick YBCO film on 60 nm thick solution LZO-buffered Ni-W substrate.

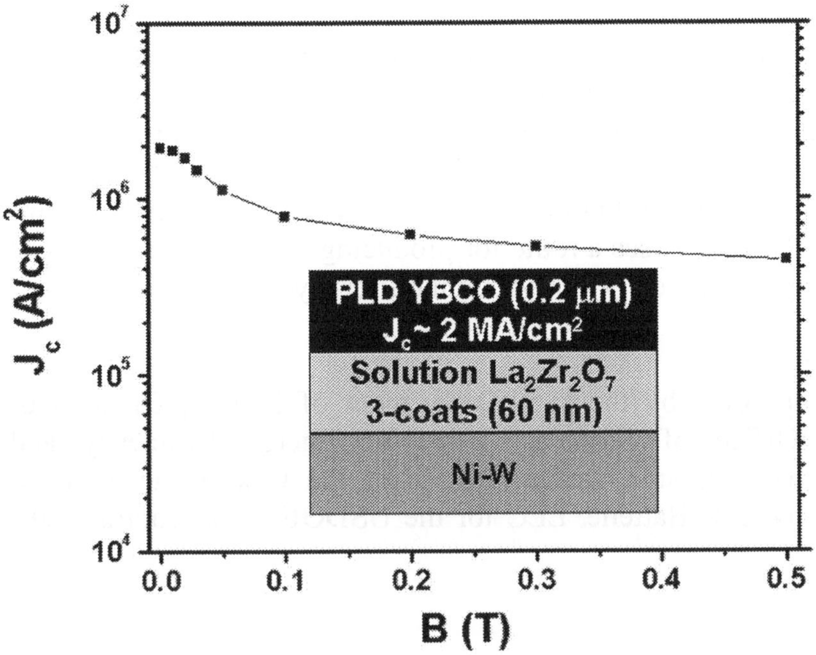

FIGURE 3. Field dependence of critical current density, J_c, for 200 nm thick PLD YBCO film on 3-coats of solution LZO buffered-Ni-W substrate.

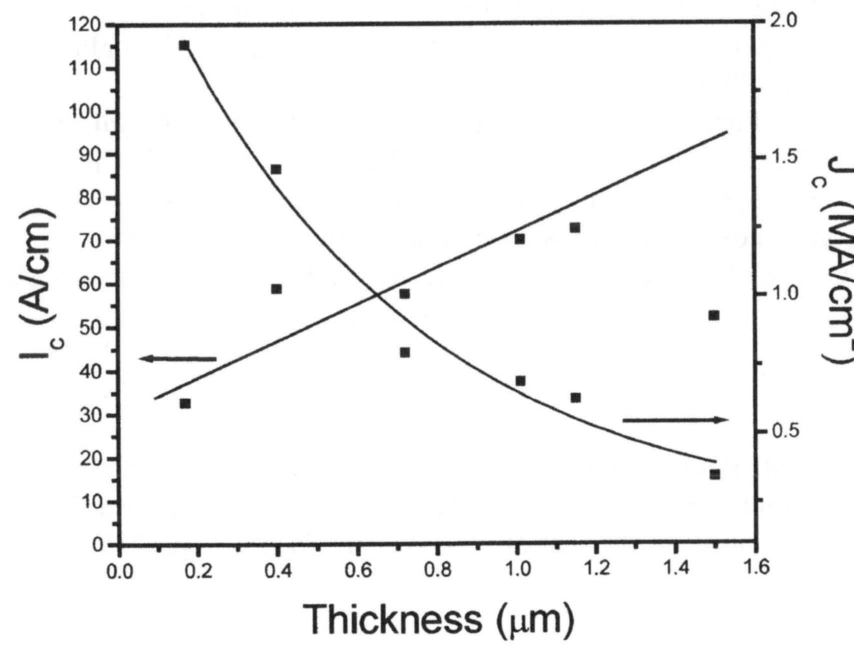

FIGURE 4. Variation of YBCO film thickness on 60 nm thick solution LZO-buffered Ni-W and critical current, I_c and critical current density, J_c.

SUMMARY

We have demonstrated that YBCO films with good superconducting properties can be grown on 3-coats of 60 nm thick solution LZO buffered Ni-W substrate. This proves that $La_2Zr_2O_7$ can be an excellent diffusion barrier for Ni. This also eliminates the need for both Ni protective layers and seed layers. For a 1.1 µm thick YBCO film on just 60 nm thick LZO/Ni-W, an I_c of over 70 A/cm-width was measured. This promises a route for producing long lengths of YBCO coated conductors using a completely non-vacuum buffer layer technology.

ACKNOWLEDGEMENTS

This work was supported by the U.S. Department of Energy, Division of Materials Sciences, Office of Science, Office of Energy Efficiency and Renewable Energy. The research was performed at the Oak Ridge National Laboratory, managed by U.T.-Battelle, LLC for the USDOE under contract DE-AC05-00OR22725.

REFERENCES

[1]D.T. Verebelyi, U. Schoop, C. Thieme, S. Annavarapu, S. Cui, X. Li, W. Zhang, T. Kodenkandath, L. Fritzemeier, Q. Li, A.P. Malozemoff, N. Nguyen, D. Buczek, J. Lynch, J. Scudiere, M. Rupich, A. Goyal, L. Heatherly and M. Paranthaman' "Uniform performance of meter long YBCO on textured Ni alloy substrates," submitted to Physica C (2002).

[2]T.G. Chirayil, M. Paranthaman, D.B. Beach, D.F. Lee, A. Goyal, R.K. Williams, X. Cui, D.M. Kroeger, R. Feenstra, D.T. Verebelyi, and D.K. Christen, "Epitaxial growth of $La_2Zr_2O_7$ thin films on rolled-Ni substrates by sol-gel process for high T_c superconducting tapes," *Physica C* **336**, 63-69 (2000).

[3]S. Sathyamurthy, M. Paranthaman, T. Aytug, B.W. Kang, P.M. Martin, A. Goyal, D.M. Kroeger, and D.K. Christen, "Chemical solution deposition of lanthanum zirconate buffer layers on biaxially textured Ni-1.7% Fe-3% W alloy substrates for coated-conductor fabrication," *J. Mater. Res.* **17**, pp. 1543-1549 (2002).

[4]S. Sathyamurthy, M. Paranthaman, H-Y. Zhai, H.M. Christen, P.M. Martin, and A. Goyal, "Lanthanum zirconate: A single buffer layer processed by solution deposition for coated conductor fabrication," *J. Mater. Res.* **17**, pp.2181-2184 (2002).

SOLUTION BUFFER LAYERS FOR YBCO COATED-CONDUCTORS

S. Sathyamurthy, M. Paranthaman, H-Y.Zhai, S.Kang, C.Cantoni, S.Cook, L. Heatherly, A. Goyal, and H.M.Christen
Oak Ridge National Laboratory, Oak Ridge, TN 37831
Md.S. Bhuiyan, and K. Salama
University of Houston, Houston, TX 77204.

A single layer of $La_2Zr_2O_7$ (LZO), deposited on textured Ni and Ni-3%W (Ni-W) tapes by a low-cost sol-gel process, is used as a buffer layer for the growth of $YBa_2Cu_3O_{7-\delta}$ (YBCO) coated conductors. It is shown for the first time that such single buffer layers can be used for the deposition of YBCO yielding critical current densities (J_c) that are comparable to those typically obtained using $CeO_2/YSZ/Y_2O_3$ tri-layers on identical substrates, i.e. in excess of 1 MA/cm^2 on nickel substrates and close to 2 MA/cm^2 on Ni-W substrates (which has a sharper texture) at 77K and self-field. Reel-to-reel deposition of LZO buffer layers in long lengths and their use in fabrication of coated conductors in long lengths has also been demonstrated. These results offer promise to the fabrication of low cost all-solution coated conductors.

INTRODUCTION

The focus of our research in the area of high-temperature superconductivity (HTS) in recent years has been on the development of the second generation wires also known as coated-conductors [1-3]. One of the leading textured-template approaches for the fabrication of coated-conductors is the rolling assisted biaxially textured substrates (RABiTS) approach [2]. In this approach, cube textured nickel or Ni alloy substrates, obtained by cold rolling and recrystallization, act as a template for the epitaxial deposition of buffer layers and the YBCO superconductor. The buffer layers, apart from providing a structural template, also acts as a chemical barrier between the metal substrate and the HTS coating. Using such architecture, sufficient biaxial texturing of the HTS layer has been obtained to avoid problems associated with weak-linked, high-angle grain boundaries [4].

To date, in the processing of high current coated conductors using RABiTS, the best results have been obtained reproducibly using three or four layer buffer architectures like $CeO_2/YSZ/Y_2O_3/Ni/NiW$. The fabrication of this multi-layered buffer architecture, however, may present significant roadblocks to the scale-up of the process to long-lengths. Typically, deposition of the buffer layer stack would involve a combination of e^- beam and rf-sputter deposition techniques coupled with the exposure of the samples to thermal cycling and ambient environment. These requirements could lead to the degradation of the individual buffer layers and add to the complexity, and cost of the over-all process. However, if a single buffer layer deposited using a scaleable technique is developed, it would significantly decrease the processing time and make the process simpler and more conducive to scale-up to long lengths.

EXPERIMENTAL

Stoichiometric quantities of lanthanum isopropoxide and zirconium-n-propoxide were dissolved in 2-methoxyethanol and refluxed to obtain the precursor solution with 0.25 M cation concentration. All-solution buffered substrates were prepared by spin coating the precursor solution on textured Ni and Ni-3at. % W (NiW) substrates, and heat treating the films at 1100°C for 1 h in $Ar/4\%H_2$ atmosphere (described in detail elsewhere [5]). The process was repeated to get thicker coating of the buffer layer. For long length samples, the buffer layer deposition was carried out in a reel-to-reel dip coating process, and the samples were also heat treated in a continuous reel-to-reel fashion. The $YBa_2Cu_3O_x$ (YBCO) was deposited using pulsed laser deposition (PLD) at 790°C in 120mTorr oxygen with average laser energy of 400-410mJ. The phase purity and texture of the samples was analyzed using X-ray diffraction (XRD). The film thickness and compositional homogeneity were analyzed using Rutherford Backscattering Spectroscopy (RBS) and the microstructural analysis of the samples was performed using a Scanning Electron Microscope (SEM). The critical current densities of the samples were measured using a standard four-point probe technique with a 1 μV/cm voltage criterion.

RESULTS AND DISCUSSION

A typical XRD of multiple coats of LZO on NiW substrates is shown in Fig.1. This figure illustrates the presence of a strong (400) reflection from the LZO film while the (222) reflection is at the background levels. It is clear from this figure the highly oriented LZO films can be processed using multiple coating and annealing cycles. On the multiple coated LZO films, YBCO was deposited by PLD. Figure 2 shows the XRD pattern of the YBCO film on LZO buffered NiW substrates. Along with a good c-axis texture, the XRD pattern also shows that there is no detectable amount of NiO in the sample. This suggests that the LZO buffer layer acts as a good barrier layer and protects the NiW substrate from oxidation. The microstructure of the multiple coated LZO films and the PLD deposited YBCO films are shown in Fig. 3. This figure shows that the LZO films are fine-grained, dense and crack-free with a smooth surface. Such a dense, smooth surface is essential for all-solution buffers to be effective. The YBCO microstructure is typical of what is observed for PLD films on metal substrates.

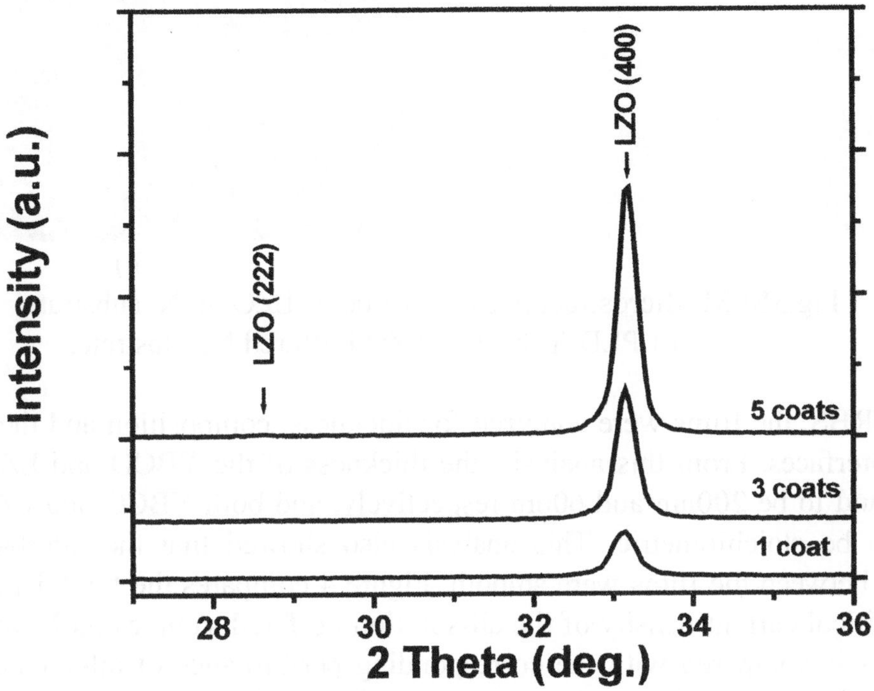

Fig.1 XRD patterns of multiple coats of LZO on textured Ni substrates.

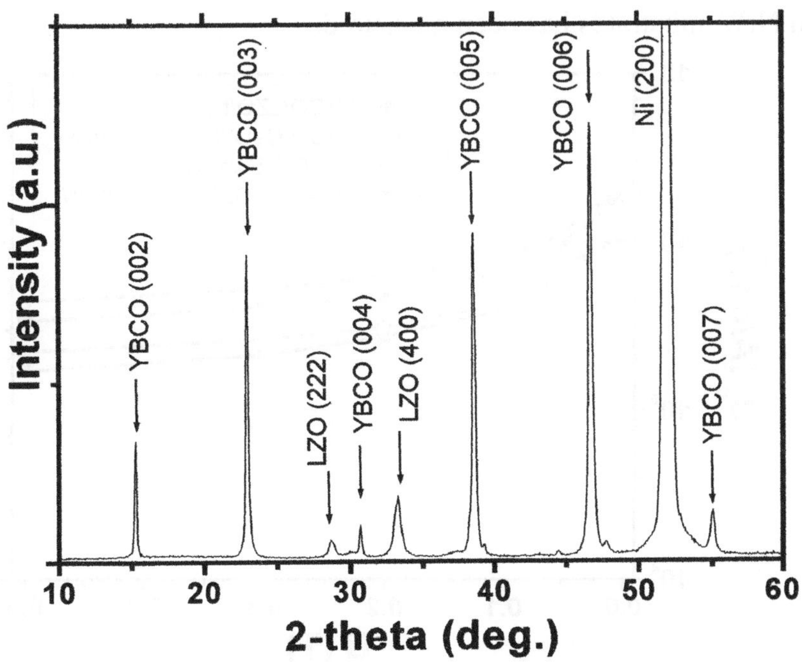

Fig.2 XRD pattern of PLD-YBCO on 60nm all-solution LZO buffered Ni substrates.

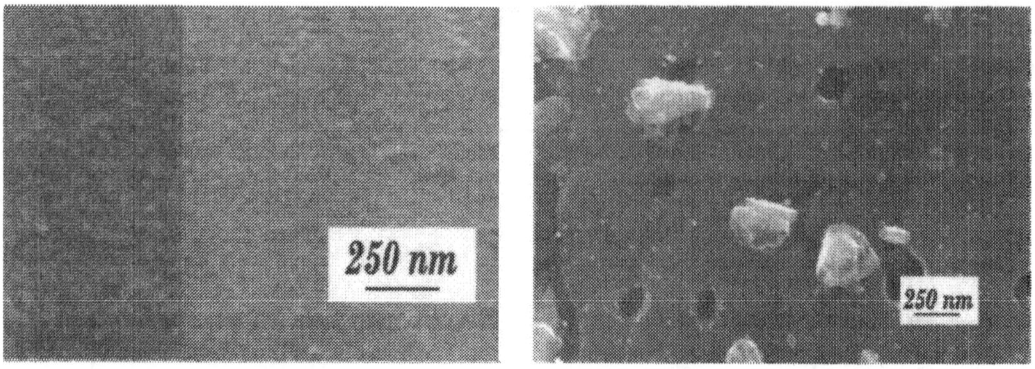

(a) (b)

Fig.3 SEM Microstructures of (a) 3 coats LZO on Ni substrate,
(b) PLD-YBCO on LZO buffered Ni substrate.

Using RBS, the films were analyzed for thickness, composition and the quality of the interfaces. From this analysis, the thickness of the YBCO and LZO films were found to be 200nm and 60nm respectively, and both YBCO and LZO were found to be stoichiometric. This analysis also showed that the interfaces and surfaces between the films were smooth. Figure 4 compares the field dependence of the critical current density of the all-solution buffered samples on Ni and NiW substrates is compared with the corresponding performance of all-vacuum three layer buffers on similar substrates. This figure clearly illustrates that the performance of all-solution LZO buffers is comparable to that of the three layer buffers. Critical current density of over 1 MA/cm^2 on Ni substrates and about 2 MA/cm^2 on NiW substrates has been measured.

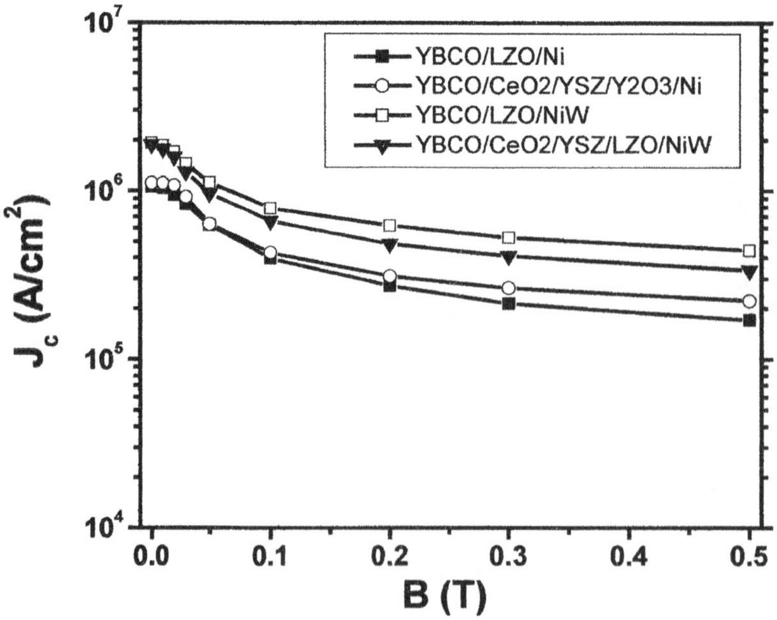

Fig. 4 Comparison of the J_c Vs B performance of YBCO on all-solution LZO buffers that of YBCO on tri-layer buffer architectures for Ni and NiW substrates.

Using a reel-to-reel dip coating process (described elsewhere [5]), multiple coats of LZO were processed over long lengths of Ni and NiW substrates. The in-plane and out-of-plane texture of 3 coats of LZO on Ni substrate and 4 coats of LZO on NiW substrate were characterized using a reel-to-reel X-ray diffractometer. The results from these measurements are shown in Fig.5. It is clear from this figure that using a reel-to-reel dip coating process, it is possible to deposit multiple coated LZO films with uniform texture over the entire length of the tape. It also illustrates that the texture of the LZO film is as good as that of the metal substrate. This offers promise to processing long lengths of solution buffered Ni and NiW tapes for coated conductor fabrication.

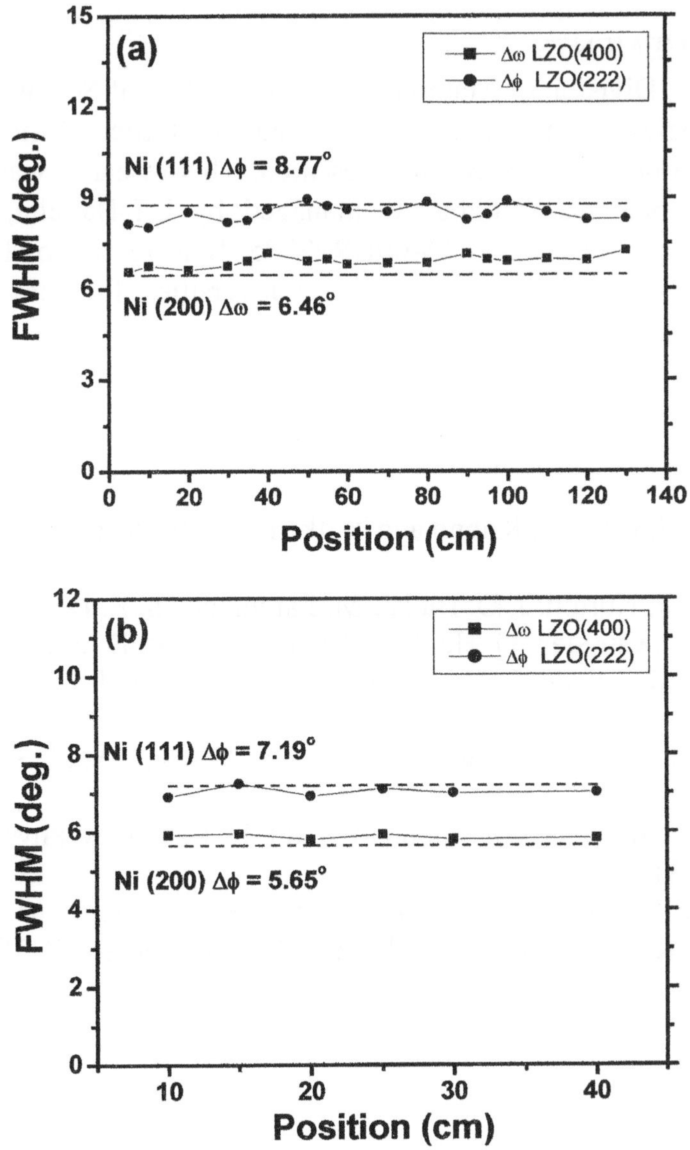

Fig. 5 Texture data from multiple coated LZO films on long length metal substrates. (a) 3-coats of LZO on long Ni substrate and (b) 4-coats of LZO on long NiW substrate.

CONCLUSIONS

A solution-based approach to buffer layer processing has been explored for coated conductors fabrication. Critical current densities over 1 MA/cm^2 on Ni substrates and up to 2 MA/cm^2 on NiW substrates have been measured using all-solution LZO buffer layers. Using XRD, it has been confirmed that the LZO films can act as good barrier layers and prevent the metal substrates from oxidation. Multiple coating of LZO has also been demonstrated on long lengths of Ni and NiW substrates with uniform texture through the length of the tape. These results are promising and give hope to the possibility of developing all-solution coated-conductors in long lengths.

ACKNOWLEDGEMENTS

The U.S. DOE, Division of Materials Sciences, Office of Science and Office of Power Technologies-Superconductivity Program, and Office of Energy Efficiency and Renewable Energy sponsored this research. This research was performed at the Oak Ridge National Laboratory, managed by UT-Battelle, LLC for the U.S.DOE under contract DE-AC05-00OR22725. Sincere acknowledgements are also extended to Oak Ridge Associated Universities for making this work possible.

REFERENCES

[1] Y. Iijima, N. Tanabe, O. Kohno and Y. Ikeno, Appl. Phys. Lett., **60**, 769 (1992).

[2] A. Goyal, D. P. Norton, J. D. Budai, M. Paranthaman, E. D. Specht, D. M. Kroeger, D. K. Christen, Q. He, B. Saffian, F. A. List, D. F. Lee, P. M. Martin, C. E. Klabunde, E. Hartfield, and V. Sikka, Appl. Phys. Lett., **69**, 1795 (1996).

[3] M. Bauer, R. Semerad, and H. Kinder, IEEE Trans. Appl. Supercond., **9**, 1502 (1999).

[4] D. Dimos, P. Choudhary, J. Mannhart, and F.K. Le Goues, Phys. Rev. Lett. **61**, 219 (1988).

[5] S. Sathyamurthy, M. Paranthaman, T. Aytug, B.W. Kang, P. M. Martin, A. Goyal, D.M. Kroeger, and D. K. Christen, Jl. Mater. Res., **17**, 1543 (2002).

SCALE UP OF HIGH PERFORMANCE HIGH TEMPERATURE SUPERCONDUCTORS

V. Selvamanickam, Y. Li, H. G. Lee, X. Xiong, Y. Qiao, J. Reeves, Y. Xie, A. Knoll, and K. Lenseth
SuperPower Inc., Schenectady, NY 12304

ABSTRACT

Second-generation High Temperature Superconducting (HTS) conductors have been fabricated by two high rate processes, Pulsed Laser Deposition (PLD), and Metal Organic Chemical Vapor Deposition (MOCVD). Ten meter long tapes have been produced with performance exceeding 100 A end to end. Also, critical currents as high as 173 A have been achieved in meter-long lengths by MOCVD.

INTRODUCTION

Second-generation HTS conductors based on RE-Ba-Cu-O (RE = rare-earths) have been developed since the early 1990's for electric power applications. The enabling technology for the second-generation HTS conductors is a textured template that can be produced by a variety of techniques. These include Ion Beam Assisted Deposition (IBAD) [1-3], Rolling Assisted Biaxially Textured Substrates (RABiTS) [4], and Inclined Substrate Deposition (ISD) [5]. Several techniques are being used to deposit HTS layers on the textured templates including PLD, MOCVD, E-beam Evaporation (BaF_2 process), and Metal Organic Deposition. Among these process PLD and MOCVD offer the advantage of high deposition rates. Deposition rates greater than 100 Angstroms/second have been demonstrated to produce YBCO films with Jc greater than 1 MA/cm^2 by both PLD [6] and MOCVD [7]. In addition to high deposition rate, MOCVD offers an additional advantage of large deposition zone. The deposition zone in MOCVD can be as long and as wide as the showerhead. Because of the combination of high deposition rate and large deposition zone, very high tape throughput can be achieved with MOCVD. We have been scaling up both high rate processes, PLD and MOCVD to fabricate second-generation HTS conductors in long lengths. In this article, we report the progress in high performance second-generation HTS conductors based on PLD and MOCVD.

EXPERIMENTAL AND RESULTS

Hastelloy C-276 substrates are polished in a custom built Chemical Mechanical Polishing facility in lengths of up to 100 m in a single pass. The surface roughness of the polished substrates is monitored on-line in the polishing facility using a laser scatterometer device. Measurements are obtained every 0.1 mm of the polished tape. Measurement obtained over a 10 m long section of a polished tape is shown in figure 1. As shown in figure 1, the average surface roughness over 10 m is about 1 nm and the standard deviation is approximately 0.08 nm. This roughness value compares well with the values we have reported on meter-long substrates [8].

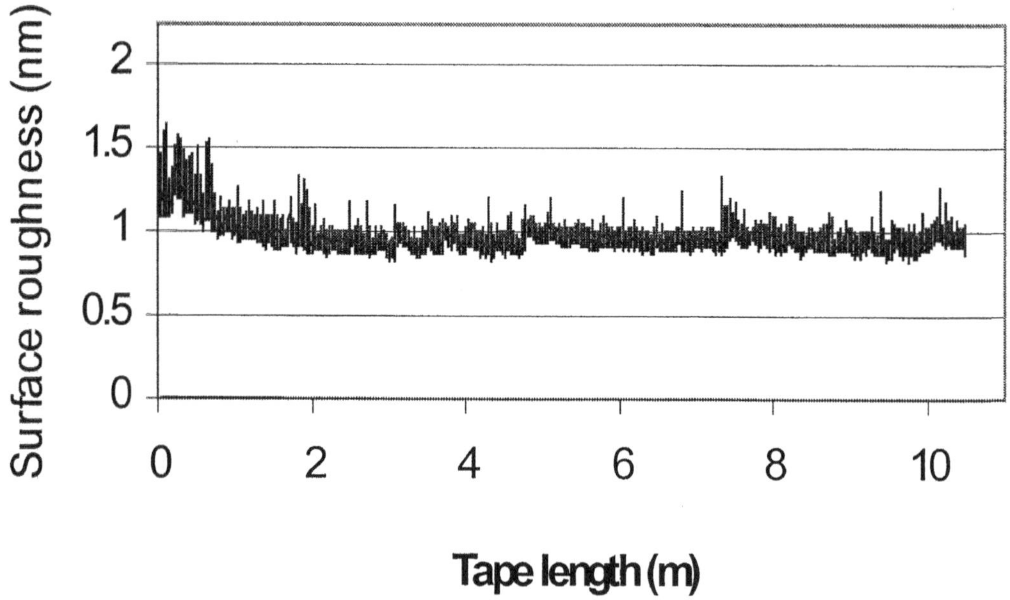

Figure 1 Surface roughness measurements obtained on-line from a 10 m long substrate tape polished by chemical mechanical polishing.

Buffer layers are then deposited on the polished substrate in a pilot IBAD facility. This facility is equipped with r.f. ion sources that can be operated continuously for several hundred hours. Ten meter long buffered substrates are routinely produced in the pilot IBAD facility. The in-plane texture of the buffered substrates is measured in a General Area Detector Diffraction System (GADDS) equipped with a spooling system. Direct polefigures of the film are obtained over long tape lengths. This avoids problems that can arise from indirectly determining the texture quality from intensity measurements. Results from direct in-plane texture measurements over a 10 m long buffered IBAD substrate are shown in figure 2. The texture measurement was obtained every meter of the tape. The average texture over 10 m is about 10.9 degrees with a standard deviation of 0.7 degrees. This compares well with the best reported bulk texture values on IBAD tapes. It also compares well with the texture values that we obtained on meter long tapes [9].

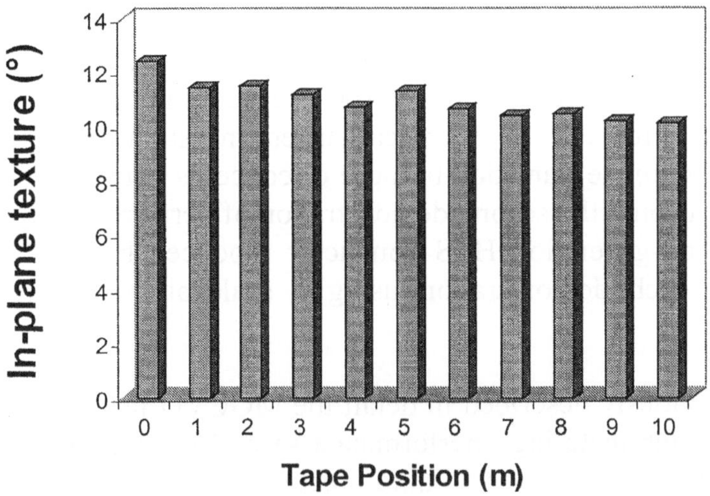

Figure 2 Distribution of in-plane texture in a 10 m long buffer tape produced by IBAD

Deposition of superconducting layers is conducted in a pilot PLD facility and a prototype MOCVD facility. The PLD facility is equipped with three, 3" diameter targets that can be rotated as well as rastered for uniform ablation. An industrial laser, Lambda Steel 670, that is rated for 24 hours/day, 7 days/week operation is used. This laser can deliver a maximum pulse energy of 670 mJ/pulse at a maximum repetition rate of 300 Hz. Pulse repetition rates of up to 200 Hz have been employed in this work. Previously, we reported achievement of critical currents over 100 A and up to 135 A with multiple meter-long tapes [10]. We also reported demonstration of 106 A over a 3 m long tape produced using PLD. In this article, we report demonstration of 106 A over 10 m long conductor. Results from critical current measurements from this tape are shown in figure 3.

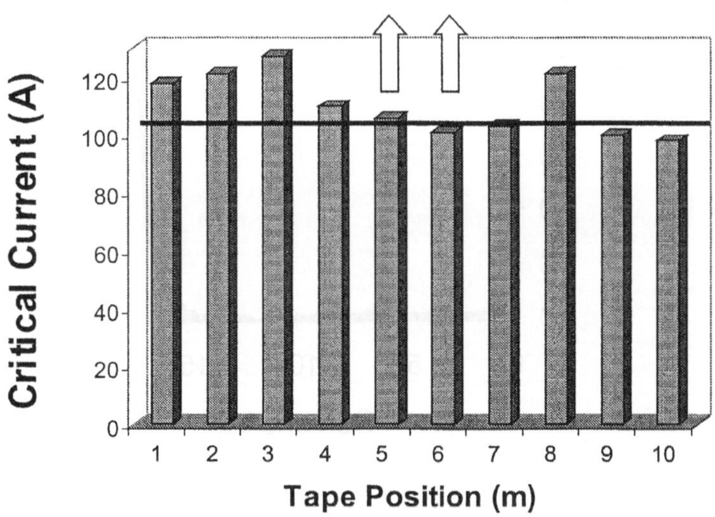

Figure 3 Critical current distribution over a 10 m PLD tape that exhibited an end-to-end critical current value of 106 A.

Critical current of the tape was measured every 1 m using a four-probe technique and a voltage criterion of 1 microvolt/cm. Every meter-long section of the tape exhibited a critical current over 100 A except the last meter (98 A). Also, as indicated by 2 arrows in fig. 3, critical current measurements were halted in 2 sections before complete transition because of concerns with tape heating. To our knowledge, this is only the second demonstration of performance over 100 A in a 10 m long second-generation HTS conductor produced by PLD. Also, this is likely the first such demonstration using a scaleable reel-to-reel deposition system.

We have previously described in detail the MOCVD facility and process that we developed to fabricate high performance second-generation HTS conductors [11]. Our MOCVD process uses a liquid precursor delivery system which has the advantages of long-term stability, high throughput, and reproducibility. We demonstrated 100 A and 1 MA/cm^2 performance in short samples of second-generation conductor using MOCVD a few years ago [11]. Recently, we scaled up our process to demonstrate multiple meter-long tapes with performance exceeding 100 A [10]. We believe that was the first demonstration of meter-long second-generation HTS conductor with performance greater than 100 A using MOCVD. In this work, we further improved the performance of meter-long conductors produced by MOCVD. Figure 4 shows a current-voltage curve obtained from a meter-long tape produced by MOCVD. As shown in the figure, a critical current of 173 A, corresponding to a Jc of 1.2 MA/cm^2 was achieved. To our knowledge there has been only one other demonstration of a meter-long second-generation HTS conductor with a better performance. That demonstration was with PLD [12].

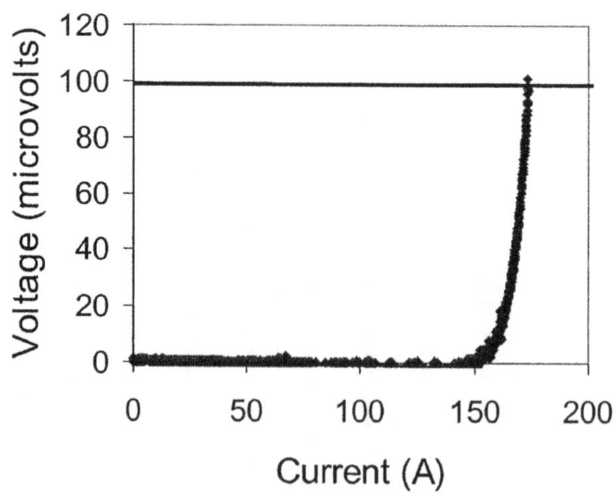

Figure 4 Current-voltage curve obtained from a 1 m long tape produced by MOCVD. A critical current value of 173 A was obtained in this tape.

Figure 5 exhibits the critical current distribution over a 2.8 m long tape produced by MOCVD. An end-to-end critical current of 105 A was achieved in this tape. Critical current measurements were obtained every 0.2 m of the tape and every section exhibited a critical current greater than 100 A except one (97 A). As shown by the arrows in figure 6, measurements had to be halted in 8 sections because of tape heating.

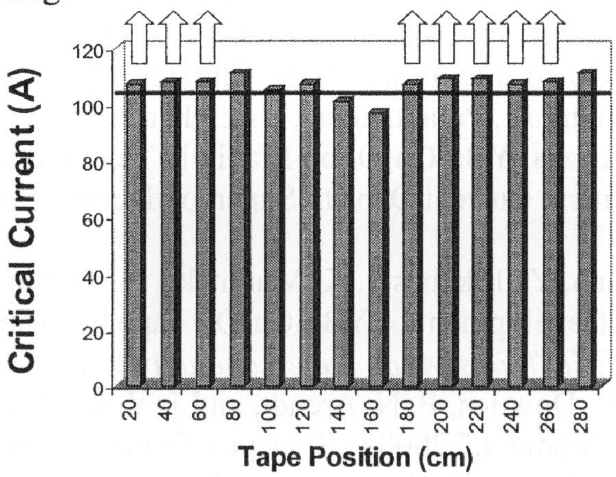

Figure 5 Critical current distribution over a 2.8 m MOCVD tape that exhibited an end-to-end critical current value of 105 A.

SUMMARY

Up to 10 m long second-generation HTS conductors have been produced. Surface roughness values of 1 nm have been achieved uniformly over 10 m long Hastelloy-C substrates by reel-to-reel chemical mechanical polishing. In-plane texture values of 10.9 degrees have been achieved uniformly over 10 m long buffer layers deposited by IBAD. End-to-end critical current value of 106 A has been demonstrated in 10 m long conductor produced by PLD. MOCVD has been used to produce second-generation conductor with critical currents as high as 173 A in meter-lengths and 105 A in 2.8 m lengths.

ACKNOWLEDGMENTS

This work was partially supported by the U.S. Department of Energy, U.S. Air Force including the Dual Use Science and Technology program and the Air Force Office of Scientific Research, and the New York State Research and Development Authority. Part of the work was also done in a Cooperative Research and Development Agreements with Los Alamos National Laboratory, Argonne National Laboratory and the Air Force Research Laboratory at Wright Patterson Air Force Base.

REFERENCES

[1]Y. Iijima, N. Tanabe, O. Kohno, and Y. Ikeno, "In-Plane Aligned $YBa_2Cu_3O_{7-x}$ Thin Films Deposited on Polycrystalline Metallic Substrates," *Appl. Phys Lett.*. **60**, 770 (1992).

[2]S. R. Foltyn, P. N. Arendt, P. C. Dowden, R. F. DePaula, J. R. Groves, J. Y. Coulter, Q. Xia, M. P. Maley, and D. E. Peterson, "High-T_c Coated Conductors – Performance of Meter-Long YBCO/IBAD Flexible Tapes, *IEEE Trans. Appl. Superconductivity* **9**, 1519 (1999).

[3]A. Usokin, J. Knoke, F. Garcia-Moreno, A. Issaev, J. Dzick, S. Sievers, and H. C. Freyhardt, "Large-area YBCO-Coated Stainless Steel Tapes with High Critical Currents," *IEEE Trans. Appl. Supercond.* **9**, 2248 (1999).

[4]D.P. Norton, A. Goyal, J.D. Budai, D. K. Christen, D. M. roeger, E. D. Specht, Q. He, B. Saffain, M. Paranthaman, C. E. Klabunde, D. F. Lee, B. C. Sales, F. A. List, "Epitaxial $YBa_2Cu_3O_7$ on Biaxially Textured (001) Ni : An Approach to High Critical Current Density Superconducting Tapes," *Science*, **274**, 755 (1996).

[5]Y. Sato, K. Matsuo, Y. Takahashi, K. Muranaka, K. Fujino, S. Hahakura, K. Ohmatsu, H. Takei, "Development of $YBa_2Cu_3O_Y$ Tape by Using Inclined Substrate Method," *IEEE Trans. Appl. Superconductivity* **11**, 3365 (2001).

[6]S. R. Foltyn, P. C. Dowden, P. N. Arendt, and J. Y. Coulter, "Coated Conductor Tape Development," 2000 U.S. Dept. of Energy Annual Peer Review, Washington D.C., July 17-19, 2000.

[7]P. Chou, Q. Zhong, Q. L. Li, K. Abazajian, A. Ignatiev, C. Y. Wang, E. E. Deal, and J. G. Chen, "Optimization of Jc of YBCO Thin Films Prepared by Photo-assisted MOCVD through Statistical Robust Design," *Physica C* **254**, 93 (1995).

[8]Y. Qiao, Y. Li, S. Sathiaraju, J. Reeves, K. Lenseth, and V. Selvamanickam, "An Overview of the Coated Conductor Progress at IGC-SuperPower," *Physica C* **382**, 48 (2002).

[9]V. Selvamanickam, H.-G.Lee, Y. Li, J. Reeves, Y. Qiao, Y.Y. Xie, K. Lenseth, G. Carota, M. Funk, K. Zdun, J. Xie, K. Likes, M. Jones, L. Hope, and D. W. Hazelton, "Scale up of High-Performance Y-Ba-Cu-O Coated Conductors," *Proc. Appl. Supercond. Conf.* Houston, Aug. 3 – 8, 2002 (to be published).

[10] V. Selvamanickam, H. G. Lee, Y. Li, X. Xiong, Y. Qiao, J. Reeves, Y. Xie, A. Knoll, and K. Lenseth, "Fabrication of 100 A Class, 1 m long Coated Conductor Tapes by Metal Organic Chemical Vapor Deposition and Pulsed Laser Deposition," *Physica C.* (*Proc. ISS*, Yokohama, Nov. 11-13, 2002) (in print)

[11] V. Selvamanickam, G. Carota, M. Funk, N. Vo, and P. Haldar, U. Balachandran, and M. Chudzik, P. Arendt, J. R. Groves, R. DePaula, B. Newnam, "High-Current Y-Ba-Cu-O Coated Conductor using Metal Organic Chemical-Vapor Deposition and Ion-Beam-Assisted Deposition," *IEEE Trans. Appl. Supercond.* **11**, 3379 (2000).

[12] A. Usokin, H. C. Freyhardt, A. Issaev, J. Dzick, J. Knoke, M. P. Oomen, H. –W. Neumueller, "Large-area YBCO-Coated Stainless Steel Tapes with High Critical Currents," *Proc. Appl. Supercond. Conf.* Houston, Aug. 3 – 8, 2002 (to be published).

INCLINED-SUBSTRATE PULSED LASER DEPOSITION OF YTTRIA-STABILIZED ZIRCONIA TEMPLATE FILM FOR YBCO COATED CONDUCTORS

B. Ma, M. Li, B. L. Fisher, R. E. Koritala, R. M. Baurceanu, S. E. Dorris, and U. Balachandran
Energy Technology Division
Argonne National Laboratory
Argonne, IL 60439

ABSTRACT

Biaxially textured films of yttria-stabilized zirconia (YSZ) were deposited on Hastelloy C276 (HC) substrates by inclined-substrate pulsed laser deposition (ISPLD). This method is promising for fabrication of YSZ templates on poly-crystalline metallic tapes for coated conductor applications. Scanning electron microscopy showed columnar grains in ISPLD-YSZ films. X-ray pole figure analysis revealed good biaxial alignment in these films. The in-plane texture was determined to be ≈16° from the full-width at half maximum (FWHM) in the YSZ (111) ϕ-scan; and the out-of-plane texture was ≈8° from the FWHM in the YSZ (002) ω-scan. Before the deposition of YBCO films by pulsed laser deposition, a thin layer of CeO_2 was deposited on the ISPLD-YSZ. The YBCO deposited on ISPLD-YSZ-buffered HC substrates were biaxially textured. T_c = 90 K and J_c = 180 kA/cm^2 at 77 K in self-field were measured.

INTRODUCTION

$YBa_2Cu_3O_{7-\delta}$ (YBCO) coated conductors are promising for high-current carrying wires and other electric power devices operating at temperatures that approach liquid nitrogen [1-3]. Textured template films or buffer layers are needed for deposition of biaxially aligned YBCO films to overcome weak links at the grain boundaries and, therefore, to achieve high critical current density (J_c) in the YBCO films on metallic substrates [4]. Several techniques, including ion-beam-assisted deposition (IBAD), rolling-assisted biaxially textured substrates (RABiTS), and inclined-substrate deposition (ISD), have been developed in recent years [5-9].

We grew biaxially textured YSZ thin films on mechanically polished Hastelloy C276 (HC) substrates by inclined-substrate pulsed laser deposition (ISPLD). Ceria cap layers and YBCO films were subsequently deposited on ISPLD-YSZ-buffered metallic substrates by pulsed laser deposition (PLD). X-ray pole figures, ϕ-scan and ω-scan were used to analyze texture. Scanning electron microscopy (SEM) and atomic force microscopy (AFM) were utilized to study morphology and surface roughness. In this paper, we discuss the growth conditions, microstructure, and crystalline texture of ISPLD YSZ buffer layers and YBCO films deposited on polished HC substrates.

EXPERIMENTAL PROCEDURE

HC coupons ($\approx$5 mm wide and 10 mm long) were mechanically polished to a mirror finish with 0.25-μm diamond paste for use as substrates. Surface roughness of $\approx$3 nm was measured by AFM. A schematic illustration of the experimental setup is shown in Fig. 1. The polished substrate was mounted on a tiltable heater stage using silver paste. The substrate inclination angle (α), substrate normal with respect to the laser plume axle, was set to a desired value (45-55°). A Lambda Physik LPX 210i excimer laser, with a Kr-F_2 gas premixture as the lasing medium, was used for ablation of targets (Superconductive Components, 99.99% pure). YSZ, CeO_2 and YBCO targets are 45 mm in diameter and 6 mm thick. The size of laser spot focused at the rotating target was $\approx$12 mm^2, which produced an energy density of $\approx$2.0 J/cm^2. The distance between the target and the substrates was 4-7 cm. The desired oxygen partial pressure was obtained by flowing ultra-high-purity oxygen through the chamber. Details of deposition conditions are listed in Table 1. Layers of CeO_2 and YBCO were subsequently deposited by PLD on the ISPLD-YSZ buffered HC substrate at zero degree inclination angle.

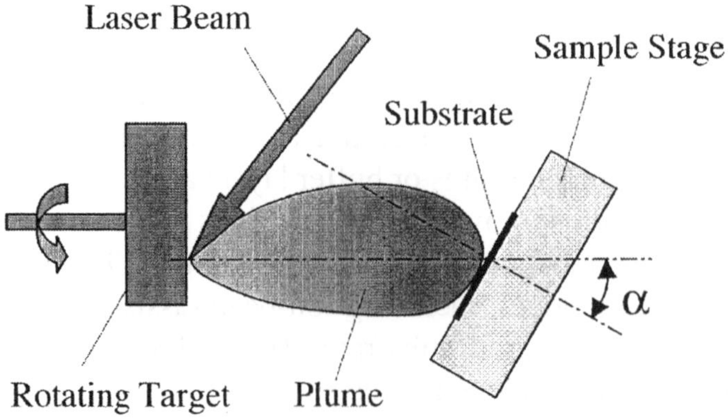

FIGURE 1. Schematic illustration of experimental setup for ISPLD system.

Table 1. PLD Conditions for Growth of Biaxially Textured YSZ Films

Laser System	Lambda Physik LPX 210i
Laser Wavelength	248 nm (KrF)
Pulse Duration	25 ns
Energy Density	1-3 J/cm^2
Repetition Rate	20-90 Hz
Substrate Temperature	400-650°C
Inclination Angle	45-55°
Oxygen Partial Pressure	1-200 mTorr
Target-to-Substrate Distance	4-7 cm

Crystalline texture was measured by X-ray diffraction pole-figure analysis using Cu-K_α radiation. In-plane texture was characterized by the FWHM of ϕ-scans for the YSZ (111) reflection, and out-of-plane texture was characterized by the FWHM of ω-scans for YSZ (002). Surface morphology was investigated by SEM using a Hitachi S-4700-II. Surface roughness was measured by AFM using a Digital Instruments D3100 scanning probe microscope (operated in tapping mode). The superconducting critical transition temperature (T_c) and J_c for the YBCO films were determined by the inductive method reported earlier [9].

RESULTS AND DISCUSSION

The X-ray diffraction 2θ scan patterns for the ISPLD YSZ films deposited at different temperature are shown in Fig. 2. Data indicate that out-of-plane orientation

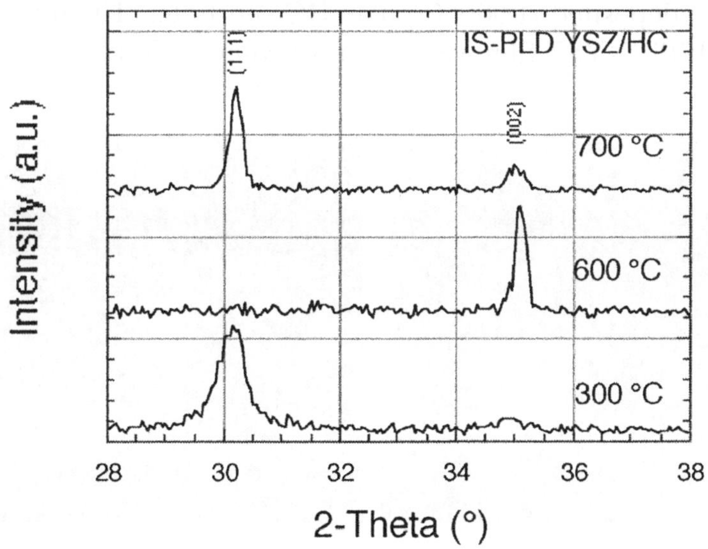

FIGURE 2. X-ray diffraction 2θ scan patterns for YSZ films deposited on HC substrate at different temperatures using ISPLD.

is dependent on deposition temperature. C-axis oriented films were obtained with a deposition temperature of ≈600°C. The (111) oriented YSZ films, whose texture was not in favor for the coated conductor applications, were obtained at higher (above 650°C) or lower (below 400°C) deposition temperatures. We will limit our discussion to c-axis oriented films in this paper.

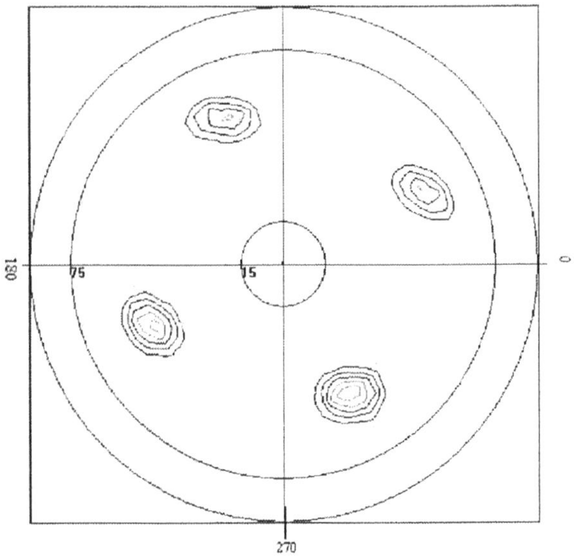

FIGURE 3. YSZ (111) pole figure for ISPLD YSZ film deposited at 600°C.

Figure 3 shows a typical X-ray diffraction pole figure of YSZ (111) for an ISPLD-YSZ film deposited at 600°C on an HC substrate. The four distinct poles indicate that the ISPLD-YSZ film was biaxially textured. In-plane texture measured

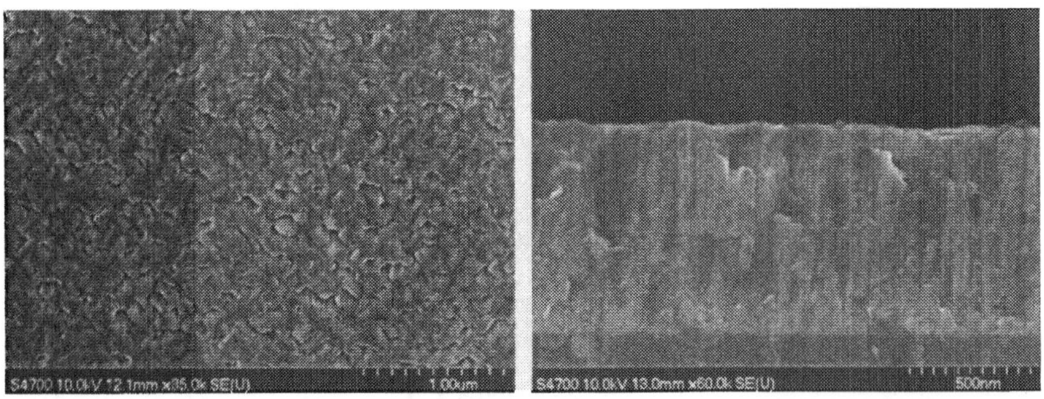

FIGURE 4. (a) Plan-view and (b) fracture cross-sectional view SEM images of ISPLD YSZ film deposited at 600°C.

from the FWHM of the YSZ (111) ϕ-scan was ≈16.0°. The c-axis of ISPLD-YSZ film was slightly tilted, with a tilt angle of ≈6°, similar to that reported by Hasegawa et al. [10]. Out-of-plane texture measured from the FWHM of the YSZ (002) ω-scan was ≈7.8°. Gaps between grains were observed from the plan-view SEM image (Fig. 4a) of an ISPLD YSZ film. Columnar grains were observed from the cross-sectional fracture surface (Fig. 4b). This funding suggests that the texturing mechanism for ISPLD-YSZ films is most likely due to grain self-shadowing effect, similar to that for the IDS MgO films [11,12]. However, unlike the ISD MgO, textured YSZ films can hardly be grown by simple inclination of substrates and using electron beam evaporation, where the momentum energy of the atoms in YSZ was not sufficent. In order to have atoms in YSZ to have a better arrangement at their sublattices, additional momentum other than thermal energy must be supplied to these atoms. In the case of ISPLD, the energy needed for preferred lattice arrangement in YSZ was provided by the pulsed laser plasma/plume. Root-mean-square (RMS) surface roughness of 16 nm was measured by tapping mode AFM on the as-deposited YSZ films.

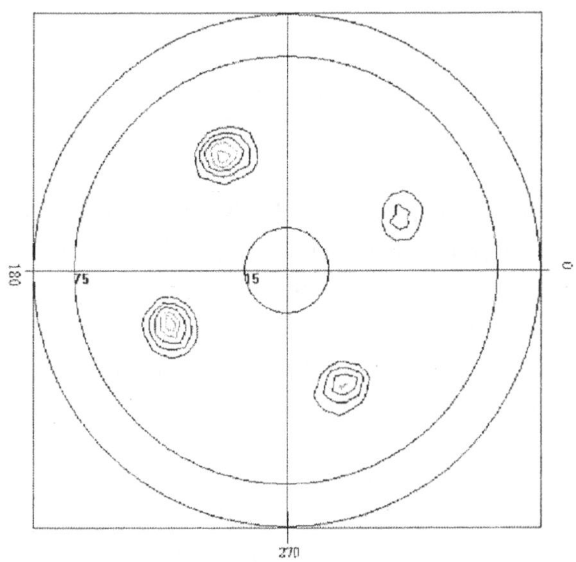

FIGURE 5. YBCO (103) pole figure of YBCO deposited on ISPLD YBZ buffered HC substrate.

YBCO deposited on ISPLD-YSZ buffered HC substrate was biaxially textured, as shown in the YBCO (103) pole figure (Fig. 5). A thin ceria layer was deposited prior to deposition of YBCO by PLD. The YBCO c-axis was parallel to the substrate normal, as illustrated by the four evenly distributed peaks in the YBCO (103) pole figure. FWHM = 18.2° was detected from the ϕ-scan of YBCO (103). T_c = 90 K and J_c = 180 kA/cm^2 at 77 K in self-field were measured.

CONCLUSIONS

Biaxially textured YSZ films were successfully grown by the ISPLD method. These films contained columnar grains. The surface roughness ≈16 nm was measured on the as-grown surface of an ISPLD YSZ thin films deposited on an HC substrate. In-plane texture measured by the FWHM in the YSZ (111) ϕ-scan was ≈16°. Out-of-plane texture measured by the FWHM in the YSZ (002) ω-scan was ≈8°. YBCO films deposited on ISPLD YSZ buffered HC substrates were biaxially textured with 18.2° FWHM in the ϕ-scan for YBCO (103). T_c = 90 K and J_c = 180 kA/cm^2 at 77 K in self-field were measured.

ACKNOWLEDGMENT

SEM analysis was performed in the Electron Microscopy Center for Materials Research at Argonne National Laboratory. This work was supported by the U.S. Department of Energy (DOE), Energy Efficiency and Renewable Energy, as part of a DOE program to develop electric power technology, under Contract W-31-109-Eng-38.

REFERENCES

1. D. K. Finnemore, K. E. Gray, M. P. Maley, D. O. Welch, D. K. Christen, and D. M. Kroeger, "Coated Conductor Development: An Assessment," *Physica C*, 320, 1-8 (1999).
2. Y. Iijima and K. Matsumoto, "High-Temperature-Superconductor Coated Conductors: Technical Progress in Japan," *Supercond. Sci. Technol.*, 13, 68-81 (2000).
3. J. O. Willis, P. N. Arendt, S. R. Foltyn, Q. X. Jia, J. R. Groves, R. F. DePaula, P. C. Dowden, E. J. Peterson, T. G. Holesinger, J. Y. Coulter, M. Ma, M. P. Maley, and D. E. Peterson, "Advance in YBCO-Coated Conductor Technology," *Physica C*, 335, 73 (2000).
4. D. Dimos, P. Chaudhari, and J. Mannhart, "Superconducting Transport Properties of Grain Boundaries in $YBa_2Cu_3O_7$ Bicrystals," *Phys. Rev. B*, 41, 4038-4049 (1990).
5. Y. Iijima, M. Kimura, T. Saitoh, and K. Takeda, "Development of Y-123-Coated Conductors by IBAD Process," *Physica C*, 335, 15 (2000).
6. C. P. Wang, K. B. Do, M. R. Beasley, T. H. Geballe, and R. H. Hammond, "Deposition of In-Plane Textured MgO on Amorphous Si_3N_4 Substrate by Ion-Beam-Assisted Deposition and Comparisons with Ion-Beam-Assisted Deposited Yttria-Stabilized-Zirconia," *Appl. Phys. Lett.*, 71, 2955-2958 (1997).
7. A. Goyal, D. P. Norton, J. D. Budai, M. Pranthaman, E. D. Specht, D. M. Kroeger, D. K. Christen, Q. He, B. Saffian, F. A. List, D. F. Lee, P. M. Martin, C. E. Klabunde, E. Hardtfield, and V. K. Sikka, "High Critical Current Density Superconducting Tapes by Epitaxial Deposition of YBCO Films on Biaxially Textured Metals," *Appl. Phys. Lett.*, 69, 1975 (1996).
8. M. Bauer, R. Semerad, and H. Kinder, "YBCO Films on Metal Substrates with Biaxially Aligned MgO Buffer Layers," *IEEE Trans. Appl. Supercond.*, 9, 1502, (1999).

9. B. Ma, M. Li, Y. A. Jee, B. L. Fisher, and U. Balachandran, "Inclined Substrate Deposition of Biaxially Textured Magnesium Oxide Films for YBCO Coated Conductors," *Physica C*, 366, 270-276 (2002).
10. K. Hasegawa, K. Fujino, H. Mukai, M. Konishi, K. Hayashi, K. Sato, S. Honjo, Y. Sato, H. Ishii, and Y. Iwata, "Biaxially Aligned YBCO Film Tapes Fabricated by All Pulsed Laser Deposition," *Appl. Supercond.*, 4, 487-493 (1996).
11. B. Ma, M. Li, R. E. Koritala, B. L. Fisher, A. R. Markowitz. R. A. Erck, R. Baurceanu, S. E. Dorris, D. J. Miller, and U. Balachandran, "Pulsed Laser Deposition of YBCO Films on ISD MgO Buffered Metal Tapes," *Supercond. Sci. Technol.*, 16, 464-472 (2003).
12. B. Ma, M. Li, B. L. Fisher, R. E. Koritala, and U. Balachandran, "Inclined Substrate Deposition of Biaxially Aligned Template Films for YBCO Coated Conductors," *Physica C*, 382, 38-42 (2002).

DEVELOPMENT OF CONDUCTIVE $La_{0.7}Sr_{0.3}MnO_3$ BUFFER LAYERS FOR Cu-BASED RABiTS

Tolga Aytug, M. Parans Paranthaman, Amit Goyal, Albert Gapud, Noel Rutter, Hong Ying Zhai and David K. Christen
Oak Ridge National Laboratory
1 Bethel Valley Road
Oak Ridge, TN 37831

ABSTRACT

For the development of $YBa_2Cu_3O_{7-\delta}$ (YBCO)-based coated conductors for electric power applications, it is desirable to electrically and thermally stabilize the high temperature superconducting (HTS) coating during its operation. In addition, use of non-magnetic substrates is an equally important consideration in conductor configurations, in order to reduce ferromagnetic hysteresis energy loss in *ac* applications. We have developed a conductive buffer layer of $La_{0.7}Sr_{0.3}MnO_3$ (LSMO) on biaxially textured non-magnetic Cu substrates to electrically couple the HTS layer to the underlying metal substrate. Property characterizations of YBCO films on LSMO/Ni/Cu multilayer architecture revealed good electrical connectivity over the entire structure.

INTRODUCTION

Potential applications of HTS coated-conductors involve areas of electric utility and power industry. For these applications, coated conductor tapes must be stabilized against thermal runaway in the event of an over-current situation. The ideal solution to avoid thermal runaway is to electrically couple the thin HTS layer to the thick underlying metal substrate through a conductive buffer layer, which reduces the overall resistivity in the event of a transient loss of superconductivity. As long as the power generated per unit surface area of conductor, due to Joule heating, remains sufficiently low, (eg. relative to the critical heat flux of boiling cryogen, or the thermal diffusivity of the system [1,2]) then event can be controlled and superconductivity should eventually be restored.

To date, much of the work reported using the rolling assisted biaxially textured substrates (RABiTS) [3] technique has utilized high purity Ni (99.99%) as the texture material. However, the ferromagnetism (FM) of pure Ni presents significant challenges in long length manufacturing where AC losses are an issue. On the other hand, Cu is a lower-cost, lower-resistivity and non-magnetic alternative substrate material for the production of long length RABiTS-based coated conductor tapes.

In this study, we assess the viability of $La_{0.7}Sr_{0.3}MnO_3$ (LSMO) as a *conductive* buffer-layer on biaxially Cu tapes. The structural, electrical and superconducting properties of these short prototype conductors are reported.

EXPERIMENTAL

Biaxially textured Cu substrates were obtained from randomly oriented high purity (99.99%) Cu bars, which are first mechanically deformed by cold-rolling and then annealed in vacuum at 800 °C for 1 hour, to obtain the desired (100) cube texture. For some samples, a thin ~ 2μm layer of Ni was first deposited. Deposition of LSMO films was conducted by *rf*-magnetron sputtering in a mixture of reducing forming gas (96%Ar + 4%H_2) and $2x10^{-5}$ Torr of H_2O. The substrate temperature was kept around 550-625 °C. The YBCO films were grown by pulsed laser deposition (PLD), using a KrF excimer laser system operated at an energy density of ≈2 J/cm^2 at 780 °C in 120 mTorr of O_2. Typical thicknesses of the YBCO films were 200 nm. The crystallographic orientation and texture of the films were characterized by X-Ray Diffraction (XRD), and surface microstructural investigations were conducted on a JOEL model, JSM-840 scanning electron microscope (SEM). A standard four-probe technique was used to evaluate the electrical properties, including T_c, and J_c of the YBCO films.

RESULTS

Typical θ-2θ XRD spectra for a 600 nm thick LSMO layer deposited directly on Cu substrate and the subsequent YBCO coating are shown in Figs. 1a and 1b, respectively. The patterns indicate only (00*l*) reflections from both the YBCO and the LSMO layers, demonstrating that the YBCO/LSMO/Cu are *c*-axis-oriented. There is no evidence of CuO/Cu_2O formation at the interface after buffer layer deposition. However, after YBCO deposition, additional peaks at around 37° and 43° appeared, indicating the formation of a large amount of copper oxide in the sample. To investigate the details of this issue, SEM analysis has been made. Figure 2a illustrates the surface morphology of the same YBCO as in Fig. 1b. Figure 2b is the higher magnification image of the boxed region. It can be clearly seen from these micrographs that the YBCO surface is covered with cracks, lying parallel to the sample, and through these cracks Cu-O diffuses in to the sample. After some detailed annealing studies, we deduce that the reason for the existence of cracks can be explained by the volume expansion (≈ 58 %) per Cu atom in Cu_2O, relative to that of the Cu in pure Cu matrix, and not because of the thermal expansion difference between the metal substrate ($\alpha = 18x10^{-6}/ °C$) and the oxide layers ($\alpha \approx 11x10^{-6}/ °C$). During the YBCO deposition, oxygen diffuses through the LSMO into the Cu substrate resulting in large amount of interfacial Cu-O

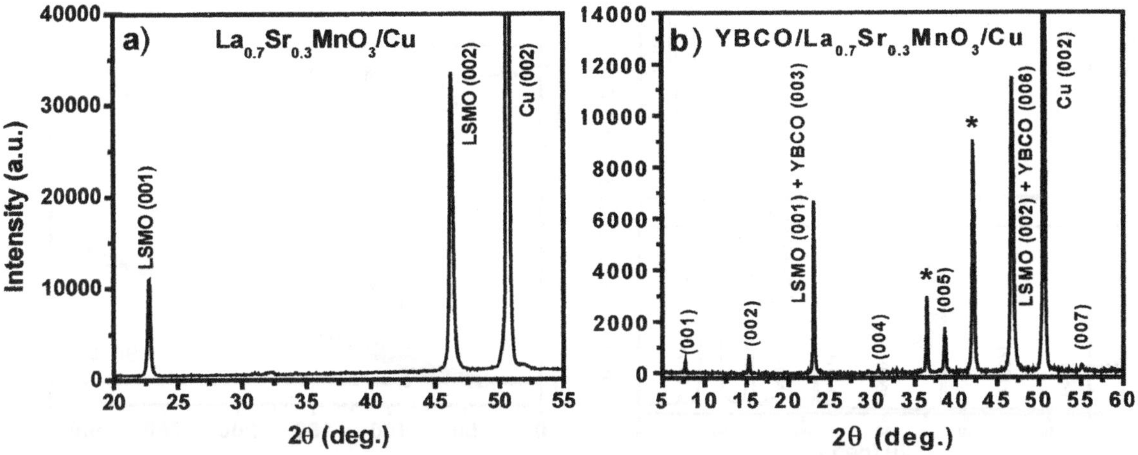

Fig. 1 Typical XRD θ-2θ scans (a) for LSMO/Ni. (b) for 200 nm YBCO film on the same LSMO/Cu multilayer. (* indicates Cu_2O impurities)

formation with associated volume expansion, leading to stress build-up on the LSMO layer, and eventually to crack development. In order to overcome this issue and to achieve an electrically conductive architecture we have chosen the

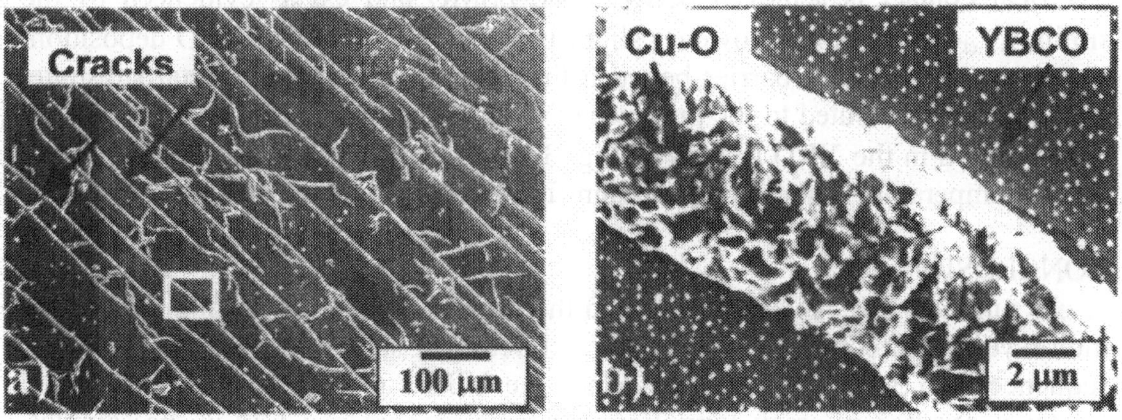

Fig. 2 SEM micrograph of (a) the surface morphology of the YBCO film grown on the LSMO/Cu substrate, and (b) the magnified image of the boxed region in part a.

alternative route using protective Ni overlayers on Cu substrates. Previously we have demonstrated high quality growth of La-based manganese oxide layers on Ni-based templates [4]. Figure 3a shows the θ-2θ XRD pattern for the LSMO/Ni/Cu, while Fig. 3b plots the pattern for the YBCO film on the same LSMO/Ni/Cu architecture. The presence of only (00*l*) reflections indicates that all layers are oriented with its *c*-axis normal to the plane of the substrate. After YBCO deposition, presence of a small amount of NiO is detected, the consequence of which is explained in Fig. 4. Note that the T_c of the sample is 90.4 K and the 77 K self-field J_c value is 2.3×10^6 A/cm^2. In this figure, we compared the net resistivity (ρ_{net}) versus temperature behavior of YBCO/LSMO/Ni/Cu

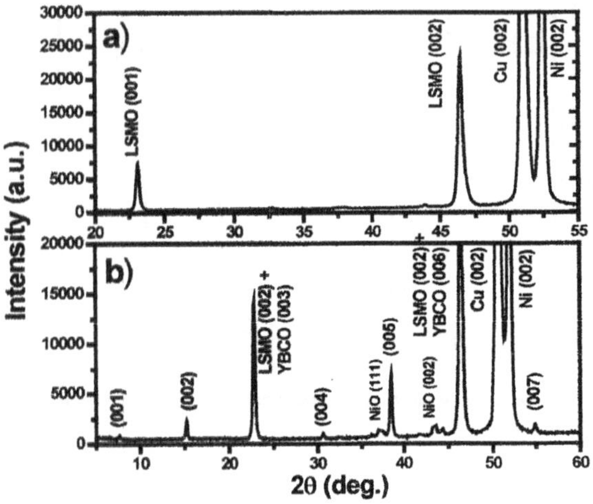

Fig. 3 XRD θ-2θ patterns (a) for LSMO/Ni/Cu (b) for a YBCO film on LSMO/Ni/Cu multilayers.

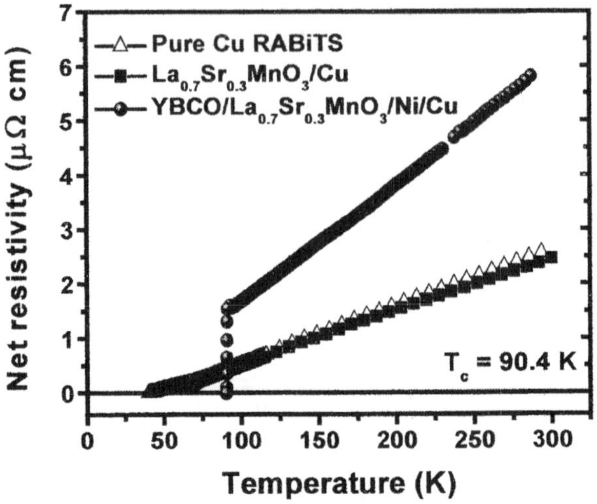

Fig. 4 shows the temperature dependent resistivities of pure Cu LSMO/Cu, and YBCO/LSMO/Ni/Cu.

sample with data for the LSMO/Cu as well as with the pure textured Cu. Clearly, electrical contact is achieved between the LSMO and Cu as evidenced by the similar ρ_{net}-T behaviors. However slight increase in ρ_{net}, after YBCO deposition, indicates some contact barrier between the oxide layers and the metal substrate, which can be attributed to the presence of NiO at the substrate-buffer interface, as was detected in the XRD spectra of Fig. 3b. Nevertheless, ρ_{net} values < 2 μΩ-cm at LN_2 temperatures should provide significant stability to the entire structure.

CONCLUSION

In summary, we have demonstrated the successful fabrication of a conductive multilayer architecture on biaxially textured Cu substrates having the layer sequence of YBCO/LSMO/Ni/Cu. Since the YBCO growth on LSMO buffered Cu tapes resulted in crack formation and large amount of Cu-O in the samples, additional Ni overlayer is used to provide a barrier for Cu oxidation and/or diffusion. Electrical property characterization revealed good electrical coupling between the YBCO coating and the LSMO/Ni/Cu.

REFERENCES

1. C. Cantoni, T. Aytug, D.T. Verebelyi, M. Paranthaman, E.D. Specht, D.P. Norton, and D.K. Christen, IEEE Trans. on Appl. Supercond. **11**, 3309 (2001).
2. Y. Fu, O. Tsuamoto, and M. Furuse, IEEE Trans. on Appl. Supercond. (2003).
3. A. Goyal, D.P. Norton, J.D. Budai, M. Paranthaman, E.D. Specht, D.M. Kroeger, D.K. Christen, Q. He, B. Saffian, F.A. List, D.F. Lee, P.M. Martin, C.E. Klabunde, E. Hatfield, and V.K. Sikka, Appl. Phys. Lett. **69**, 1795 (1996).
4. T. Aytug, M. Paranthaman, H.Y. Zhai, H.M. Christen, S. Sathyamurthy, D.K Christen, and R.E. Ericson, J. Mater. Res. **17**, 2194 (2002).

ANALYTICAL TRANSMISSION ELECTRON MICROSCOPY OF THICK $YBa_2Cu_3O_{7-\delta}$ FILMS ON RABITS

Keith J. Leonard, Sukill Kang, Byeongwon Kang*, Amit Goyal
and Donald M. Kroeger
Oak Ridge National Laboratory, Metals and Ceramics Division,
Superconducting Materials Group, One Bethel Valley Road
Oak Ridge, TN 37831-6116.

ABSTRACT

Microstructural changes related to the critical current density dependence on film thickness in superconducting $YBa_2Cu_3O_{7-\delta}$ (YBCO) films have been investigated through electron microscopy. Pulsed laser deposited YBCO films ranging in thickness from 0.19 to 6.4 μm were grown on rolling assisted biaxially textured substrates (RABiTS) having two different buffer layer architectures. Remarkable improvements in the YBCO microstructure were observed in samples deposited on Ni-3%W, with a thick Y_2O_3 seed layer. Films grown on this structure showed little random oriented grain formations along with minimal interfacial reactions. The causes of reduced performance with increased film thickness will be addressed for both RABiTS architectures.

INTRODUCTION

Difficulties in depositing or growing thick, high quality, c-axis oriented $YBa_2Cu_3O_{7-\delta}$ (YBCO) films whose properties do not become degraded with increasing film thickness, has been an elusive goal in producing commercially viable superconducting films. Typical problems in producing thick YBCO films can include: the development of misoriented YBCO grains or porosity,[1-5] formation of non-superconducting phases,[6-9] oxygen deficiencies[10] and changes in stoichiometry.[11] In addition, the choice of substrate can have a significant effect on the defects and microstructures generated within the YBCO film.[12] In this investigation, transmission electron microscopy was used to examine the causes for the reduction in J_c with film thickness for pulsed laser deposited (PLD) YBCO

* Byeongwon Kang, is now with the National Creative Research Initiative Center for Superconductivity, Department of Physics, Pohang University of Science and Technology, Pohang, South Korea.

films on two different architectures of rolling assisted biaxially textured substrates (RABiTS). The architecture of the two substrates investigated were as follows:

RABiTS1: Ni / CeO_2 / yttrium-stabilized zirconia (YSZ) / CeO_2 / YBCO

RABiTS2: Ni-3at.%W / Y_2O_3 / YSZ / CeO_2 / YBCO

The use of Ni-3at.% W, hereby written as Ni-W, posses a sharper texture over earlier pure Ni tapes and is a recent improvement in the development of RABiTS materials.[13] The use of a Y_2O_3 seed layer instead of CeO_2 is the result of another upgrade in the RABiTS materials, due to concerns over crack formation in the CeO_2 seed layer and subsequent NiO growth into the upper layers of the coated conductor stack during processing.

EXPERIMENTAL

The 12 nm thick CeO_2 and 150 nm thick Y_2O_3 seed layers of the two substrates were deposited through electron beam evaporation of pure metals under a partial pressure of water onto the {100} <100> biaxially textured Ni and Ni-W substrates, respectively. Additional 150 to 200 nm thick YSZ and 20 nm thick CeO_2 layers were deposited by RF magnetron sputtering. Details in the processing conditions and substrate properties are covered elsewhere.[14-17]

YBCO films ranging in thickness from 0.19 to 6.4 μm were deposited by pulsed laser deposition (PLD) using a XeCl excimer laser (λ=308nm) on 2.5 cm x 0.5 cm samples. Deposition was done at 790°C under an oxygen partial pressure of 120 mTorr. Laser energy density was 4 J/cm^2 with a film growth rate between 5 to 13 Å/s. Following deposition, the films were cooled to room temperature at a rate of 5°C/min, under an oxygen partial pressure of 550 Torr.

Cross-section samples were prepared for transmission electron microscopy (TEM), either by tri-pod polishing or conventional ion-milling methods, details of which are described elsewhere.[9,18] The samples were investigated using a Philips Technai 20 (LaB_6, 200kV) and CM200 (FEG, 200kV) microscopes, equipped with energy dispersive spectrometry (EDS) units. The CM200 TEM was used in both conventional and high-resolution scanning modes.

RESULTS AND DISCUSSION

Properties of the Thick YBCO Films

From previous work,[16-17] the critical current densities (J_c) of the thick YBCO films along with the film textures determined through x-ray diffraction (XRD) are listed in Tables I and II for the two substrate types. The YBCO films characterized on RABiTS1 by TEM were 0.19, 0.5, 1.7 and 3.0 μm thick, and were similar in electrical and texture properties to the samples shown in Table I. The RABiTS2 samples characterized by TEM are those listed in Table II. The values of J_c were calculated from the four-point probe measurements of critical current (I_c) conducted at 77K in self-field without micro-bridge patterning. Limitations in characterizing the film properties to 120 amps prevented the full determination of I_c in films thicker than 1.0 μm of the RABiTS2 materials under self-field. For these films a zero-field J_c was calculated from in-field measurements at 0.5 Tesla, assuming that a factor of 4 to 5 drop in J_c occurs from self-field to that of 0.5 Tesla.[16] Therefore, a range of calculated values are presented for the 2.9 and 4.3 μm thick films in Table II. During measurement of the 6.4 μm thick film on RABiTS2, a crack was generated across the sample upon reaching 60 amps, believed to have been the result of sample heating.

Table I. Measured electrical and texture properties of thick YBCO films on the RABiTS1 material, from ref. [17].

YBCO thickness (μm)	J_c (MA/cm^2)	$\Delta\phi^*$	$\Delta\omega^{**}$	a-axis fraction $Ia_\perp(200) / Ic_\perp(002)$	% Cube
0.19	2.6	1.07	0.73	0.0	98.8
0.43	1.4	1.04	0.70	0.0	90.4
1.6	0.59	1.03	0.76	0.03	89.7
3.0	0.45	1.08	0.73	0.05	79.9

* In-plane texture: ratio of the FWHM of x-ray intensities of (113)YBCO to (111)Ni/Ni-W.

** Out-of-plane texture: ratio of (005)YBCO to (200)Ni/Ni-W.

Table II. Measured electrical and texture properties of thick YBCO films on the RABiTS2 material, from ref. [16].

YBCO thickness (μm)	J_c (MA/cm^2)	$\Delta\phi$	$\Delta\omega$	a-axis fraction $Ia_\perp(200) / Ic_\perp(002)$	% Cube
1.0	1.18	0.90	0.79	0.0	92.5
2.9	0.9 - 1.1	1.03	0.78	0.1	93.5
4.3	0.65 - 0.81	1.09	0.89	0.2	87.9
6.4		0.96	0.91	0.1	95.8

The fraction of a-axis ($a_\perp$) to c-axis ($c_\perp$) oriented grains within the films showed an increase with film thickness for both substrates, with that of the RABiTS2 materials being significantly higher. Although the larger $a_\perp$ concentration observed was substantially less than that appearing in YBCO films deposited on single crystal substrates under the same conditions.[17]

While the a-axis fraction of the Ni-W RABiTS2 showed a higher value than that of the Ni RABiTS1 substrates, the percent cube texture was significantly better despite the random variation in results measured. A slight increase in $\Delta\omega$ with thickness was also observed for the RABiTS2 materials.

Microstructural Changes with Film Thickness for RABiTS1.

The 0.19 μm YBCO film examined in cross-section (Fig. 1(a)) showed complete epitaxial c-axis orientation, with no secondary phase formation or porosity. The YBCO film consisted of c-axis oriented grain columns, separated by anti-phase boundaries whose contrast is visible due to the displacement of the (001) planes by a fraction of the unit cell dimension. Created either by ledge steps at the CeO_2 interface,[19] or by a variation in the initiating atomic stacking sequence of YBCO.[20] Deviations in $c_\perp$ columns across the anti-phase boundaries, measured as splitting of YBCO electron diffraction intensities, was negligible in the thinner films, but was found to increase with either film thickness or aging time.[9] Investigation of the YBCO/CeO_2 interface showed no evidence of reaction between the layers in the 0.19 μm film, nor the presence of any disordered perovskite layer within the YBCO as reported to occur in PLD and ex-situ grown films on single crystal and metal substrates.[20-21]

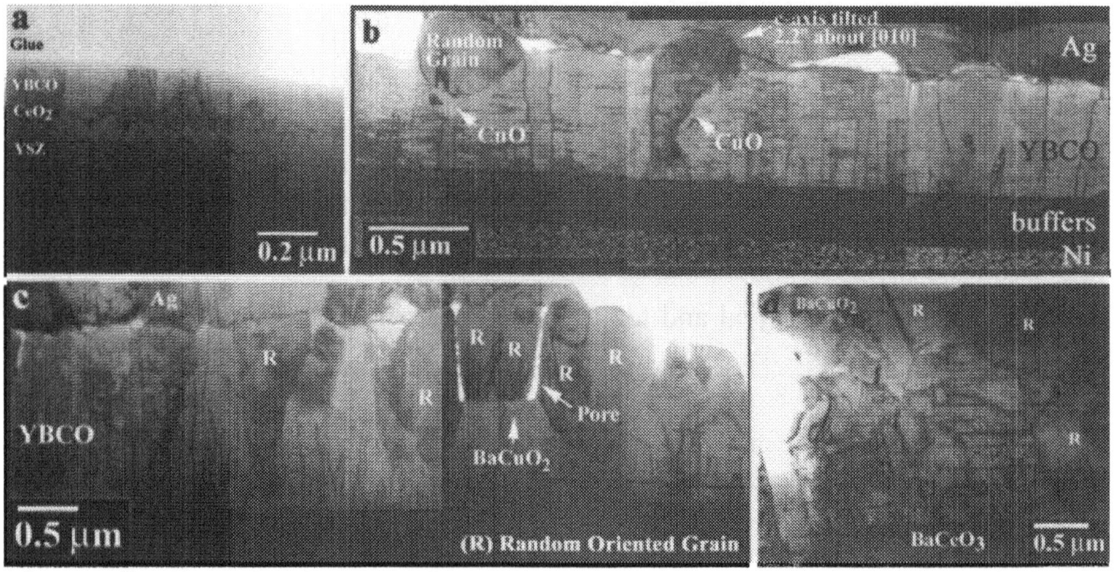

Fig. 1. Cross-section images of: (a) 0.19 μm, (b) 0.5 μm, (c) 1.7 μm, and (d) 3.0 μm thick YBCO films on RABiTS1.

Examination of the 0.5 μm thick YBCO sample on the RABiTS1 substrate (Fig. 1(b)) also showed $c_\perp$ columns extending the entire film thickness and no porosity. However, the film did show the formation of random oriented grains near the top surface beginning at a thickness of 0.45 μm, contributing to the drop in cube texture from 98.8% in the 0.19 μm film to 90.4%.

Observed within the 0.5 μm film was an outgrowth grain of YBCO, characterized as having an ~2.2° tilt in its c-axis, rotated about the [010] direction, relative to the surrounding film. The outgrowth originated 150 nm above the CeO_2 cap layer, and was not found to be associated with any defects within the buffer layer or the formation of secondary phases. A detailed explanation of the types of surface outgrowths observed in PLD films is provided elsewhere,[9] but it is believed to have occurred in this film as a result of a strain release associated the lattice mismatch between the growing YBCO film and CeO_2 layer.

As a result of reaction between YBCO and CeO_2 cap layer, $BaCeO_3$ particles of 10 to 20 nm in size were randomly observed along the interface (Fig. 2(a)).

Their formation occurred following the initial nucleation and growth of the YBCO, as no $BaCeO_3$ was observed in the 0.19 μm thick YBCO film. Due to the excess Y and Cu released from the formation of $BaCeO_3$, occasional particles of Y_2O_3 were observed within the film in addition to CuO formation along the interfaces between the misoriented and $c_\perp$ grains. Investigation of possible orientation relationships between CuO and YBCO revealed multiple growth directions possible for CuO.[9] It was found that while some CuO particles held a common relationship to the $c_\perp$ grains, others shared one with the random oriented grains. Suggesting that the formation of CuO followed random grain growth, and was not responsible for the misoriented YBCO grains.

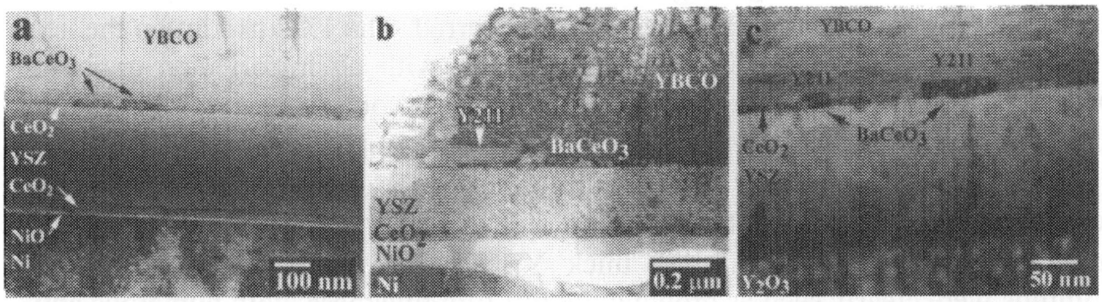

Fig. 2. Interfacial reactions between YBCO and CeO_2 in (a) 0.5 μm and (b) 3.0 μm thick YBCO films on RABiTS1, and in the (c) 6.4 μm thick RABiTS2 film.

The microstructure of the 1.7 μm film showed relatively undisturbed $c_\perp$ grains in the lower portion of the YBCO film (Figure 1(c)), however, the misoriented grains were more numerous near the top of the film surface. Unlike YBCO films grown by PLD on IBAD YSZ buffered substrates,[10] the misoriented grains do not completely cover the top of the film. The majority of the misoriented grains were formed in groups that typically shared a common twin boundary. The presence of porosity within the microstructures was also beginning to be seen, but was usually associated with the large mismatches between the $c_\perp$ and random oriented grains. These regions, however, may have contained secondary phases that were removed during sample preparation. CuO particles at the interfaces between the two types of grain orientations were also present, in addition to the formation of $BaCuO_2$.

The thickest YBCO film on RABiTS1, which suffered from the largest degradation in cube texture, showed a very rough, highly misaligned upper portion in its cross-section (Fig. 1(d)). Pores were found more frequently within the YBCO, which are believed to be associated with either the formation of secondary phases or the removal of quenched in vacancies.

Formation of CuO along the interfaces of the YBCO grains were again present within the 3.0 μm film, but a dramatic increase in the amount of $BaCuO_2$ was observed at locations both near the top surface of the film as well as at intermediate depths.

The extent of the reaction between the YBCO and CeO_2 cap layer appeared to be complete within the 3.0 μm sample (Fig. 2(b)). Investigation of the interface region through both diffraction and composition analysis could not identify any CeO_2 layer remaining. Y-enrichment resulting from the formation of $BaCeO_3$ did produce Y_2BaCuO_5 (Y211) grain formations along the top of the $BaCeO_3$ reaction layer rather than Y_2O_3 particles. While Y_2O_3 can nucleate homogeneously in YBCO and is more frequently observed in YBCO films despite being less thermodynamically stable than Y211,[22-23] the high energy interfacial sites associated with the $BaCeO_3$ formation allowed for easier Y211 nucleation.

Microstructural Changes with Film Thickness for RABiTS2

Immediately, it can be seen by comparison between the 1.0 μm YBCO film on the RABiTS2 material (Fig. 3(a)) to that of the films on RABiTS1 (Fig. 1), that a significant improvement in film quality has occurred. The 1.0 μm film on the new substrates showed completed c-axis orientation, with no misoriented grains or porosity observed within the examined sample. Only contrast from stacking faults and the $c_{\perp}$ column boundaries are visible within the composite image.

The YBCO films greater than 1.0 μm on RABiTS2, were found to be similar to one another. The thickest of the films studied and the most remarkable of the microstructures, was the 6.4 μm thick YBCO film (Fig. 3(b)). The c-axis columns were clearly seen as growing the full film thickness, with numerous anti-phase boundaries observed which originate and terminate within the thickness of the YBCO layers. The mismatch across these boundaries was considered negligible through electron micro-diffraction.

The thicker YBCO samples also showed tubular or elongated porosity within the film, not observed in the 1.0 μm thick film. The triangular shape of the porosity imaged in Fig. 3(b), is the result of the cross-section having been cut at an angle relative to the normal of the film. Porosity was observed at all depths in the thick YBCO layers, and did not appear to be either connected to each other or long enough to reach the free surface of the YBCO film.

Further revealed under higher magnification was the presence of secondary phase particles within the YBCO layer, observable in the dark-field image of Fig. 3(c). The particles identified as CuO and Y_2O_3 through electron diffraction, formed as the result of the coalescence of excess Y and Cu oxide along stacking faults. The monoclinic CuO particles (Fig. 3(d)) were indexed as having a CuO [010] // YBCO [100] orientation relationship[18] were much larger in size than the Y_2O_3 particles (Fig. 3(e)) and grew as long continuous particles. The Y_2O_3 particles which held the Type A,[25-26] Y_2O_3 [110] // YBCO [100] relationship, were smaller and discontinuous in the film, typically forming separate from other Y_2O_3 or CuO particles. Staking faults were no longer observed in the thicker films as they had been in the 1.0 μm film. Volume changes associated with the formation of CuO and Y_2O_3 particles are believed to be responsible for the creation of the porosity observed within the thicker RABiTS2 films.

Reaction between the YBCO and CeO_2 layers within the RABiTS2 samples was also limited (Fig. 2(c)). Formations of both $BaCeO_3$ and Y211 were identified

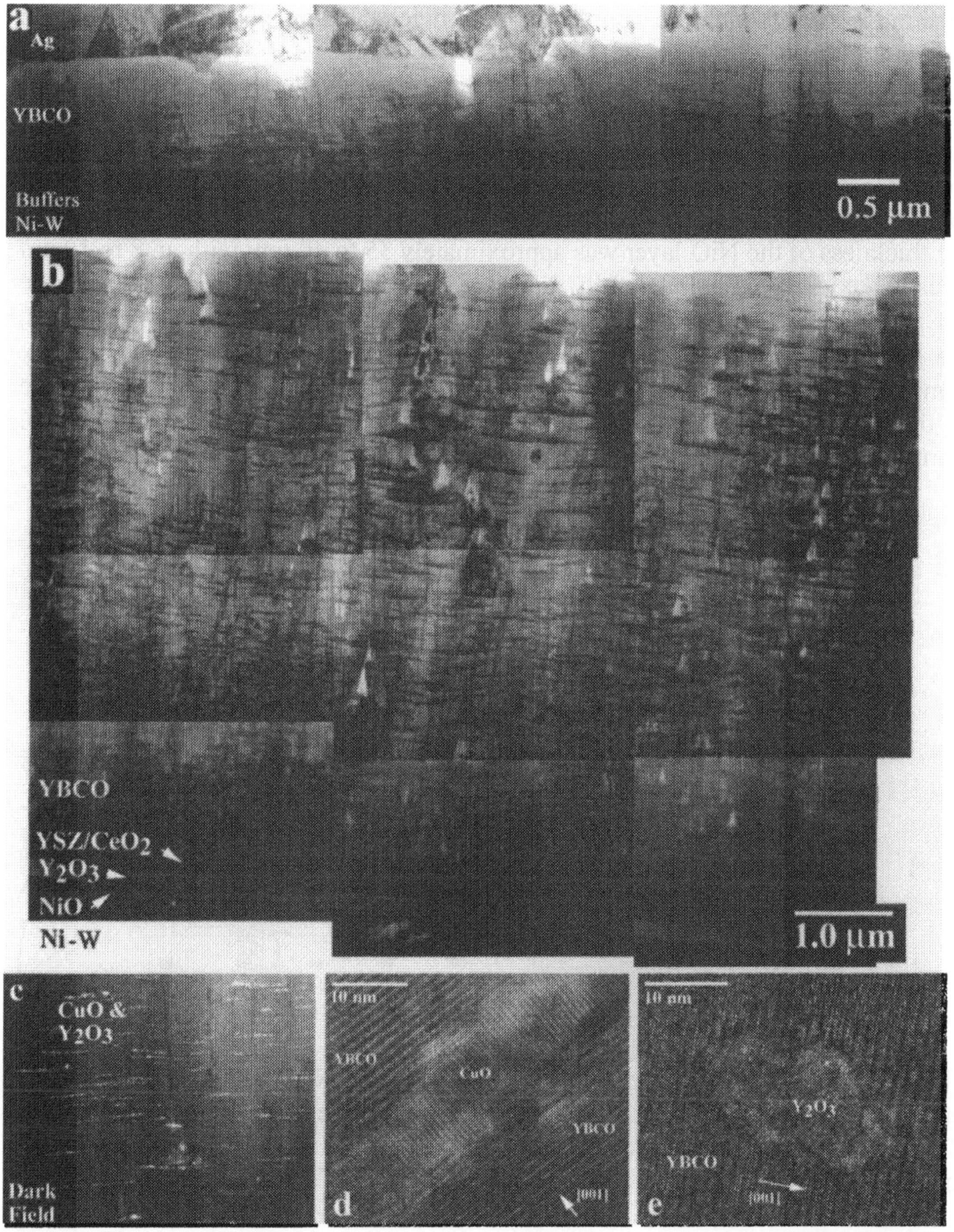

Fig. 3. Cross-section images of (a) 1.0 µm and (b) 6.4 µm thick YBCO films on RABiTS2. (c) Dark-field image of 6.4 µm thick film, and high-resolution images showing (d) CuO and (e) Y_2O_3 particles.

at the boundary of the 6.4 µm thick film, but the CeO_2 layer was found still intact. By comparison, the 6.4 µm thick YBCO film on the newer RABiTS2 structure showed much less interface reaction than that of the 3.0 µm thick film of the older substrates, even though the sample had seen twice the deposition time at elevated temperatures.

At the boundary between the Ni-W and Y_2O_3 seed layer, a NiO reaction layer was formed. While this also occurred for the RABiTS1 materials, it was found that the growth of NiO was limited within the Ni-W substrates. In the NiO layer formed within the RABiTS2 materials, a tungsten oxide layer was identified through high resolution EDS line scans (Fig. 4). For all the RABiTS2 samples, the thickness of the NiO layer was approximately 35 to 40 nm, with an 8 to 10 nm thick W-oxide layer and was fully developed or continuous in the NiO within the 6.4 μm sample. Where no W-oxide layer was found in the RABiTS1 materials, the NiO layer grew from 10 nm to over 150 nm for the 0.19 and 3.0 μm thick films, respectively. The tungsten oxide layer may have prevented further damage to the upper layers of the conductor stack by preventing the diffusion of Ni. Contributing to the enhanced performance of the YBCO on the newer substrates.

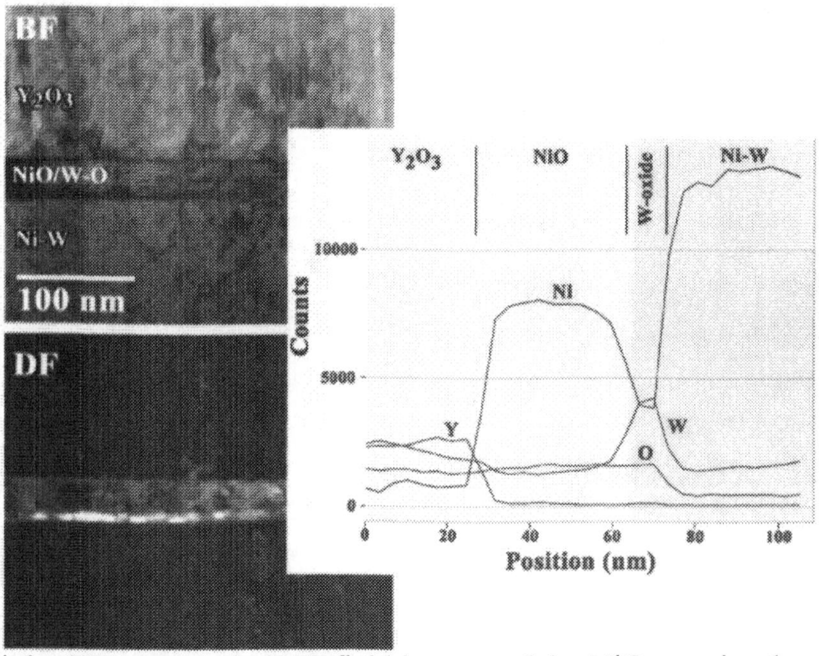

Fig. 4. Bright-(BF) and dark-(DF) field images of the NiO reaction layer between the Y_2O_3 buffer and Ni-W substrate showing the tungsten-oxide layer formed. An EDS nano-probe line scan across the interfaces shown in the images, illustrating the tungsten-oxide layer.

CONCLUSIONS

In combining microstructural investigation of the YBCO films with the x-ray characterization data, the leading cause of J_c loss for the RABiTS1 films was the degradation of cube texture. The formation of randomly oriented grains near the top of the YBCO film began around a thickness of 0.45 μm under the given PLD conditions for the RABiTS1 architecture. The amount of non c-axis grain formations increased with film thickness, along with the formation of a-axis oriented grains beginning with the 1.7 μm thick film. While the random grain formations dominated the upper portion of the film, they did not create a dead layer preventing c-axis growth up to the surface of the YBCO film.

The development of secondary phases along the $YBCO/CeO_2$ interface also played a significant role in the reduction of YBCO film properties for the RABiTS1 films. Transforming up to 0.1 μm of the YBCO film near the buffer surface into a non-conducting phase within the 3.0 μm thick film. This was not the case for RABiTS2 materials, which showed little reaction at the $YBCO/CeO_2$ interface.

The RABiTS2 samples showed only a modest decrease in J_c with film thickness. In the thicker YBCO films on RABiTS2, micro-porosity associated with the formation of Y_2O_3 and CuO was observed, though its effects on electrical properties of the film are believed to be minimal. In addition, the formation of a tungsten oxide layer within the NiO reaction layer of the RABiTS2 samples appeared to have acted as a Ni diffusion barrier, further protecting the YBCO layer.

ACKNOWLEDGEMENT

Research sponsored by the United States Department of Energy, Office of Energy Efficiency and Renewable Energy, Office of Distributed Energy and Electric Reliability- Superconductivity Program. This research was performed at the Oak Ridge National Laboratory, managed by UT-Battelle, LLC for the United States Department of Energy under contract No. DE-AC05-00OR22725. The authors (K.J. Leonard, B.W. Kang, and S. Kang) are grateful to the Oak Ridge Associated Universities for their support.

REFERENCES

[1]S.R. Foltyn, Q.X. Jia, P.N. Arendt, L. Kinder, Y. Fan and J.F. Smith, *Appl. Phys. Lett.*, **75** [23] 3692-3694 (1999).

[2]S. Sievers, F. Mattheis, H.U. Krebs and H.C. Freyhardt, *J. Appl. Phys.*, **78** [9] 5545-5548 (1995).

[3]X.F. Zhang, H.H. Kung, S.R. Foltyn, Q.X. Jia, E.J. Peterson and D.E. Peterson, *J. Mater. Res.*, **14** [4] 1204-1211 (1999).

[4]J.H. Park and S.Y. Lee, *Physica C*, **314**, 112-116 (1999).

[5]T.G. Holesinger, S.R. Foltyn, P.N. Arendt, Q.X. Jia, P.C. Dowden, R.F. DePaula and J.R. Groves, *IEEE Trans. On Appl. Superconductivity*, **11** [1] 3359-3364 (2001).

[6]K.D. Develos, H. Yamasaki, A. Sawa and Y. Nakagawa, *Physica C*, **361,** 121-129 (2001).

[7]K.D. Develos, H. Yamasaki, A. Sawa, Y. Nakagawa, S. Oshima, and M. Mukaida, *Physica C*, **357**, 1353-1357 (2001).

[8]T.G. Holesinger, S.R. Foltyn, P.N. Arendt, H. Kung, Q.X. Jia, R.M. Dickerson, P.C. Dowden, R.F. DePaula, J.R. Groves and J.Y. Coulter, *J. Mater. Res.,* **15** [5] 1110-1119 (2000).

[9]K.J. Leonard, B.W. Kang, A. Goyal, D.M. Kroeger, J.W. Jones, S. Kang, N. Rutter, M. Paranthaman, and D.F. Lee, *J. Mater. Res.*, **18** [5] 1109-1122 (2003).

[10]O. Eibl and B. Roas, *J. Mater. Res.*, **15** [11] 2620-2632 (1990).

[11]S.R. Foltyn, E.J. Peterson, J.Y. Coulter, P.N. Arendt, Q.X. Jia, P.C. Dowden, M.P. Maley, X.D. Wu and D.E. Peterson, *J. Mater. Res.*, **12** [11] 2941-2946 (1997).

[12]H.Y. Zhai, I. Rusakova, R. Fairhurst and W.K. Chu, Phil. Mag. Lett., **81** [10] 683-690 (2001).

[13]A. Goyal, R. Feenstra, M. Paranthaman, J.R. Thompson, B.Y. Kang, C. Cantoni, D.F. Lee, P.M. Martin, E. Lara-Curzio, C. Stevens, D.M. Kroeger, M. Kowalewski, E.D. Specht, T. Aytug, S. Sathyamurthy, R.K. Williams and R.E. Ericson, *Physica C*, **382**, 251-262 (2002).

[14]A. Goyal, D.P. Norton, J.D. Budai, M. Paranthaman, E.D. Specht, D.M. Kroeger, D.K. Christen, Q. He, B. Saffian, F.A. List, D.F. Lee, P.M. Martin, C.E. Klabunde, E.C. Hatfield, and V.K. Sikka, *Appl. Phys. Let.*, **69** [12] 1795-1797 (1996).

[15]J.D. Budai, D.K. Christen, A. Goyal, Q. He, D.M. Kroeger, D. F. Lee, F.A. List, D.P. Norton, M. Paranthaman, B.C. Sales and E.D. Specht, U.S. Patent 5 968 877 (1999).

[16]S. Kang, A. Goyal, N.A. Rutter, K.J. Leonard, M. Paranthaman, S. Sathyamurthy, and D.M. Kroeger, *J. Mater. Res.*, in review (2003).

[17]B.W. Kang, A. Goyal, D.F. Lee, J.E. Mathis, E.D. Specht, P.M. Martin, D.M. Kroeger, M. Paranthaman, and S. Sathyamurthy, *J. Mater. Res.*, **17** [7] 1750-1757 (2002).

[18]K.J. Leonard, S. Kang, A. Goyal, K.A. Yarborough and D.M. Kroeger, *J. Mater. Res.*, in review (2003).

[19]J.G. Wen, C. Traeholt and H.W. Zandbergen, *Physica C*, **205**, 354-362 (1993).

[20]S. Bals, G. Van Tendeloo, G. Rijnders, D.H.A. Blank, V. Leca and M. Salluzzo, *Physica C*,**372**, 711-714 (2002).

[21]D.M. Hwang, T. Venkatesan, C.C. Chang, L. Nazar, X.D. Wu, A. Inam and M.S. Hegde, *Appl. Phys. Lett.*, **54** [17] 1702-1704 (1989).

[22]J.F. Hamet, B. Mercey, M. Hervieu, G. Poullain and B. Raveau, *Physica C*, **198**, 293-302 (1992).

[23]U. Scotti di Uccio, F. Miletto Granozio, A. Di Chiara, F. Tafuri, O.I. Lebedev, K. Verbist and G. van Tendeloo, *Physica C*, **321**, 162-176 (1999).

[24]P.R. Broussard. L.H. Allen, V.C. Cestone and S.A. Wolf, *J. Appl. Phys.*, **74** [1] 446 (1993).

[25]K. Verbist, A.L. Vasiliev and G. Van Tendeloo, *Appl. Phys. Lett.*, **66** [11] 1424-1426 (1995).

[26]A. Catana, R.F. Broom, J.G. Bednorz, J. Mannhart and D.G. Schlom, *Appl. Phys. Lett.*, **60** [8] 1016-1018 (1992).

THICKNESS DEPENDENCE OF J_CS IN YBCO, TL-2212 AND HG-1212 THICK FILMS

J.Z. Wu, R. Emergo and X. Wang
Department of Physics and Astronomy, University of Kansas, Lawrence, KS 66045

Abstract
Epitaxial growth of thick (typically a few microns) high-T_c superconducting (HTS) films on metal substrates is desired for coated conductor applications in order to carry high electrical current of several thousands amperes per centimeter width. Serious degradation of the critical current (J_c) at large film thickness in $YBa_2Cu_3O_7$ (YBCO) films has raised a question on the mechanism related. To understand such mechanisms, we have investigated the thickness dependence of J_cs in several different types of HTS thick films including YBCO, $Tl_2Ba_2CaCu_2O_{7+\delta}$ (Tl-2212), and $HgBa_2CaCu_2O_{6+\delta}$ (Hg-1212). By comparing materials with different anisotropy that may affect the thickness dependence of J_cs, and microstructures that are resulted from different processing, we intend to pinpoint the mechanisms that are responsible for the degradation of the J_c in HTS thick films.

1. Introduction

The second-generation high-temperature superconducting (HTS) wires employ the "coated conductor" approach, where long lengths of highly grain-oriented HTS coatings are epitaxially deposited on metal tapes [1,2]. This approach is key to overcoming the problem of "weak-link" current transport across high-angle grain boundaries (GBs) in HTS films and may offer an attractive alternative to the first-generation Bi-HTSs powder-in-tube tapes. The ultimate requirement for coated conductors is to fabricate tapes in kilometers which can carry large currents in the order of a few hundreds to thousands Ampere per centimeter width. Since typical YBCO thin films with their thickness less than 0.3 μm can carry J_cs of 4-5 MA/cm^2 at 77 K and zero applied magnetic field, the capability to carry critical current of a few hundreds to thousands Ampere per centimeter width will require YBCO films with thickness in the order of a few micrometers or thicker. Unfortunately, superconducting critical current density (J_c) drops drastically in $YBa_2Cu_3O_{7-\delta}$ (YBCO) thick films at large thickness. Several groups reported monotonic decrease of J_cs with increasing film thickness whether on single crystals or on metal substrates [3,4]. It is generally true that J_cs drop by a factor of 3-5 from several MA/cm^2 to about 1 MA/cm^2 when the film thickness increases from ~0.2 μm to ~1 μm. At larger thickness, J_cs continue to decrease but at a much slower pace. This pace, however, is different for YBCO films grown on single crystals and coated conductors. A much larger drop in J_c has been observed on coated conductors.

An interesting question arises on the mechanisms of the observed J_c vs. thickness behaviors: whether they are intrinsic to HTS materials due to, for example anisotropy of their crystalline and electronic structures (see Fig. 1), or extrinsic due to microsctructures formed via different processing. To answer this question, we have made a comparative study on the thickness dependence of J_c in $YBa_2Cu_3O_7$ (YBCO), $Tl_2Ba_2CaCu_2O_{7+\delta}$ (Tl-2212), $HgBa_2CaCu_2O_{6+\delta}$ (Hg-1212) films. In this paper, we describe our experimental results.

2. Sample preparation

Pulsed laser deposition (PLD) was employed to fabricate YBCO thick films on $SrTiO_3$ (STO) substrates. The (100) axis of the substrate is aligned with one of the edges of the substrates. A Lambda Physik KrF excimer laser was used for PLD with λ=248 nm and pulse duration of 25 ns. The laser energy density was estimated to be 1-3 J/cm^2 and the deposition rate was 0.06 nm/pulse at 5 Hz repetition rate. The substrates were silver pasted to the heater and deposition was made at 810 °C in 200 millitorr oxygen partial pressure. After the film deposition, the samples were annealed *in situ* at 520 °C and 350 torr oxygen partial pressure. Most films of the same thickness on STO of different vicinal angles were made in the same run by placing the substrates very close to each other to ensure a uniform deposition condition.

Tl-2212 films were made in crucible process in two different ways. $LaAlO_3$ (LAO) and MgO substrates were used. The thin Tl-2212 films with a thickness of 0.25—0.30 μm were prepared using the standard crucible technique [5]. The thicker Tl-2212 films with a thickness of 2.5—3.0 μm were made in a Tl-loss process [6]. Some of these Tl-2212 films were then used as the precursor films for Hg-1212 films and the conversion from Tl-2212 to Hg-1212 was made through Hg-vapor annealing them to replace Tl cations with Hg cations via cation exchange [7]. The details of the cation exchange process can found in [8,9]. Briefly, the Tl-2212 precursor films were torch sealed in an evacuated quartz tube together with an $HgBa_2Ca_2Cu_3O_x$ pellet as Hg vapor source and a $Ba_2Ca_2Cu_3O_x$ pellet as Hg absorber. The whole sample assembly was annealed in a furnace at 780°C for 4-9 hours. In order to optimize the oxygen content in these Hg-1212 films, post annealing in flowing oxygen at 300°C for 4 hours was applied to most

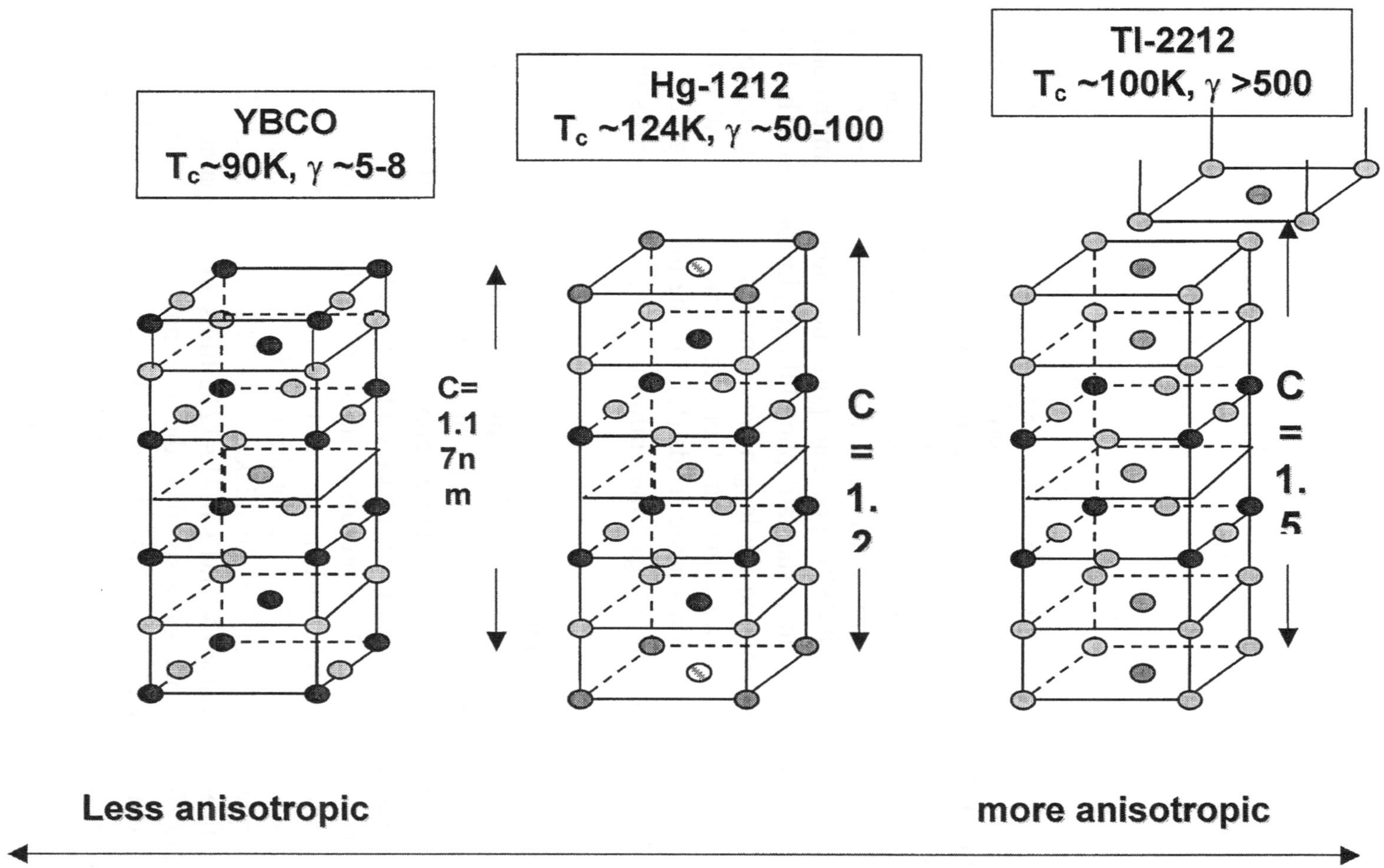

Fig. 2 Schematics showing variation of lattice structures and anisotropy in YBCO, Hg-1212 and Tl-2212.

Hg-1212 films. It should be mentioned the c-axis lattice constant for Tl-2212 is 1.48 nm, about 14% larger than that of Hg-1212. The conversion from Tl-2212 to Hg-1212 resulted in 14% thickness reduction in the c-axis as confirmed earlier in x-ray diffraction (XRD) θ-2θ scans and thickness measurements on Hg-1212 and their Tl-2212 precursor films [7]. The thickness of Hg-1212 thin and thick films used was, respectively, in the range of 0.21—0.26 μm and 2.1—2.6 μm.

3. Experimental results

The structures and phase purity of the films were characterized using x-ray diffraction (XRD) θ-2θ scans and pole figures on a Bruker x-ray diffractometer with four-circle goniometer. . The surface morphology of the films was examined using a LEO 1550 field-emission scanning electron microscope (SEM) and a Digital Instrument atomic force microscope (AFM). Both magnetic and electrical transport techniques were applied for investigation of the superconducting properties. The magnetic measurement was carried out on a Quantum Design SQUIDs magnetometer (MPMS7). For transport measurement, micro-bridges were defined on the films in a standard photolithography process. The bridge width was 20 μm or 40 μm and the length was 3 mm. Both resistivity (ρ) and critical current density (J_c) were measured on these micro-bridges at different temperatures (T) and applied magnetic fields (H). On the thicker Hg-1212 films, ion milling was applied to investigate the uniformity of ρ and J_c across the film thickness.

3.1. Structure and phase purity

XRD θ-2θ spectra were taken in most films studied in this experiment and all of them showed that the films have the c-axis orientation. XRD pole figures were acquired on some films to confirm the epitaxy. For example, (103) poles were acquired on Hg-1212 films and (105) poles, on the Tl-2212 films. The four (103) poles [or (105) poles for Tl-2212 precursor films] are 90 degree apart in φ angles and aligned with the (100) [or (010)] axes of the substrate. The full-width-half-maximum (FWHM) of the (103) poles for thinner Hg-1212 films was in the range of 0.5-0.8 degree while that for the thicker Hg-1212 films, 0.6-1.1 degree. It should be noticed that these FWHM's in Hg-1212 films are comparable to that of their Tl-2212 films. This suggests that the original epitaxial structure was transferred from the Tl-2212 precursor films to Hg-1212 films of thickness up to 2.6 μm and the texture degradation during the cation exchange was minimal [9].

The phase purity of these HTS films can also be estimated from the XRD θ-2θ spectra. The amount of impurity phases on all these three types of films was negligibly small. However, certain amount of unconverted Tl-2212 phase was visible in the XRD θ-2θ spectra for Hg-1212 thick films (thickness > 0.5 μm). The volume portion of the Hg-1212 can be estimated from *Hg-1212 left in the film (%)=* $I_{Hg}/(I_{Hg}+I_{Tl})$. Here I_{Hg} and I_{Tl} are respectively the intensity of the strongest peak for Hg-1212 [(005) peak] and Tl-2212 [(0012) peak] in their XRD θ-2θ spectra. For thin films, the volume portion of Hg-1212 above 90% can be routinely obtained and the highest so far observed is above 99%. If processed at more or less the same conditions, more remaining Tl-2212 phase was observed in thicker films [10]. This suggests that much longer processing time may be necessary to convert thicker Tl-2212 films to Hg-1212 films. To confirm this, we processed 3.0 μm thick Tl-2212 films at 780 °C for different periods in cation exchange process. It was found that the volume portion of the Hg-1212 film increases monotonically with the annealing time and exceeds ~92% when the Hg-vapor annealing time is 12 hours. Although the Hg-1212 volume portion still increases linearly with the cation exchange time up to 12 hours, longer processing time up to 24 hours did not yield the complete conversion. The mechanism remains unclear. A possibility is that the Hg-vapor pressure inside the quartz ampoule may begin to decrease after certain period, reducing the efficiency of the cation exchange afterwards. In any cases, the Hg-1212 films used in this experiments all have the volume portion of Tl-2212 less than 10%.

3.2. Surface morphology

Fig. 2 compares surface morphology of YBCO, Tl-2212 and Hg-1212 films at three different thickness: thin film regime around 0.2 μm, medium one of 1.3-1.5 μm, and thick one of 2.6-3.0 μm. All three thin films have smooth surface morphology with either particulates (for YBCO) or secondary phases-Tl-rich particles (on Tl-2212 and Hg-1212) visible on the surface of the film. Notice the scale for the YBCO films is slightly different from that for Tl-2212 and Hg-1212 films. With increasing thickness, for example at 1.5 and 3.0 μm thickness, the morphology of the films experienced considerable changes as shown in Fig. 2. On YBCO films, the surface became much rougher and the size of the particulates increased significantly. On Tl-2212 films, on the other hand, uniform porous structures were seen in the film thickness range from 0.6 to 3.0 μm. Some of the 3.0 μm thick Tl-2212 films were thinned using ion milling and SEM pictures were acquired at different depths [11]. Similar morphology was confirmed at different depths, suggesting a uniform structure across the thickness of the Tl-2212 thick films. The surface of the Hg-1212 films is dense but rougher with many voids of dimensions in the range of sub-μm to several μm's. Such voids have been observed on most Hg-1212 films converted from Tl-2212 films, despite the thickness and processing conditions employed [9]. Since no voids were observed on Hg-1212 films converted from Tl-1212 precursor films, we suspect

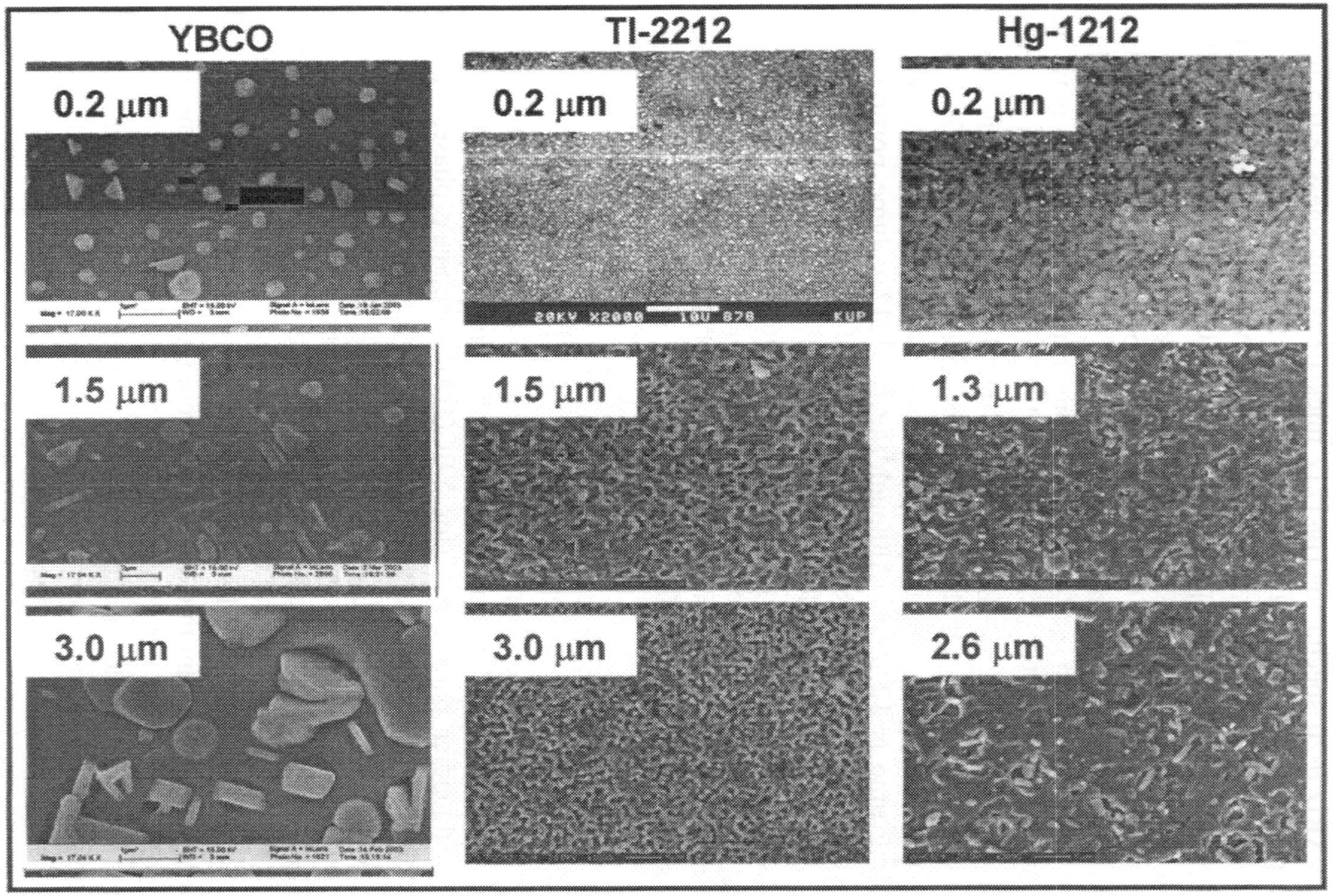

Fig. 2 SEM pictures of the surface morphology of YBCO (left column); Tl-2212 (middle column); and Hg-1212 (right column) films of a 0.2 μm, 1.3-1.5 μm, 2.6-3.0 μm thickness. Notice the scale for YBCO films is 2.5 times smaller than that for Tl-2212 and Hg-1212 films.

that the extra Tl cations in Tl-2212, by making a way out, are the major reason for the formation of the voids.

3.3. T_c and J_c (T,H)

The T_cs of the HTS films were measured using a Quantum Design SQUID magnetometer as well as using the standard four-probe electrical transport technique. The results of the magnetic and transport measurements were consistent. Fig. 3 depicts the zero-field-cooled magnetic moment (M) of three YBCO films [Fig. 3(a)] and three each of Tl-2212 and Hg-1212 films [Fig. 3(b)] as a function of temperature (T) in a 10 Oe magnetic field (H) applied along the c-axis of the film. The three samples for each material had different thickness: thin film regime around 0.2-0.6 μm, medium one of 1.3-1.5 μm, and thick one of 2.6-3.0 μm. The T_c's of three YBCO films were about 87-91 K. The T_cs of the Tl-2212 films were 100-102 K, and that for Hg-1212 films, 120-122 K, about 20 K higher than that of the original Tl-2212 precursor films. Overall, sharp transition near Tc was observed on most films studied in this experiment. The ρ measurement made on some films confirmed linear ρ vs. T curves above T_c's. ρ at room temperature was typically less than 300 μΩ cm for thin films while it was 20-50% higher in thick films.

The J_c of the Hg-1212 film can be estimated from the M-H hysteresis loops using the Bean model. Fig. 4 plots J_c vs. thickness curves for YBCO, Tl-2212 and Hg-1212 films at 77 K and self field. It should be mentioned that for each thickness, many films were made and characterized and the J_cs in Fig. 4 are the representative J_cs of samples showing consistent behaviors. As shown in Fig. 4(a), J_c's of the three different types of films experienced a similar decrease with increasing thickness. This J_c decrease is much bigger initially when the film thickness is smaller or near 1μm. For example, at 77 K, the J_c of the 0.25μm thick Hg-1212 films is around 3.8 MA/cm^2 and it reduces by a factor of four when the film thickness is increased to 1.3 μm. When the film thickness is further increased to about 2.6 μm, the J_c remains nearly the same. This J_c vs. thickness behavior demonstrated in Fig. 4(a) resembles what has been reported before on YBCO films [3,4]. This is not surprising since the same PLD technique was employed. The reduction of J_c with thickness is more or less uniform at different temperatures. For example, at 100 K, the J_c of 1.39 MA/cm^2 for the 0.25 μm thick Hg-1212 film reduces to 0.41 MA/cm^2 for the 2.6 μm thick films--a factor of 3.4, while at 77 K and 5 K, this ratio is 4.1 and 3.3, respectively. Fig. 4(b) depicts the same J_c vs. thickness curves for YBCO, Tl-2212 and Hg-1212 films on the normalized scale. It is clearly shown that Tl-2212 films have the smallest thickness dependence while Hg-1212, the largest. This seems to suggest the thickness dependence may not have a direct correlation with the anisotropy of the

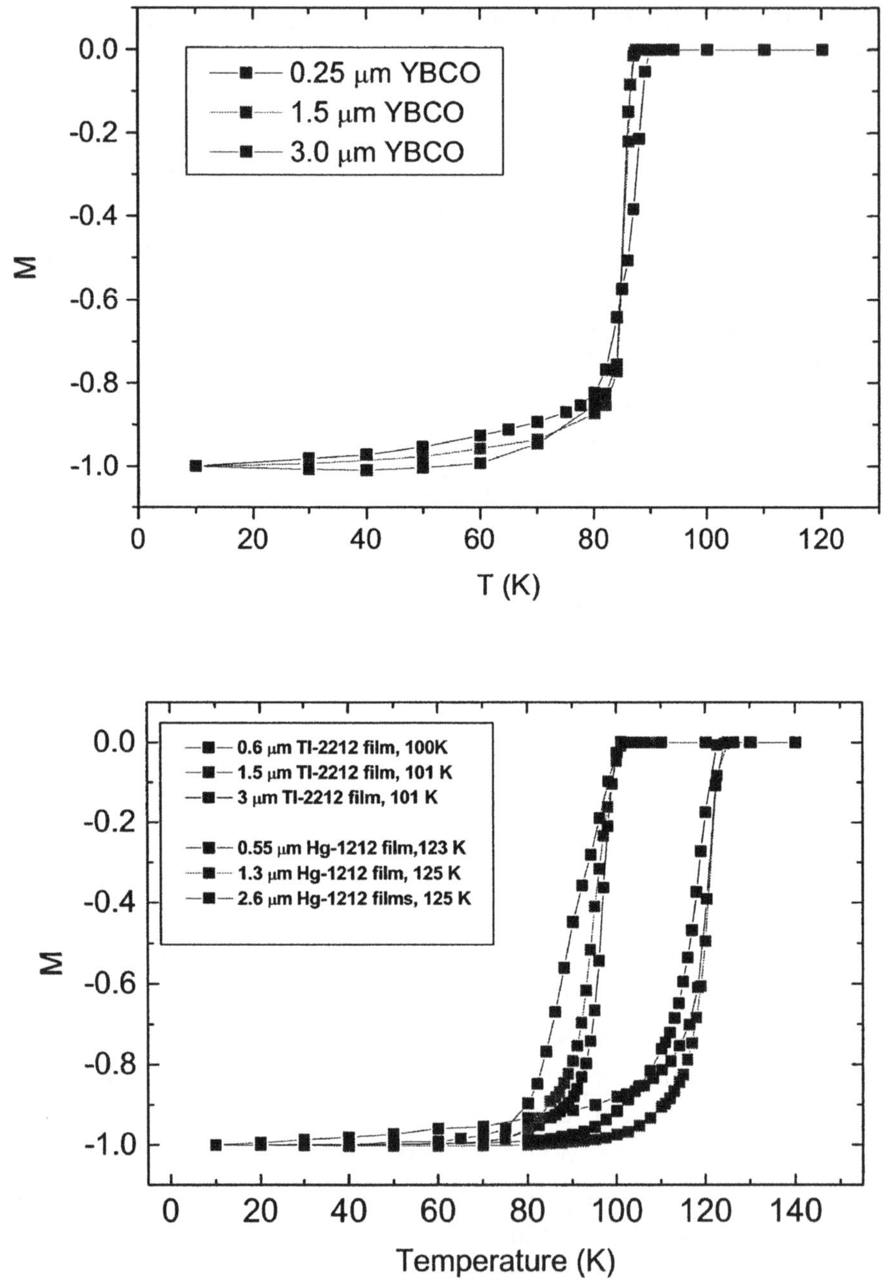

Fig. 3 Zero-field-cooled M-T curves of (a) three YBCO films of 0.25, 1.5 and 3.0 μm thickness; and (b) three Tl-2212 films of 0.6, 1.5 and 3.0 μm thickness and the three Hg-1212 films converted from these Tl-2212 films. H=10 Oe.

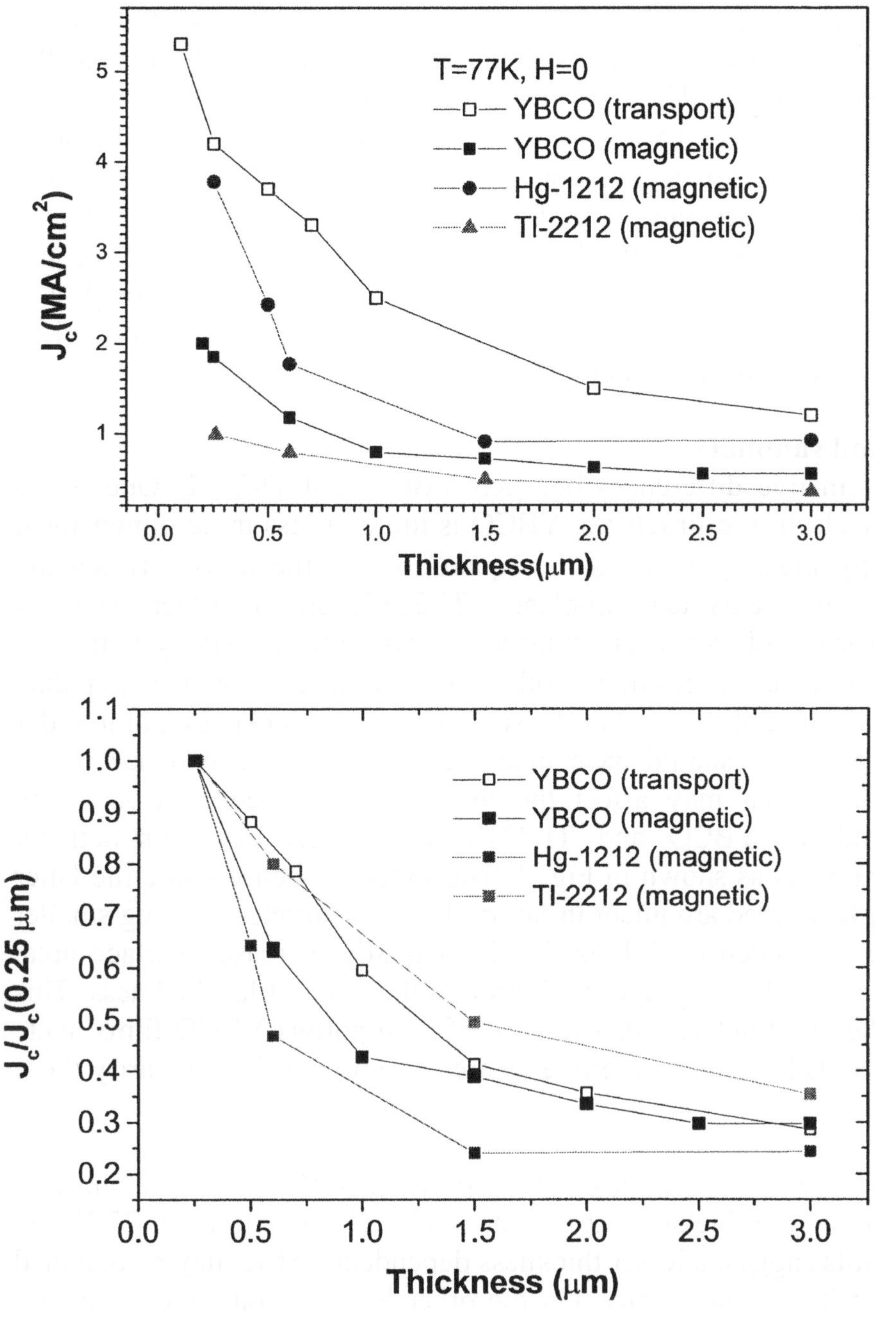

Fig. 4 A comparison of thickness dependence of J_cs of YBCO, Tl-2212 and Hg-1212 films in (a) original and (b) normalized scale at 77 K and self field.

material. However, our recent study on the depth profile of J_cs in 2.6-3.0 μm thick Tl-2212 and Hg-1212 films revealed considerable degradation of films near the substrate/film interface [10]. The thickness of the degraded layer was over 1.0 μm and it reduced to less than 0.5 μm for films of 1.5 μm thickness. On the other hand, the near interface film degradation was confirmed minimal in our PLD YBCO films. Taking this account, the J_c for Tl-2212 and Hg-1212 will be raised at large thickness, which means that the J_c vs. thickness curve for Tl-2212 and Hg-1212 in Fig. 4 will be flatter if degradation problem is solved. Although it is difficult to visualize where the J_c-thickness curve for Hg-1212 will be, it is clear that the curve for Tl-2212 is much flatter than that for YBCO.

4. Discussions and summary

There are two distinctive differences between YBCO and Tl-2212. One is the anisotropy in their electronic structures. YBCO is more or less three-dimensional with the anisotropy index $\gamma<10$ (γ is the square-root of the ratio between the effective mass along c-axis and ab-plane). Tl-2212, on the other hand, is extremely anisotropy and has been considered two-dimensional system with $\gamma>500$. Consequently, the penetration depth of the Tl-2212 is much larger than that of YBCO. If J_cs are only carried in the surface layer of thickness on the order of penetration depth, they should have much less thickness dependence in Tl-2212. Another factor that may affect the thickness dependence of J_c is the microstructure. Indeed, YBCO and Tl-2212 thick films have distinctively different microstructures as shown in Fig. 2. The former is dense while the latter is porous. A simple minded argument in favor of porous structure having smaller or none thickness dependence of J_c is the dramatically increased surface area, allowing penetration of the magnetic field through the whole film thickness. This is supported by the less thickness dependence of J_c in porous YBCO films made in solution process [12]. Further experiments are certainly needed to clarify these mechanisms.

In summary, we have fabricated thick films of YBCO, Tl-2212 and Hg-1212 and investigated the thickness dependence of J_c in these films at 77 K and self field. Our results suggest a lesser thickness dependence of Jc may be obtained in HTS films with higher anisotropy. On the other hand, porous microstructure may be favorable for higher J_cs in thicker HTS films.

Acknowledgement

JZW acknowledges supports from NSF and DOE for this work. RE and XW were supported by AFOSR.

REFERENCES

1. D.K. Finnemore, K.E. Gray, M.P. Maley, D.O. Welch, D.K. Christen, and D.M. Kroeger, Physica C 320, 1 (1999).
2. M. Paranthaman, A. Goyal, R. Feenstra, T. Izumi and V. Selvamanickam, "2002 MRS Workshop on Processing and Applications of Superconductors", Gatlinburg, TN, July 31-Aug. 2, 2002.
3. X.D. Wu, S.R. Foltyn, P.N. Arendt, W. R. Blumenthal, I. H. Campbell, J. D. Cotton, J. Y. Coulter, W. L. Hults, M. P. Maley, H. F. Safar, J. L. Smith, Appl. Phys. Lett. **67**, 2397 (1995).
4. S.R. Foltyn, Q.X. Jia, P.N. Arendt, L. Kinder, Y. Fan, and J.F. Smith, Appl. Phys. Lett. 75, 3692 (1999).
5. M. P. Siegal, E. L. Venturini, B. Morosion, and T. L. Aselage, J. Mater. Res. **12**, 2825 (1997).
6. B. Hammond, G.V. Negrete, L.C. Bourne, D.D. Strother, A.H. Cardona, and M.M. Eddy, App Phys. Lett. **57**, 825 (1990).
7. J.Z. Wu, S.L. Yan, and Y.Y. Xie, Appl. Phys. Lett. **74**, 1469 (1999).
8. S.L. Yan, Y.Y. Xie, J.Z. Wu, T. Aytug, A.A. Gapud, B.W. Kang, L. Fang, M. He, S.C. Tidrow, K.W. Kirchner, J.R. Liu and W.K. Chu, Appl. Phys. Lett. **73**, 2989(1998).
9. S.L. Yan, J.Z. Wu, L. Fang, Y.Y. Xie, T. Aytug, A.A. Gapud, and B.W. Kang, J. Appl. Phys **93**, 1666 (2003).
10. Z.W. Xing, Y.Y. Xie, J.Z. Wu and A. Cardona, to appear in Physica C.
11. X. Wang, R. Emergo, and J.Z. Wu, unpublished.
12. R. Finnstra, M. Feldmann, T. Holesinger, DOE superconductivity program peer review, Washington DC, July 2003.

DAMAGE EFFECT IN HTS IRRADIATED BY U FISSION FRAGMENTS

Alberto Gandini,
Physics Department and TCSAM
University of Houston
Houston, TX, 77204-5005, USA

Roy Weinstein,
Physics Department and TCSAM
University of Houston
Houston, TX, 77204-5005, USA

ABSTRACT

We discuss the effect that the volume of irradiation-induced damages has on the transport current properties of superconductors. We argue that irradiation-damages do not only act as pinning centers, hence increase J_c, but do also reduce the area through which current percolate, hence reduce I_c. The reduction of the percolating current area is discussed within the framework of percolation theory. Among the several irradiation processes, the U/n process is a leading method to increase I_c in HTS. In the U/n process, damages are produced by uranium fission fragments. The induced defects in this case are particular, as they change in size and morphology along the fission fragment range. Then, in here, we particularly focus on size and morphology of the U/n fission fragment defects, and on their effect on the percolating current area.

BACKGRAOUND

To achieve high critical current in high temperature superconductors (HTS) pinning centers must be introduced within the HTS matrix [1]. Pinning centers are not-superconducting regions. Theoretical works as well as experimental data indicate that the best pinning centers have dimensions of the order of the superconducting coherence length [1 - 2]. Pinning centers may be introduced by chemical doping [3 - 4], and by irradiation [5 - 6]. Irradiation-induced damages have been proven to be the most effective types of pinning center; however, their full potential has not been explored. Indeed, most of the studies in the last decade have been focused only on the low pinning density regime, where critical current, I_c, is observed to rapidly increase with the pinning-center density. Studies are discontinued at pinning-center density beyond which I_c begins to fall-off. Little attention has been paid to this effect [7 – 8] (i.e., the I_c fall-off at large density.)

The I_c fall-off is usually attributed mostly to the reduction of T_c, and the fall-off has been considered as an inevitable secondary effect [5]. For this reason, most of the studies have been carried at low radiation fluence, below the peak of I_c attainable.

In recent works [7 - 8], the mechanisms of the I_c fall-off at large fluence have been addressed. The driving idea is that in order to full explore the potential of irradiation-induced pinning centers the mechanisms, which limit the I_c enhancement, should be fully understood. As the mechanisms are identified and their effects quantified, a way to reduce them may be found as well. The main result of these works was to show that defects "collectively" result in a reduction of the percolating current area (this is also called the flow-area.)

The reduction of the flow-area, as the defect density increases, is most naturally discussed within the framework of percolation theory. Percolation theory describes the effects on the conductivity of a conductor as holes are randomly inserted in the sample [9 – 10]. Typically, it is applied to normal metal; though a priori the theory shall apply also to superconductors [11]. As one may see, a similar scenario occurs in irradiated HTS. In fact, irradiation damages are randomly distributed in non-superconducting amorphous regions. Hence, from a current conduction point of view those damages behave as holes. It is as if the damaged regions were removed from the HTS matrix, and they were not contributing to the current transport.

Weinstein et al. [7] investigated the effect of the size of columnar pinning centers on the magnetic properties of HTS. In [7], it has been shown that the volume of columnar pinning centers limits the maximum pinned field, and the percolating current area is severely reduced at a point which current can no longer flow within the sample, even at a relative low columnar pinning center density.

Gandini et al. [8], using a phenomenological model that makes use of the reduction-of-flow-area idea, showed that the I_c enhancement is limited by the volume of the defects that collectively reduce the area through which current flows. The model was tested against data on irradiated Ag/Bi-2223 tapes, and it was shown to fit experimental data well. In particular, the model indicates an interesting feature: if the volume of a defect is reduced by a factor of two, then I_c becomes larger up to a factor of two may be achieved [12 – 7].

We begin to realize that the size of a pinning center is important not only because the pinning potential depends on it, but also because the volume of the defects collectively limits severely the maximum number of pinning centers that can be added, which limits the maximum critical current achievable.

Two independent experiments [13 - 14] also showed that irradiation is mostly effective when it creates amorphous damages within the HTS. As it is discussed in the next section, not all forms of radiation create amorphous regions, although all forms of radiation induce vacancy-type damage (in here we consider only protons, neutrons and heavy ions irradiation.) [13 - 14] show that the vacancy-type damages are much less effective as a pinning center than the amorphous defects.

However, just as the amorphous damages, so do the vacancy-type damages reduce the flow-area. A deeper understanding of the volume-effect of the irradiation-induced damage, in particular of the vacancy-type damages as they reduce the flow-area while only marginally increase I_c, is then of great interest in the process of increasing the critical current.

In the following, size and morphology of U/n fission fragment [15] is investigated, and its effect on I_c is discussed.

THE U/n PROCESS

The U/n process has been proven as one of the most effective process for increasing I_c [15 - 16]. In the U/n process, U is added to the HTS precursor powder. The doped powder is processed following the standard HTS preparation methods, then samples are irradiated with thermal neutrons. Thermal neutrons fission some of the U atoms. The product of the fission is two heavy ions; in the following we refer to those as fission fragments. Fission fragments kinetic energy is discussed in [15]. As discussed in the introduction, the volume of the damage has a large effects on the maximum achievable I_c. In this paper, we then discussed in detail the damage due to the U/n process fission fragment, as this process is one of the leading processes to increase I_c.

ENERGY LOSS AND DAMAGES BY HEAVY ION IN HTS

As a swift ion travels through matter, it transfers energy to the target atoms. The energy of the ion is transferred via two different mechanisms, which occurs simultaneously. Ion may lose energy via Coulomb interaction with the electron [17]. In this case, the energy loss per unit of length is commonly referred in the literature as S_e. Ions may also lose energy via Coulomb force with the target atom nuclei, the energy loss per unit of distance in this case is called S_n [17]. Typically, the ion and energies used to irradiate HTS are such that: $S_e \sim 1000$ larger then S_n.

As the ion travels through matter and its energy is transferred to the target atoms, damage in the target lattice may form. Size and density of defects vary with the energy-loss mechanisms [17 - 19].

Defect Production Via S_e

The size and the morphology of the S_e-induced defects vary significantly with S_e. For $S_e < 8$ KeV/nm, no defects are formed. Whereas for $S_e > 8$ KeV/nm amorphous regions begin to form, in YBCO as well as in BiSCCO. The physics is that the amount of energy transferred below 8 KeV/nm is not enough to locally melt the HTS, hence to form an amorphous region [17 - 18]. The diameter and the length of these amorphous regions vary with S_e. A collection of data of defect diameter versus S_e is presented in [7]. At low S_e, an ion track looks like a string of beads separated by gaps of undamaged regions. As S_e increases the gaps between the amorphous damages diminishes. At $S_e > 35$ KeV/nm, continuous amorphous columnar defects form. The diameter of the damage has been extensively studied, whereas the length of the damaged segment, relative to the ion range, has not; though some experiments have been done [18]. An experiment done with gold and silver ions showed that for $S_e \sim 35$ KeV/nm full columns are formed, while for $S_e \sim$ 25 KeV/nm amorphous regions form only along ~ 50% of the ion range.

Defect Production Via S_n

Energy loss via S_n leads to the formation of defects provided the energy transferred to the nucleus, E_t, is larger than the target displacement energy, E_d. The target displacement energy is the energy that must be given to a target atom to remove it from its equilibrium position. E_d is typically of the order of 20-25 eV. For $E_t > E_d$ vacancies and clusters of vacancies may form. Typically, vacancies are too small to be observed by SEM [17], though cluster of vacancies have been observed to be ~ 3 nm in diameter [20]. The S_n-induced damage increases the normal state resistivity as well as it decreases the critical temperature T_c. The change of resistivity and of T_c has been shown to scale with the defect density, expressed in term of number of displacements per atom, *dpa*. For example, Hensel et al. [17] showed that for several ions, with various energies and different values of S_n, the change of resistivity is equal for equal *dpa* amount.

ENERGY LOSS AND DAMAGE OF FSSION FRAGMENT

S_e for fission fragment has been calculated using the TRIM program [21]. The result is shown in figure 1. It is seen that as the ion travel ~ 3 μm S_e falls below the 8 KeV/nm threshold to form amorphous regions. A collection of data from the literature in combination with the TRIM results shown in figure 1 indicate: (i) along the first ~ 3 μm, fission fragment tracks consist of broken column followed by string of beads; (ii) fission fragment tracks have an amorphous core of radius ~ 1.8 nm (at the beginning of the track).

As we are interested in the net volume of the defect, which is subtracted from the current transport area, we shall notice two important aspects of irradiation damage. i) The amorphous core is surrounded by a damaged region (due to the shock wave) twice as large as the amorphous core itself [18]. ii) Because superconductivity is only fully restored into the matrix away from a non-superconducting region over a distance of the order of the coherence length, ξ, the volume over which superconductivity is destroyed is larger than the actual geometric volume of the damage. Then the volume of the first part of a fission fragment track is $\sim \pi(r + \xi)^2 l \sim 0.54\times10^{-15}$ cm^3, where r ~ 3.6 nm is the radius of the damaged area (amorphous core + shock wave), ξ ~ 4 nm at 77 K in Bi-2223 [22], and l ~ 3 μm is the length of the amorphous part.

As a fission fragment passes the 3μm, although most of the energy is still lost via S_e (see figure 1), no amorphous regions are formed since S_e is below the 8 KeV/nm amorphization-threshold. However, after the first ~ 3 μm a fission fragment still transfers enough energy via S_n to the target nuclei to displace some of them. In fact, the transferred energy, E_t, to the target atoms is $\sim 11\times S_n$, as on

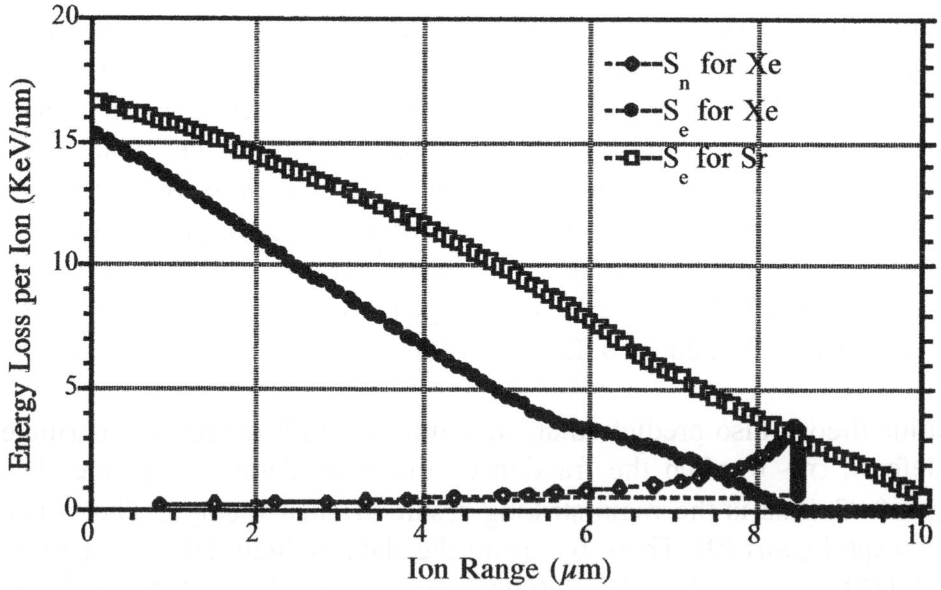

Figure 1. Energy loss per unit of length (KeV/nm) versus ion (i.e., fission fragment) range, for two examples of fission fragments. Fission fragment Xe has initial energy of 65 MeV. Fission fragment Sr has initial energy of 96 MeV. Close circles and open squares are energy loss via S_e for Xe and Sr, respectively. Open diamonds are energy loss via S_n for Xe; energy loss via S_n for Sr is similar to S_n for Xe, hence it is not shown in figure.

average fission fragments encounter 1 target atom every 1.1 nm. Then, it follows from Figure 1 that E_t is larger then E_d along most part of the fission fragment range, hence a fission fragment produces vacancy-type defect along most part of its range. TRIM [21] calculation shows that each fission fragment creates several vacancies along its path. The number of vacancies depends on the particular pair of fission fragments form during the U fission [15]. In fact, as U fissions, several different ions-pairs (e.g., fission fragment pair) may form, e.g., Xe and Sr. On average, the number of vacancies per fission fragment per cm, P, is ~ 0.42×10^8 vac/(cm*ff*) and ~ 1.2×10^8 vac/(cm*ff*) for Xe and Sr, respectively. The pair Xe-Sr is just one of the many possible U fission fragment pairs. The above averages are calculated along the second part of the fission fragment track, where S_e is too low to create damage.

The volume of damage along the second part of the fission fragment track is more complicated to evaluate, and the result has a larger error as the size of these damages is too small to be studied systematically with SEM. However, we think that a good estimate of the S_n-induced damage can be obtained by performing the following percolation theory-based analysis. Hensel et al. [17] showed that conductivity in YBCO as the ions lose energy via S_n (while the S_e is below the amorphous threshold) varies with *dpa*. In particular, they show that at low *dpa*, $\sigma \sim 1/(1400 \times dpa+1)$, whereas at high *dpa* decreases more rapidly, and $\sigma \sim 0$ at $dpa \sim 0.025$. Here, we use Hensel et al. [17] data and re-interpret them in term of percolation theory. Accordingly to the percolation theory, the conductivity decreases proportionally to the product: $V_{Sn} \times n$, where: 1) n is the number of defects per unit of volume, and it is related to the *dpa* by the simple expression: $n \equiv (\#atoms/cm^3) \times dpa$, with $(\#atoms/cm^3) \sim 7.07 \times 10^{22}$; 2) V_{Sn} is the volume of a S_n-induced defect. Then, it follows that $V_{Sn} \times n = V_{Sn} \times (\#atoms/cm^3) \times dpa = 1400 \times dpa$. Hence, the volume of a S_n-induced defect is $\sim 1.98 \times 10^{-20}$ cm^3, which gives a radius of 1.7 nm.

Percolation theory also predicts that, in conductor with randomly distributed spherical defects, $\sigma \sim 0$ when the fraction of the total damaged volume, $V_c \equiv V_{damage}/V_{tot}$, ~ 0.03 (this is the main striking result of such theory). The volume fraction is $\sim \exp(-V_{Sn} \times n)$ [9]. Then, by using the data at high defect density by Hensel et al. [17], i.e., $\sigma \sim 0$ at $dpa \sim 0.025$, we get that $V_{Sn} \sim 1.98 \times 10^{-21}$ cm^3, hence: r ~ 0.8 nm.

Our reinterpretation, within the framework of percolation theory, of Hensel's data [17] shows that S_n-induced damages may be treated as spherical defects with radius ~ 0.8 – 1.7 nm, the reduction of conductivity is due to the loss of percolation area.

Hensel et al. data [17], as well as other experiments in the early '90's, were done at $T > T_c$. To support our analysis we search the literature for similar data obtained at $T < T_c$. Weaver et al. [23] show an interesting behavior of J_c. It is shown that the ratio of J_c after irradiation to J_c before irradiation, $J_c(\Phi)/J_c(\Phi=0)$, decreases as $\sim exp(-7.7 \times 10^{-17} \times \Phi)$, where Φ is the proton fluence. We calculate, using the TRIM program, the *dpa* per unit of proton fluence, then by carrying the same analysis as above, we obtain the radius of the proton damage to be ~ 1.8 nm . This is also in agreement with the estimate by Civale [20]. We should notice at this time that the proton used in Weaver et al. [23] experiment had $S_e \sim 4$ eV/nm, much lower than the threshold for amorphization, and $S_n \sim 2.7 \times 10^{-3}$ eV/nm.

In conclusion, Hensel et al. [17] and Weaver et al. [23] data, re-interpreted in terms of percolation theory, indicate that the change in σ, by ions which have S_e below the amorphization threshold but S_n large enough to create vacancy-type defects, is due to the reduction of percolation area. Furthermore, the damage produced via S_n is composed of spherical defects of radius ~ 0.8 –1.8 nm. We shall shortly comment on the magnitude of the radius found. Since S_n damages are vacancy-type damages, one should indeed expect them to be at most of the order of few unit cells (unit cell size ~ 1 nm), in agreement with the above found value of ~ 1.5 –1.7 nm. This is also in agreement with Civale [20] observation.

We have now enough information to estimate the volume damage along the second part of the track. We use an average radius for the S_n-induced vacancy-type damage of ~ 1.6 nm, and an average number of vacancies per fission fragment

, P, of ~ 0.8×10^8 vacancies/cm. Then, $V_{Sn} = (4\pi/3)r^3 xPxl$, where r ~ 1.6 nm is the radius of a spherical S_n-induce defect, and l ~ 6 μm is the average length of the second half of the fission fragment track. Hence, V_{Sn} ~ 0.82×10^{-15} cm^3. Surprisingly, V_{Sn} is comparable to the amorphous damage along the first part of the track.

The total volume of a fission fragment defects, V_d, is given by: $V_d = V_{Se} + V_{Sn}$ ~ 1.37×10^{-15} cm^3. The effect of this damage is to reduce the flow-area by a factor of ~ exp ($-1.37\times10^{-15} xff$). The combine volume of the damaged induced via S_e with the damage volume via S_e turn out to be on the same order as the volume empirically found ~ exp ($-4.1\times10^{-15} xff$).

SUMMARY

We have discussed the effects of damage volume on critical current, and estimated the value of the U/n fission fragment volume. We have shown that fission fragment track is composed of two regions. Along the first ~ 3 μm of the their range fission fragments form amorphous quasi-columnar defects of ~ 3.6 nm in radius; along the last ~ 6 μm of their range fission fragments instead produce vacancy-type spherical defects of radius ~ 1.6 nm. The density of the vacancy-type defects was calculated with the TRIM program to be ~ 0.82×10^8 #vacancies/cm per fission fragment. We estimated that the volume of the damage along the first part of the track is comparable to the volume along the second part. Hence, both parts contribute in equal manner to decrease the current percolating area. However, as it has been discussed, the amorphous part of the fission fragment damage is an efficient pinning center, while the vacancy-type damage has a low pinning efficiency. In other word, the second part of the fission fragment while reduces the current percolating does not increase J_c. We finally argue that a reduction of the volume of second part of the track may lead to a smaller reduction of the current percolating area, hence it may lead to a larger I_c enhancement as now achievable [24]. The vacancy-type damage along the second part of the fission fragment range may be reduced by annealing [25]. A preliminary experiment has been done in which U/n processed Ag/Bi-2223 tapes have been annealed at 400 Celsius. The first results show that I_c increases after annealing by ~ 20%.

ACKNOWLEDGMENT

This work was supported in part by the U.S. Army Research Office, the Welch Foundation, the State of Texas via the Texas Center for Superconductivity and Advance Material (TCSAM).

REFERENCES

[1]L. Civale, "Irradiation Processing for Flux Pinning Enhancement in High-T_c superconductor", Processing and Properties of High-Tc Superconductors, Vol 1, Chap. 8, World Scientific.

[2]D. Nelson, V. Vinokur, Phys. Rev B 48 13,060 (1993).

[3]R. Weinstein and Ravi-Persad Sawh, "A Class of Chemical Pinning Centers Including Two Elements Forein to HTS", Physics C 383, 438-444 (2003).

[4]Y. Ren, R. Weinstein and Ravi-Persad Sawh, "New Chemical Pinning Ceneter from Uranium Compound in Melt-Textured YBCO", Physics C 282-287 (1997) 2275-2276.

[5]R. C. Budhani, "Studies of Vortex Localization Columnar Defects in High Temperature Superconductor", Studies of High Temperature Superconductor, Vol. 33, Ed. Anant Narlikar, 2000 Nova Science Publishers.

[6]L. Civale, "Vortex Pinning and Creep in High-Temperature Superconductors with Columnar defects", Supercod. Sci. Technol. 10, A11-A28 (1997).

[7]R. Weinstein, A. Gandini, Ravi-Persad Sawh, D. Parks, and B. Mayes, "Improved Pinning Regime by Energetic Ions Using Reduction of Pinning Potential", Physics C 387 (2003) 391-405.

[8]A. Gandini, R. Weinstein, D. Parks, R. P. Sawh, and S.X. Dou, "On the limiting Mechanism of Irradiation Enhancement of I_c", Presented at the 2002 Applied Superconductivity Conference (ASC 2002), Agust 4-9, 2002 Houston, Texas, USA. Accepted for publication on IEEE Transaction on Applied Superconductivity.

[9]D. Stauffer, and A. Aharony, Introduction to Percolation Theory 2nd edn (1992), (Bristol, Pa: Taylor & Francis).

[10]M.D. Rintoul, "Precise Determination of their Void Percolation Threshold for Two Distributions of Overlapping Spheres', Phys. Rev. E, (62) 68-725 (2000).

[11]B. Zeimetz, B.A. Glowacki, and J.E. Evetts, "Applivation of Percolation Theory to Current Transport in Superconductors", Eur. Phys. J. B 29, 359-367 (2002).

[12]A. Gandini, "Improvement of Critical Current, Irreversibility Field, and Anisotropy in BiSCCO Tape by the U/n Method", Ph.d. Dissertation, University of Houston (2001).

[13]R. Weinstein, Ravi-Persad Sawh, J. Liu, D. Parks, Y. Ren, V. Obot, C. Foster, "Threshold for Creation of Ionization Pinning Centers in YBCO by Heavy Ions", Physics C 357 (2001) 743-746.

[14]Ravi-Persad Sawh, R. Weinstein, Y. Ren, V. Obot, and H. Weber, "Uranium Fission Fragment Centers in Melt-Textured YBCO", Physics C 341-348, 2441-2442 (2000).

[15]R. Weinstein, Invited Paper, "An Overview of U/n Processing," Proc. of 12th International Symposium of Superconductivity (ISS-99) Morioka, Japan, Oct. 17-19, 1999, Advances in Superconductivity XII, Springer-Verlag, 521-526 (2000).

[16]S. Tönies, H. W. Weber, Y. C. Guo, Shi X. Dou, R-P. Sawh, R. Weinstein, "Improved In-field Behavior of Uranium Doped BiSCCO Tapes by Enhanced Flux Pinning," IEEE Transactions on Applied Superconductivity, Vol. 11 (March 2001), pg.3904-3907. (ASC-2000, Virginia Beach, Sept 17-22, 2000).

[17]R. Hensel, B. Roas, S. Henke, R. Hopfengartner, M. Lippert, J.P. Strobel, M. Vildic, and G. Saemann-Ischenko, "Ion Irradiation of Epitaxial $YBa_2Cu_3O_{7-X}$ Films: Effects of Electronic Energy Loss", Phys. Rev. B 48, 4135 (1990).

[18]Y. Zhu, Z. X. Cal, R. C. Budhani, M. Suenaga, and D. O. Welch, " Structure and Effects of Radiation Damage in Cuprate Superconductors Irradiated with Several-Hundred-MeV Heavy Ions", Phys. Rev. B 48, 6436 (1993).

[19]D.X. Huang, Y. Sasaki, and Y. Ikuhara, "Influence of Ion Velocity on the Damage efficiency in the Single Ion-Target Irradiation System: Au-$Bi_2Sr_2CaCu_2O_x$", Phys. Rev. B (59) 3862-3869 (1999).

[20]L. Civale et al., "Defect Independence of the Irreversibility Line in Proton-Irradiated Y-Ba-Cu-O Crystal", Phys. Rev Lett. (65) 1164-1167 (1990).

[21]J.F. Ziegler, J.P. Biersack, V. Littmork, "The Stopping and Range of Ions in Solids", Vol. 1, Pergamon Press, Oxford, 1985.

[22]P. Poole, "Handbook of Superconductivity", Academic Press, 2000

[23]B.D. Weaver, M.E. Reeves, D.B. Chrisey, G.P. Summer, W.L. Olson, M.M. Eddy, T.W. James, and E.J. Smith, "Critical Current enhancement in Proton-Irradiated $Tl_2CaBa_2Cu_2O_8$", J. Appl. Phys. 69 (2), 1119-1121 (1991).

[24]A. Gandini, R. Weinstein, Y. R. Ren, R. P. Sawh, D. Parks, Y. C. Guo, B. Zeimetz, S. X. Dou, S. Tönies, C. Klein and H. W. Weber, " Critical Current Enhancement in $(Bi,Pb)_2Sr_2Ca_2Cu_3O_{10}$ Tapes Via Isotropic Quasi-Columnar Defects, Induced by Fission Products". Physica C 341-348,1453-1454 (2000).

[25]B. M. Vlcek et al., "Enhancements of the Critical Current of YBCO Single Crystals by Neutron and Proton Irradiation", IEEE Transaction on Appl. Supercon., Vol. 3 No. 1 (1993).

EFFECT OF Y_2BaCuO_5 MORPHOLOGY AND SIZE IN SEMISOLID MELT ON GROWTH RATE OF $YBa_2Cu_3O_{7-x}$ SINGLE CRYSTALS

Oratai Jongprateep
Department of Ceramic Engineering
University of Missouri-Rolla
Rolla, MO 65401

Fatih Dogan
Department of Ceramic Engineering
University of Missouri-Rolla
Rolla, MO 65401

ABSTRACT

To control the growth rate of melt textured $YBa_2Cu_3O_{7-x}$ (Y123) single crystals, the effects of starting powder composition and heating rate of the compacts on the morphology and size of Y_2BaCuO_5 (Y211) in semisolid melt were studied. The processing route of the samples involved the mixture of Y_2O_3 and liquid phase composition (barium cuprate and copper oxide) as starting compounds. The correlations between the heating rate, the morphology of Y211, and the growth rate of Y123 ware examined. Observations from crystal size and microstructural analysis revealed that lower heating rate of compacts led to formation of needle shaped Y211 particles, which consequently resulted in faster growth rate of melt textured Y123 single crystal.

INTRODUCTION

The process of top seeded melt texture growth reduces weak-link problems, and greatly enhances the critical current density of Y123 samples through grain alignment. While samples with large critical current density can be fabricated, obtaining large size and high quality Y123 crystals can be limited. To optimize the size of a single crystal, it is crucial to understand growth mechanism and factors that affect the crystal growth in the fabrication process of melt textured Y123. One of the crucial factors that plays an important role in the kinetics of Y123 crystal growth is the size and morphology of Y211 inclusion in the melt. It has been shown that the controlled Y211 morphology leads to the enhancement of Y123 crystal growth [1-5].

However, the effect of some other factors such as heating rate along with powder-processing route, have not been well established. In this study, the effect of heating rate along with a composite powder processing technique impacting on Y211 morphology and crystal growth of Y123, will be presented.

EXPERIMENTAL

The Y123 pellets were prepared by solid liquid melt growth process. Samples were prepared by mixing commercially available Y_2O_3, CuO and barium cuprate (prepared by combustion synthesis). The starting compositions were pure Y123, Y123+16.7 mol% excess Y211 and Y123+23.1 mol% excess Y211. Very fine platinum powder in the amount of 0.1 wt% was added to all of the powders. The mixtures of powder were uniaxially pressed to obtain 1-inch diameter pellets. $SmBa_2Cu_3O_{7-x}$ (Sm123) was used as the seed on each pellet.

Each set of pellets was heated at a different heating rate—47°C/hr, 94°C/hr and 172°C/hr—from room temperature to 1050°C. Samples were subsequently held at 1050°C for 1 hr and cooled down rapidly to 1010°C. The cooling rate from 1010°C to 990°C remained constant at 0.286°C/hr for all sets. Samples were held at 990°C for 10 hrs prior to cooling to room temperature. Y211 morphology and size were observed using quenched samples in semisolid melt.

Crystal size of samples was determined by measuring a distance from corner to corner of the crystal growth region (Fig. 1a). For samples with the crystal region extended to the edge of the pellet, the diameter of the pellet was used as the crystal size (Fig. 1b). The microstructure of the samples were studied by JEOL-T330A scanning electron microscope (SEM).

RESULTS

Effect of Heating Rate on Crystal Growth:

For the sample with the smallest growth region, the crystal size was 0.4 cm by 0.4 cm. The crystal growth regions accounted for more than 80% of the area of the pellet surface in most of the pellets. Some samples demonstrated the growth region extended to the edge of the pellet.

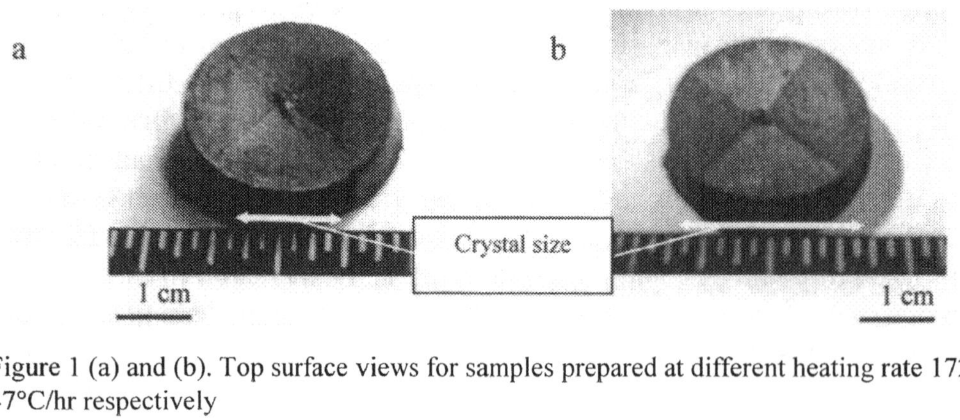

Figure 1 (a) and (b). Top surface views for samples prepared at different heating rate 172°C/hr and 47°C/hr respectively

Fig. 1 (a) and (b), reveals the effect of heating rates on the crystal size. For the samples with identical amount of Y211 inclusions, the lower heating rate tended to favor the crystal growth.

As shown in Figs. 2 and 3, the crystal sizes of samples were in the range of 4 mm–21 mm. The smaller crystal sizes were observed in samples with low amounts of Y211, heated at high heating rate. The larger crystal sizes were observed in samples with high Y211 content, heated at the lowest heating rate.

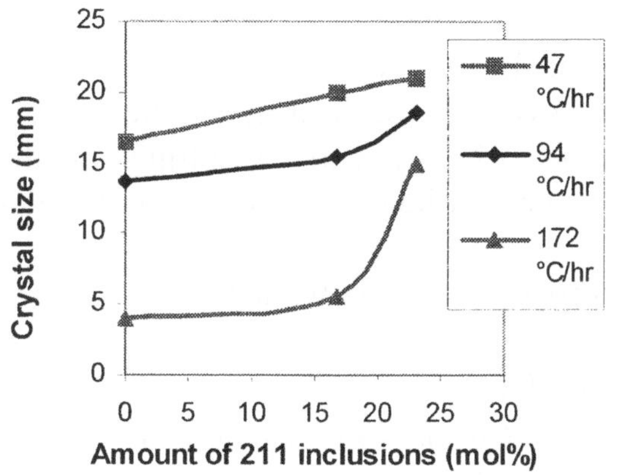

Figure 2. Growth dimension (crystal size) of Y123 as a function of Y211 inclusion

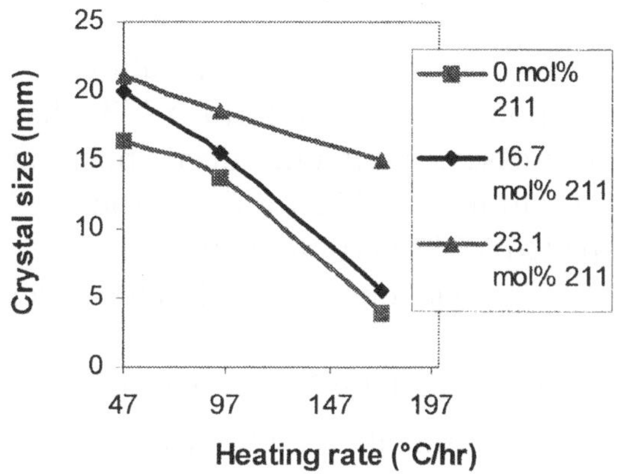

Figure 3. Growth dimension (crystal size) of Y123 as a function of heating rate

Microstructural Development:

From Fig. 4, it was observed that different heating rates resulted in different morphology of Y211 particles in semisolid melt. The shape of Y211 particles in samples with higher heating rates appeared to be platelet-like (4a), while needle-like particles were observed in the samples with lower heating rate (4b).

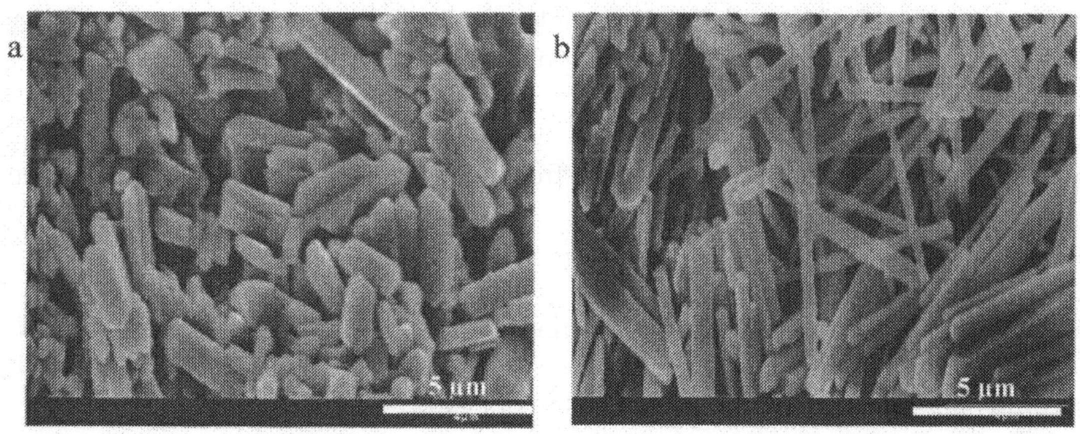

Figure 4 (a) and (b). SEM micrographs of Y123 samples with 23.1 mol% of Y211 inclusion, heated with heating rate 172°C/hr and 47°C/hr respectively

DISCUSSION

Mechanisms in the formation of platelet and needle-shaped 211 particles in this experiment are proposed as follows:

During the heating of powder compacts with a high heating rate, the starting compounds bypass the formation of Y123, which is the stable phase at temperatures below 1000°C. Instead, the formation of Y211 particles occurs in the melt. The direct formation of Y211 from Y_2O_3 phase results in the coarsened platelet-shaped Y211 particles (Fig. 4a).

For samples with low heating rate, an additional intermediate reaction occurs below the peritectic temperature. During the heating of samples, the starting compounds initially form Y123 phase. However, as the heating of the sample progresses, peritectic decomposition of the Y123 occurs, resulting in the formation of Y211 particles and the liquid phase. The process leads to needle-shaped Y211 particles (Fig. 4b).

The reactions involved in the described processes are shown as:

High heating rate: $Y_2O_3 + BaCuO_2 + CuO \rightarrow Y_2BaCuO_5 + L$
Low heating rate: $Y_2O_3 + BaCuO_2 + CuO \rightarrow 2YBa_2Cu_3O_{7-x} \rightarrow Y_2BaCuO_5 + L$

The advantage of the needle-shaped Y211 particles is that they can redissolve relatively fast in the liquid phase as compared to bulky platelet shape Y211. This may lead to a more rapid supply of yttrium ions in the liquid phase for the formation of Y123 resulting in faster crystal growth.

SUMMARY

The relationship between the heating rate, microstructure and crystal growth of Y123 was investigated using Y_2O_3, CuO and barium cuprate as starting powders. It was observed that samples with a low heating rate resulted in a needle-like morphology of Y211 particles in semisolid melt and in a higher solidification rate during melt textured growth of Y123 single crystals.

ACKNOWLEDGEMENT

This work was supported by a grant from the Materials Research Center at the University of Missouri-Rolla.

REFERENCES

[1]M. Murakami, A. Kondoh, H. Fujimoto, K. Takamuku, N. Nakamura, and N. Koshizuka, "A Comparison of Melt Processes to Prepare $YBa_2Cu_3O_{7-x}$ with High J_c," *Advances in Superconductivity*, **4** 451–54 (1991).

[2]C.Varanasi and P.J. McGinn, "Y_2BaCuO_5 Particle Coarsening During Melt Processing of $YBa_2Cu_3O_{7-x}$," *Journal of Electronic Materials*, **22** [10] 1251–57 (1993).

[3]I. Monot, J. Wang, M.P. Delamare, J. Provost, and G. Desgardin, "Influence of Different Melt Process upon the Microstructure and Critical Current of Textured Y123," *Physica C*, **267** 173–82 (1996).

[4]Y. Shiohara, and A. Endo, "Crystal Growth of Bulk High-T_c Superconducting Oxide Materials," *Materials Science and Engineering*, **R19** [1-2] 1-86 (1997).

[5]M. L. Griffith, J. W. Halloran, and R. T. Huffman, "Formation and Coarsening behavior of Y_2BaCuO_5 from Peritectic Decomposition of $YBa_2Cu_3O_{7-x}$," *Journal of Materials Research*, **9** [7] 1633 (1994).

FLUX PINNING and PROPERTIES OF SOLID-SOLUTION $(Y,Nd)_{1+x}Ba_{2-x}Cu_3O_{7-\delta}$ SUPERCONDUCTORS PROCESSED IN AIR and PARTIAL OXYGEN ATMOSPHERES

T. J. Haugan, J. M. Evans, J. C. Tolliver, I. Maartense, P. N. Barnes
Air Force Research Laboratory, 2645 Fifth St. Ste. 13, Wright-Patterson AFB, OH 45433-7919

W. Wong-Ng, L. P. Cook, R.D. Shull
National Institute of Standards and Technology, Materials Science and Engineering Laboratory, Gaithersburg, MD 20899-8520

ABSTRACT

The effect of chemical composition substitutions on the flux pinning and physical properties of $(Y,Nd)_{1+x}Ba_{2-x}Cu_3O_{7-\delta}$ superconductors was studied in powders processed by solid-state reaction and equilibrated in air at 910°C. The powders were subsequently processed in 1% O_2 atmosphere at < 920°C to increase the superconducting transition temperature (T_c) and critical current density (J_c). After processing in air, the powders were nearly pure single-phase compositions as determined by X-ray diffraction. Powders were finally annealed in 100% O_2 atmosphere at temperatures < 500 °C to maximize T_c. The T_cs of the powders were measured by ac susceptibility and dc magnetization methods. Annealing powders with a final step in 1% O_2 atmosphere compared to processing in air significantly enhanced T_c from 65-90 K to > 92 K for all compositions tested, and also increased J_c from about ~10^3-10^5 A/cm^2 to ~10^6 A/cm^2. The flux pinning properties varied depending on exact composition, and the intrinsic behaviors changed with the final 1% O_2 annealing treatment.

INTRODUCTION

High transition temperature (Rare-Earth)$Ba_2Cu_3O_{7-z}$ (RE123) superconductors are being considered for applications including thin film coated conductors and bulk devices because of their high superconducting transition temperatures (T_c) > 90 K, and enhanced critical current density (J_c) at 77 K in useful magnetic fields. While these materials have desirable qualities at 77 K, it is of interest to increase the $J_c(H)$ properties even further by increasing the flux pinning properties of the superconductor. Different methods can be considered to introduce flux pinning defects into the superconductors, including irradiation and addition of second-phase defects or precipitates. One method being considered for bulk applications is to substitute rare-earth cations into the RE123 compound. Different variations of this theme have been studied, including substitution in (Y,R)123 with R = Ho, Dy, Gd, Eu, and Pr, and various other combinations of rare-earths such as (Gd,Sm,Eu)123 [1-16]. Possible mechanisms by which such substitutions increase the flux pinning include: **(a)** addition of second-phase defects by precipitation or composition changes, **(b)** formation of finely distributed lower T_c components from the mixed solubility of RE with Ba and intersolubility of RE or other mechanisms, or **(c)** randomly distributed oxygen-deficient zones which have lower T_cs [7,8]. The finely distributed lower T_c components are suggested as a cause of the so-called 'fishtail' effect, where as the magnetic field is increased, the J_c increases as the lower T_c components transition to normal behavior before the J_c decreases at much higher applied magnetic fields [7,8]. The peak of J_c maximum in these materials typically occurs at applied fields of about 2 T to 3 T.

Studies of the system $(Y_{1-x}Nd_x)Ba_2Cu_3O_{7-\delta}$ have, to our knowledge, been limited thus far to melt-processed or single crystal materials [9-14], and powders [15,16]. Single crystals in this system demonstrated very high J_c and the 'fish-tail' effect for Nd content varying from 0.1 to 0.4 after varying oxygenation treatments [13]. Recently our group published initial studies on $(Y_{1-x}Nd_x)Ba_2Cu_3O_{7-\delta}$ powders processed in air [16].

In this paper, previous studies on powders in this system [16] are continued by adding an additional processing step of annealing in reduced oxygen atmospheres to enhance T_c and J_c. The T_c in particular was noted to be quite low for some of the Nd-rich and Ba-poor compositions after annealing in air [16]. Subsequent processing in reduced oxygen atmospheres is well known to sharpen the T_c transition of Nd123 type compositions and improve superconducting properties [15]. How effective this is for processing the mixed compositions and their effect on flux pinning is studied herein.

In this and previous work, the effect of composition changes in $(Y,Nd)_{1+x}Ba_{2-x}Cu_3O_{7-\delta}$ was studied in solid-state powders annealed in air to achieve chemical equilibrium [16]. Powder compositions are expected to have different properties than (Y,Nd)123 melt-processed or crystals in previous studies, where non-equilibrium processes can affect the physical properties. In melt-processed materials it is almost impossible to eliminate the formation of second-phase defects such as RE211, which can affect pinning. With powder processing, it is possible to completely eliminate second-phase defects, and investigate other causes of pinning more related to the intrinsic properties of the crystal structure.

For the studies herein, powders were annealed in 1% oxygen atmosphere, which is expected to enhance site occupation of Nd on the RE rather than Ba site [15,17]. This is expected to sharpen and increase the T_c transition, however it could decrease flux pinning by reducing the formation of lower T_c solid-solutions [8,15,16].

In this paper, only solid-solution single-phase compositions are considered that preserve the single-crystal structure however perturb the superconductor in localized regions surrounding the chemically substituted volumes. It's expected that with Nd substitution on about 5 % or 10% of the Y or Ba sites, the defect density of the locally-perturbed areas can be high enough for effective pinning ~ 2-3 $x10^{11}$ cm^{-2} equivalent to a fluxon density of about 3-6 T. How effectively this can increase the flux pinning must be determined, however.

EXPERIMENTAL**

Experimental methods for this work are described in the following, using procedures identical to previous work [16]. Superconducting powders were prepared by the solid-state reaction method, using starting reactants of Nd_2O_3, Y_2O_3, $BaCO_3$, and CuO (≥ 99.95% purity). The powders were dehydrated at 450 °C prior to weighing. The powders were mixed and ground with mortar and pestle, calcined by slow heating 650 °C to 850 °C at 25 °C/h, and subsequently annealed with intermediate grinding at 880 °C and 910 °C. Powders were annealed at 910 °C with intermediate grinding until phase equilibrium was reached (3-4 annealings), as determined by X-ray diffraction (XRD). The powders were reacted in ~1 cm diameter pellets (0.5 g to 1 g batches), formed by lightly pressing ($5\text{-}10x10^6$ Pa) in molds. X-ray diffraction was performed with a Rigaku diffractometer. A step size of 0.03° was used for the θ-2θ scans.

After single-phase compositions were obtained, an additional processing step in 1% O_2 atmosphere was performed; the annealing temperature was at 820°C to 920°C increasing in 20°C increments as the Nd content increased from 0 to 1.0, and pellets were reacted on mixed (Nd,Y)211 pellets to minimize reactions with the substrate. After annealing in 1% O_2 atmosphere, the samples were checked with XRD to determine if the single phase composition was preserved or whether some slight (melting or decomposition) reaction might have occurred to cause formation of Y_2BaCuO_5 and $BaCuO_2$ phases (for example). A final annealing step was performed in 100% O_2 atmosphere at 460°C to 260°C, decreasing in 40°C increments as the Nd content increased from 0 to 1.0; this assumed a linear gradient in temperature between optimized annealing temperatures for Y123 and Nd123 compositions [18]. The optimal annealing temperature in 100% O_2 atmosphere for the mixed RE123 compounds to maximize oxygen content and T_c is unknown yet, therefore this approximation was used.

Superconducting properties of powders were measured with a SQUID magnetometer (Quantum Design, MPMS/$MPMS^2$). Magnetization-applied field (M-H) hysteresis loops at different temperatures were made by heating samples to 100 K and zero-field cooling (ZFC) to the measurement temperature. The magnetic J_c was estimated using the extended Bean critical current model $J_c = 15(\Delta M)/R$ where ΔM is the volume magnetization, and R is the radius of the superconducting volume roughly approximated as 0.00005 cm for the finely reacted and ground powders [19].

Field-cooled (FC) Meissner and zero-field-cooled (ZFC) measurements were performed from 5 K to 100 K [19]. The SQUID magnet was reset to zero before any measurements. The superconducting volume percentages were calculated using $\chi(\%) = 4\pi\chi_v/(1\text{-}D*4\pi\chi_v)$, where $\chi_v = M/H_{appl}$ is the measured magnetic susceptibility, and $D = 1/3$ is the demagnetization factor assuming a spherical particle distribution [19]. The applied magnetic field was $H_{appl} = 796$ A/m - H_{rem}, where H_{rem} is the remnant field of the magnet after resetting to zero, determined for each sample by measuring M(H) from 796 A/m to -398 A/m at 79.6 A/m intervals and plotting when $M = 0$ ($\pm$ 8 A/m accuracy). The transition temperature of the largest volume fraction of powder was determined by finding the temperature when $(d^2\chi/dT^2) \cong 0$ upon cooling from 100 K. The standard uncertainty of T_c measurements was 0.5 K, as determined from both ZFC and FC curves and multiple measurements. The standard uncertainty of Meissner volume fractions was ± 5%, as determined from multiple measurements of one sample. The standard uncertainty of J_cs was measured using differences of ΔM for both positive and negative applied fields. Standard uncertainties for normalized $J_c(H)$ curves were calculated by measuring errors of ΔM for positive and negative applied fields, and averaging the ΔM errors over the range of approximately 0.1 T to 4.8 T.

The superconducting transition temperature (T_c) was also measured using an AC susceptibility technique with the amplitude of the magnetic sensing field, h, varied from 0.025 Oe to 2.2 Oe, at a frequency of approximately 4 kHz. The AC susceptibility technique provides information about primary and secondary transitions of the bulk samples. T_cs were measured on palletized samples with very small thickness. Samples were mounted onto the end of a sapphire rod and measured as the samples were warmed through the transition region at very slow rate of ~ 0.06 K/min. The T_c measurements were accurate within ≤ 0.1 K at three calibration points: liquid He at 4.2 K, liquid N_2 at 77.2 K, and room temperature. Different methods of measuring powders were tested including mounting pellets and powders placed in silicone grease, and the different methods gave the same T_c values within about 0.1 K.

RESULTS

The range of single-phase compositions determined in previous work is shown in Figure 1 [16]. All of the compositions were nominal single-phase as determined by XRD, in agreement with previous results [15], except for a small region (Ba = 2.0 and Y = 0.3 to 0.4) where sluggish reactions were suspected [16].

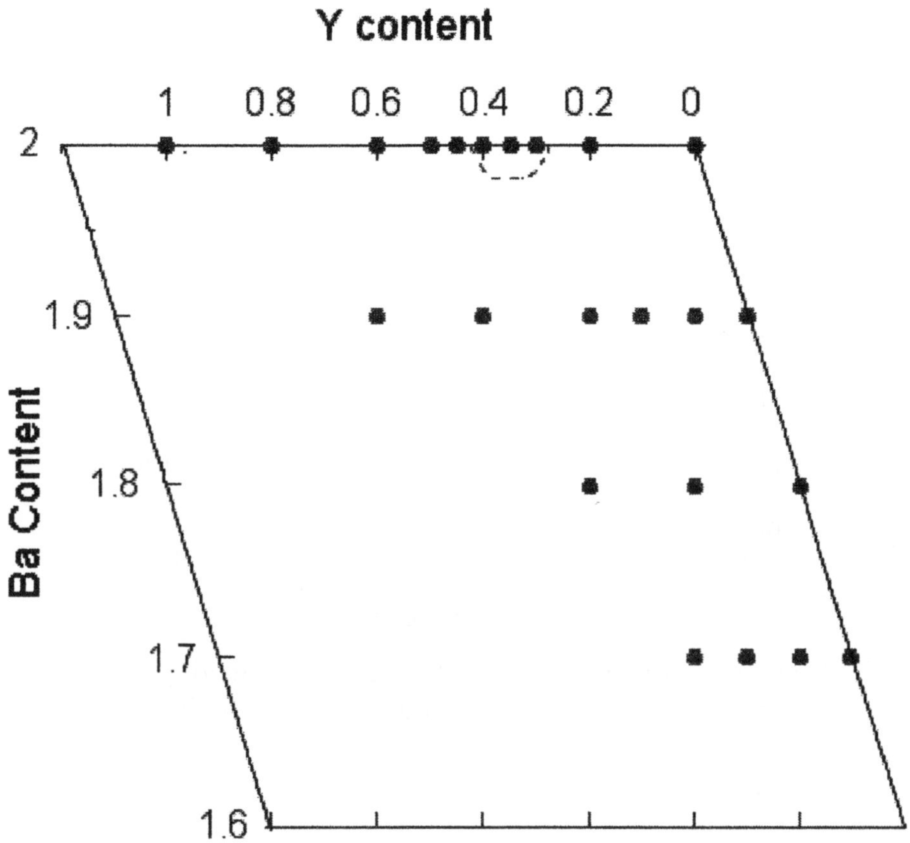

Figure 1. Compositions of $(Y,Nd)_{1+x}Ba_{2-x}Cu_3O_{7-\delta}$ that were nominal single-phase, excluding Ba = 2 and Y = 0.3 to 0.4 [16].

The superconducting transitions of Ba = 2.0 and 1.9 compositions annealed only in air determined from previous work are shown in Figure 2 [16]. The transitions were broadened for Nd > 0.4, and the transition for the largest volume fraction of the powder was reduced to ~ 73 K for the Nd = 1.0 composition. A reduction and broadening of T_c for the Nd = 1.0 composition is usually observed without further processing in reduced oxygen partial pressures [15,17]. With increasing Nd and reduced Ba content, the T_c of the bulk powder is decreased, in agreement with trends reported previously for $(Y,Nd)_{1+x}Ba_{2-x}Cu_3O_{7-\delta}$ and single-phase $Nd_{1+x}Ba_{2-x}Cu_3O_{7-\delta}$ [15,17].

After annealing these compositions in 1% reduced oxygen atmosphere, the T_cs sharpened considerably and the transition onsets were greatly increased to > 92K on average, as shown in Figure 3. The increase and sharpening of T_c for the Nd-rich compositions in reduced O_2 atmosphere is consistent with previous work [7,8,15,17]. Of special note in Figure 3 is that the T_cs of the Ba = 1.9 and Nd-rich compositions were greatly increased sufficiently high so that they can be considered for practical applications.

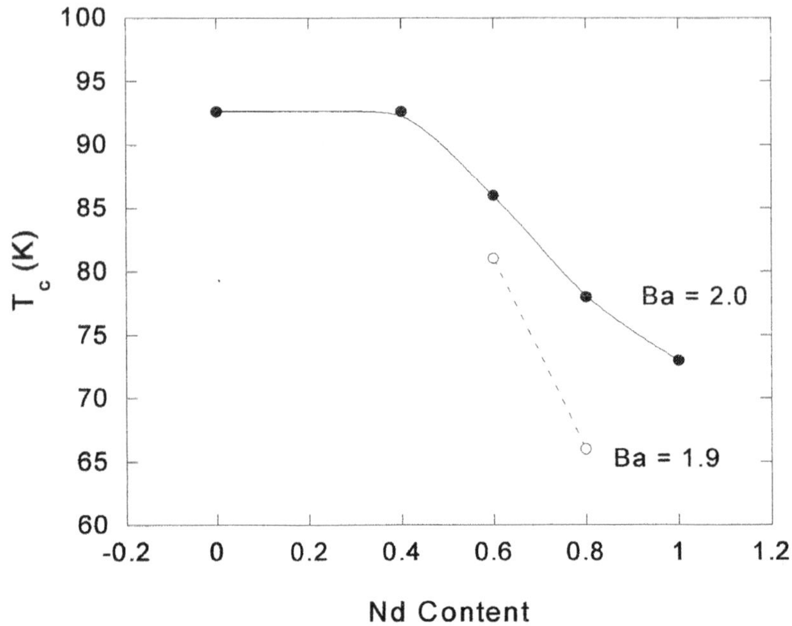

Figure 2. Transition temperature of the bulk superconducting volume fraction of powders processed in air measured by dc magnetization [16].

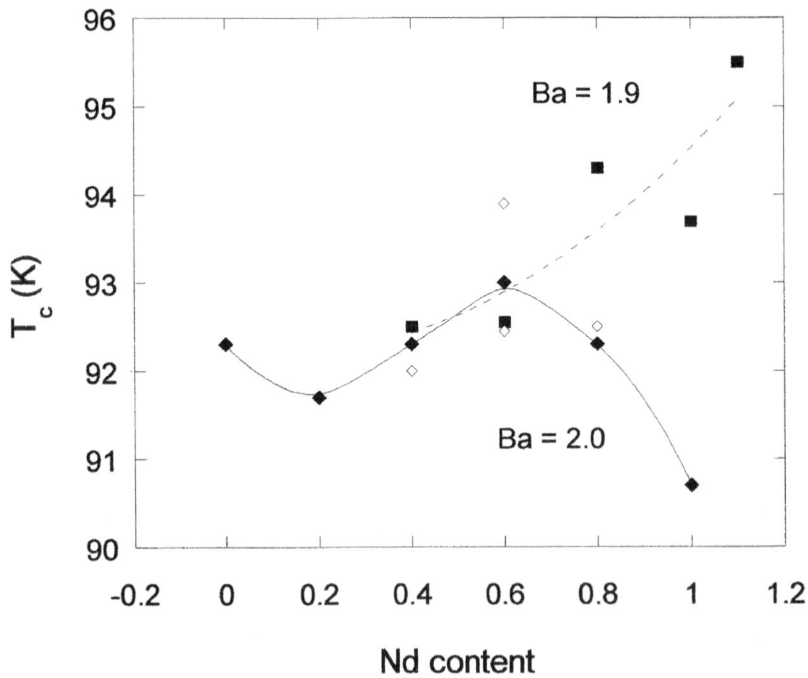

Figure 3. Transition temperature of powders processed with an additional annealing step in 1% O_2 atmosphere; Ba = 1.9 powders measured by ac susceptibility χ' onset (■) and for Ba = 2.0 powders measured by DC magnetization (◊) or ac susceptibility χ' onset (♦).

Figure 4 plots the maximum J_cs of the powders for different composition and temperature annealed only in air. In general, the J_cs decreased with increasing Nd substitution and decrease of Ba, consistent with the decrease of T_c shown in Figure 2. Only the Y123 compound had reasonably high J_c to be considered for practical applications. The normalized $J_c/J_{c\text{-max}}$ curves for the same powders is shown in Figure 5, where $J_{c\text{-max}}$ is the maximum J_c measured in the range of H_{appl} ~ (0.06 to 0.1) T. More intrinsic differences in the pinning can be observed in Figure 5.

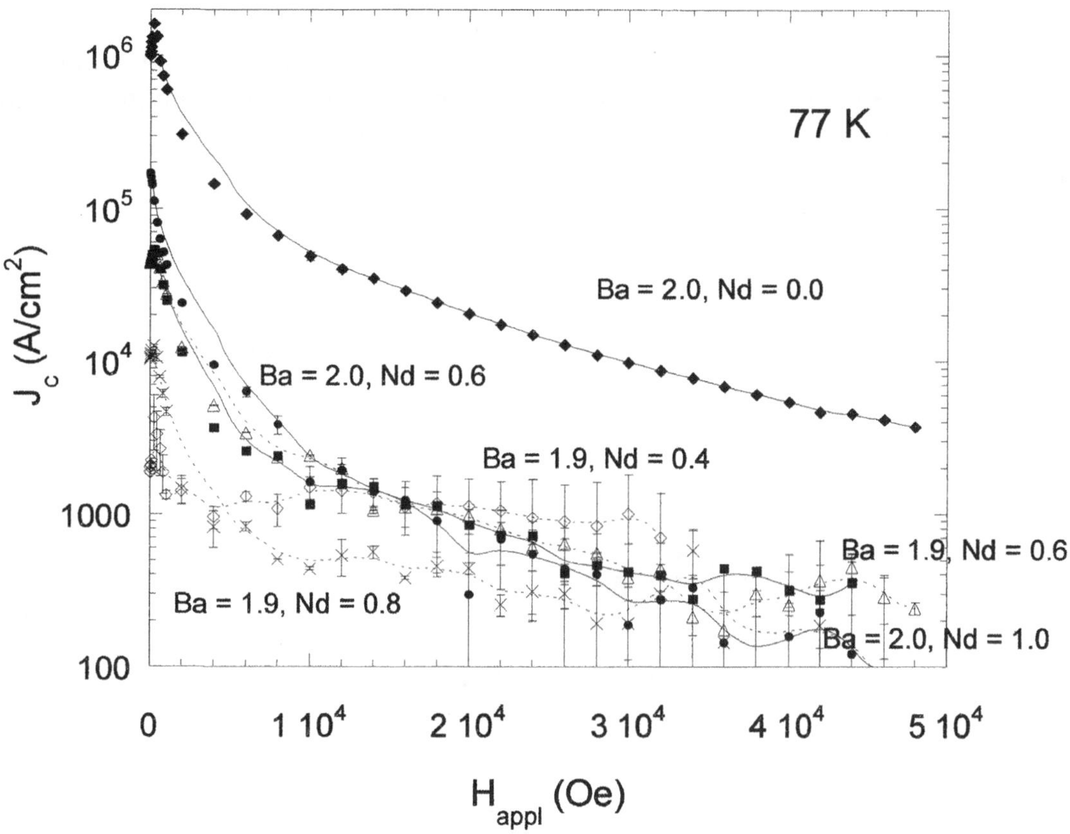

Figure 4. Critical current density as a function of applied magnetic field (H_{appl}) of powders annealed in air estimated from M-H loops for varying Nd content and Ba = 1.9 (dashed lines, open symbols) and Ba = 2.0 (solid lines, filled symbols).

The effect of the additional annealing step in 1% O_2 atmosphere is shown in Figure 6, where $J_c(H)$ curves are compared for powders processed with and without the additional step. The increase of J_c with the additional processing in 1% O_2 atmosphere can be clearly seen in Figure 6, with the J_c increased more than an order of magnitude for any applied field and all compositions. The increase is sufficiently high so that the J_cs of the Ba 1.9 compositions are about equal or better than values for Y123 material (Ba 2.0, Nd = 0), as shown in Figure 4. The

J_c of the Ba = 2.0, Nd = 0.6 compound was higher by about 50% than Y123 especially for H_{appl} > 3 T. Therefore these compositions might be considered for additional processing steps or for processing in thin film form to determine additional effects on flux pinning.

In Figure 7 the normalized J_c curves from Figure 6 are plotted, which gives indication of more intrinsic changes of pinning. For one composition (Ba = 1.9, Nd = 0.6), additional annealing in 1% O_2 atmosphere significantly changed the flux pinning behavior by enhancing the lower field J_cs (H< 2T) and reducing the higher field J_cs (H > 2T). The intrinsic pinning is better than Y123 as shown in Figure 5, which suggests that with additional optimization of J_c(0T) values, the absolute values of J_c could be increased substantially also. For another composition (Ba = 1.9, Nd = 0.4), the J_c was quite low for processing in air, so the normalized curve showed more usual behavior after processing in 1% O_2 atmosphere.

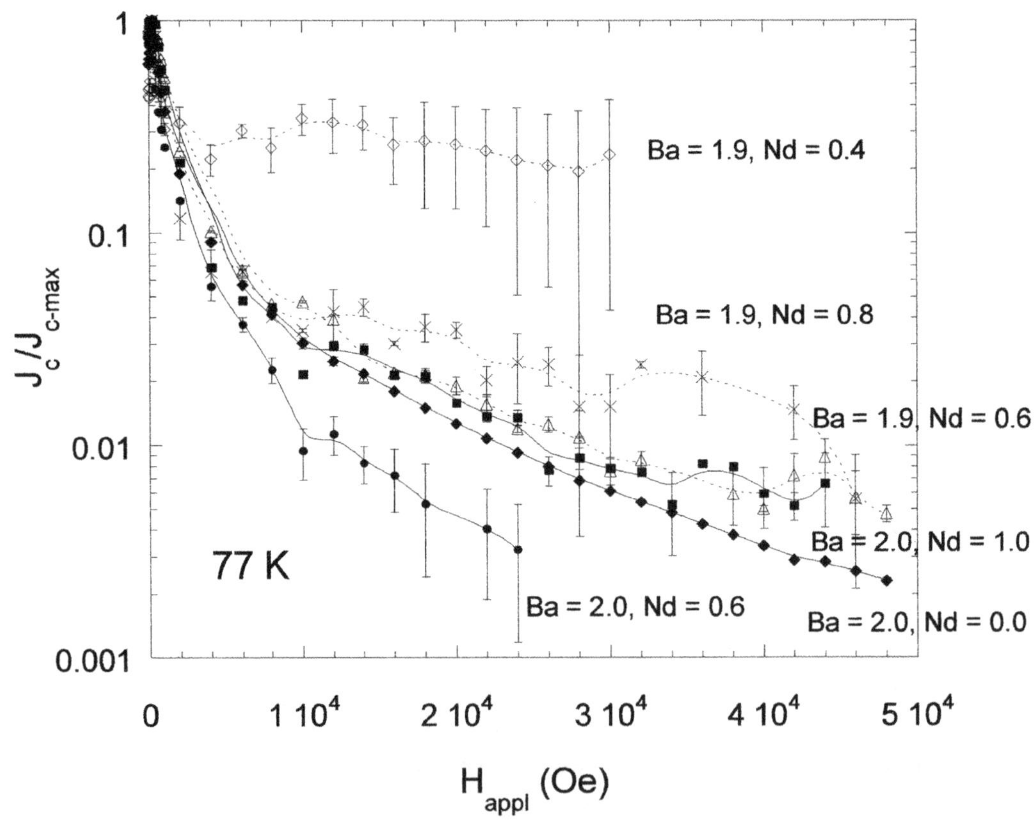

Figure 5. Normalized J_c as a function of H_{appl} for films from Figure 4; maximum J_c ($J_{c\text{-}max}$) was measured for H_{appl} ~ (0.06-0.1) T.

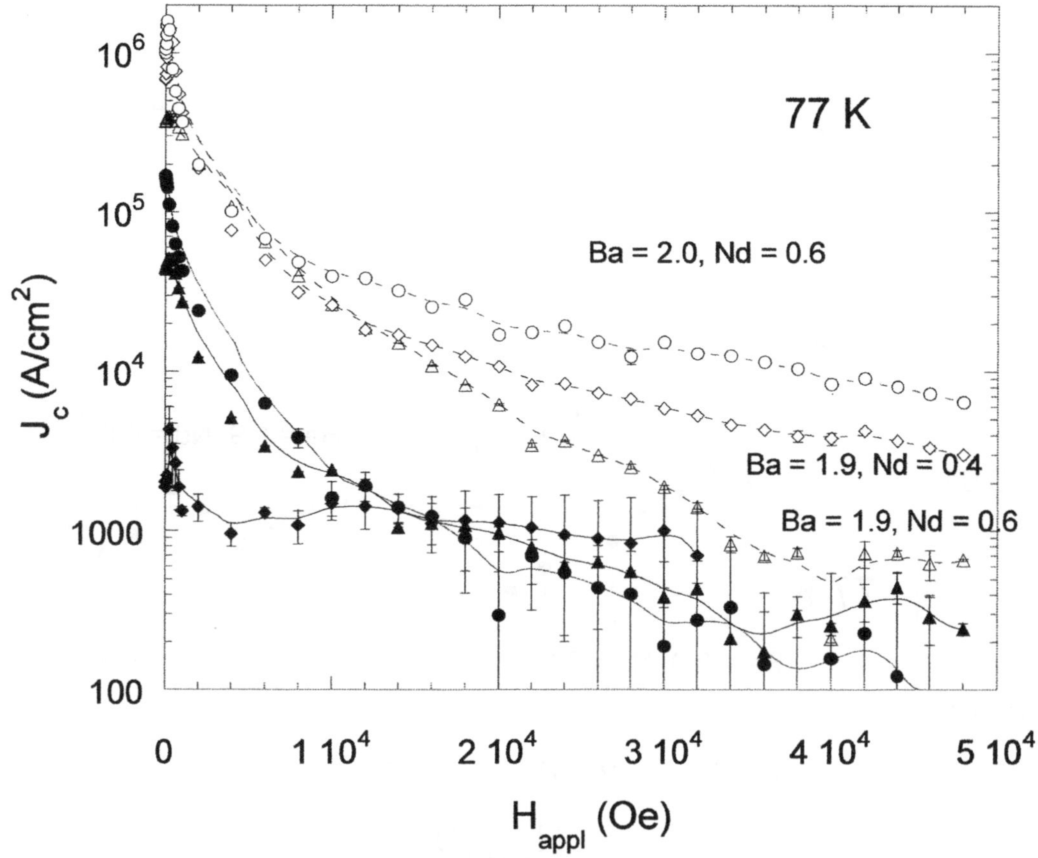

Figure 6. $J_c(H_{appl})$ comparison of powders annealed with additional step in 1% O_2 atmosphere (dashed lines, open symbols), to the same compositions annealed only in air (solid lines, closed symbols).

For all compositions 77 K, the powders showed a very small fish-tail effect, with $J_{c\text{-}max}$ occurring at 0.06 T to 0.1 T depending on composition. By comparison, the peak of maximum J_c in oxygenated melt-processed materials usually occurs around 2 to 3 T [7,8]. This suggests the powder compositions in this work are homogeneous in nature without the oxygen-deficient regions thought to cause the fish-tail effect [7,8]. A homogenous distribution of oxygen is expected as a consequence of the small size of the powders, and the near-equilibrium conditions used for processing.

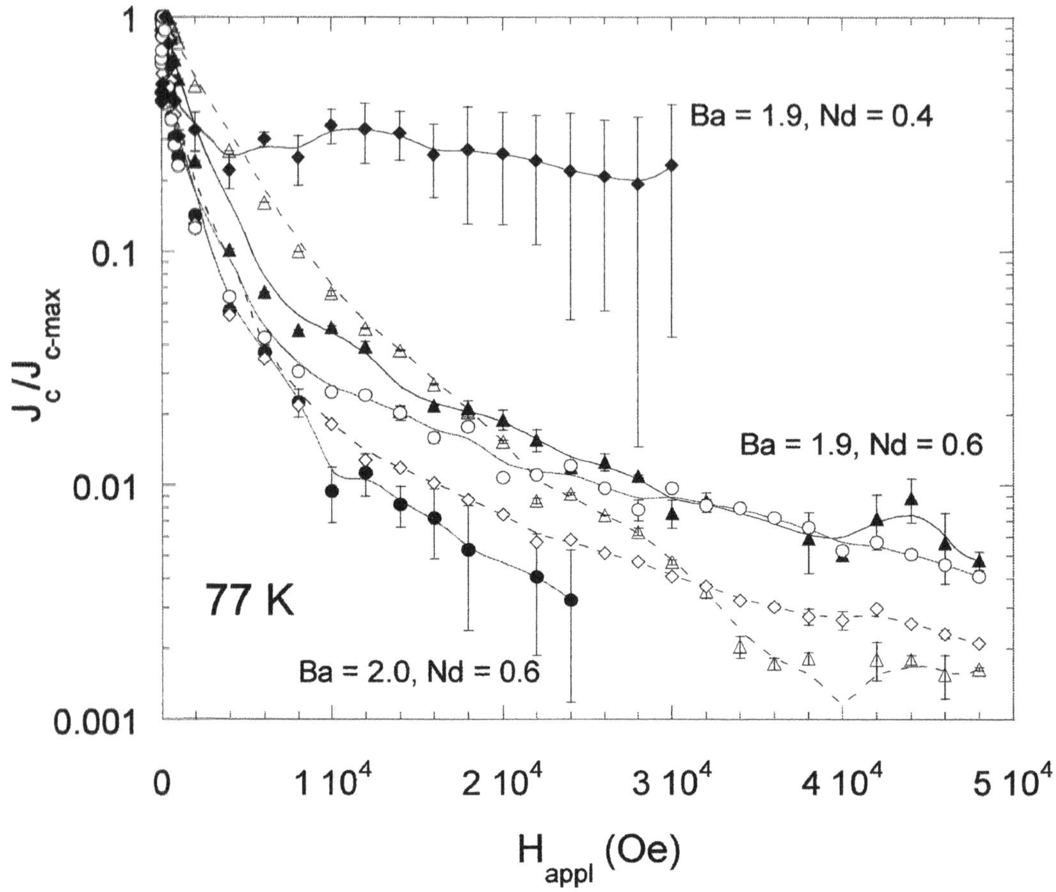

Figure 7. Normalized J_c as a function of H_{appl} for films from Figure 6; maximum J_c ($J_{c\text{-}max}$) was measured for $H_{appl} \sim (0.06\text{-}0.1)$ T. Powders were treated in 1% O_2 atmosphere (open symbols, dashed lines), or only treated in air (filled symbols, solid lines).

CONCLUSIONS

The flux pinning and superconducting properties of $(Y,Nd)_{1+x}Ba_{2-x}Cu_3O_{7-\delta}$ powders were studied for varying compositions, and compared for different final high temperature processing steps in air or 1% O_2 atmosphere. Processing of powders in air was found to form single-phase structures, however the superconductivity and T_c were much less than optimal as evidenced by low and wide T_c transitions, and J_cs were in the range of 10^3-10^5 A/cm^2 for most compositions. Subsequent processing of powders in 1% O_2 atmosphere at reduced temperatures < 920°C increased and narrowed the T_c transitions and raised J_c values to the range of 10^6 A/cm^2, where J_c values were estimated assuming a uniform distribution of spherical particles with diameter ~1.0 micron. Compositions including Ba = 2.0, 1.9 and Nd = 0.4-0.6 had sufficiently high J_c (actual values and normalized for maximum J_c) that they could be considered for further investigation. Annealing in 1% O_2 atmosphere was observed to change the normalized J_c pinning, e.g. by increasing J_c for H < 2 T and reducing J_c for H > 2 T for one composition (Ba = 1.9, Nd = 0.6)

For the range of single-phase compositions studied, additional processing in 1% O_2 atmosphere significantly increased T_c > 92 K for all compositions tested, including Ba = 1.9. By comparison, T_cs of the same powders processed in air were only in the range of 65-90 K, which was too low to consider for applications at 77 K. Additional processing in 1% O_2 atmosphere significantly changed the T_cs and J_cs and therefore the range of compositions than could be considered for possible application in coated conductors.

ACKNOWLEDGEMENTS

The authors would like to thank T. Spry, M.E. Fowler, and K. Fields at AFRL for assistance with sample preparation, and R. Drew of NIST-Gaithersburg for SQUID measurements.

REFERENCES

[1]Y. Feng, A. K. Pradhan, Y. Zhao, Y. Wu, N. Koshizuka, and L. Zhou, "Influence of Ho substitution for Y on flux pinning in melt-processed YBCO superconductors," *Physica C* **357-360** 799-802 (2001).

[2]A. R. Devi, V. S. Bai, P. V. Patanjali, R. Pinto, N. H. Kumar, and S. K. Malik, "Enhanced critical current density due to flux pinning from lattice defects in pulsed laser ablated $Y_{1-x}Dy_xBa_2Cu_3O_{7-\delta}$ thin films," *Supercond. Sci. Technol.* **13** 935-939 (2000).

[3]H. H. Wen, Z. X. Zhao, R. L. Wang, H. C. Li, and B. Yin, "Evidence for lattice-mismatch-stress-field induced flux pinning in $(Gd_{1-x}Y_x)Ba_2Cu_3O_{7-\delta}$ thin films," *Physica C* **262** 81-88 (1996).

[4]Y. Li, G. K. Perkins, A. D. Caplin, G. Cao, Q. Ma, L. Wei and Z X Zhao, "Study of the pinning behaviour in yttrium-doped Eu-123 superconductors" *Supercond. Sci. Technol.* **13** 1029-1034 (2000).

[5]H. H. Wen, Z. X. Zhao, Y. G. Xiao, B. Yin, and J. W. Li, "Evidence for flux pinning induced by spatial fluctuation of transition temperatures in single domain $(Y_{1-x}Pr_x)Ba_2Cu_3O_{7-\delta}$ samples," *Physica C* **251** 371-378 (1995).

[6]E. S. Reddy, P. V. Patanjali, E. V. Sampathkumaran, R. Pinto, "Fabrication and superconducting properties of ternary $REBa_2Cu_3O_y$ thin films," *Physica C* **366** 123-128 (2002).

[7]M. R. Koblischka, M. Muralidhar, M. Murakami, "Flux pinning sites in melt-processed $(Nd_{0.33}Eu_{0.33}Gd_{0.33})Ba_2Cu_3O_y$ superconductors," *Physica C* **337** 31-38 (2000).

[8]M. Jirsa, M. R. Koblischka, T. Higuchi, M. Muralidhar, M. Murakami, "Comparison of different approaches to modelling the fishtail shape in RE-123 bulk superconductors," *Physica C* **338** 235-245 (2000).

[9]C. Varanasi, P. J. McGinn, H. A. Blackstead, and D. B. Pulling, "Nd Substitution in Y/Ba Sites in Melt Processed $YBa_2Cu_3O_{7-\delta}$ Through Nd_2O_3 Additions," *Journal of Electronic Materials,* **24** [12] 1949-1953 (1995).

[10]D. N. Matthews, J. W. Cochrane, G. J. Russell, "Melt-textured growth and characterization of a $(Nd/Y)Ba_2Cu_3O_{7-\delta}$ composite superconductor with very high critical current density," *Physica C* **249** 255-261 (1995).

[11]P. Schätzle, W. Bieger, U. Wiesner, P. Verges and G. Krabbes, "Melt Processing of (Nd,Y)BaCuO and (Sm,Y)BaCuO composites," *Supercond. Sci. and Technol.* **9** 869-874 (1996).

[12]A. S. Mahmoud and G. J. Russell, "Large crystals of the composite Y/Nd(123) containing various dopants grown by melt-processing in air," *Supercond. Sci. and Technol.* **11** 1036-1040 (1998).

[13]X. Yao, E. Goodilin, Y. Yamada, H. Sato and Y. Shiohara, "Crystal growth and superconductivity of $Y_{1-x}Nd_xBa_2Cu_3O_{7-\delta}$ solid solutions," *Applied Superconductivity* **6** [2-5] 175-183 (1998).

[14]D. K. Aswal, T. Mori, Y. Hayakawa, M. Kumagawa, "Growth of $Y_{1-z}Nd_zBa_2Cu_3O_x$ single crystals," *Journal of Crystal Growth* **208** 350-356 (2000).

[15]H. Wu, K. W. Dennis, M. J. Kramer, and R. W. Mccallum, "Solubility Limits of $LRE_{1+x}Ba_{2-x}Cu_3O_{7+\delta}$," *Applied Superconductivity* **6** [2-5] 87-107 (1998).

[16] T. J. Haugan, M. E. Fowler, J. C. Tolliver, P. N. Barnes, W. Wong-Ng, L. P. Cook, "Flux Pinning and Properties of Solid-Solution $(Y,Nd)_{1+x}Ba_{2-x}Cu_3O_{7-\delta}$ Superconductors", in *Processing of High Temperature Superconductors*, Ceramic Transactions Vol. 104, edited by A. Goyal, W. Wong-Ng, M. Murakami, J. Driscoll (American Ceramic Society, Westerville OH, 2003), p. 299 - 307.

[17]R. W. McCallum, M. J. Kramer, K. W. Dennis, M. Park, H. Wu, and R. Hofer, "Understanding the Phase Relations and Cation Disorder in $LRE_{1+x}Ba_{2-x}Cu_3O_{7+\delta}$," *J. of Electr. Mater.* **24** [12] 1931-1935 (1995).

[18] J. Shimoyama, Univ. of Tokyo, private communication.

[19]*Magnetic Susceptibility of Superconductors and Other Spin Systems*, Plenum Press, New York, 1991.

PHASE RELATIONS IN THE $BaO-R_2O_3-CuO_x$ SYSTEMS

W. Wong-Ng, L.P. Cook, and J. Suh
Ceramics Division
Materials Science and Engineering Laboratory
National Institute of Standards and Technology
Gaithersburg, MD 20899

ABSTRACT

Flexible second-generation coated conductors are based on barium-yttrium-copper-oxide and barium-lanthanide-copper-oxide materials. Processing of these conductors requires phase diagram data collected under controlled-atmosphere conditions. A special experimental procedure was used for preparing BaO starting material and for the handling and heat-treatment of samples. The phase diagrams of the carbonate-free $BaO-R_2O_3-CuO_x$ (where R=Nd, Sm, and Y) were constructed under atmospherically-controlled conditions. A discussion of the crystal chemistry and phase relationships of selected phases found in these systems is presented. A trend as a function of the size of R was observed.

INTRODUCTION

Phase diagrams are 'road maps' for processing of $Ba_2RCu_3O_{6+z}$ materials (R-213, where R=lanthanides and Y). Phase diagrams are especially important for processing routes based on the melt-growth method [1, 2]. A number of studies pertaining to the $BaO-R_2O_3-CuO_x$ phase equilibria have been reported. A majority of the data was collected using $BaCO_3$ as a starting reagent [1, 3-6]. Other reported Ba-containing starting reagents include BaO_2 [7], BaO and $Ba(NO_3)_2$ [8]; however, most of the reported data was not collected entirely under controlled-atmosphere conditions. As a result, the presence of carbonate or hydroxide in the samples has rendered phase diagram determinations in the high-BaO region tentative, and the reported diagrams are incomplete.

This paper is a part of a continuing effort to understand the phase equilibria and crystal chemistry of $AO-R_2O_3-CuO_x$ systems (A = alkaline-earths). We will concentrate on three systems (A=Ba, and R = Nd, Sm and Y) that were prepared under controlled atmosphere using BaO-derived starting materials. Particular emphasis is placed on the region near the BaO corner, and on the solid solution formation in $Ba_{2-x}R_{1+x}Cu_3O_{6+z}$ and its tie-line relationships with neighboring compounds. This work has utilized special experimental procedures to minimize the presence of carbonate and minimize exposure to atmospheric CO_2 and moisture. Experimental work was carried out at two oxygen pressures, namely, at p_{O2} = 100 Pa (0.1 % O_2 in Ar volume fraction) and at p_{O2} = 21 kPa (21 % O_2 in Ar volume fraction). The former p_{O2} was selected to approximate ion beam assisted deposition (IBAD) and rolling assisted biaxially textured substrate (RABiTS) processing conditions.

EXPERIMENTAL[1]

The primary issue in our sample preparation procedure was controlling the atmosphere to avoid CO_2 and moisture. A reducing atmosphere during heat-treatment is also important to match the deposition conditions of the R-213 films. Since it is critical to use pure BaO as a starting material, we prepared BaO by decomposing $BaCO_3$ (99.99% purity, metals basis) at 1300 °C in a vacuum furnace. Sample weighing and handling were performed inside a glove box, and sample transference between the glove box and the atmosphere-controlled furnace was accomplished using a sealed transfer vessel.

A solid-state reaction technique was used for sample preparation. Since the commercial Nd_2O_3 and Sm_2O_3 powder used for this study (99.9% purity, metals basis) react readily with atmospheric moisture to form $Nd(OH)_3$ and $Sm(OH)_3$ respectively, a preliminary heat-treatment at 450 °C was necessary to convert hydroxide into oxide prior to sample preparation. After weighing and homogenization of individual compositions prepared from BaO, R_2O_3 (R=Nd, Sm and Y) and CuO (99.99 % purity, metals basis), pellets were pressed and placed inside individual MgO crucibles, and then transferred to a controlled atmosphere annealing furnace using a transfer vessel and an interlock system. Experiments were conducted at p_{O2} = 100 Pa and at p_{O2} = 20 kPa for R=Nd and Y, and at 100 Pa for R=Sm. Oxygen partial pressures were maintained by a mass flow-meter connected to cylinders of O_2 and Ar, and were monitored by a zirconia oxygen sensor at both the gas inlet and the outlet of the furnace. Heat-treatments were performed at 810 °C. Several intermediate grindings and pressings took place until the content of the phase assemblages did not change, as determined from the results of powder x-ray diffraction. A total of 281 samples were prepared for this study (142 for the Nd-systems, 35 for the Sm-system and 104 for the Y-systems).

For x-ray phase analysis, specimens were loaded into a sealed cell designed by Ritter [9]. The process of sample loading was performed inside an Ar-filled

[1] Certain trade names and company products are mentioned in the text or identified in illustrations in order to adequately specify the experimental procedure and equipment used. In no case does such identification imply recommendation or endorsement by National Institute of Standards and Technology.

glove box. X-ray powder diffraction was used to identify the phases synthesized and to confirm phase purity. Data were collected using a computer-controlled automated diffractometer equipped with a theta-compensation slit; CuK_α radiation was used at 45 kV and 40 mA. The radiation was detected by a scintillation counter and a solid state amplifier. A Siemens diffraction software package was used for diffractometer control and data acquisition. Phase identification was accomplished using reference x-ray diffraction patterns from the ICDD Powder Diffraction File (PDF) [10].

DISCUSSION

In the following, discussion about the phase diagrams of the BaO-½R_2O_3-CuO_x systems (R=Nd, Sm,Y) and a comparison of the diagrams prepared under 100 Pa oxygen partial pressure will be given. The notation adopted for lanthanide oxides is ½ R_2O_3, so that the composition ratio is the same as the cation ratio, as it is for the other end members of the system.

BaO-½Nd_2O_3-CuO_x Systems

(1) *Phase diagrams*

Figures 1a and 1b give the ternary phase diagrams of BaO-½Nd_2O_3-CuO_x prepared at p_{O2} = 21 kPa (930 °C), and p_{O2} = 100 Pa (810 °C), respectively. The diagram prepared at p_{O2} = 21 kPa is similar to that reported by Abbittasta et al. which was prepared under pure O_2 [8]. Both the solid solution $(Ba,Nd)_2CuO_{3+z}$ and the $Ba_4Nd_2Cu_2O_9$ phase were identified in the BaO-rich domain. However, many of the phases previously reported in the literature, when $BaCO_3$ was used, were not observed. For example, $Ba_3RCu_2O_z$, $Ba_4RCu_2O_z$, $Ba_5RCu_3O_z$ [11-12], $Ba_4RCu_3O_z$, $Ba_8R_3Cu_5O_z$ [13] and $Ba_6RCu_3O_z$ [6, 14] were not observed regardless of the oxygen pressure. Fu et al. [15] reported that Ba_2NdCuO_z and the "brown phase" $BaNd_2CuO_z$ are part of the same solid solution series, rather than distinct phases.

(2) *$Ba_{2-x}Nd_{1+x}Cu_3O_{6+z}$ Solid Solution*

At p_{O2} = 21 kPa, the solid solution range 'x' of the superconductor solid solution, $Ba_{2-x}Nd_{1+x}Cu_3O_{6+z}$, was confirmed to extend nominally from x = 0.0 to x = 0.95. X-ray diffraction patterns of $Ba_{2-x}Nd_{1+x}Cu_3O_{6+z}$, as quenched using flowing liquid nitrogen-cooled helium, are shown in Fig. 2. These patterns illustrate that single phase Nd-213 was obtained. Lattice parameter determination using the least-squares refinement program LSQ-NBS92 [16] indicated that samples with $0.7 > x \geq 0$ are tetragonal. Peak splitting in the x-ray patterns appears for samples with $x > 0.7$. We were able to index these patterns using the supercell reported by Abbattista et al. [8, 17] and Petrykin et al. [18], namely, orthorhombic Ammm, with $a'=a$, $b'=2a$ and $c'=2c$ [19]. A plot of the unit cell

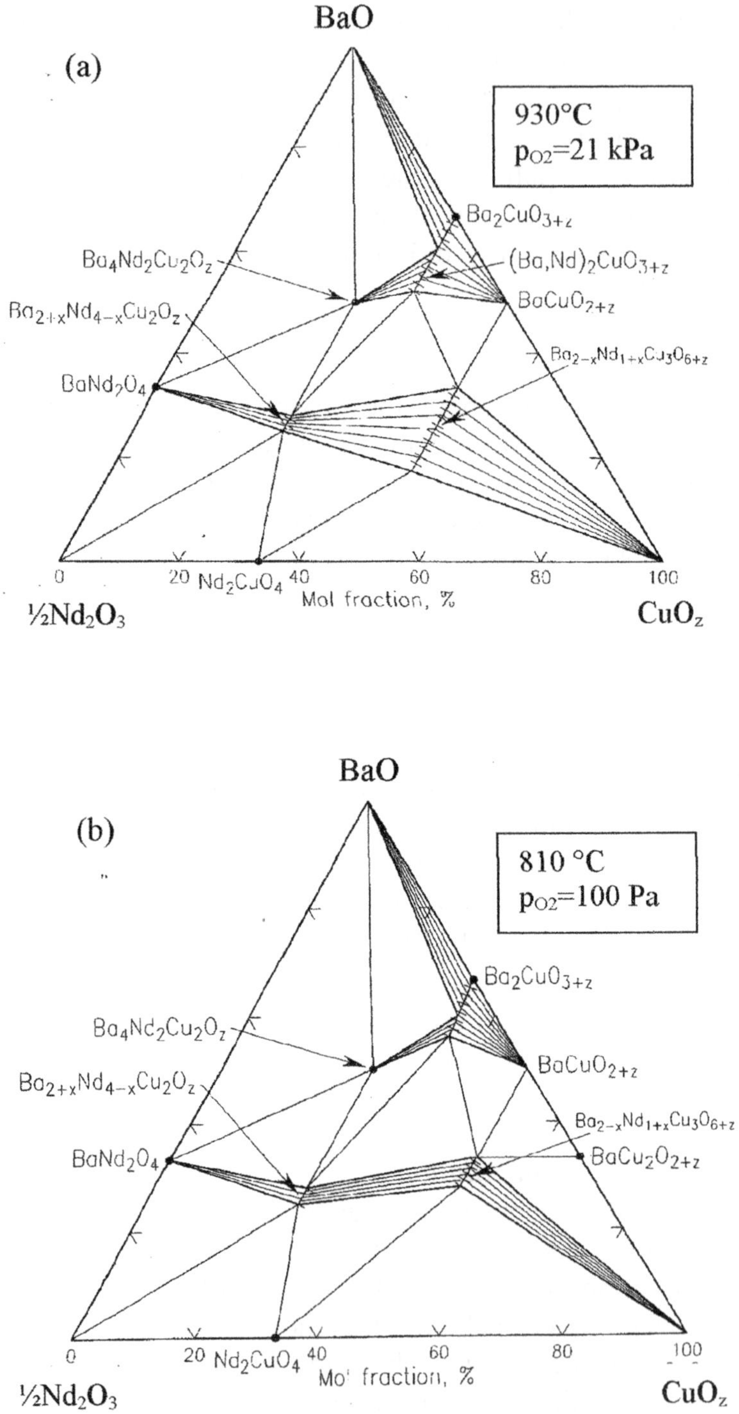

Fig. 1. Phase Diagrams for the BaO- $\frac{1}{2}Nd_2O_3$-CuO_x system under (a) 21 kPa and (b) 100 Pa oxygen partial pressure.

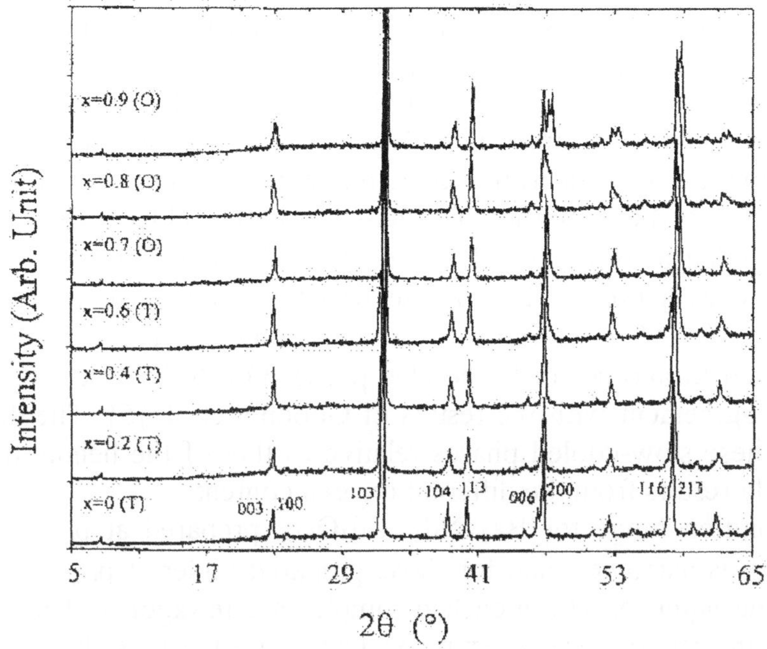

Fig. 2. X-ray diffraction patterns of the $Ba_{2-x}Nd_{1+x}Cu_3O_{6+z}$ series after Annealing at $p_{O2} = 21$ kPa at 930 °C and quenching into helium gas cooled by liquid nitrogen.

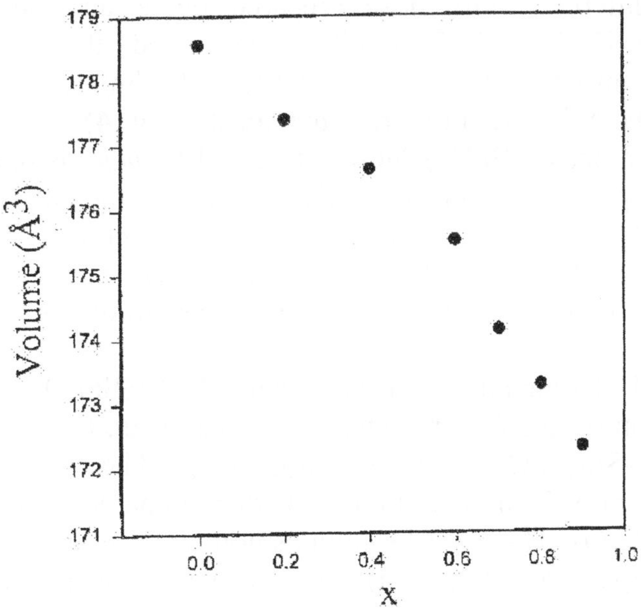

Fig. 3. Plots of unit cell volume against x in $Ba_{2-x}Nd_{1+x}Cu_3O_{6+z}$ (experiments at p_{O2}=21 kPa). For $0.95 \geq x \geq 0.7$, V of the tetragonal subcell is used.

volume, V as a function of x in $Ba_{2-x}Nd_{1+x}Cu_3O_{6+z}$ is illustrated in Fig. 3. For the orthorhombic phases, the approximate subcell volumes (one-fourth of the volume of the supercell) are used. Since the size of Nd^{3+} is smaller than that of Ba^{2+} [20], a decrease of cell volume as a function of Nd substitution is observed. Two lines with different slope are inferred from Fig. 3, one corresponding to the tetragonal phase and the other to the orthorhombic phase. For $Ba_{2-x}Nd_{1+x}Cu_3O_{6+z}$ samples that were slow-cooled at p_{O2} = 21 kPa from 900 °C, three different phases are observed in the x-ray patterns [21]. Phase transformations occurred from the orthorhombic (O), ($0<x<0.2$, *Pmmm*), to the tetragonal (T) phase, ($0.2<x<0.7$, *P4/mmm*), and then to another orthorhombic phase (O'), ($0.95 \geq x \geq 0.7$). This observation is in agreement with the results of Goodilin et al. [5]. The difference in symmetry of these slow-cooled phases relative to that of the helium-quenched samples apparently results from the different oxygen content.

The solid solution range for $Ba_{2-x}Nd_{1+x}Cu_3O_{6+z}$, prepared at p_{O2} = 100 Pa, with $0.3 > x \geq 0$, is narrower than for those prepared under at p_{O2} = 21 kPa or under oxygen. The liquid N_2/He quenched samples are tetragonal. The narrower solid solution range occurs because extra oxygen is needed to stabilize the solid solution members that are Nd-rich, and the oxygen partial pressure is not sufficient to stabilize such members [22].

<u>BaO-½Sm_2O_3-CuO_x system</u>

(1) *<u>Phase Diagram</u>*

The diagram of the BaO-½Sm_2O_3-CuO_x system prepared under 100 Pa 0.1% O_2 is substantially different from those prepared under air and from those prepared using $BaCO_3$ [7, 23-25]. These differences include the phase formation and the tie-line relationships. The diagram of the BaO-Sm_2O_3-CuO_x system is somewhat similar to the Nd-system reported previously (Fig. 4), except that in the Sm-system, the "brown phase" $BaNd_2CuO_5$ (tetragonal *I4/mcm*) is replaced by the orthorhombic "green phase" $BaSm_2CuO_5$. Also, because the difference of the ionic size between Ba^{2+} and Sm^{3+} is greater than that between Ba^{2+} and Nd^{3+}, the extent of solid solution also in the Sm system is smaller. The structure of the BaR_2CuO_5 "green phase" (*Pnma* space group) has been studied extensively [26-28].

There are a total of four phases in the ternary BaO-½Sm_2O_3-CuO_x system. Among them, two form solid solutions. These four phases are $Ba_{2-x}Sm_{1+x}Cu_3O_{6+z}$, $(Ba_{2-x}Sm_x)CuO_{3+z}$, $BaSm_2CuO_5$, and $Ba_4Sm_2Cu_2O_{9-z}$. The perovskite-related phase, $Ba_6SmCu_3O_z$ (Sm-613), although it exists when prepared in air [7], was not observed under 100 Pa O_2 (The Sm-613 composition that we prepared showed a mixture of $Ba_4Sm_2Cu_2O_{9-z}$ and $(Ba, Sm)_2CuO_{3+x}$). The cubic perovskite-related phase, $Ba_4SmCu_3O_{8.5+z}$ phase, which was reported to exist under oxygen, was also

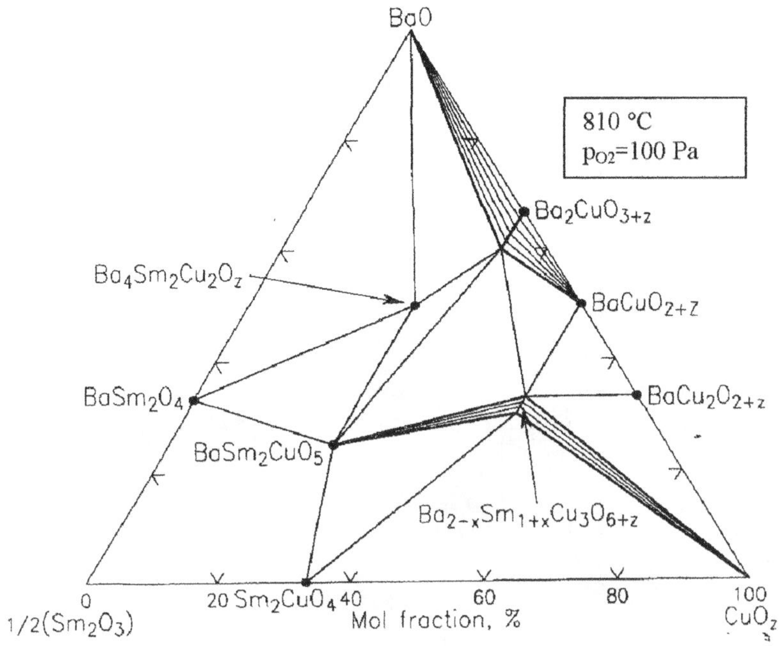

Fig. 4. Phase Diagram for the BaO-½Sm_2O_3-CuO_x system under 100 Pa oxygen partial pressure.

found to be unstable under 100 Pa O_2. Apparently when the size of R is smaller, such as that of Y, the 413 phase is stable under 100 Pa O_2 [7].

(2) *Structure of $Ba_4Sm_2Cu_2O_{9-z}$*

Under 100 Pa p_{O2}, $Ba_4Sm_2Cu_2O_{9-z}$ was found to be orthorhombic with a space group *Pnn2* (a subgroup of *P4n2*, No. 34). The (001) projection of the $Ba_4Sm_2Cu_2O_{9-z}$ structure is shown in Fig. 5a and the one-dimensional chains of CuO_5 are shown in Fig. 5b. The lattice parameters determined using powder x-ray diffraction are a = 11.9718(7)X, b = 11.8884(7) X, c = 3.8465 (2) X, and V = 547.47X^3). This semi-conductor phase is one of the few cuprate phases reported which possesses the unusual 1-dimensional chains of CuO_5 units [29]. These distorted and isolated CuO_5 units consist of square pyramids with four almost equal basal Cu-O distances (1.921(6) Å to 1.935(5) Å) and one much longer axial Cu-O bond of 2.473(14) Å. These square pyramids form corner-shared infinite chains running parallel to the c-axis. The CuO_5 chains are connected via Ba^{2+} and Sm^{3+} cations. The SmO_7 adopts a distorted mono - capped trigonal prismatic

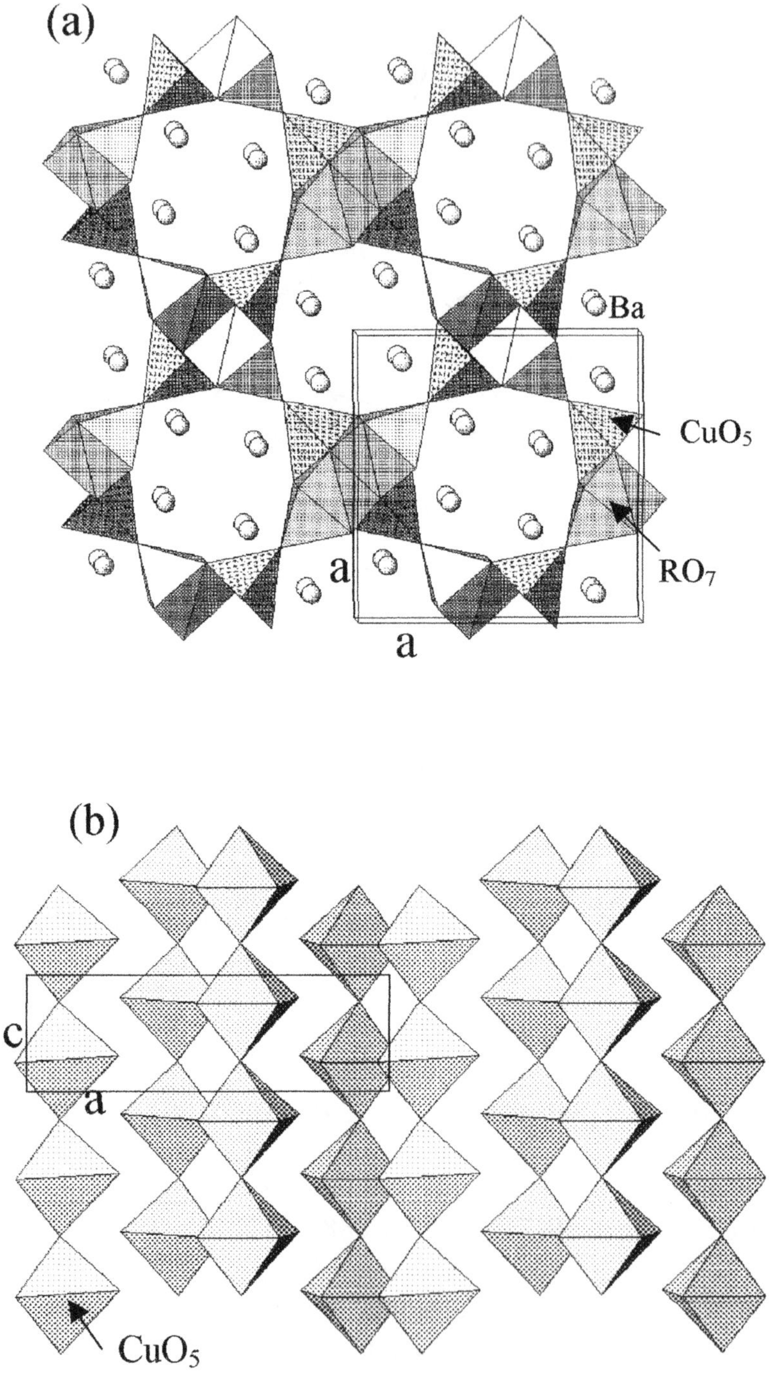

Fig. 5. Crystal Structure of $Ba_4R_2Cu_2O_x$ (R=Nd and Sm).
(a) The (001) projection, and (b) one-dimensional chains of CuO_5

configuration. The environment of Sm^{3+} in Sm-422 is similar to that of the 'green phase' $BaSm_2CuO_5$ [23]. The SmO_7 and CuO_5 units form a 3-dimensional network, and wide octagonal tunnels are found throughout the structure along the *c*-axis inside which Ba ions are found.

BaO-½Y_2O_3-CuO_x system

Figures 6a and 6b show the phase diagrams of the BaO-Y_2O_3-CuO_x system prepared at p_{O2} = 21 kPa (900 °C) and at p_{O2} = 100 Pa (810 °C). The phases formed under these two conditions are similar, except for those in the binary BaO-CuO_x system. For example, the reduced phase $BaCu_2O_{2+x}$ is stable when prepared under 0.1% O_2, but not under purified air. There are four ternary oxides in the BaO-Y_2O_3-CuO_x system, namely, BaY_2CuO_5 (2:1:3), $Ba_4YCu_3O_x$ (4:1:3), $Ba_6YCu_3O_x$ (6:1:3) and BaY_2CuO_5 (1:2:1). The compatibility relationships of these four phases are similar in these two diagrams, but differ from those prepared using $BaCO_3$ [7, 11, 30], particularly in the Ba-rich region. The occurrence of the compounds $Ba_4YCu_3O_x$ and $Ba_6YCu_3O_x$ in the BaO-rich part of the diagram agrees with that reported by Abbattista et al. [31] and Osamura and Zhang [7]. The $Ba_2YCu_3O_{6+x}$ phase is tetragonal P4/mmm when quenched from 900 °C, and 810 °C.

A tie-line was found between $Ba_4YCu_3O_x$ and $Ba_2YCu_3O_x$, which is in contrary to the literature reported tie-line between $BaCuO_{2+x}$ and BaY_2CuO_5. In the relatively Cu-poor regions, while a tie-line was found between BaY_2CuO_5 and $Ba_6YCu_3O_x$ under pure air, $Ba_3Y_4O_9$ and $Ba_4YCu_3O_x$ were found to be compatible with each other under 0.1% O_2.

Comparison of the BaO-½R_2O_3-CuO_x (R=Nd, Sm and Y) diagrams

As a comparison, Figs. 7(a) to 7(c) give the three diagrams of the Ba-R-Cu-O systems prepared under 100 Pa. A trend in phase formation is observed. Systems with larger size of R show a greater number as well as a larger range of solid solutions. For example, the R=Nd and Sm diagrams show solid solutions such as $Ba_{2-x}Nd_{1+x}Cu_3O_{6+z}$, and $(Ba_{2-x}Sm_x)CuO_{3+z}$, while in the R=Y system, the size of Y is smaller and no solid solution is observed. The 422 phase that can be prepared in the Nd and Sm systems is absent in the Y system. In the Y_2O_3-CuO_x binary system, the $Y_2Cu_2O_5$ phase is formed instead of R_2CuO_4 as found in the ½Nd_2O_3-CuO_x and ½Sm_2O_3-CuO_x systems. While the literature reports a tie-line between $BaCuO_{2+x}$ and BaR_2CuO_5 [7,11,23], this tie-line is not found in the carbonate-free BaO-R_2O_3-CuO_x systems.

ACKNOWLEDGEMENTS

Partial financial support from the US Department of Energy is acknowledged. The authors would like to thank Nils Swanson for his graphical assistance.

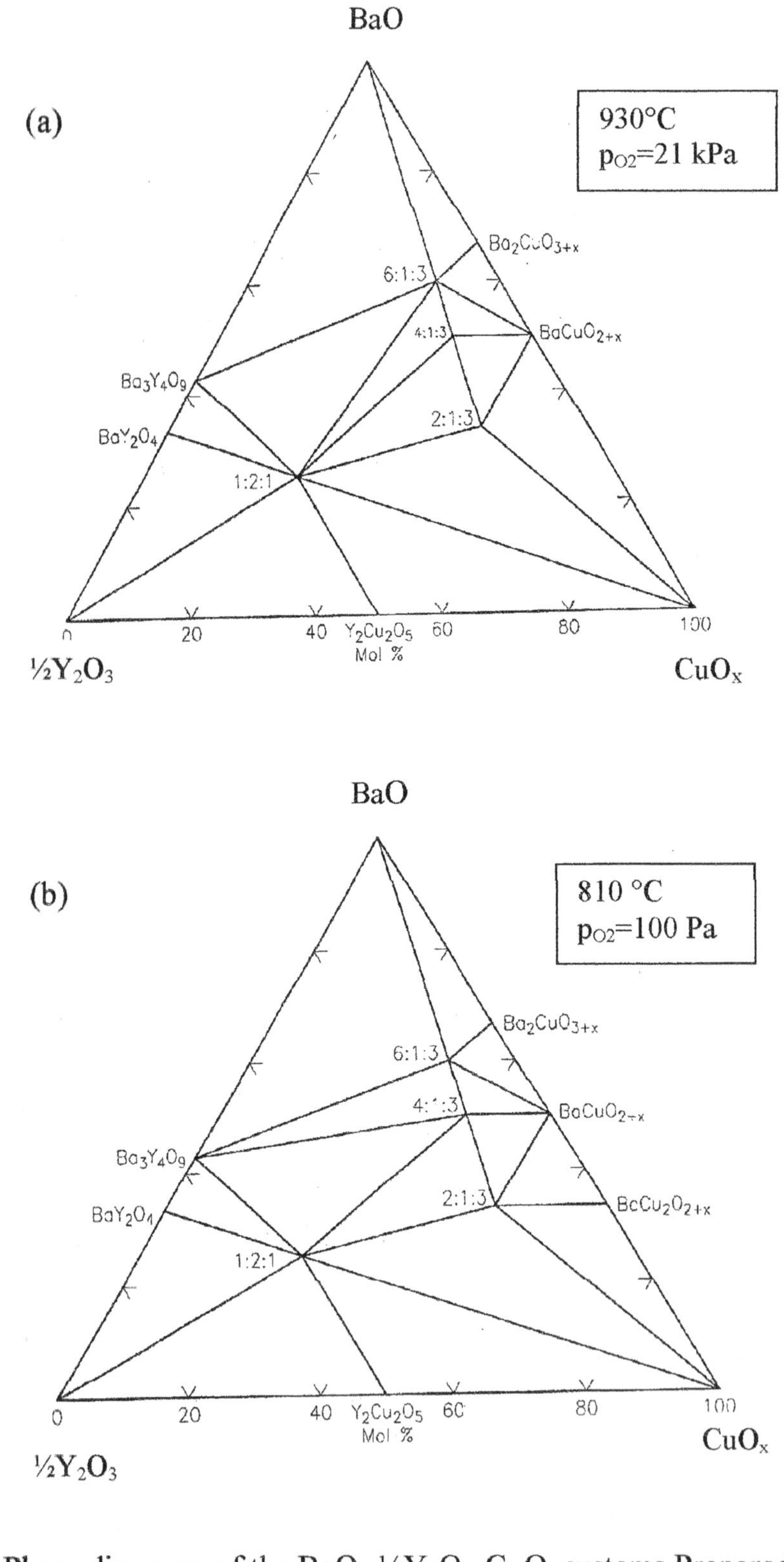

Fig 6. Phase diagrams of the BaO- ½Y_2O_3-CuO_x systems Prepared under 21 KPa and 100 Pa oxygen partial pressure.

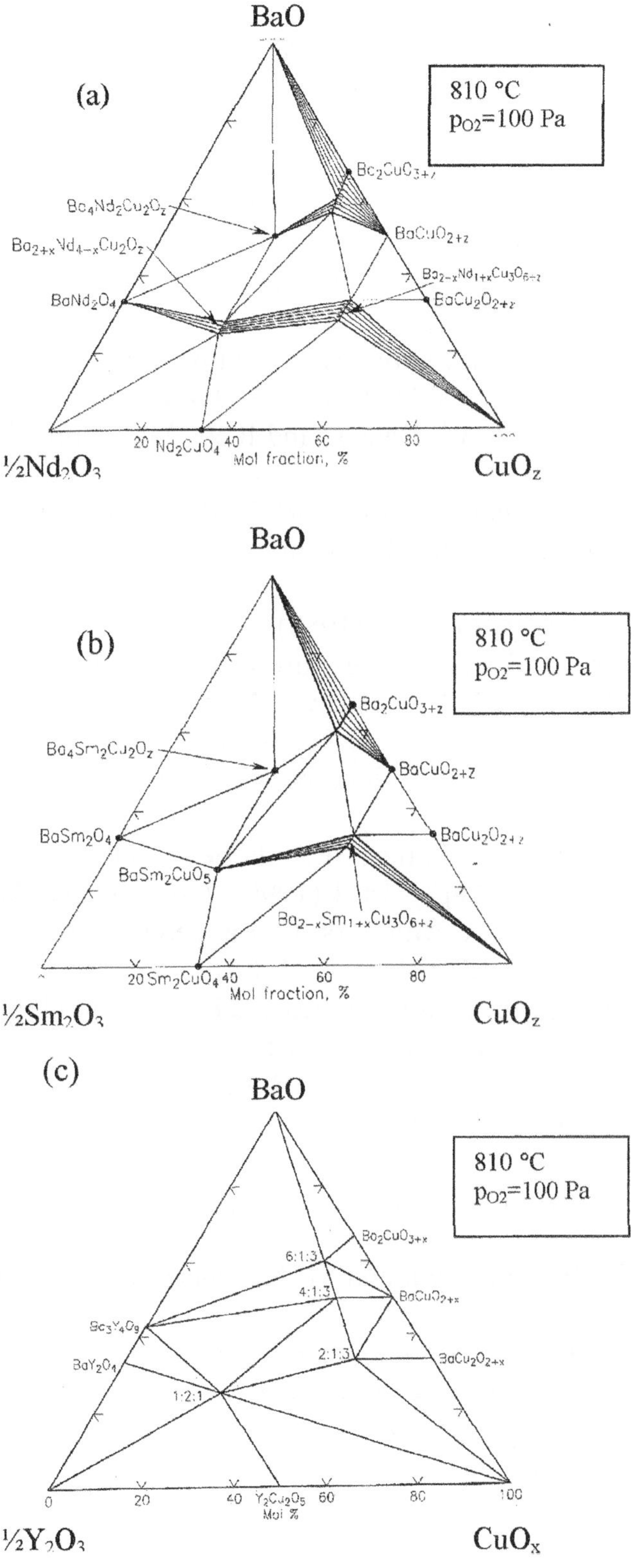

Fig. 7. Phase diagrams of the BaO- ½ R_2O_3-CuO systems, R=(a) Nd, (b) Sm and (c)Y. A trend of phase formation is observed.

REFERENCES

[1]M. Kambara, T. Umeda, M. Tagami, X. Yao, E.A. Goodilin, and Y. Shiohara, *J. Am. Ceram. Soc.* **81** (8), 2116 (1998).

[2] K. Oka, M. Saito, M. Ito, K. Nakane, K. Murata, Y. Nishihara, and H. Unoki, *Jpn. J. Phy. Lett.* **28** (2) L219 (1989)

[3]E. Goodilin, M. Kambara, T. Umeda and Y. Shiohara, *Physica C* **289**, 251 (1997).

[4]W. Wong-Ng, L.P. Cook, B. Paretzkin, M. D. Hill and J.K. Stalick, *J. Am. Ceram. Soc.* 77(9), 2354 (1994).

[5]E.A. Goodilin, N.N. Oleynikov, E.V. Antipov, R.V. Shpanchenko, G. Yu. Popov, V.G. Balakirev, and Yu. D. Tretyakov, *Physica C* **272**, 65 (1996).

[6]H. Wu, M.J. Kramer, K.W. Dennis, and R.W. McCallum, *Physica C* **290**, 252 (1997).

[7]K. Osamura and W. Zhang, *Z. Metallkd.* **84** (8) 522 (1993).

[8]F. Abbattista, D. Mazza and M. Vallino, *Eur. J. Solid State Inorg. Chem.* **28**, 649 (1991).

[9] J.J. Ritter, *Powd. Diff.* **3** (1), 30 (1988).

[10]PDF, Powder Diffraction File, produced by International Centre for Diffraction Data (ICDD), 12 Campus Blvd., Newtown Square, PA. 19073-3273.

[11]R.S. Roth, C.J. Rawn, F. Beech, J.D. Whitler, and J.O. Anderson; pp. 13-26 in *Ceramic Superocnductors II.* Edited by M.F. Yan, American Ceramic Society, Wersterville, OH, (1988).

[12]S.N. Koshcheeva, V.A. Fotiev, A.A. Fotiev, and V.G. Zubkov, *Izv. Akad. Nauk SSSR, Neorg. Mater.* **26**[7], 1491 (1990), *Inorg. Mater.* **26** [7] 1267 (1990).

[13]D.M.Deleeuw, C.A.H.A. Mutsaers, C. Langereis. H.C.A. Smoorenburg, and P.J. Rommers, *Physica C,* **152** [1] 39-49 (1988).

[14]S.I.Yoo and R.W. McCallum, *Physica C* **210** 147-156 (1993).

[15]S.J. Fu and S.S. Xie, *Chin. Sci. Bull.* **35** [10] 816-820 (1990).

[16]D.E. Appleman and H.T. Evans, Jr., Report PB216188, U.S. Department of Commerce, National Technical Information Service,5285 Port Royal Rd., Springfield, VA 22151.

[17]F. Abbattista, C. Brisi, M. Lucco-Borlera, and M. Vallino, *Nuovo Cimento* **10** 611 (1988).

[18]V.V. Petrykin, P. Berastegui and M. Kakihana, *Chem. Mater.* **11**, 3445-3451 (1999).

[19]W. Wong-Ng, L.P. Cook, J. Suh, R. Coutts, J.K. Stalick, I. Levin, and Q. Huang, J. Solid State Chem. **173** 476 (2003).

[20]R.D. Shannon, *Acta Crystallogr.* **A32**, 751 (1976).

[21]W. Wong-Ng, C.K. Chiang, B. Paretzkin, and E.R. Fuller, Jr. *Powd. Diffr.* **5** (1), 26-32 (1990).

[22]Y. Shiohara and E.A. Goodilin, Handbook on the Physics and Chemistry of Rare Earths, **30,** edited by K.A. Gschneidner, Jr., L. Eyring and M.B. Maple, Elsevier Science B.V., Ch. 189, pp. 67-227 (2000).

[23]W. Wong-Ng, B. Paretzkin, and E. R. Fuller, Jr., *J. Solid State Chem.* **85**, 117 (1990).

[24]Z.Y. Qiao, X.R. Xing, W.X. Yuan, S.K. Wei, X.L. Chen, J.K. Liang, and S.S. Xie, *J. Alloys Compd.*, **202** [1-2] 77-80 (1993).

[25]J. Czerwonka and H.A. Eick, *J. Solid State Chem.*, **90** [1] 69-78 (1991).

[26]R.M. Hazen, L.W. Finger, R.J. Angel, C.T. Prewitt, N. L. Ross, H. K. Mao, & C.G. Hadidiaco, P.H. Hor, R.L. Meng, C.W. Chu, *Phys. Rev.* B. **35** 7238 (1987).

[27]S.F. Watkins, F.R. Fronczek, K.S. Wheelock, R.G. Goodrich, W.O. Hamilton, and W.W. Johnson, *Acta Cryst.* **C44** 3 (1988).

[28]W. Wong-Ng, M.A Kuchinski, H.F. McMurdie and B. Paretzkin, *Powd. Diff.* **4** 1 (1989).

[29]B. Domenges, F. Abbattista, C. Michel, M. Vallino, L. Barbey, N. Nguyen, and B. Raveau, *J. Solid State Chem.* **106** 271 (1993).

[30]W. Reichelt, H. Wilhelm, G. Foersterling, and H. Oppermann, *Crysta. Res. Technol.* **24** [2] K26 (1989).

[31]F. Abbattista, M. Vallino, and D. Mazza, *Mater. Chem. Phys.* **21** 521 (1989).

Printed and bound by CPI Group (UK) Ltd, Croydon, CR0 4YY
15/09/2026
14962415-0002